APPLIED FUNCTIONAL ANALYSIS

J. Tinsley Oden
Leszek F. Demkowicz
Texas Institute for Computational
and Applied Mathematics
The University of Texas at Austin

CRC Press
Boca Raton New York London Tokyo

Library of Congress Cataloging-in-Publication Data

Catalog record is available from the Library of Congress.

Direct all inquiries to CRC Press, Inc., 2000 Corporate Blvd., N.W., Boca Raton, Florida 33431.

International Standard Book Number 0-8493-2551-X
Printed in the United States of America 1 2 3 4 5 6 7 8 9 0
Printed on acid-free paper

To
John James and Sara Elizabeth Oden
and
Wiesława and Kazimierz Demkowicz

CRC Series in

COMPUTATIONAL MECHANICS and APPLIED ANALYSIS

Series Editor: J.N. Reddy
Texas A&M University

New and Forthcoming Titles

APPLIED FUNCTIONAL ANALYSIS
J. Tinsley Oden and Leszek F. Demkowicz

Preface

Worldwide, in many institutions of higher learning, there has emerged in recent years a variety of new academic programs designed to promote interdisciplinary education and research in applied and computational mathematics. These programs, advanced under various labels such as computational and applied mathematics, mathematical sciences, applied mathematics, and the like, are created to pull together several areas of mathematics, computer science, and engineering and science which underpin the broad subjects of mathematical modeling and computer simulation. In all such programs, it is necessary to bring students of science and engineering quickly to within reach of modern mathematical tools, to provide them with the precision and organization of thought intrinsic to mathematics, and to acquaint them with the fundamental concepts and theorems which form the foundation of mathematical analysis and mathematical modeling. These are among the goals of the present text.

This book, which is the outgrowth of notes used by the authors for over a decade, is designed for a course for beginning graduate students in computational and applied mathematics who enter the subject with backgrounds in engineering and science. The course purports to cover in a connected and unified manner an introduction to the topics in functional analysis important in mathematical modeling and computer simulation; particularly, the course lays the foundation for futher work in partial differential equations, approximation theory, numerical mathematics, control theory, mathematical physics, and related subjects.

Prerequisites for the course for which this book is written are not extensive. The student with the usual background in calculus, ordinary differential equations, introductory matrix theory, and, perhaps, some background in applied advanced calculus typical of courses in engineering mathematics or introductory mathematical physics, should find much of the book a logical and, we hope, exciting extension and abstraction of his knowledge of these subjects.

It is characteristic of such courses that they be paradoxical, in a sense, because on the one hand they presume to develop the foundations of algebra and analysis from the first principles, without appeal to any previous prejudices toward mathematical methods; but at the same time, they call upon undergraduate mathematical ideas repeatedly as examples or as illustrations of purpose of the abstractions and extensions afforded by the abstract theory. The present treatment is no exception.

We begin with an introduction to elementary set theoretics, logic, and general abstract algebra, and with an introduction to real analysis in Chapter 1. Chapter 2 is devoted to linear algebra in both finite and infinite-dimensions. These two chapters could be skipped by many readers who have an undergraduate background in mathematics. For engineering graduate students, the material is often new and should be covered. We have provided numereous examples throughout the book to illustrate concepts, and many of these, again, draw from undergraduate calculus, matrix theory, and ordinary differential equations.

Chapter 3 is devoted to measure theory and integration and Chapter 4 covers topological and metric spaces. In these chapters, the reader encounters the fundamentals of Lebesgue integration, L^p spaces, the Lebesgue dominated convergence theorem, Fubini's theorem, the notion of topologies, filters, open and closed sets, continuity, convergence, Baire categories, the contraction mapping principle, and various notions of compactness.

In Chapter 5, all of the topological and algebraic notions covered in Chapters 1–4 are brought together to study topological vector spaces and, particularly, Banach spaces. This chapter contains introductions to many fundamental concepts, including the theory of distributions, the Hahn-Banach theorem and its corollaries, open mappings, closed operators, the closed graph theorem, Banach's theorem and the closed range theorem. The main focus is on properties of linear operators on Banach spaces and, finally, the solution of linear equations.

Chapter 6 is devoted to Hilbert spaces and to an introduction to the spectral theory of linear operators. There some applications to boundary-value problems involving partial differential equations of mathematical physics are discussed in the context of the theory of linear operators on Hilbert spaces.

Depending upon the background of the entering students, the book may be used as a text for as many as three courses: Chapters 1 and 2 provide a course on real analysis and linear algebra; Chapters 3 and 4, a text on integration theory and metric spaces, and Chapters 5 and 6 an introductory course on linear operators and Banach spaces. We have frequently taught all six chapters in a single semester course, but then we have been very selective of what topics were or were not taught. The material can be covered comfortably in two semesters, Chapters 1–3 and, perhaps, part of 4 dealt with in the first semester and the remainder in the second.

As with all books, these volume reflects the interests, prejudices, and experience of its authors. Our main interests lie in the theory and numerical analysis of boundary- and initial-value problems in engineering science and physics, and this is reflected in our choice of topics and in the organization of this work. We are fully aware, however, that the text also provides a foundation for a much broader range of studies and applications.

The book is very much based on the text with the same title by the first author and, indeed, can be considered as a new, extended and revised version of it. It draws heavily from other monographs on the subject, listed in the References, as well as from various old personal lecture notes taken by the authors when they themselves were students. The second author would like especially to acknowledge the privilege of listening to unforgettable lectures of Prof. Stanisław Łojasiewicz at the Jagiellonian University in Cracow, from which much of the text on integration theory has been borrowed.

We wish to thank a number of students and colleagues who made useful suggestions and read parts of the text during the preparation of this work: Waldek Rachowicz, Andrzej Karafiat, Krzysztof Banaś, Tarek Zohdi and others. We thank Ms. Judith Caldwell for typing a majority of the text.

J. Tinsley Oden and Leszek F. Demkowicz

Austin, September 1995

Contents

1

Preliminaries

Elementary Set Theory

1.1 Sets and Preliminary Notations, Number Sets

An axiomatic treatment of algebra, as with all mathematics, must begin with certain primitive concepts that are intuitively very simple but that may be impossible to define very precisely. Once these concepts have been agreed upon, true mathematics can begin—structure can be added, and a logical pattern of ideas, theorems, and consequences can be unraveled. Our aim here is to present a brief look at certain elementary, but essential, features of mathematics, and this must begin with an intuitive understanding of the concept of a *set*.

The term *set* is used to denote a collection, assemblage, or aggregate of objects. More precisely, a set is a plurality of objects that we treat as a single object. The objects that constitute a set are called the *members* or *elements* of the set. If a set contains a finite number of elements, we call it a *finite set*; if a set contains an infinity of elements, we call it an *infinite set*. A set that contains no elements at all is called an *empty*, *void*, or *null set* and is generally denoted $\emptyset$.

For convenience and conciseness in writing, we should also agree here on certain standard assumptions and notations. For example, any collection of sets we consider will be regarded as a collection of subsets of some mathematically well-defined set in order to avoid notorious paradoxes concerned with the "set of all sets," etc. The sets to be introduced here will always be well defined in the sense that it will be possible to determine if a given element is or is not a member of a given set. We will denote sets by uppercase Latin letters such as A, B, C, ... and elements of sets by lowercase Latin letters such as a, b, c, ... The symbol $\in$ will be used to denote membership of a set. For example, $a \in A$ means "the element a belongs to the set A" or "a is a member of A." Similarly, a stroke through $\in$ negates membership; that is, $a \notin A$ means "a does *not* belong to A."

Usually, various objects of one kind or another are collected to form a set because they share some common property. Indeed, the commonality or the characteristic of its elements serves to define the set itself. If set A has a small finite number of elements, the set can be defined simply by displaying all its elements. For example, the set of natural (whole) numbers greater than 2 but less than 8 is written

$$A = \{3, 4, 5, 6, 7\}$$

However, if a set contains an infinity of elements, it is obvious that a more general method must be used to define the set. We shall adopt a rather widely used method: Suppose that every element of a set A has a certain property P; then A is defined using the notation

$$A = \{a : \ a \text{ has property } P\}$$

Here a is understood to represent a typical member of A. For example, the finite set of whole numbers mentioned previously can be written

$$A = \{a : \ a \text{ is a natural number; } 2 < a < 8\}$$

Again, when confusion is likely, we shall simply write out in full the defining properties of certain sets.

Sets of primary importance in calculus are the number sets. These include:

the set of *natural* (whole) *numbers*

$$\mathbb{N} = \{1, 2, 3, 4, \ldots\}$$

the set of *integers* (this notation honors Zermelo, a famous Italian mathematician who worked on number theory)

$$\mathbb{Z} = \{\ldots, -3, -2, -1, 0, 1, 2, 3, \ldots\}$$

the set of *rational numbers* (fractions)

$$\mathbb{Q} = \left\{\frac{p}{q} : p \in \mathbb{Z}, \ q \in \mathbb{N}\right\}$$

the set of *real numbers* $\mathbb{R}$

the set of *complex numbers* $\mathbb{C}$

We do not attempt here to give either axiomatic or constructive definitions of these sets. Intuitively, once the notion of a natural number is adopted, $\mathbb{Z}$ may be constructed by adding zero and negative numbers and $\mathbb{Q}$ is the set of fractions with integer numerators and natural (in particular different from zero) denominators. The real numbers may be identified with their decimal representations and complex numbers may be viewed as pairs of real numbers with a specially defined multiplication.

The block symbols introduced above will be used hereinafter to denote the number sets.

1.2 Subsets and Equality of Sets

If A and B are two sets, A is said to be a *subset* of B if and only if *every* element of A is also an element of B. The subset property is indicated by the symbolism

$$A \subset B$$

which is read "A is a subset of B" or, more frequently, "A is contained in B." Alternately, the notation $B \supset A$ is sometimes used to indicate that "B contains A" or "B is a 'superset' of A."

It is clear from this definition that every set A is a subset of itself. To describe subsets of a given set B that do not coincide with B, we use the idea of *proper subsets*: A set A is a proper subset of B if and only if A is a subset of B and B contains one or more elements that do not belong to A. Occasionally, to emphasize that A is a subset of B but possibly not a proper subset, we may write $A \subseteq B$ or $B \supseteq A$.

We are now ready to describe what is meant by equality of two sets. It is tempting to say that two sets are "equal" if they simply contain the same elements; but this is a little too imprecise to be of much value in proofs of certain set relations to be described subsequently. Rather, we use the equivalent idea that equal sets must contain each other: Two sets A and B are said to be *equal* if and only if $A \subset B$ and $B \subset A$. If A is equal to B, we write

$$A = B$$

In general, to prove equality of two sets A and B, we first select a typical member of A and show that it belongs to the set B. Then, by definition, $A \subset B$. We then select a typical member of B and show that it also belongs to A, so that $B \subset A$. The equality of A and B then follows from the definition.

1.3 Set Operations

Some structure can be added to the rather loose idea of a set by defining a number of so-called set operations. We shall list several of these here. As a convenient conceptual aid, we also illustrate these operations by means of Venn diagrams in Fig. 1.1; there an abstract set is represented graphically by a closed region in the plane. In this figure, and in all of the definitions listed below, set A, B, etc., are considered to be subsets of some fixed master set U called the *universal set*; the universal set contains all elements of a type under investigation.

Union. The *union* of two sets A and B is the set of all elements x that belong to A or B. The union of A and B is denoted by $A \cup B$ and, using the notation introduced previously,

$$A \cup B = \{x : x \in A \text{ or } x \in B\}$$

Thus an element in $A \cup B$ may belong to either A or B or to both A and B. The union of a number of sets $A_1, A_2, \dots, A_n$ is, of course, the set of elements x belonging to A_1 or A_2 or $A_3 \dots$ or A_n; we use the notation

$$A_1 \cup A_2 \cup \cdots \cup A_n = \bigcup_{k=1}^{n} A_k$$

Intersection. The *intersection* of two sets A and B is the set of elements x that belong to both A and B. The symbolism $A \cap B$ is used to denote the intersection of A and B:

$$A \cap B = \{x : x \in A \text{ and } x \in B\}$$

The intersection of a number of sets $A_1, A_2, \dots, A_n$ is the set of elements common to all n sets and is denoted

$$A_1 \cap A_2 \cap \cdots \cap A_n = \bigcap_{k=1}^{n} A_k$$

Disjoint Sets. Two sets A and B are *disjoint* if and only if they have no elements in common. Then their intersection is the empty set $\emptyset$ described earlier:

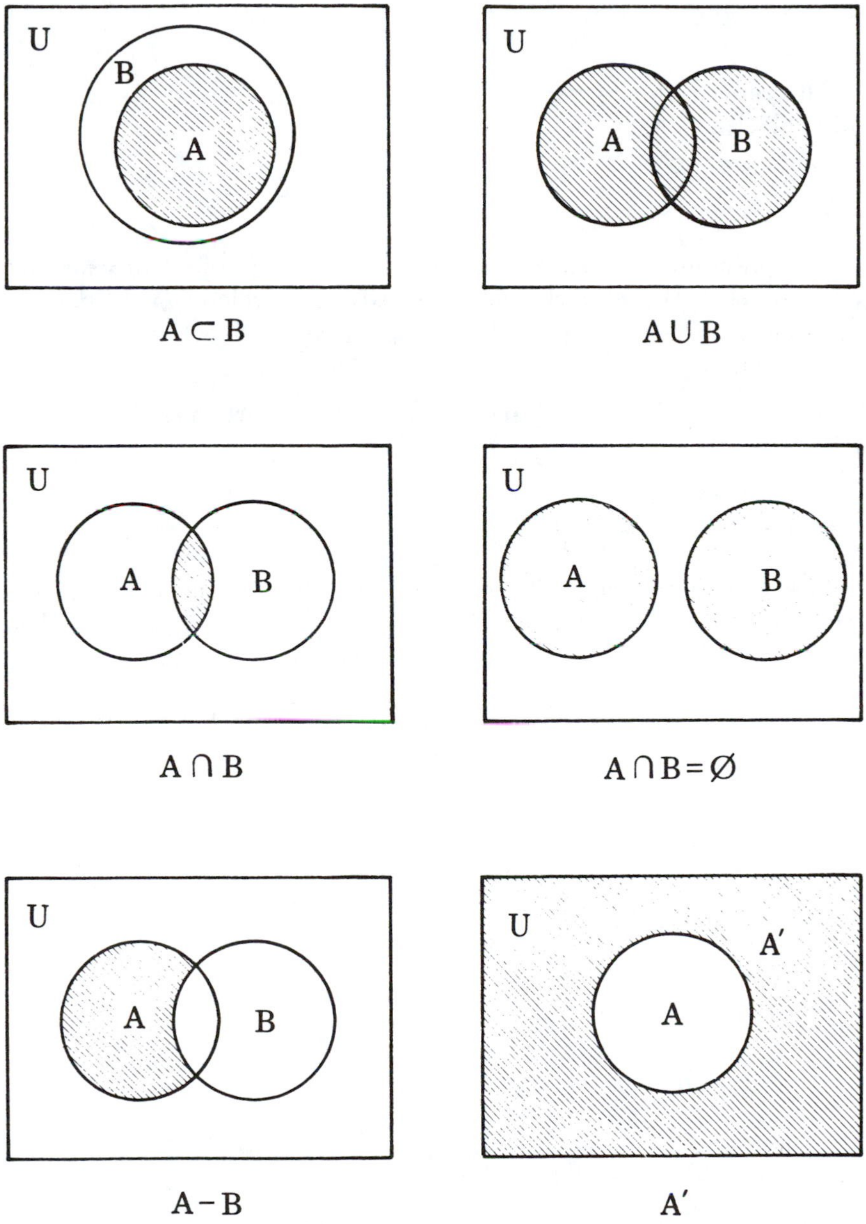

Figure 1.1
Venn diagrams.

$$A \cap B = \emptyset$$

Difference. The *difference* of two sets A and B, denoted $A - B$, is the set of all elements that belong to A but not to B.

$$A - B = \{x : x \in A; \quad x \notin B\}$$

Complement. The *complement* of a set A (with respect to some universal set U) is the set of elements, denoted by A', which do not belong to A.

$$A' = \{x : x \in U; \quad x \notin A\}$$

In other words, $A' = U - A$ and $A' \cup A = U$. In particular, $U' = \emptyset$ and $\emptyset' = U$.

Example 1.3.1

Suppose U is the set of all lowercase Latin letters in the alphabet, A is the set of vowels ($A = \{a, e, i, o, u\}$), $B = \{c, d, e, i, r\}$, $C = \{x, y, z\}$. Then the following hold:

$$A' = \{b, c, d, f, g, h, j, k, l, m, n, p, q, r, s, t, v, w, x, y, z\}$$

$$A \cup B = \{a, c, d, e, i, o, r, u\}$$

$$(A \cup B) \cup C = \{a, c, d, e, i, o, r, u, x, y, z\}$$

$$= A \cup (B \cup C)$$

$$A - B = \{a, o, u\}$$

$$B - A = \{c, d, r\}$$

$$A \cap B = \{e, i\} = B \cap A$$

$$A' \cap C = \{x, y, z\} = C \cap A'$$

$$U - ((A \cup B) \cup C) = \{b, f, g, h, j, k, l, m, n, p, q, s, t, v, w\}\ , \text{ etc.}$$

□

Classes. We refer to sets whose elements are themselves sets as *classes.* Classes will be denoted by script letters $\mathcal{A}$, $\mathcal{B}$, $\mathcal{C}, \ldots$, etc. For example, if A, B and C are the sets

$$A = \{0,\ 1\}, \qquad B = \{a, b, c\}, \qquad C = \{4\}$$

the collection

$$\mathcal{A} = \{\{0,1\}, \{a,b,c\}, \{4\}\}$$

is a class with elements A, B and C.

Of particular interest is the *power set* or *power class* of a set A, denoted $\mathcal{P}(A)$. Based on the fact that a finite set with n elements has 2^n subsets, including $\emptyset$ and the set itself (see Exercise 1.3.4), $\mathcal{P}(A)$ is defined as the class of all subsets of A. Since there are 2^n sets in $\mathcal{P}(A)$, when A is finite, we sometimes use the notation

$$\mathcal{P}(A) = 2^A$$

Example 1.3.2

Suppose $A = \{1,2,3\}$. Then the power class $\mathcal{P}(\mathcal{A})$ contains $2^3 = 8$ sets:

$$\mathcal{P}(A) = \{\emptyset, \{1\}, \{2\}, \{3\}, \{1,2\}, \{1,3\}, \{2,3\}, \{1,2,3\}\}$$

It is necessary to distinguish between, e.g., element 1 and the single-element set $\{1\}$ (sometimes called *singleton*). Likewise, $\emptyset$ is the null set, but $\{\emptyset\}$ is a nonempty set with one element, that element being $\emptyset$. □

Infinite Unions and Intersections. Notions of union and intersection of sets can be generalized to the case of infinite families of sets. Let $\mathcal{B}$ be a class of sets (possibly infinite). The *union* of sets from class $\mathcal{B}$ is the set of all elements x that belong to *some* sets from $\mathcal{B}$. We use the notation

$$\cup\{B : \ B \in \mathcal{B}\} = \{x : \text{ there exists a } B \in \mathcal{B} \text{ such that } x \in B\}$$

Similarly, the *intersection* of sets from class $\mathcal{B}$ is the set of all elements x that belong to *every* set from $\mathcal{B}$. We have

$$\cap\{B : \ B \in \mathcal{B}\} = \{x : \text{ for every } B \in \mathcal{B},\ x \in B\}$$

Example 1.3.3

Suppose $\mathbb{R}$ denotes the set of all real numbers and $\mathbb{R}^2$ the set of *ordered pairs* (x, y) of real numbers (we make these terms precise subsequently). Then the set $A_m = \{(x, y) : y = mx\}$ is equivalent to the set of points on the straight line $y = mx$ in the euclidean plane. The set of all such lines is the class

$$\mathcal{A} = \{A_m :\ m \in \mathbb{R}\}$$

In this case,

$$\cap A_m = \{(0, 0)\}$$

$$\cup A_m = \mathbb{R}^2 - \{(0, y) : |y| > 0\}$$

That is, the only point common to all members of the class is the origin $(0, 0)$, and the union of all such lines is the entire plane $\mathbb{R}^2$, excluding the y-axis, except the origin, since $m = \infty \notin \mathbb{R}$. $\square$

Exercises

1.3.1. If $\mathbb{Z} = \{\dots, -2, -1, 0, 1, 2, \dots\}$ denotes the set of all integers and $\mathbb{N} = \{1, 2, 3, \dots\}$ the set of all natural numbers, exhibit the following sets in the form $A = \{a, b, c, \dots\}$:

(i) $\{x :\ x \in \mathbb{Z},\ x^2 - 2x + 1 = 0\}$

(ii) $\{x :\ x \in \mathbb{Z},\ 4 \leq x \leq 10\}$

(iii) $\{x :\ x \in \mathbb{N},\ x^2 < 10\}$

1.3.2. Of 100 students polled at a certain university, 40 were enrolled in an engineering course, 50 in a mathematics course, and 64 in a physics course. Of these, only 3 were enrolled in all three subjects, 10 were enrolled only in mathematics and engineering, 35 were enrolled only in physics and mathematics, and 18 were enrolled only in engineering and physics.

(i) How many students were enrolled only in mathematics?

(ii) How many of the students were not enrolled in any of these three subjects?

1.3.3. List all of the subsets of $A = \{1, 2, 3, 4\}$. (Note: A and $\emptyset$ are considered to be subsets of A.)

1.3.4. Verify that a set A with a finite number n of elements has 2^n subsets.

1.4 Set Relations—Algebra of Sets

The set operations described in the previous sections can be used to construct a sort of algebra of sets that is governed by a number of basic laws. We list several of these as follows:

Idempotent Laws

$$A \cup A = A; \qquad A \cap A = A$$

Commutative Laws

$$A \cup B = B \cup A; \qquad A \cap B = B \cap A$$

Associative Laws

$$A \cup (B \cup C) = (A \cup B) \cup C$$

$$A \cap (B \cap C) = (A \cap B) \cap C$$

Distributive Laws

$$A \cup (B \cap C) = (A \cup B) \cap (A \cup C)$$

$$A \cap (B \cup C) = (A \cap B) \cup (A \cap C)$$

Identity Laws

$$A \cup \emptyset = A; \qquad A \cap U = A$$

$$A \cup U = U; \qquad A \cap \emptyset = \emptyset$$

Complement Laws

$$A \cup A' = U; \qquad A \cap A' = \emptyset$$

$$(A')' = A; \qquad U' = \emptyset, \quad \emptyset' = U$$

There are also a number of special identities that often prove to be important. For example:

De Morgan's Laws

$$A - (B \cup C) = (A - B) \cap (A - C)$$

$$A - (B \cap C) = (A - B) \cup (A - C)$$

All of these so-called *laws* are merely theorems that can be proved by direct use of the definitions given in the preceding section and, in particular, use of the idea of equality of sets.

Example 1.4.1
(Proof of Associative Laws)

To prove the first associative law, consider a typical element x in the set $A \cup (B \cup C)$. By definition,

$$x \in A \text{ or } (B \cup C), \text{ which means that}$$

$$x \in A \text{ or } B \text{ or } C, \text{ which means that}$$

$$x \in (A \cup B) \cup C$$

Consequently, a typical member of $A \cup (B \cup C)$ is also a member of $(A \cup B) \cup C$. From this fact, we conclude that

$$A \cup (B \cup C) \subseteq (A \cup B) \cup C \tag{a}$$

Now consider a typical element y of $(A \cup B) \cup C$. By definition,

$$y \in A \cup B \text{ or } C, \text{ which means that}$$

$$y \in A \text{ or } B \text{ or } C, \text{ which means that}$$

$$y \in A \text{ or } (B \cup C), \text{ which means that}$$

$$(A \cup B) \cup C \subseteq A \cup (B \cup C) \tag{b}$$

Recalling the definition of equality of sets and comparing (a) and (b), we arrive at the first associative law. The second law is proved in a similar manner. $\square$

Example 1.4.2
(Proof of De Morgan's Laws)

If $x \in A - (B \cup C)$, then $x \in A$ but $x \notin (B \cup C)$. Consequently, $x \notin B$ and $x \notin C$, so that $x \in A - B$ and $A - C$. But this implies that $x \in (A - B) \cap (A - C)$; hence $A - (B \cup C) \subseteq (A - B) \cap (A - C)$. Similarly, if $y \in (A - B) \cap (A - C)$, then $y \in (A - B)$ and $(A - C)$, which implies that $y \in A$ but not B and $y \in A$ but not C. It follows that $y \in A$, but $y \notin B$ or C. Therefore, $(A - B) \cap (A - C) \subseteq A - (B \cup C)$. This completes the proof of the first of De Morgan's laws; the second is obtained by similar reasoning. $\square$

De Morgan's laws can be generalized to the case of unions and intersections of arbitrary classes of sets (in particular infinite).

$$A - \bigcup_{B \in \mathcal{B}} B = \bigcap_{B \in \mathcal{B}} (A - B)$$

$$A - \bigcap_{B \in \mathcal{B}} B = \bigcup_{B \in \mathcal{B}} (A - B)$$

When the universal set U is taken for A, we may use the notion of the complement of a set and write De Morgan's laws in the more concise form

$$\left(\bigcup_{B \in \mathcal{B}} B \right)' = \bigcap_{B \in \mathcal{B}} B'$$

$$\left(\bigcap_{B \in \mathcal{B}} B \right)' = \bigcup_{B \in \mathcal{B}} B'$$

De Morgan's laws express a *duality effect* between the notions of union and intersection of sets and sometimes they are called the *duality laws.* They are a very effective tool in proving theorems. We shall return to the proof of them in the Section 1.6.

Exercises

1.4.1. Construct Venn diagrams to illustrate the idempotent, commutative, associative, distributive, and identity laws. (Note: some of these are trivially illustrated.)

1.4.2. Construct Venn diagrams to illustrate De Morgan's laws.

1.4.3. Prove the distributive laws.

1.4.4. Prove the identity laws.

1.4.5. Prove the second of De Morgan's laws.

1.4.6. Prove that $(A - B) \cap B = \emptyset$.

1.4.7. Prove that $B - A = B \cap A'$.

Elementary Logic

1.5 Statements, Statement Operations, Truth Tables

Before we turn to more complicated notions like relations or functions, we would do well to examine briefly some elementary concepts in logic so that we may have some idea about the meaning of a proof.

We are not interested here in examining the foundations of mathematics, but in formalizing certain types of thinking people have used for centuries to derive meaningful conclusions from certain premises. Millenia ago, the ancient Greeks learned that a deductive argument must start somewhere. In other words, a certain set of statements which are supposed to be true, must be assumed and, then, by reasonable arguments, new "true" statements are derived. The notion of "truth" in mathematics may thus have nothing to do with concepts of "truth" (whatever the term may mean) discussed by philosophers. It is merely the starting point of an exercise in which new true statements are derived from old ones by certain fixed rules of logic. We expect that there is general agreement among knowledgeable specialists that this starting point is acceptable and that the consequences of the choices of truth agree with our experiences.

Typically, a branch of mathematics is constructed in the following way.

A small number of statements, called *axioms,* is assumed to be true. To signify this, we may assign the letter "t" to them. Then there are various ways to construct new statements and some specific rules are prescribed to assign the value "t" or "f" (false) to them. Each of the new statements must be assigned only one of the two values. In other words, no situation can be accepted in which a statement could be simultaneously true and false. If this happens, it will mean that the set of axioms is *inconsistent* and the whole theory should be abandoned (at least from the mathematical point of view; there are many inconsistent theories in engineering practice and they are still in operation).

For a consistent set of axioms, the statements bearing the "t" value are called *theorems*, *lemmas*, *corollaries* and *propositions.* Though many inconsistencies in using these words are encountered, the following rules may be suggested:

- a *theorem* is an important true statement;
- a *lemma* is a true statement, serving, however, as an auxiliary tool to prove a certain theorem or theorems;
- a *proposition* is, in fact, a theorem which is not important enough to be called a *theorem.* This suggests that the name *theorem* be used rather rarely to emphasize especially important key results;
- finally, a *corollary* is a true statement, derived as an immediate consequence of a theorem or proposition with little extra effort.

Lowercase letters will be used to denote statements. Typically letters p, q, r, s are preferred. Recall once again that a statement p is a sentence for which only one of the two values "true" or "false" can be assigned. In the following, we shall list the fundamental operations on statements allowing us to construct new statements and we shall specify precisely the way to assign the "true" and "false" values to those new statements.

Negation

$\sim q$, to be read : not q

If $p = \sim q$ then p and q always bear opposite values; p is false when q is true and, conversely, if q is false then p is true. Assigning value 1 for "true" and 0 for "false" we may illustrate this rule using the so-called truth table

q	$\sim q$
1	0
0	1

Alternative

$p \vee q$, to be read : p or q

The alternative $r = p \vee q$ is true whenever at least one of the two component statements p or q is true. In other words, r is false only when both p and q are false. Again we can use the truth table to illustrate the definition:

p	q	$p \vee q$
1	1	1
1	0	1
0	1	1
0	0	0

Note in particular the nonexclusive character of the alternative. The fact that $p \vee q$ is true does not indicate that only one of the two statements p or q is true; they *both* may be true. This is somewhat in conflict with the everyday use of the word "or."

Conjunction

$p \wedge q$, to be read : p and q

The conjunction $p \wedge q$ is true only if both p and q are true. We have the following truth table:

p	q	$p \wedge q$
1	1	1
1	0	0
0	1	0
0	0	0

Implication

$p \Rightarrow q$, to be read in one of the following ways:

- p implies q
- q if p
- q follows from p
- if p then q
- p is a sufficient condition for q
- q is a necessary condition for p

It is somewhat confusing, but all these sentences mean exactly the same thing. The truth table for implication is as follows:

p	q	$p \Rightarrow q$
1	1	1
1	0	0
0	1	1
0	0	1

Thus, the implication $p \Rightarrow q$ is false only when "true" implies "false". Surprisingly, a false statement may imply a true one and the implication is still considered to be true.

Equivalence

$p \Leftrightarrow q$, to be read: p is equivalent to q.

The truth table is as follows:

p	q	$p \Leftrightarrow q$
1	1	1
1	0	0
0	1	0
0	0	1

Thus the equivalence $p \Leftrightarrow q$ is true (as expected) when both p and q are simultaneously true or false.

All theorems, propositions, etc., are formulated in the form of an implication or an equivalence. Notice that in proving a theorem in the form of implication $p \Rightarrow q$, we typically assume that p is true and attempt to show that q must be true. We do not need to check what will happen if p is false. No matter which value q takes on, the whole implication will be true.

Using the five operations on statements, we may build new combined operations and new statements. Some of them always turn out to be true no matter which values are taken on by the initial statements. As an example, let us study the fundamental statement showing the relation between the implication and equivalence operations.

$$(p \Leftrightarrow q) \Leftrightarrow ((p \Rightarrow q) \wedge (q \Rightarrow p))$$

One of the very convenient ways to prove that this statement is always true is to use the truth tables. This is illustrated in the table below.

p	q	$p \Leftrightarrow q$	$p \Rightarrow q$	$q \Rightarrow p$	$(p \Rightarrow q) \wedge (q \Rightarrow p)$	$(p \Leftrightarrow q) \Leftrightarrow ((p \Rightarrow q) \wedge (q \Rightarrow p))$
1	1	1	1	1	1	1
1	0	0	0	1	0	1
0	1	0	1	0	0	1
0	0	1	1	1	1	1

The law just proven is very important in proving theorems. It says that whenever we have to prove a theorem in the form of the equivalence $p \Leftrightarrow q$, we need to show that both $p \Rightarrow q$ and $q \Rightarrow p$. The fact is commonly expressed by replacing the phrase "p is equivalent to q" with "p is a necessary and sufficient condition for q."

Another very important law, fundamental for the methodology of proving theorems, is as follows:

$$(p \Rightarrow q) \Leftrightarrow (\sim q \Rightarrow \sim p)$$

Again, the truth table method can be used to prove that this statement is always true (see Exercise 1.5.1). This law lays down the foundation for the so-called *proof by contradiction.* In order to prove that p implies q, we negate q and show that this implies $\sim p$.

Example 1.5.1

As an example of the proof by contradiction, we shall prove the following simple proposition:

> If $n = k^2 + 1$, and k is natural number, then n cannot be a square of a natural number.

Assume, contrary to the hypothesis, that $n = \ell^2$. Thus $k^2 + 1 = \ell^2$ and, consequently,

$$1 = \ell^2 - k^2 = (\ell - k)(\ell + k)$$

a contradiction, since $\ell - k \neq \ell + k$ and 1 is divisible only by itself. $\square$

In practice there is more than one assumption in a theorem which means that statement p in the theorem $p \Rightarrow q$ is not a simple statement but rather a collection of many statements. Those include all of the theorems (true statements) of the theory being developed, not necessarily listed as explicit assumptions. Consider for example the proposition:

> $\sqrt{2}$ is not a rational number

It is somewhat confusing that this proposition is not in the form of an implication (nor equivalence). It looks to be just a single (negated) statement. In fact the proposition should be read as follows:

> If all the results concerning the integers and the definition of rational numbers hold, then $\sqrt{2}$ is not a rational number.

We may proceed now with the proof as follows. Assume, to the contrary, that $\sqrt{2}$ is a rational number. Thus $\sqrt{2} = \frac{p}{q}$, where p and q are integers and may be assumed, without loss of generality (why?), to have no common divisor. Then $2 = \frac{p^2}{q^2}$, or $p^2 = 2q^2$. Thus p must be even. Then p^2 is divisible by 4, and hence q is even. But this means that 2 is a common divisor of p and q – a contradiction of the definition of rational numbers.

Exercises

1.5.1. Construct truth tables to prove the following identities:

$$(p \Rightarrow q) \Leftrightarrow (\sim q \Rightarrow \sim p)$$
$$\sim (p \Rightarrow q) \Leftrightarrow (\sim q \wedge p)$$

1.5.2. Construct truth tables to prove De Morgan's laws in logic:

$$\sim (p \wedge q) \Leftrightarrow (\sim p) \vee (\sim q)$$
$$\sim (p \vee q) \Leftrightarrow (\sim p) \wedge (\sim q)$$

1.5.3. Construct truth tables to prove the associative laws in logic:

$$p \vee (q \vee r) \Leftrightarrow (p \vee q) \vee r$$
$$p \wedge (q \wedge r) \Leftrightarrow (p \wedge q) \wedge r$$

1.6 Algebra of Sets—Revisited

There is an intrinsic relation between the fundamental notions of logic and set theory. First of all, let us notice that we have implicitly used the operations on statements when defining the set operations. We have, for instance:

$$x \in A \cup B \Leftrightarrow (x \in A \vee x \in B)$$

$$x \in A \cap B \Leftrightarrow (x \in A \wedge x \in B)$$

$$x \in A' \Leftrightarrow \sim (x \in A)$$

Thus, the notions like union, intersection and complement of sets correspond to the notions of alternative, conjunction and negation in logic. The situation is even more evident when we use the laws of logic (statements which are always true) to prove the laws of algebra of sets. As an example, let us try to prove the first De Morgan's law

$$(A \cup B)' = A' \cap B'$$

This may be equivalently stated as follows:

$$\sim (x \in A \cup B) \Leftrightarrow ((x \in A') \text{ and } (x \in B'))$$

or

$$\sim (x \in A \text{ or } x \in B) \Leftrightarrow \sim (x \in A) \text{ and } \sim (x \in B)$$

Thus, it is sufficient to prove that the following statement is always true:

$$\sim (p \vee q) \Leftrightarrow (\sim p) \wedge (\sim q)$$

This is also called the first De Morgan's law in logic. It can be easily proven using the truth table technique (cf. Exercise 1.5.2).

Similarly, the second De Morgan's law in set algebra

$$(A \cap B)' = A' \cup B'$$

corresponds to the second De Morgan's law in logic

$$\sim (p \wedge q) \Leftrightarrow (\sim p) \vee (\sim q)$$

1.7 Open Statements. Quantifiers

Suppose that $S(x)$ is an expression which depends upon a variable x. One may think of variable x as the name of an unspecified object from a certain given set X. In general it is impossible to assign the "true" or "false" value to such an expression unless a specific value is substituted for x. If, after such a substitution, $S(x)$ becomes a statement then $S(x)$ is called an *open statement.*

Example 1.7.1

Consider the expression

$$x^2 > 3 \qquad \text{with} \qquad x \in \mathbb{N}$$

Then "$x^2 > 3$" is a open statement which becomes true for x bigger than 1 and it is false for $x = 1$. $\square$

Thus, having an open statement $S(x)$ we may obtain a statement by substituting a specific variable from its domain X. Another way is to add to $S(x)$ one of the two so-called *quantifiers*:

$\forall x \in X$, to be read: for all x, for every x, etc. (belonging to X)

$\exists x \in X$, to be read: for some x, there exists x such that, etc. (belonging to X)

The first one is called the *universal quantifier* and the second the *existential quantifier.* Certainly, by adding the universal quantifiers to the open statement from Example 1.7.1. we get the false statement:

$$\forall x \in \mathbb{N} \qquad x^2 > 3$$

(every natural number, when squared is greater than 3), while by adding the existential qualifier we get the true statement:

$$\exists x \in \mathbb{N} \qquad x^2 > 3$$

(there exists a natural number whose square is greater than 3).

Naturally, the quantifiers may be understood as generalizations of the alternative and conjunction. First of all, due to the associative law in logic (recall Exercise 1.5.3)

$$p \vee (q \vee r) \Leftrightarrow (p \vee q) \vee r$$

we may agree to define the alternative of these statements

$$p \vee q \vee r$$

by either of the two statements above. Next, this can easily be generalized to the case of the alternative of the finite class of statements

$$p_1 \vee p_2 \vee p_3 \vee \ldots \vee p_N$$

Note that this statement is true whenever *there exists* a statement p_i, for some i which is true. Thus, for finite sets $X = \{x_1, \ldots, x_N\}$, the statement

$$\exists x \in X \qquad S(x)$$

is equivalent to the alternative

$$S(x_1) \vee S(x_2) \vee \ldots \vee S(x_N)$$

Similarly, the statement

$$\forall x \in X \qquad S(x)$$

is equivalent to

$$S(x_1) \wedge S(x_2) \wedge \ldots \wedge S(x_N)$$

Negation Rules for Quantifiers. We shall adopt the following *negation rule* for the universal quantifier

$$\sim (\forall x \in X,\ S(x)) \Leftrightarrow \exists x \in X \quad \sim S(x)$$

Observe that this rule is consistent with De Morgan's law

$$\sim (p_1 \wedge p_2 \wedge \ldots \wedge p_N) \Leftrightarrow (\sim p_1 \vee \sim p_2 \vee \ldots \vee \sim p_N)$$

Substituting $\sim S(x)$ for $S(x)$ and negating both sides we get the negation rule for the existential quantifier

$$\sim (\exists x \in X,\ S(x)) \Leftrightarrow \forall x \in X \sim S(x)$$

which again corresponds to the second De Morgan's law

$$\sim (p_1 \vee p_2 \vee \ldots \vee p_N) \Leftrightarrow (\sim p_1 \wedge \sim p_2 \wedge \ldots \wedge \sim p_N)$$

Example 1.7.2

The negation rules for quantifiers must be used when proving the De Morgan's laws for infinite unions and intersections in set algebra. Indeed, the equality of sets

$$\Big(\bigcap_{B \in \mathcal{B}} B\Big)' = \bigcup_{B \in \mathcal{B}} B'$$

is equivalent to the statement

$$\sim (\forall B \in \mathcal{B},\ x \in B) \Leftrightarrow \exists B \in \mathcal{B} \sim (x \in B)$$

and, similarly, the second law

$$\Big(\bigcup_{B \in \mathcal{B}} B\Big)' = \bigcap_{B \in \mathcal{B}} B'$$

corresponds to the second negation rule

$$\sim (\exists B \in \mathcal{B}\; x \in B) \Leftrightarrow \forall B \in \mathcal{B} \sim (x \in B)$$

□

Example 1.7.3
(Principle of Mathematical Induction)

Using the proof by contradiction concept and the negation rules for qualifiers we can easily prove the *principle of mathematical induction.*

Let $T(n)$ be an open statement for $n \in \mathbb{N}$. Suppose that

1. $T(1)$ (is true)

2. $T(k) \Rightarrow T(k+1) \qquad \forall k \in \mathbb{N}$

then

$$T(n) \quad \forall n \quad \text{(is true)}$$

PROOF Assume, to the contrary, that the statement $T(n)$ $\forall n$ is not true. Then, by the negation rule, there exists a natural number, say k, such that $T(k)$ is false. This implies that the set

$$A = \{k \in \mathbb{N} \; : \; T(k) \text{ is false}\}$$

is not empty. Let ℓ be the minimal element of A (for a definition of this term, see Section 1.9). Then $\ell \neq 1$ since, according to the assumption, $T(1)$ is true. Thus ℓ must have a predecessor $\ell - 1$, for which $T(\ell - 1)$ holds. But according to the second assumption this implies that $T(\ell)$ is true as well, a contradiction. ∎

It is easy to generalize the notion of open statements to more than one variable; for example,

$$S(x,y) \qquad x \in X,\; y \in Y$$

then the two negation rules may be used to construct more complicated negation rules for many variables, e.g.,

$$\sim (\forall x \in X\; \exists y \in Y\; S(x,y)) \Leftrightarrow \exists x \in X\; \forall y \in Y\; \sim S(x,y)$$

This is done by negating one quantifier at a time.

$$\sim (\forall x \in X \ \ \exists y \in Y \ \ S(x,y))$$

$$\Leftrightarrow \sim (\forall x \in X \ \ (\exists y \in Y \ \ S(x,y)))$$

$$\Leftrightarrow \exists x \in X \ \sim (\exists y \in Y \ \ S(x,y))$$

$$\Leftrightarrow \exists x \in X \ \ \forall y \in Y \ \ \sim S(x,y)$$

We shall frequently use this type of technique throughout this book.

Relations

1.8 Cartesian Products, Relations

We are accustomed to the use of the term "relation" from elementary algebra. Intuitively, a *relation* must represent some sort of rule of correspondence between two or more objects—for example, "Bob is related to his brother Joe" or "real numbers are related to a scale on the x-axis." One of the ways to make this concept more precise is to recall the notion of the open statement from the preceding section.

Suppose we are given an open statement of two variables

$$R(x,y), \qquad x \in A, \qquad y \in B$$

We shall say that "a is related to b" and we write $a\,R\,b$ whenever $R(a,b)$ is true, i.e., upon the substitution $x = a$ and $y = b$ we get the true statement.

There is another equivalent way to introduce the notion of the relation by means of the set theory. First, we must introduce the idea of *ordered pairs* of mathematical objects and then the concept of the *product set*, or the *Cartesian product* of two sets.

Ordered Pairs. By an ordered pair (a,b) we shall mean the set $(a,b) = \{\{a\},\{a,b\}\}$. Here a is called the first member of the pair and b the second member.

We now have:

Cartesian Product. The Cartesian product of two sets A and B, denoted $A \times B$, is the set of all ordered pairs (a, b), where $a \in A$ and $b \in B$:

$$A \times B = \{(a, b) : \ a \in A \text{ and } b \in B\}$$

We refer to the elements a and b as *components* of the pair (a, b).

Two ordered pairs are equal if their respective components are equal, e.g.,

$$(x, y) = (a, b) \Leftrightarrow x = a \text{ and } y = b$$

Note that, in general,

$$A \times B \neq B \times A$$

More generally, if $A_1, A_2, \ldots, A_k$ are k sets, we define the Cartesian product $A_1 \times A_2 \times \ldots \times A_k$ to be the set of all ordered k-tuples $(a_1, a_2, \ldots, a_k)$, where $a_i \in A_i$, $i = 1, 2, \ldots, k$.

Example 1.8.1

Let $A = \{1, 2, 3\}$ and $B = \{x, y\}$. Then

$$A \times B = \{(1, x), (1, y), (2, x), (2, y), (3, x), (3, y)\}$$

$$B \times A = \{(x, 1), (x, 2), (x, 3), (y, 1), (y, 2), (y, 3)\}$$

□

Suppose now that we are given an open statement $R(x, y)$, $x \in A$, $y \in B$ and the corresponding relation R. With each such open statement $R(x, y)$ we may associate a subset of $A \times B$, denoted by R again, of the form

$$R = \{(a, b) \in A \times B : a \, R \, b\} = \{(a, b) \in A \times B : R(a, b) \text{ holds}\}$$

In other words, with every relation R we may identify a subset of the Cartesian product $A \times B$ of all the pairs in which the first element is related to the second by R. Conversely, if we are given an arbitrary subset $R \subset A \times B$, then we may define the corresponding open statement as

$$R(x, y) = \{(x, y) \in R\}$$

which in turn implies that

$$a \, R \, b \Leftrightarrow (a, b) \in R$$

Thus there is the one-to-one correspondence between the two notions of relations which let us identify relations with subsets of the Cartesian products. We shall prefer this approach through most of this book.

More specifically, the relation $R \subseteq A \times B$ is called a *binary relation* since two sets A and B appear in the Cartesian product of which R is a subset. In general, we may define a "k-ary" relation as a subset of $A_1 \times A_2 \times \ldots \times A_k$.

The *domain* of a relation R is the set of all elements of A that are related by R to at least one element in B. We use the notation "dom R" to denote the domain of a relation. Likewise, the *range* of R, denoted "range R", is the set of all elements of B to which at least one element of A is related by R. Thus

$$\text{dom } R = \{a : \ a \in A \text{ and } a\, R\, b \text{ for some } b \in B\}$$

$$\text{range } R = \{b : \ b \in B \text{ and } a\, R\, b \text{ for some } a \in A\}$$

We see that a relation, in much the same way as the common understanding of the word, is a rule that establishes an *association* of elements of a set A with those of another set B. Each element in the subset of A that is from R is associated by R with one or more elements in range R. The significance of particular relations can be quite varied; for example, the statement "Bob Smith is the father of John Smith" indicates a relation of "Bob Smith" to "John Smith," the relation being "is the father of." Other examples are cited below.

Example 1.8.2

Let $A = \{1, 2, 3\}$ and $B = \{\alpha, \beta, \gamma, \delta\}$, and let R be the subset of $A \times B$ that consists of the pairs $(1, \alpha), (1, \beta), (2, \delta), (3, \beta)$. Then

$$\text{dom } R = \{1, 2, 3\} = A$$

$$\text{range } R = \{\alpha, \beta, \delta\} \subset B$$

We see that R establishes a *multivalued correspondence* between elements of A and B. It is often instructive to represent relations such as this by diagrams; this particular example is indicated in Fig. 1.2 (a). Fig. 1.2(b) depicts the relation R and "sending" or "mapping" certain elements of A into certain elements of B. $\square$

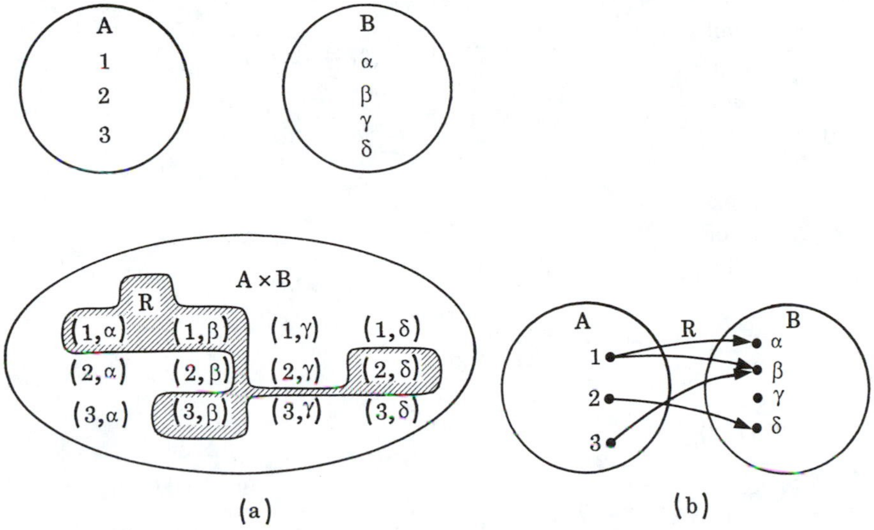

Figure 1.2
Graphical representation of a relation R from a set A to a set B.

Example 1.8.3

Let $P = \{a, b, c, \ldots\}$ be the set of all people in a certain school, and let $T = \{a, b, c\}$ denote the set of teachers at the school. We may consider relations on P of the type a "is a teacher of" d. For example, if a is a teacher of d, e, f, g; b is a teacher of h, i; and c is a teacher of k, l, we use R to mean "is a teacher of" and write the relations

$$a\,R\,d,\ a\,R\,e,\ a\,R\,f,\ a\,R\,g$$
$$b\,R\,h,\ b\,R\,i,\ \ c\,R\,k,\ c\,R\,l$$

□

Example 1.8.4

Let $X = \{2, 3, 4, 5\}$ and R mean "is divisible by on $\mathbb{N}$." Then

$$X \times X = \{(2,2), (2,3), (2,4), (2,5), (3,2), (3,3), (3,4), (3,5),$$
$$(4,2), (4,3), (4,4), (4,5), (5,2), (5,3), (5,4), (5,5)\}$$

Then $2\,R\,2$, $3\,R\,3$, $4\,R\,2$, $4\,R\,2$, $4\,R\,4$, and $5\,R\,5$; i.e.,

$$R = \{(2,2),(3,3),(4,2),(4,4),(5,5)\}$$

□

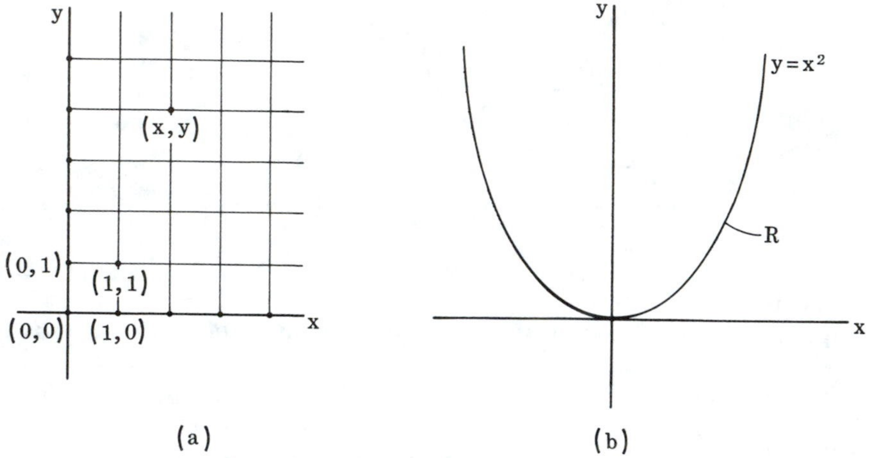

Figure 1.3

The relation $y = x^2$ from $\mathbb{R}$ into $\mathbb{R}$.

Example 1.8.5

Let $\mathbb{R}$ denote the set of real numbers and $\mathbb{R} \times \mathbb{R}$ the set of ordered pairs (x, y) of real numbers. Descartes exploited the fact that $\mathbb{R} \times \mathbb{R}$ could be represented geometrically as a plane, with an origin $(0,0)$ and each element $(x,y) \in \mathbb{R} \times \mathbb{R}$ a point with *Cartesian coordinates* (x, y) measured off in perpendicular distances x and then y according to some preselected directions (the x and y coordinate axes) and some preselected scale. This is illustrated in Fig. 1.3 (a).

Now consider the relation

$$R = \left\{(x,y) : \; x, y \in \mathbb{R}, \; y = x^2\right\}$$

that is, R is the set of ordered pairs such that the second member of the pair is the square of the first. This relation, of course, corresponds to the sets of all points on the parabola. The rule $y = x^2$ simply identifies a special subset of the Cartesian product (the plane) $\mathbb{R} \times \mathbb{R}$. □

We now list several types of relations that are of special importance. Suppose R is a relation on a set A, i.e., $R \subseteq A \times A$; then R may fall into one of the following categories:

1. *Reflexive.* A relation R is *reflexive* if and only if for every $a \in A$, $(a, a) \in R$; that is, $a\,R\,a$, for every $a \in A$.
2. *Symmetric.* A relation R is *symmetric* if and only if $(a, b) \in R \Longrightarrow (b, a) \in R$; that is, if $a\,R\,b$, then also $b\,R\,a$.
3. *Transitive.* A relation R is *transitive* if and only if $(a, b) \in R$ and $(b, c) \in R \Longrightarrow (a, c) \in R$; that is, if $a\,R\,b$ and if $b\,R\,c$, then $a\,R\,c$.
4. *Antisymmetric.* A relation R is *antisymmetric* if and only if for every $(a, b), \in R$, $(b, a) \in R \Longrightarrow a = b$; that is, if $a\,R\,b$ and $b\,R\,a$, then $a = b$.

The next two sections are devoted to a discussion of two fundamental classes of relations satisfying some of these properties.

Exercises

1.8.1. Let $A = \{\alpha, \beta\}$, $B = \{a, b\}$, and $C = \{c, d\}$. Determine

(i) $(A \times B) \cup (A \times C)$

(ii) $A \times (B \cup C)$

(iii) $A \times (B \cap C)$

1.8.2. Let R be the relation $<$ from the set $A = \{1, 2, 3, 4, 5, 6\}$ to the set $B = \{1, 4, 6\}$.

(i) Write out R as a set of ordered pairs.

(ii) Represent R graphically as a collection of points in the x, y-plane, $\mathbb{R} \times \mathbb{R}$.

1.8.3. Let R denote the relation $R = \{(a, b), (b, c), (c, b)\}$ on the set $R = \{a, b, c\}$. Determine whether or not R is (a) reflexive, (b) symmetric, or (c) transitive.

1.8.4. Let R_1 and R_2 denote two nonempty relations on set A. Prove or disprove the following:

(i) If R_1 and R_2 are transitive, so also is $R_1 \cup R_2$.

(ii) If R_1 and R_2 are transitive, so also is $R_1 \cap R_2$.

(iii) If R_1 and R_2 are symmetric, so also is $R_1 \cup R_2$.

(iv) If R_1 and R_2 are symmetric, so also is $R_1 \cap R_2$.

1.9 Partial Orderings

One of the most important kinds of relations is that of *partial ordering*. If $R \subset A \times A$ is a relation, then R is said to be a *partial ordering* of A, iff R is

(i) transitive

(ii) reflexive

(iii) antisymmetric

We also may say that A is *partially ordered* by the relation R. If, additionally, every two elements are comparable by R, i.e., for any $a, b \in A$, $(a, b) \in R$ or $(b, a) \in R$, then the partial ordering R is called the *linear ordering* (or *total ordering*) of set A and A is said to be *linearly (totally) ordered* by R.

Example 1.9.1

The simplest possible example of partial ordering is furnished by any subset A of the real line $\mathbb{R}$ and the usual $\leq$ (less than or equal) inequality relation. In fact, since every two real numbers are comparable, A is totally ordered by $\leq$. □

Example 1.9.2

A nontrivial example of partial ordering may be constructed in $\mathbb{R}^2 = \mathbb{R} \times \mathbb{R}$. Let $\boldsymbol{x} = (x_1, x_2)$ and $\boldsymbol{y} = (y_1, y_2)$ be two points of $\mathbb{R}^2$. We shall say that

$$\boldsymbol{x} \leq \boldsymbol{y} \quad \text{iff} \quad x_1 \leq y_1 \quad \text{and} \quad x_2 \leq y_2$$

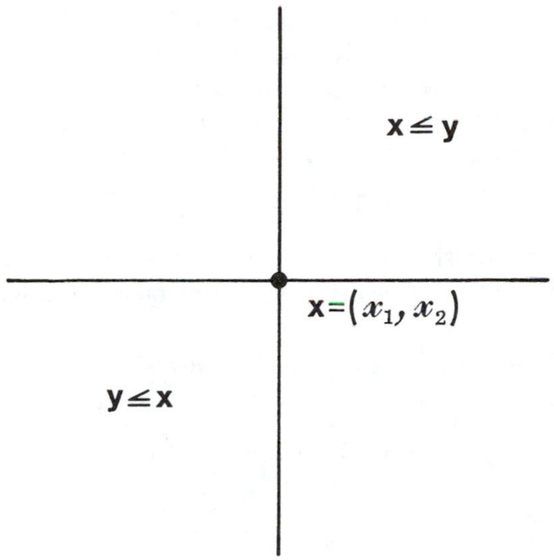

Figure 1.4

Illustration of a partial ordering in $\mathbb{R}^2$.

Note that we have used the same symbol $\leq$ to define the new relation as well as to denote the usual "greater or equal" correspondence for the numbers (coordinates of $\boldsymbol{x}$ and $\boldsymbol{y}$). The reader will easily verify that the relation is transitive, reflexive and antisymmetric, and therefore it is a partial ordering of $\mathbb{R}^2$. It is *not,* however, a total ordering of $\mathbb{R}^2$. To visualize this let us pick a point $\boldsymbol{x}$ in $\mathbb{R}^2$ and try to specify points $\boldsymbol{y}$ which are "greater or equal" (i.e., $\boldsymbol{x} \leq \boldsymbol{y}$) than $\boldsymbol{x}$ and those which are "smaller or equal" (i.e., $\boldsymbol{y} \leq \boldsymbol{x}$) than $\boldsymbol{x}$.

Drawing horizontal and vertical lines through $\boldsymbol{x}$ we subdivide the whole $\mathbb{R}^2$ into four quadrants (see Fig. 1.4). It is easy to see that the upper-right quadrant (including its boundary) corresponds to the points $\boldsymbol{y}$ such that $\boldsymbol{x} \leq \boldsymbol{y}$, while the lower-left quadrant contains all $\boldsymbol{y}$ such that $\boldsymbol{y} \leq \boldsymbol{x}$. In particular, all the points which do not belong to the two quadrants are neither "greater" nor "smaller" than $\boldsymbol{x}$. Thus $\leq$ is not a total ordering of $\mathbb{R}^2$.

$\square$

Example 1.9.3

Another common example of a partial ordering is furnished by the *inclusion* relation for sets. Let $\mathcal{P}(A) = 2^A$ denote the power class of the set A and define a relation R on $\mathcal{P}(A)$ such that $(C, B) \in R$ iff $C \subset B$, where $C,\ B \in \mathcal{P}(A)$. Then R is a partial ordering. R is transitive: if $C \subset B$ and $B \subset D$, then $C \subset D$. R is reflexive, since $C \subset C$. Finally, R is antisymmetric, for if $C \subset B$ and $B \subset C$, then $B = C$. □

The notion of a partial ordering makes it possible to give a precise definition to the idea of the greatest and least elements of a set.

(i) An element of $a \in A$ is called a *least element* of A, iff $a\,R\,x$, for every $x \in A$.

(ii) An element $a \in A$ is called a *greatest element* of A, iff $x\,R\,a$, for every $x \in A$.

(iii) An element $a \in A$ is called a *minimal element* of A, iff $x\,R\,a \Longrightarrow x = a$, for every $x \in A$.

(iv) An element $a \in A$ is called a *maximal element* of A, iff $a\,R\,x \Longrightarrow x = a$, for every $x \in A$.

While we used the term "a is a least element," it is easy to show that when a exists, it is unique. Indeed, if a_1 and a_2 are two least elements of A, then in particular $a_1\,R\,a_2$ and $a_2\,R\,a_1$ and therefore $a_1 = a_2$.

Similarly, if the greatest element exists, then it is unique. Note also, that in the case of a totally ordered set, every two elements are comparable and therefore the notions of the greatest and maximal as well as the least and minimal elements coincide with each other (a maximal element is the greatest element of all elements comparable with it). In the general case, of a set only partially ordered, if the greatest element exists, then it is also the unique maximal element of A (the same holds for the least and minimal elements). The notion of maximal (minimal) elements is more general in the sense that there may not be a greatest (least) element, but still maximal (minimal) elements, in general not unique, may exist. Examples 1.9.4–1.9.6 illustrate the difference.

The notion of a partial ordering makes it also possible to define precisely the idea of *bounds* on elements of sets. Suppose R is a partial ordering on a set B and $A \subset B$. Then, continuing our list of properties, we have

(v) An element $a \in B$ is an *upper bound of A* iff $x\,R\,a$, for every $x \in A$.

(vi) An element $a \in B$ is a *lower bound of A* iff $a\,R\,x$, for every $x \in A$.

(vii) The *least upper bound of* A (i.e., the *least* element of the set of all upper bounds of A), denoted $\sup(A)$, is called the *supremum* of A.

(viii) The *greatest lower bound* of A, denoted $\inf(A)$, is called the *infimum* of A.

Note that if A has the *greatest* element, say a, then $a = \sup(A)$. Similarly, the *smallest* element of a set coincides with its *infimum*.

Example 1.9.4

The set $\mathbb{R}$ of real numbers is totally ordered in the classical sense of real numbers (indeed, if $x \leq y$, then $x+z \leq y+z \; \forall \; z \in \mathbb{R}$ and if $x \geq 0$ and $y \geq 0$ then $xy \geq 0$; see Section 1.17). Let $A = [0,1) = \{x \in \mathbb{R} : \; 0 \leq x < 1\} \subset \mathbb{R}$ be the interval "closed on its left end and open on the right." Then:

- A is bounded from above by every $y \in \mathbb{R}$, $y \geq 1$.
- $\sup A = 1$.
- There is neither a greatest nor a maximal element of A.
- A is bounded from below by every $y \in \mathbb{R}$, $y \leq 0$.
- $\inf A =$ the least element of $A =$ the minimal element of $A = 0$.

□

Example 1.9.5

Let $\mathbb{Q}$ denote the set of all rational numbers and A be a subset of $\mathbb{Q}$ such that for every $a \in A$, $a^2 < 2$. Then:

- A is bounded from above by every $y \in \mathbb{Q}$, $y > 0$, $y^2 > 2$.
- There is not a least upper bound of A ($\sup A$) and therefore there is neither a greatest nor a maximal element of A.
- Similarly, no $\inf A$ exists.

□

We remark that set A is referred to as *order complete* relative to a linear ordering R if and only if every nonempty subset of A that has an upper bound also has a least upper bound. This idea makes it possible to dis-

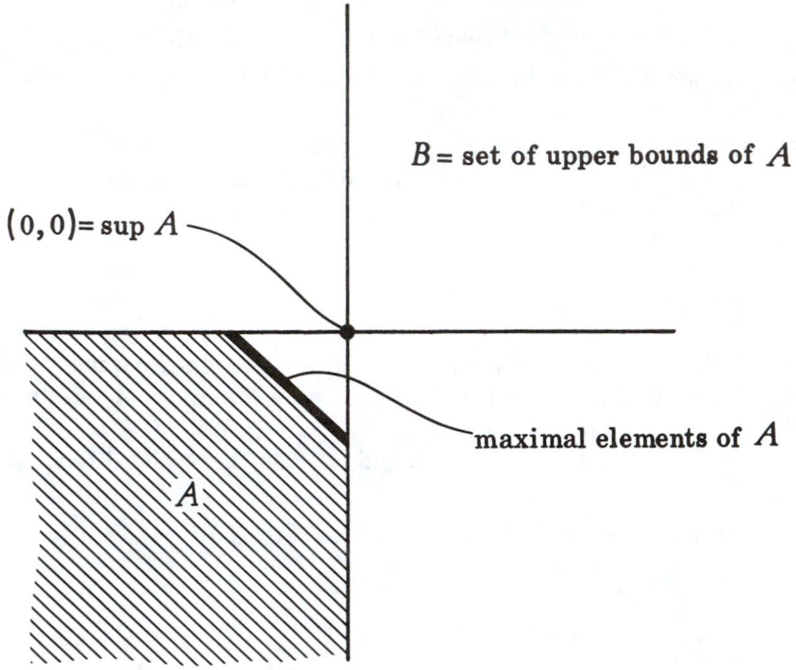

Figure 1.5
Illustration of the notion of an upper bound, supremum, maximal and the greatest elements of a set.

tinguish between real numbers (which are order complete) and rational numbers (which are not order complete).

Example 1.9.6

Let $A \subset \mathbb{R}^2$ be the set represented by the shaded lower-left area in Fig. 1.5, including its boundary, and consider the partial ordering of $\mathbb{R}^2$ discussed in Example 1.9.2. Then:

- The first (upper-right) quadrant of $\mathbb{R}^2$, denoted B, (including its boundary) consists of all *upper bounds* of A.
- The origin $(0,0)$ is the *least* element of B and therefore is the *supremum* of A.
- All the points belonging to the "outer" corner (see Fig. 1.5) are *maximal elements* of A.

- There is not a *greatest* element of A.

The Kuratowski-Zorn Lemma and the Axiom of Choice. The notion of maximal elements leads to a fundamental mathematical axiom known as the Kuratowski-Zorn lemma:

The Kuratowski-Zorn Lemma. Let A be a nonempty partially ordered set. If every linearly ordered subset of A contains an upper bound, then A contains at least one maximal element. □

Kuratowski-Zorn's lemma asserts the existence of certain maximal elements without indicating a constructive process for finding them. It can be shown that Kuratowski-Zorn's lemma is equivalent to the axiom of choice:

Axiom of Choice. Let $\mathcal{A}$ be a collection of disjoint sets. Then there exists a set B such that $B \subset \cup A$, $A \in \mathcal{A}$ and, for every $A \in \mathcal{A}$, $B \cap A$ has exactly one element.

The Kuratowski-Zorn lemma is an essential tool in many existence theorems covering *infinite-dimensional vector spaces* (e.g., the existence of Hamel Bases, proof of the Hahn-Banach theorem, etc.).

Exercises

1.9.1. Consider the partial ordering of $\mathbb{R}^2$ in Example 1.9.2. Construct an example of a set A which has many minimal elements. Can such a set have the least element?

1.9.2. Consider the following relation in $\mathbb{R}^2$.

$$\boldsymbol{x} \, R \, \boldsymbol{y} \quad \text{iff } (x_1 < y_1 \text{ or } (x_1 = y_1 \text{ and } x_2 \le y_2))$$

(i) Show that R is a *linear (total) ordering* of $\mathbb{R}^2$.

(ii) For a given point $\boldsymbol{x} \in \mathbb{R}^2$ construct the set of all points $\boldsymbol{y}$ "greater than or equal" to $\boldsymbol{x}$, i.e., $\boldsymbol{x} \, R \, \boldsymbol{y}$.

(iii) Does the set A from Example 1.9.6 have the greatest element with respect to this partial ordering?

1.9.3. Consider a contact problem for the simply supported beam shown in Fig. 1.6. The set K of all *kinematically admissible deflections* $w(x)$ is defined as follows

$$K = \{w(x) : w(0) = w(l) = 0 \text{ and } \\ w(x) \le g(x), x \in (0, l)\}$$

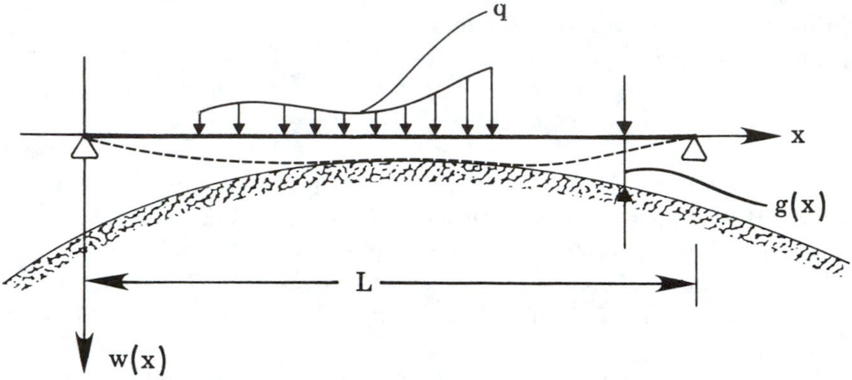

Figure 1.6
A contact problem for a beam.

where $g(x)$ is an initial gap function specifying the distance between the beam and the obstacle.
Let V be a class of functions defined on $(0, l)$ including the gap function $g(x)$.
For elements $w \in V$ define the relation

$$w\,R\,v (w \leq v) \text{ iff } w(x) \leq v(x) \text{ for every } x \in (0, l)$$

(i) Show that R is a partial ordering of V.

(ii) Show that R is not a linear ordering of V.

(iii) Show that the set K can be rewritten in the form

$$K = \{w(x) : \; w(0) = w(l) = 0 \text{ and } w \leq g\}$$

1.9.4. Let $\mathcal{P}(A)$ denote the power class of a set A: then $\mathcal{P}(A)$ is partially ordered by the inclusion relation (see Example 1.9.3). Does $\mathcal{P}(A)$ have the smallest and greatest elements?

$\square$

1.10 Equivalence Relations, Equivalence Classes

Equivalence Relations. A relation $R \subset A \times A$ is called an *equivalence relation on* A iff it is

(i) reflexive

(ii) transitive

(iii) symmetric

Such relations may serve to generalize the familiar notion of equality $(=)$ in the set of real numbers. Numerous other examples can be cited.

Example 1.10.1

Let C denote the set of male children living in a certain residential area. We may introduce a relation "is a brother of" on $C \times C$. If we accept that a given male child can be the brother of himself, then this relation on C is reflexive. Also, if a is the brother of b, then, of course, b is the brother of a. Moreover, if b is also the brother of c, then so is a. It follows that "is the brother of" is an equivalence relation on C. $\square$

Example 1.10.2

Let L denote the set of all straight lines in the euclidean plane. The rule "is parallel to" defines a relation R on L; indeed, R is an equivalence relation: R is transitive, since if line a is parallel to line b and b is parallel to c, then a is parallel to c ($a\,R\,b$ and $b\,R\,c \Longrightarrow a\,R\,c$). R is symmetric, since $a\,R\,b$ and $b\,R\,a$ and R is reflexive, if we admit that every line is parallel to itself. $\square$

Example 1.10.3

Let $\mathbb{Z}^+ = \{0, 1, 2, \ldots\}$ denote the nonnegative integers and R be a relation on $\mathbb{Z}^+$ such that $(a, b) \in R$ if $a - b$ is divisible by 3 on $\mathbb{Z}$. Then R is an equivalence relation. It is reflexive, since $a - a = 0$ is divisible by 3; it is transitive, for if $a - b = 3r$ and $b - c = 3s$, where r, $s \in \mathbb{Z}$, then $a - c = 3(r + s)$. Finally, R is symmetric, because $b - a$ is divisible by 3 if $a - b$ is divisible by 3. $\square$

Equivalence Classes. Let A be a set and R be an equivalence relation defined on A. If $a \in A$, the elements $x \in A$ satisfying $x \, R \, a$ constitute a subset of A, denoted $R[a]$, called an *equivalence class of* a. That is,

$$R[a] = \{x : \; x \in A, x \, R \, a\}$$

Example 1.10.4

Let $\mathbb{Z}^+$ be the set of nonnegative integers and define a relation R on $\mathbb{Z}^+$ such that $(a, b) \in R$ iff $a - b$ is divisible by 3 on $\mathbb{Z}$. Consider the equivalence class $R[1]$. By definition, $R[1]$ is the set of elements x in $\mathbb{Z}^+$ such that $x \, R 1$; i.e., $x - 1$ is divisible by 3. Hence,

$$R[1] = \{1, 4, 7, 10, \ldots\}$$

Similarly,

$$R[2] = \{2, 5, 8, 11, \ldots\}$$

$$R[3] = \{0, 3, 6, 9, 12, \ldots\}$$

$$R[4] = \{1, 4, 7, 10, \ldots\}$$

and so forth. Notice that $4 \in R[1]$ and (as a consequence) $R[4] = R[1]$.
□

Example 1.10.5

The rational numbers $\mathbb{Q}$ are equivalence classes on pairs of integers. Consider the relation R on $\mathbb{Z} \times (\mathbb{Z} - \{0\})$ defined by $\{(p, q), (r, s)\} \in R$ iff $ps = qr$. Then $R[(p, q)]$ is a set of pairs $(r, s) \in \mathbb{Z} \times \mathbb{Z}$ such that $(r, s) R (p, q)$; i.e., $r/s = p/q$. Of course, instead of always writing out this elaborate equivalence class notation, we prefer to simply denote

$$R[(p, q)] = \frac{p}{q}$$

□

Partition of a Set. A class $\mathcal{B} \subset \mathcal{P}(A)$ of nonempty subsets of a set A is called a *partition of* A iff

(i) $\cup\{B : B \in \mathcal{B}\} = A$

(ii) every pair of distinct subsets of $\mathcal{B}$ is disjoint, i.e., if $B,\ C \in \mathcal{B}$, $B \neq C$ then $B \cap C = \emptyset$

We are approaching the important idea of an equivalence relation R partitioning a set into equivalence classes. For example, if A is the set of all triangles, the relations "is congruent to" or "has the same area as" are equivalence relations that segregate triangles into certain equivalence classes in A. To make this assertion precise, we first need:

LEMMA 1.10.1

Let R denote an equivalence relation on a set A and $R[a]$ an equivalence class for $a \in A$. If $b \in R[a]$, then $R[b] = R[a]$.

PROOF By definition, $b \in R[a] = \{x :\ x \in A, x\, R\, a\} \Longrightarrow bRa$; similarly, $x \in R[b] \Longrightarrow x\, R\, b$. Since R is transitive, $x\, R\, a$, and, therefore, $R[b] \subseteq R[a]$. A repetition of the argument assuming $x \in R[a]$ yields $R[a] \subseteq R[b]$, which means that $R[a] = R[b]$ and completes the proof. ∎

Example 1.10.6

Recall the previous example in which R was a relation on $\mathbb{Z}^+$ such that $(a, b) \in R \Longrightarrow a - b$ was divisible by 3. Observe that $R[1] = R[4]$. □

LEMMA 1.10.2

If $R[a] \cap R[b] \neq \emptyset$, then $R[a] = R[b]$.

PROOF Suppose that $R[a] \cap R[b] = \{\alpha, \beta, \ldots\} \neq \emptyset$. Then $\alpha \in R[a]$ and $\alpha \in R[b]$. By Lemma 1.10.1, this means that $R[a] = R[b]$. ∎

LEMMA 1.10.3

If R is an equivalence relation on A and $R[a]$ an equivalence class for $a \in A$, then

$$\cup\{R[x] :\ x \in A\} = A$$

PROOF Let $Y = \cup\{R[x] :\ x \in A\}$. Then each $y \in Y$ belongs to a $R[x]$

for some $x \in A$, which means that $Y \subseteq A$. Consequently, $\cup\{R[x] : x \in A\} \subset A$. Now take $z \in A$. Since R is reflexive, $z\,R\,z$ and $z \in R[z]$. Therefore $A \subseteq \cup\{R[x] : x \in A\}$. This completes the proof. ∎

Finally, we have

PROPOSITION 1.10.1

An equivalence relation R on a set A effects a partitioning of A into equivalence classes. Conversely, a partitioning of A defines an equivalence relation on A.

PROOF Let $R[a], R[b], R[c], \dots$ denote equivalence classes induced on A by R, with $a, b, c, \dots \in A$. By Lemma 1.10.3, $R[a] \cup R[b] \cup R[c] \cup \dots = A$, so that property (i) of a partitioning of A is satisfied. Also (ii) is satisfied, because if $R[a]$ and $R[b]$ are distinct equivalence classes, they are disjoint by Lemma 1.10.2.

To prove the converse, let $\mathcal{B}$ be any partition of A and define a relation R on A such that $a\,R\,b$ iff there exists a set B in the partition such that $a, b \in B$. Clearly, such a relation is both reflexive and symmetric: for every $a \in A$ there is a set B in the partition such that $a \in B$; formally, $a, a \in B$, also $a, b \in B$ implies that $b, a \in B$.

Now suppose $a\,R\,b$ and $b\,R\,c$. This implies that $a, b \in B$ for some B and $b, c \in C$ for some C. Consequently, $b \in B \cap C$, which, according to the definition of partition, implies that $B = C$. Thus both a and c belong to the same set $B = C$ and therefore $a\,R\,c$, which means that R is transitive. This completes the proof that R is an equivalence relation. ∎

Quotient Sets. The collection of equivalence classes of A is a class, denoted by A/R, called the *quotient* of A by R: $A/R = \{R[a] : a \in A\}$. According to Proposition 1.10.1, the quotient set A/R is, in fact, a partition of A.

Example 1.10.7

Let $\mathcal{B}$ be a collection of circles in $\mathbb{R}^2$, centered at the origin

$$B(\mathbf{0}, \varepsilon) = \left\{ \boldsymbol{x} \in \mathbb{R}^2 : \text{ distance from } \mathbf{0} \text{ to } \boldsymbol{x} = \varepsilon \right\}$$

Let us identify the origin $\mathbf{0}$ with the circle of zero radius. Obviously,

$$\mathcal{B} = \{B(\mathbf{0}, \varepsilon) \mid \varepsilon \geq \mathbf{0}\}$$

is a partition of $\mathbb{R}^2$.

Defining an equivalence relation R by $\boldsymbol{x}\, R\, \boldsymbol{y}$ iff the distance from $\boldsymbol{x}$ to $\mathbf{0}$ = distance from $\boldsymbol{y}$ to $\mathbf{0}$, we may identify the circles with equivalence classes of points in $\mathbb{R}^2$ and the partition with the quotient set $\mathbb{R}^2/R$. □

We shall return to the important issue of equivalence classes and quotient sets in the context of vector spaces in Chapter 2.

Exercises

1.10.1. (i) Let T be the set of all triangles in the plane $\mathbb{R}^2$. Show that "is similar to" is an equivalence relation on T.

(ii) Let P be the set of all polygons in the plane $\mathbb{R}^2$. Show that "has the same number of vertices" is an equivalence relation on P.

(iii) For part (i) describe the equivalence class $[T_0]$, where T_0 is a (unit) right, isoceles triangle with unit sides parallel to the x- and y-axes.

(iv) For part (ii) describe the equivalence class $[P_0]$, where P_0 is the unit square

$$\{(x,y):\ 0 \leq x \leq 1,\ 0 \leq y \leq 1\}.$$

(v) Give examples of quotient sets corresponding to the relations in (i) and (ii).

1.10.2. Let $A = \mathbb{N} \times \mathbb{N}$, where $\mathbb{N}$ is the set of natural numbers. The relation $(x,y)R(u,v) \iff x + v = u + y$ is an equivalence relation on A. Determine the equivalence classes $[(1,1)]$, $[(2,4)]$, $[(3,6)]$.

Functions

1.11 Fundamental Definitions

A *function* from a set A into a set B, denoted $f : A \to B$, is a relation $f \subset A \times B$ such that

(i) for every $x \in A$ there exists a $y \in B$ such that $x \, f \, y$

(ii) for every $x \in A$ and y_1, $y_2 \in B$, if $x \, f \, y_1$ and $x \, f \, y_2$, then $y_1 = y_2$

In other words, for every $x \in A$ there exists a *unique* $y \in B$ such that $x \, f \, y$. We write

$$y = f(x)$$

If $f : A \to B$, A is called the *domain* of f, denoted $\operatorname{dom} f$, and B is called the *co-domain* of f. Clearly, a function f from A into B is a relation $R \subset A \times B$ such that for every $(a, b) \in R$, each first component a appears only once. Thus a function may be thought of as a "single-valued relation" since each element in $\operatorname{dom} f$ occurs only once in f. We also note that $\operatorname{dom} f \neq \emptyset$. The element $y \in B$ in $f(x) = y$ is called the *image* of $x \in A$, or *the value of the function* at x.

The *range* of a function $f : A \to B$, denoted $\mathcal{R}(f)$, is the set of elements in B that are images of elements in A; i.e., $\mathcal{R}(f)$ is the set of all images of f:

$$\mathcal{R}(f) = \{f(a) : a \in A\}$$

$\mathcal{R}(f)$ is also sometimes called the *image set.*

Once the relations are identified with subsets of the appropriate Cartesian products, functions are identified with their *graphs.* The *graph* of function $f : A \to B$ is the set

$$\textit{graph}\; f = \{(x, f(x)) : x \in A\}$$

It is customary to use the terms *function*, *mapping*, *transformation*, and *operator* synonymously. Thus, if $f : A \to B$, we say that "f maps A into B" or "f is a transformation from A into B" or "f is an operator from A into B" (in some works the term "operator" is reserved for functions whose domains are subsets of *spaces*; we consider these later).

Clearly, this generalization of the elementary idea of a function is in complete accord with our first notions of functions: Each pair $(a, b) \in$

f associates an $a \in A$ with an element $b \in B$. The function thereby establishes a correspondence between elements of A and those of B that appear in f.

For example, we have often encountered expressions of the form

$$y = f(x) = x^2$$

that we read as "y is a function of x." Technically, we consider a set of (say) real numbers $\mathbb{R}$ to which there belong the elements x, and another set $\mathbb{R}^+$ of nonnegative numbers $y \in \mathbb{R}^+$. The particular subset of $\mathbb{R} \times \mathbb{R}^+$, which is the function f under consideration, is identified by the *rule* $y = x^2$. We may define the *function* f by

$$f = \{(x, y) : x \in \mathbb{R},\ y \in \mathbb{R}^+,\ y = x^2\}$$

Example 1.11.1

Let $\mathbb{R}$ be the real numbers and consider the relation

$$R = \left\{(x, y) : x, y \in \mathbb{R},\ x^2 + \left(\frac{y}{2}\right)^2 = 1\right\}$$

Obviously, R defines the points on the ellipse shown in Fig. 1.7(a). Remarkably, R is *not* a function, since elements $x \in \mathbb{R}$ are associated with *pairs* of elements in $\mathbb{R}$. For example, both $(0, +2)$ and $(0, -2) \in R$. □

Example 1.11.2

The relation $R = \{(x, y) : x,\ y \in \mathbb{R},\ y = \sin x\}$ is shown in Fig. 1.7(b). This relation *is* a function. Its domain is $\mathbb{R}$, the entire x-axis, $-\infty < x < \infty$. Its co-domain is also $\mathbb{R}$, i.e., the y-axis. Its range is the set $\{y : y \in \mathbb{R},\ -1 \leq y \leq 1\}$. Notice that specific values of $y \in \mathcal{R}(R)$ are the images of infinitely many points in the domain of R. Indeed, $y = 1$ is the image of $\pi/2$, $5\pi/2$, $9\pi/2$, ... □

An arbitrary function $f : A \to B$ is said to map A *into* B and this terminology suggests nothing special about the range of f or the nature of its values in B. To identify special properties of f, we use the special nomenclature listed below:

1. *Surjective (Onto) Functions.* A function $f : A \to B$ is *surjective*, or from A *onto* B if every $b \in B$ is the image of some element of A.

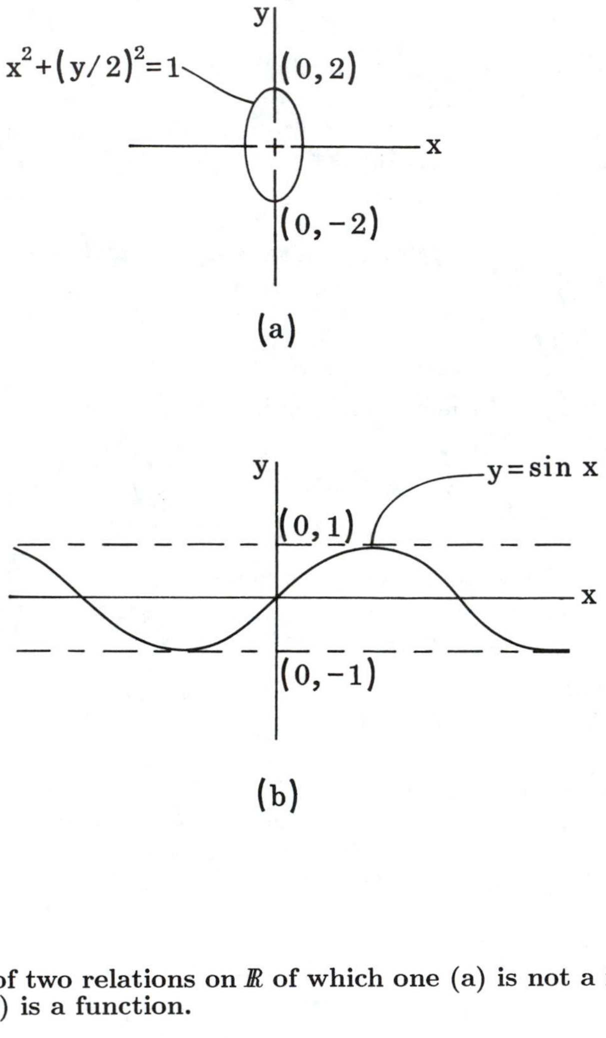

Figure 1.7
Examples of two relations on $\mathbb{R}$ of which one (a) is not a function and one (b) is a function.

2. *Injective (One-to-One) Functions.* A function $f: A \to B$ is said to be *injective* or *one-to-one* (denoted 1:1) from A into B iff, for every $b \in \mathcal{R}(f)$, there is exactly one $a \in A$ such that $b = f(a)$.

3. *Bijective (One-to-One and Onto) Functions.* A function $f: A \to B$ is *bijective*, or *one-to-one and onto*, iff it is both injective and surjective, i.e., iff every $b \in B$ is the unique image of some $a \in A$.

Figure 1.8 illustrates geometrically the general types of functions. The correspondence indicated in Fig. 1.8(a) is a relation, but is not a function, because elements of A do not have distinct images in B. That in Fig. 1.8(d) is one-to-one, but not onto, because the element b_3 is not an image of an element of A.

Example 1.11.3

Let $\mathbb{R}$ denote the set of real numbers and $\mathbb{R}^+$ the set of nonnegative real numbers. Let f denote the rule $f(x) = x^2$. Then consider the following functions:

1. $f_1: \mathbb{R} \to \mathbb{R}$. This function is not one-to-one, since both $-x$ and $+x$ are mapped into x^2. It is not onto, since the negative real numbers are in the co-domain $\mathbb{R}$, but are not images.

2. $f_2: \mathbb{R} \to \mathbb{R}^+$. This function is not one-to-one, but it is onto.

3. $f_3: \mathbb{R}^+ \to \mathbb{R}$. This function is one-to-one, but it is *not* onto.

4. $f_4: \mathbb{R}^+ \to \mathbb{R}^+$. This function is bijective; it is both one-to-one and onto.

Note that although the rule $f(x) = x^2$ defining each function f_1, f_2, f_3, and f_4 is the same, the four are quite different functions. □

Direct and Inverse Images. The set

$$f(C) = \{f(a) : a \in C \subseteq A\}$$

is called the *direct image* of C. Obviously,

$$f(C) \subset \mathcal{R}(f)$$

Likewise, suppose $f: A \to B$ and D is a subset of B. Then the set

$$f^{-1}(D) = \{a : f(a) \in D\}$$

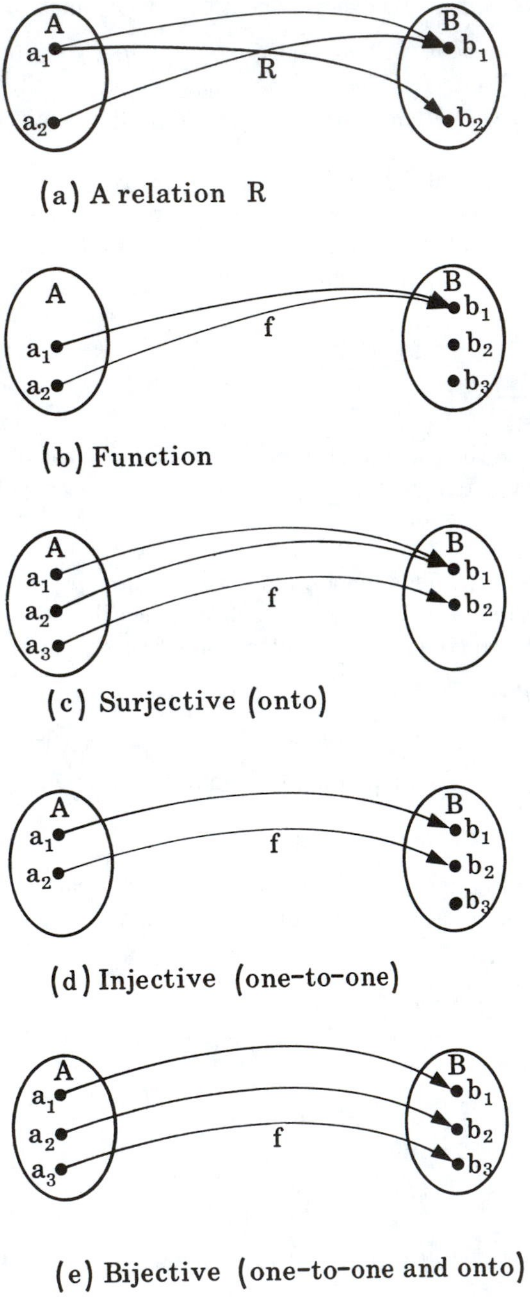

Figure 1.8
Classification of functions $f:\ A \to B$.

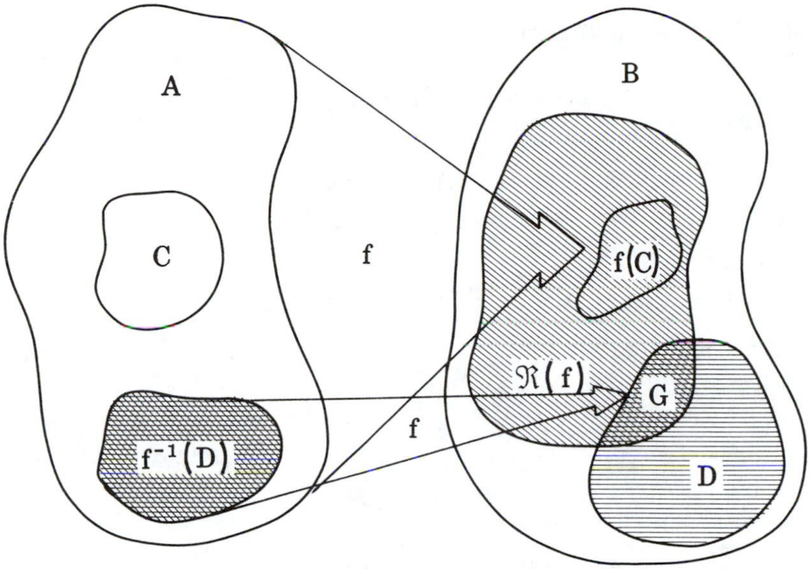

Figure 1.9

Illustration of $f(C)$, $\mathcal{R}(f)$, and $f^{-1}(D)$ for $f: A \to B$.

is called the *inverse image* of D *under* f. Clearly,

$$f^{-1}(D) \subset A$$

i.e., $f^{-1}(D)$ is a subset of the domain of f.

These ideas are illustrated symbolically in Fig. 1.9, where the set G denotes the intersection of $\mathcal{R}(f)$ and a subset $D \subset B$. It is clear that $f^{-1}(D)$ consists of those elements in A that have images in $G \subset D$; in other words, not all of D need consist of images of elements of A.

We now list several properties involving functions. Let $f: X \to Y$, $A \subset X$, $B \subset X$, $D \subset Y$, and $F \subset Y$. Then the following hold:

1. $f(A \cup B) = f(A) \cup f(B)$
2. $f(A \cap B) \subset f(A) \cap f(B)$
3. $f^{-1}(D \cup F) = f^{-1}(D) \cup f^{-1}(F)$
4. $f^{-1}(D \cap F) = f^{-1}(D) \cap f^{-1}(F)$

Example 1.11.4

Let $A = \{-1, -2\}$, $B = \{1, 2\}$, $f(x) = x^2$. Then

$$A \cap B = \emptyset$$

$$f(A \cap B) = \emptyset$$

However,

$$f(A) = \{1, 4\}, \qquad f(B) = \{1, 4\}$$

Consequently,

$$f(A) \cap f(B) = \{1, 4\} \neq \emptyset$$

$$\neq f(A \cap B)$$

However, $f(A \cup B) = f(A) \cup f(B)$ always. $\square$

Example 1.11.5
(Proof of $f(A \cup B) = f(A) \cup f(B)$)

Let $y \in f(A \cup B)$. Then there exists an $x \in A$ or B such that $y = f(x)$. If $x \in A$, then $y = f(x) \in f(A)$; if $x \in B$, then $y = f(x) \in f(B)$. Consequently, $y \in f(A) \cup f(B)$, which proves that $f(A \cup B) \subseteq f(A) \cup f(B)$. Conversely, let $w \in f(A) \cup f(B)$. Then w is the image of an $x \in A$ or an $x \in B$; i.e., $w = f(x)$, $x \in A \cup B$. Hence $w \in f(A \cup B)$ and, therefore, $f(A) \cup f(B) \subseteq f(A \cup B)$. $\square$

Example 1.11.6
(Proof of $f^{-1}(D \cup F) = f^{-1}(D) \cup f^{-1}(F)$)

Suppose that $x \in f^{-1}(D \cup F)$. Then there exists a $y \in D$ or F such that $y = f(x)$; i.e., $f(x) \in D \cup F$. If $f(x) \in D$, $x \in f^{-1}(D)$ and if $f(x) \in F$, $x \in f^{-1}(F)$, so that $x \in f^{-1}(D) \cup f^{-1}(F)$ and $f^{-1}(D \cup F) \subseteq f^{-1}(D) \cup f^{-1}(F)$. Following the reverse procedure, we can show that $f^{-1}(D) \cup f^{-1}(F) \subseteq f^{-1}(D \cup F)$, which completes the proof. $\square$

It is important to note that $f(x) \in D \Longrightarrow x \in f^{-1}(D)$, but $f(x) \in f(C) \not\Longrightarrow x \in C$, because f need not be injective. This is illustrated in the diagram shown in Fig. 1.10. Consider, for example, $f : \mathbb{R} \to \mathbb{R}$, $f(x) = x^2$, and let

$$A = \{1, 2\} \qquad D = \{-1, -2, -3\}$$

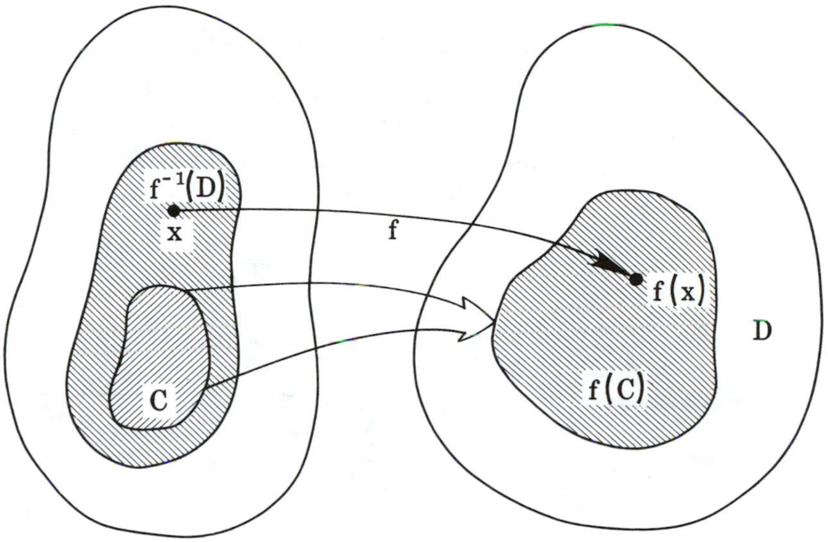

Figure 1.10

Illustration of the fact that $f(x) \in D \Longrightarrow x \in f^{-1}(D)$ **but** $f(x) \in f(C) \not\Longrightarrow x \in C$**.**

Then, since $f^{-1}(D)$ is the set in $\mathbb{R}$ for which $x^2 = -1, -2$, or -3, $f^{-1}(D) = \emptyset$. However, $f^{-1}(f(A)) \supset A$. In fact, $f(A) = \{1, 4\}$ and $f^{-1}(f(A)) = f^{-1}(\{1, 4\}) = \{1, -1, 2, -2\}$.

We conclude this section with a list of some important types of functions. Let $f: A \to B$. Then

1. f is a *constant function* iff there exists a $b_0 \in B$ such that for every $a \in A$, $b_0 = f(a)$.
2. The function $i_A : A \to A$ such that for every $a \in A$, $i_A(x) = x$ is called the *identity function* for A.
3. If $f: X \to Y$ and $A \subset X$, the function $f|_A: A \to Y$ is called the *restriction* of f to A if $f|_A(x) = f(x)$ for every $x \in A$.
4. If $f: A \to B$, $A \subset X$, and if there exists a function $g: X \to B$ such that $g|_A = f$, then g is called an *extension of f to X*.

Let now $f_1 : A_1 \to B_1$ and $f_2 : A_2 \to B_2$ be two functions. Then

5. The function denoted $f_1 \times f_2$ from the Cartesian product $A_1 \times A_2$ into the Cartesian product $B_1 \times B_2$ defined by

$$(f_1 \times f_2)(x_1, x_2) = (f_1(x_1), f_2(x_2))$$

is called the *Cartesian product of functions* f_1 and f_2.
Similarly, if $f_1 : A \to B_1$ and $f_2 : A \to B_2$ are defined on the same set A, we define the *composite function* of functions f_1 and f_2, denoted $(f_1, f_2) : A \to B_1 \times B_2$, as

$$(f_1, f_2)(x) = (f_1(x), f_2(x))$$

Exercises

1.11.1. Prove that $f^{-1}(D \cap H) = f^{-1}(D) \cap f^{-1}(H)$.

1.11.2. Prove or disprove that $f(A - B) = f(A) - f(B)$.

1.11.3. Prove or disprove that $f^{-1}(A) - f^{-1}(B) = f^{-1}(A - B)$.

1.11.4. Prove that if C is an arbitrary set, $f^{-1}(\mathcal{R}(f) \cap C) = f^{-1}(C)$.

1.12 Compositions. Inverse Functions

Compositions or Product Functions. Let $f : X \to Y$ and $g : Y \to Z$. Then f and g define a *product function*, or *composition*, denoted $g \circ f$ (or sometimes simply gf), from X into Z, $g \circ f : X \to Z$. We define $g \circ f$ by saying that for every $x \in X$,

$$(g \circ f)(x) = g(f(x))$$

Example 1.12.1

$f : \mathbb{R} \to \mathbb{R}$, $f(x) = x^2$ and $g : \mathbb{R} \to \mathbb{R}$, $g(x) = 1 + x$. Then

$$(gf)(x) = 1 + x^2$$

$$(fg)(x) = (1 + x)^2$$

$\square$

Note that if $f : X \to Y$ is defined on X and $g : Y \to Z$ is defined on Y, then it does not make sense to speak about the composition $f \circ g$. The preceding example shows that even in the case of functions prescribed on the same set into itself, when it does make sense to speak about both compositions, in general

$$fg \neq gf$$

Inverses. Let $R \subset X \times Y$ denote a relation. A relation

$$\check{R} = \{(y, x) \in Y \times X : (x, y) \in R\}$$

is called the *converse* of R.

It follows from the definition that

$$\text{(i) domain } \check{R} = \text{range } R$$

$$\text{(ii) range } \check{R} = \text{domain } R$$

$$\text{(iii) } (\check{R})\check{} = R$$

In general, if R is a function f, its converse $\check{f}$ may not be a function. If it happens that $\check{f}$ is also a function, then it is called the *inverse* of f and is denoted f^{-1}. We also then say that f is *invertible.* In other words, $f : X \to Y$ is invertible iff there exists a function $g : Y \to X$ such that for every $x \in X$ if $y = f(x)$ then $x = g(y)$ and for every $y \in Y$ if $x = g(y)$ then $y = f(x)$.

The concept of the inverse function is illustrated in Fig. 1.11. The element x is set forth into the element y by function f and then back from y into x again by the inverse $g = f^{-1}$. Similarly, starting with y, we prescribe $x = g(y)$ and taking $f(x) = f(g(y))$ we arrive at x again. We can express this algebraically writing:

$$f^{-1}(f(x)) = x \qquad \text{and} \qquad f\left(f^{-1}(y)\right) = y$$

or, equivalently,

$$f^{-1} \circ f = i_X \qquad \text{and} \qquad f \circ f^{-1} = i_Y$$

where i_X and i_Y are the identity functions on X and Y, respectively.

In other words,

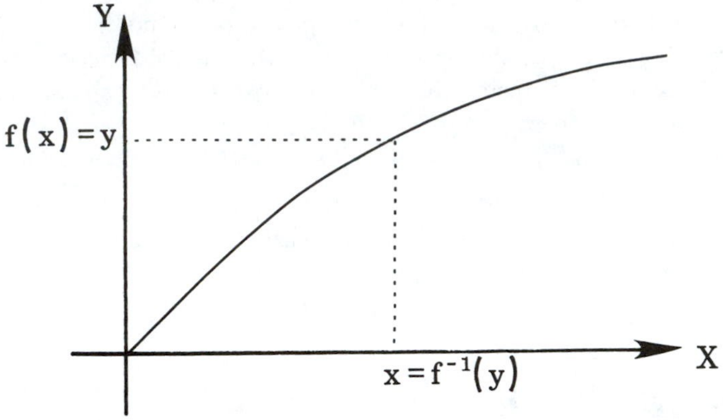

Figure 1.11
Concept of the inverse functions.

a function $f: X \to Y$ is *invertible* iff there exists a function $g: Y \to X$ such that

$$g \circ f = i_X \qquad \text{and} \qquad g \circ f = i_Y$$

Note that $g = f^{-1}$ as the *converse* of f is unique.

This suggests the following definitions.

A function $f: X \to Y$ is said to be *left invertible* if there exists a function $g: Y \to X$ such that

$$g \circ f = i_X$$

The function g is called a *left–inverse* of f.

A function $f: X \to Y$ is said to be *right invertible* iff there exists a function $g: Y \to X$ such that

$$f \circ g = i_Y$$

The function g is called a *right-inverse* of f.

Thus, if function f is invertible then it is both left- and right-invertible and its inverse is a left- and a right-inverse as well. It turns out that the converse is also true. We need the following:

LEMMA 1.12.1

Let $f : X \to Y$ and $g : Y \to X$ be two functions such that

$$g \circ f = i_X$$

Then f is injective and g is surjective.

PROOF Pick an arbitrary $x \in X$ and set $y = f(x)$. We have

$$g(y) = g(f(x)) = x$$

which proves that g is surjective.

Also, if $f(x_1) = f(x_2)$ for some x_1, x_2 then

$$x_1 = g(f(x_1)) = g(f(x_2)) = x_2$$

which implies that f is injective. ∎

We have the immediate corollaries.

COROLLARY 1.12.1

(i) Every left-invertible function is injective.

(ii) Every right-invertible function is surjective.

(iii) Every left- and right-invertible function is bijective.

Finally, we arrive at the following important result.

PROPOSITION 1.12.1

Let $f : X \to Y$ be a function. The following conditions are equivalent to each other:

(i) f is invertible.

(ii) *f is both left- and right-invertible.*

(iii) *f is bijective.*

PROOF (ii) follows from (i) by definition. We have just shown in Corollary 1.12.1(iii) that (ii) implies (iii). Thus it remains to prove that every bijective function f is invertible. But this is trivial because bijective maps establish a one-to-one correspondence between *all* elements of X and *all* elements of Y. In other words, for every $y \in Y$ (f is surjective) there exists a unique $x \in X$ (f is injective) such that $y = f(x)$. Set by definition

$$g(y) = x$$

Thus g is a function and $g(f(x)) = x$ as well as $f(g(y)) = y$, which ends the proof. ∎

The notion of the inverse f^{-1} of a function $f : X \to Y$ should not be confused with the inverse image set $f^{-1}(B)$, for some $B \subset Y$. The latter is a set which exists for *every* function f and the prior is a *function* which exists only when f is bijective. Note, however, that the direct image of the inverse function f is equal to the inverse image of f and therefore it is not necessary to distinguish between the symbols $(f^{-1})(B)$ and $f^{-1}(B)$ (cf. Exercise 1.12.7).

Example 1.12.2

Let $f : \mathbb{R} \to \mathbb{R}^+$, $\mathbb{R} = \operatorname{dom} f =$ the set of real numbers and $\mathbb{R}^+ =$ range $f = \{y : y \in \mathbb{R},\ y \geq 0\}$. Suppose f is defined by the rule $f(x) = x^2$, i.e., $f = \left\{(x, y) : x,\ y \in \mathbb{R},\ y = x^2\right\}$. Then f does *not* have an inverse since it is clearly not one-to-one. □

Example 1.12.3

Let $\operatorname{dom} f = \{x : x \in \mathbb{R},\ x \geq 0\}$ and range $f = \{y : y \in \mathbb{R},\ y = x^2\}$. That is, $f = \{(x, y) : x, y \in \mathbb{R},\ x \geq 0,\ y = x^2\}$. Clearly, f is one-to-one and onto. Also, f has an inverse f^{-1} and $y = f(x) = x^2$ if and only if $x = f^{-1}(y)$. The inverse function f^{-1}, in this case, is called the *positive square root function* and we use the notation $f^{-1}(y) = \sqrt{y}$. (Likewise, if $f_1 = \left\{(x, x^2) : x \in \mathbb{R},\ x \leq 0\right\}$, $f_1^{-1}(y) = -\sqrt{y}$ is the inverse of f_1 and is called the *negative square root function*, etc.) □

Example 1.12.4

The sine function, $f(x) = \sin x$, is, of course, not one-to-one ($\sin 0 = \sin \pi = \sin 2\pi = \cdots = 0$). However, if $\mathbb{R}_{\pi/2} = \{x : x \in \mathbb{R}, -\pi/2 \le x \le \pi/2\}$, the restriction $f\,|_{\mathbb{R}_{\pi/2}}$ is one-to-one and onto and has an inverse function, called the *inverse sine function*, denoted by $f^{-1}(y) = \arcsin(y)$ or $\sin^{-1}(y)$.
□

When a function $f : X \to Y$ is not invertible, it may still have a left- or right-inverse. We have already learned from Lemma 1.12.1 that injectivity and surjectivity are the necessary conditions for the left- and right-invertibility, respectively. It turns out that they are also sufficient.

PROPOSITION 1.12.2

Let $f : X \to Y$ be a function. Then

(i) f is left-invertible iff f is injective;

(ii) f is right-invertible iff f is surjective.

PROOF

(i) Let f be an injective map. Restricting its co-domain to its range $R(f)$, we get a bijective function (it becomes surjective by definition) which, according to Proposition 1.12.1, is invertible with an inverse g defined on $R(f)$. Let G be any extension of g to Y. Then for every $x \in X$,

$$G(f(x)) = g(f(x)) = x$$

and therefore f is left-invertible.

(ii) Let f be surjective. For every $y \in Y$, consider the inverse image set

$$f^{-1}(\{y\})$$

Since f is a function, $f^{-1}(\{y_1\}) \cap f^{-1}(\{y_2\}) = \emptyset$, for different y_1 and y_2.
Thus

$$\{f^{-1}(\{y\}),\ y \in Y\}$$

is a partition of X and, by the *axiom of choice*, for every $y \in Y$ one can choose a corresponding representative $x_y \in f^{-1}(\{y\})$. Consider the relation $g \subset Y \times X$, $ygx \Leftrightarrow x = x_y$, i.e., $(y, x_y) \in g$. It is clear that g is a function from Y to X with the property that

$$g(y) = x_y$$

But then

$$f(g(y)) = y$$

which means that g is a right-inverse of f. The fact that a right-invertible f is surjective was established in Corollary 1.12.1 ∎

Example 1.12.5

Let $A = \{1, 2, 3\}$ and $B = \{x, y\}$. Consider the correspondence

$$f : \begin{array}{l} 1 \rightarrow x \\ 2 \rightarrow y \\ 3 \rightarrow x \end{array}$$

Clearly, f is onto. Hence it should have a right-inverse. Indeed, consider the function g_1 and g_2 from B to A:

$$g_1 : \begin{array}{l} x \rightarrow 1 \\ y \rightarrow 2 \end{array} \qquad g_2 : \begin{array}{l} x \rightarrow 3 \\ y \rightarrow 2 \end{array}$$

We see that

$$f \circ g_1 : \begin{array}{l} x \rightarrow 1 \rightarrow x \\ y \rightarrow 2 \rightarrow y \end{array} \qquad f \circ g_2 : \begin{array}{l} x \rightarrow 3 \rightarrow x \\ y \rightarrow 2 \rightarrow y \end{array}$$

Hence both g_1 and g_2 are right-inverses of f. □

This example shows that when f is onto but not one-to-one, it can have more than one right-inverse. If f is neither onto nor one-to-one, no inverses of any kind exist. The invertibility properties of a function $f : A \rightarrow B$ can be summarized in the network diagram shown in Fig. 1.12.

Example 1.12.6
(The Motion of a Continuous Medium)

One bijective map that is fundamental to physics is the primitive notion of *motion* that can be defined for general continua in terms of an invertible

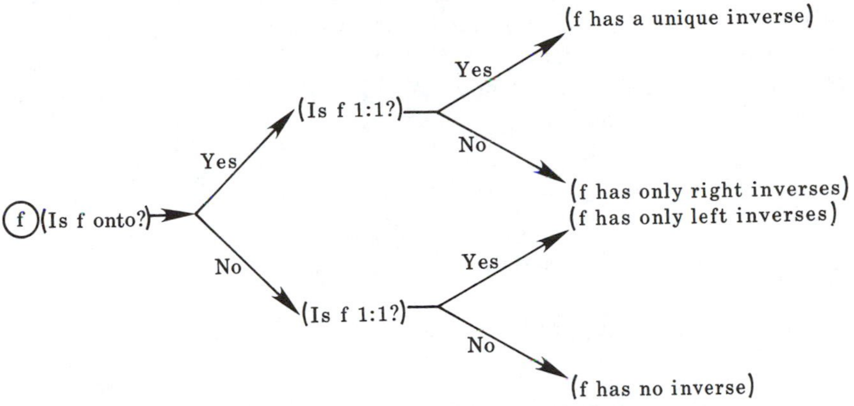

Figure 1.12
A diagram illustrating the invertibility properties of functions.

map of points in one region of Euclidean space into points in another. Since we may wish to use some concepts of continuum mechanics as examples of various mathematical ideas, we shall describe motion as a special invertible function.

Consider a material body $\mathcal{B}$ in motion under the action of external forces. The body $\mathcal{B}$ may be considered to be a nondenumerable set consisting of material points P. Such a body is also endowed with additional structures—for example, every physical body has mass, and, mathematically, this is manifested as a measure m on $\mathcal{B}$ (see Chapter 3). Conceptually, $\mathcal{B}$ is viewed as an actual piece of physical material set in motion by certain prescribed forces.

To describe this motion, we establish a fixed (inertial) frame of reference described by the Cartesian coordinates x_i, which are called *spatial coordinates* because they identify points in space as opposed to particles of material. We observe the motion of $\mathcal{B}$ by watching it assume various places in euclidean space $\mathbb{R}^3$ at each time t. These places that the body occupies are called its *configurations*. Thus, if P denotes a material particle in $\mathcal{B}$ and $\boldsymbol{x} = \sum_{k=1}^{3} x_k \boldsymbol{i}_k$ is the spatial position vector, then the relation

$$\boldsymbol{x} = \kappa(P)$$

defines the configuration κ of $\mathcal{B}$. We refer to the functions κ of $\mathcal{B}$ into $\mathbb{R}^3$ as *configuration maps*, or also simply *configurations*. The motion of the body is observed relative to some fixed configuration κ_0, known as the *reference configuration.* Generally, the reference configuration is chosen as the location of $\mathcal{B}$ at some convenient time when its geometrical features are known, and the deformation of $\mathcal{B}$ relative to this natural state is to be determined. The images X of material points $P \in \mathcal{B}$ under κ_0 are called *material coordinates*, and we use the notation

$$\boldsymbol{Z} = \kappa_0(P)$$

where $\kappa_0 : \mathcal{B} \to E_0 \subset \mathbb{R}^3$. Then the composition

$$\boldsymbol{x} = \kappa(P) = \kappa(\kappa_0^{-1}(\boldsymbol{Z})) \equiv \chi(\boldsymbol{Z})$$

is called a *deformation* of E_0 into $\mathbb{R}^3$, and the one-parameter family of deformations,

$$\boldsymbol{x} = \chi(\boldsymbol{Z}, t), \qquad t \geq 0$$

is called the *motion* of the body $\mathcal{B}$ relative to the reference configuration. In essence, the motion of the body defines a one-parameter family of deformations. $\square$

Example 1.12.7

The reader familiar with matrices will appreciate the following example of left- and right-inverses (we discuss matrices and linear equations in Chapter 2). Suppose $\mathbb{R}$ is the set of real numbers. Let $A = \mathbb{R}^3 = \mathbb{R} \times \mathbb{R} \times \mathbb{R}$ denote the set of ordered triples $A = \{(a_1, a_2, a_3) : a_1, a_2, a_3 \in \mathbb{R}\}$ and let $B = \mathbb{R}^2 = \mathbb{R} \times \mathbb{R} = \{(b_1, b_2) : b_1, b_2 \in \mathbb{R}\}$ denote the set of ordered pairs of real numbers. Consider the mapping $f : A \to B$ defined by the matrix equation

$$\begin{bmatrix} b_1 \\ b_2 \end{bmatrix} = \begin{bmatrix} 1 & 1 & 2 \\ -1 & 1 & 0 \end{bmatrix} \begin{bmatrix} a_1 \\ a_2 \\ a_3 \end{bmatrix}$$

Clearly, $\mathcal{R}(f) = B$; i.e., f is onto. Therefore, f has a right-inverse. Indeed, the mapping $g : B \to A$ defined by

$$\begin{bmatrix} a_1 \\ a_2 \\ a_3 \end{bmatrix} = \begin{bmatrix} 2 & -\frac{9}{2} \\ 2 & -\frac{7}{2} \\ -\frac{3}{2} & 4 \end{bmatrix} \begin{bmatrix} b_1 \\ b_2 \end{bmatrix}$$

is a right-inverse of f. In fact, the identity mapping for B is obtained by the composition

$$\begin{bmatrix} 1 & 1 & 2 \\ -1 & 1 & 0 \end{bmatrix} \begin{bmatrix} 2 & -\frac{9}{2} \\ 2 & -\frac{7}{2} \\ -\frac{3}{2} & 4 \end{bmatrix} = \begin{bmatrix} 1 & 0 \\ 0 & 1 \end{bmatrix}$$

since

$$\begin{bmatrix} b_1 \\ b_2 \end{bmatrix} = \begin{bmatrix} 1 & 0 \\ 0 & 1 \end{bmatrix} \begin{bmatrix} b_1 \\ b_2 \end{bmatrix}$$

Note that this right-inverse is not unique; indeed, the matrix

$$\begin{bmatrix} 0 & -1 \\ 0 & 0 \\ \frac{1}{2} & \frac{1}{2} \end{bmatrix}$$

is also a right-inverse. $\square$

Exercises

1.12.1. If F is the mapping defined on the set $\mathbb{R}$ of real numbers by the rule $y = F(x) = 1 + x^2$, find $F(1)$, $F(-1)$, and $F(\frac{1}{2})$.

1.12.2. Let $\mathbb{N}$ be the set of all natural numbers. Show that the mapping $F: \ n \to 3 + n^2$ is an injective mapping of $\mathbb{N}$ into itself, but it is not surjective.

1.12.3. Consider the mappings $F: \ n \to n+1$, $G: \ n \to n^2$ of $\mathbb{N}$ into $\mathbb{N}$. Describe the product mappings FF, FG, GF and GG.

1.12.4. Show that if $f: \ \mathbb{N} \to \mathbb{N}$, and $f(x) = x + 2$, then f is one-to-one but not onto.

1.12.5. If f is one-to-one from A onto B and g is one-to-one from B onto A, show that $(fg)^{-1} = g^{-1} \circ f^{-1}$.

1.12.6. Let $A = \{1, 2, 3, 4\}$ and consider the sets

$$f = \{(1,3), (3,3), (4,1), (2,2)\}$$

$$g = \{(1,4), (2,1), (3,1), (4,2)\}$$

(i) Are f and g functions?

(ii) Determine the range of f and g.

(iii) Determine $f \circ g$ and $g \circ f$

1.12.7. Let $f : X \to Y$ be a bijection and f^{-1} its inverse. Show that

$$(f^{-1})(B) = f^{-1}(B)$$

$$\begin{array}{cc}\text{(direct image of } B & = \text{(inverse image of } B \\ \text{through inverse } f^{-1}) & \text{through } f)\end{array}$$

1.12.8. Let $A = \{x : x \in \mathbb{R}, -1 \leq x \leq 1\}$ and let $f_i : A \to A$ be defined by

(i) $f_1(x) = \sin x$;

(ii) $f_2(x) = \sin \pi x$;

(iii) $f_3(x) = \sin(\pi x/2)$.

Classify each f_i as to whether or not it is surjective, injective, or bijective.

Cardinality of Sets

1.13 Fundamental Notions

The natural idea of counting which lets us compare finite sets (two sets are "equivalent" if they have the same number of elements) may be generalized to the case of infinite sets. Every set may be assigned a symbol, called its "cardinal number", which describes its "number of elements" in the sense that, indeed, in the case of a finite set, its cardinal number is equal to its number of elements.

To make this idea precise, we introduce the following relation for sets: two sets A and B are said to be *equivalent*, denoted $A \sim B$, if there exists a bijective map which maps A onto B. In other words, there is a one-to-one correspondence between all elements of A and all elements of B. It is easy to prove that, given a universal set U and its power set $\mathcal{P}(U)$ consisting of all subsets of U, the relation $\sim$ on $\mathcal{P}(U)$ is an *equivalence relation.* As a consequence $\mathcal{P}(U)$ may be partitioned into equivalence classes and every such class may be assigned a symbol, called its *cardinal number*, that is, *cardinality* is a property that all sets equivalent to each other have in common.

To see that the notion of equivalent sets generalizes the idea of counting, let us notice that two finite sets have the same number of elements if and only if they are equivalent to each other. More precisely, a set A is *finite* iff there exists an $n \in I\!N$ such that $A \sim \{1, 2, \ldots, n\}$. If $A \sim B$ then also $B \sim \{1, 2, \ldots, n\}$ and the class of sets equivalent to A is assigned the cardinal number n equal to the number of elements of A.

We say that a set A is *infinite* if it is *not* finite, i.e., no natural number n exists such that $A \sim \{1, \ldots, n\}$. It is obvious that the theory of cardinal numbers is mainly concerned with infinite sets. The simplest infinite sets are those which can be denumerated with natural numbers, that is, we can represent them in a sequential form.

We say that a set A is *denumerable* iff $A \sim I\!N$. The symbol $\aleph_0$ (aleph-naught) is used to denote the cardinal number of denumerable sets, $\aleph$ being the Hebrew letter aleph. Sets which are either finite or denumerable bear a common name of *countable* sets, i.e., a set A is *countable* iff it is finite or denumerable.

There are numerous properties of countable sets. Some of them are listed below:

(i) $A \subset B$, B countable implies that A is countable, i.e., every subset of a countable set is countable.

(ii) If $A_1, \ldots, A_n$ are countable then $A_1 \times \ldots \times A_n$ is countable.

(iii) If A_i, $i \in I\!N$ are countable then $\cup A_i$ is countable.

Example 1.13.1

To prove property (ii) of countable sets it is sufficient to prove that $I\!N \times I\!N$ is denumerable. Indeed, if A and B are countable then $A \times B$ is equivalent to a subset of $I\!N \times I\!N$, then by property (i), $A \times B$ is countable. Since $A_1 \times \ldots, \times A_{n-1} \sim (A_1 \times \ldots \times A_{n-1}) \times A_n$, the property may be generalized to any finite family of sets.

To see that $I\!N \times I\!N$ is denumerable, consider the diagram shown in Fig. 1.13. By letting 1 correspond to the pair (1,1), 2 correspond to (2,1), 3 to (1,2) and continuing to follow the path indicated by the arrows, it is easily

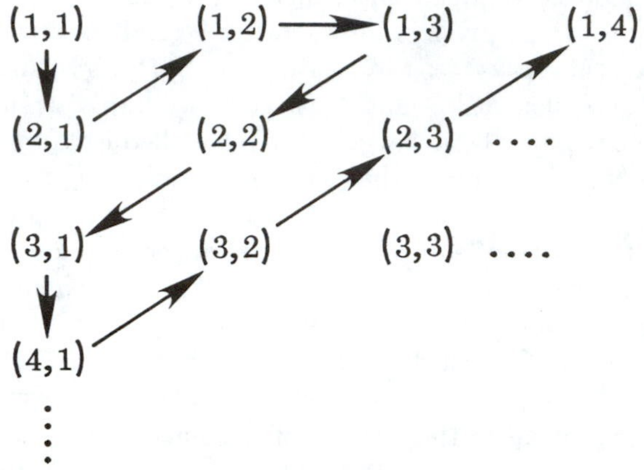

Figure 1.13
A diagram showing the equivalence of sets $\mathbb{N}$ and $\mathbb{N} \times \mathbb{N}$.

seen that the elements of $\mathbb{N} \times \mathbb{N}$ are in one-to-one correspondence with those of $\mathbb{N}$ itself. $\square$

Example 1.13.2

As an example of application of properties (i)–(iii) we may prove that the set $\mathbb{Q}$ of rational numbers is denumerable. To see this, recall that a rational number is identified with an equivalence class of all fractions of the form p/q where $p \in \mathbb{Z}$, $q \in \mathbb{N}$ and two fractions p_1/q_1 and p_2/q_2 are equivalent to each other iff $p_1 q_2 = q_1 p_2$.

Since the set of all integers is countable (explain why), the rationals may be identified with a subset of the Cartesian product $\mathbb{Z} \times \mathbb{N}$ and, therefore, by properties (i) and (ii), are denumerable.

At this moment we might ask a natural question: are all infinite sets denumerable? In other words, do infinite sets exist which could *not* be represented in a sequential form? The famous Cantor theorem brings an answer.

$\square$

THEOREM 1.13.1
(Cantor)

The power set $\mathcal{P}(A)$ of a set A is not equivalent to A.

In other words, the sets A and $\mathcal{P}(A)$ have different cardinal numbers. In particular, the power set $\mathcal{P}(\mathbb{N})$ is not equivalent to $\mathbb{N}$ and therefore is *not* denumerable.

At this point it is very important to realize that we have to deal in mathematics with infinite sets which cannot be represented in the sequential form. In particular, it can be proved that the set of real numbers $\mathbb{R}$ is equivalent to the power set of the set of natural numbers, i.e., $\mathbb{R} \sim \mathcal{P}(\mathbb{N})$. The small gothic letter "**c**" is used to denote the cardinal number assigned to $\mathbb{R}$.

Exercises

1.13.1. Show that if U is a universal set, the relation $\sim$ is an equivalence relation on $\mathcal{P}(U)$.

1.13.2. Explain why every subset of a countable set must be countable.

1.14 Ordering of Cardinal Numbers

If the cardinal numbers are supposed to be a generalization of the natural numbers then the next natural thing to do will be to establish an ordering allowing one to compare cardinal numbers. The ordering should generalize the usual "greater or equal" relation for numbers.

Let A and B be two subsets of a universal set U and $\#A$ and $\#B$ denote their corresponding cardinal numbers. We say that the *cardinal number* $\#B$ is *greater than or equal to the cardinal number* $\#A$, denoted $\#A \leq \#B$, if and only if there exists a one-to-one map T_{AB} from A into B. In other words, every element in A has its counterpart in B but not necessarily conversely; there may be some elements in B which have not been assigned to elements of A. This intuitively corresponds to the fact that B has "more" elements than A.

Obviously, if $A_1 \sim A$, $B_1 \sim B$ and $A \leq B$ then also $A_1 \leq B_1$. Thus $\leq$ is a well-defined relation on the quotient set $\mathcal{P}(U)/_\sim$.[1]

[1]Elements of $\mathcal{P}(U)/_\sim$ can in fact be identified as the cardinal numbers themselves in the same way that natural numbers $n \in \mathbb{N}$ can be identified with finite sets consisting of precisely n elements. The relation $\leq$ is defined for equivalence classes $[A]_\sim$ and $[B]_\sim$ from $\mathcal{P}(U)/_\sim$ through their representatives A and B. By saying that the relation is well-defined we mean

To prove that the relation $\leq$ between the cardinal numbers is indeed a linear ordering on $\mathcal{P}(U)/_\sim$ we have to show that it is reflexive, transitive, antisymmetric, and that every two equivalence classes from $\mathcal{P}(U)/_\sim$ are comparable to each other. Two of the four properties listed above can be easily shown. Using the identity map from a set A onto itself we easily show that

$$\#A \leq \#A$$

Also, to prove that the relation is transitive, i.e., that $\#A \leq \#B$, $\#B \leq \#C$ implies $\#A \leq \#C$, it is enough to define the mapping T_{AC} as

$$T_{AC} = T_{BC} \circ T_{AB}$$

where T_{AB} and T_{BC} are the two mappings corresponding to pairs (A, B) and (B, C), respectively. As a composition of two injective mappings, T_{AC} is injective, too. It is also defined on the whole set A, which altogether proves that $\#A \leq \#C$.

The third condition is much more difficult. It is the famous *Cantor-Berstein theorem.*

THEOREM 1.14.1
(Cantor-Berstein)

The cardinal numbers relation is antisymmetric, i.e., if $\#A \leq \#B$ and $\#B \leq \#A$ then $\#A = \#B (A \sim B)$.

The proof of this theorem exceeds considerably the scope of this book. We shall show, however, as an example of application of the Kuratowski-Zorn lemma, the proof of the last property.

PROPOSITION 1.14.1

Let U be a universal set and A, $B \in \mathcal{P}(U)$. Then either $\#A \leq \#B$ or $\#B \leq \#A$.

PROOF Consider a family $\mathcal{F}$ consisting of the following triples:

$$(X, Y, T_{XY})$$

where X is a subset of A, Y is s subset of B and $T_{XY}: X \rightarrow Y$ is a one-to-one mapping from X into Y. The family $\mathcal{F}$ is certainly not empty (explain

---

that it is *independent* of the choice of the representatives.

why). Next we introduce a relation $\leq$ in $\mathcal{F}$ by setting

$$(X_1, Y_1, T_{X_1Y_1}) \leq (X_2, Y_2, T_{X_2Y_2})$$

iff $X_1 \subset X_2$, $Y_1 \subset Y_2$ and restriction of $T_{X_2Y_2}$ to the set X_1 coincides with $T_{X_1Y_1}$. It can be easily shown (see Exercise 1.14.1) that the relation is a partial ordering of $\mathcal{F}$. We shall prove now that $\mathcal{F}$ with this partial ordering satisfies the assumptions of the Kuratowski-Zorn lemma. To do so, we have to show that every *linearly ordered* subset $\mathcal{L}$ of $\mathcal{F}$ has an upper bound in $\mathcal{F}$.

Assuming that $\mathcal{L}$ consists of triples (X, Y, T_{XY}) define

$$W = \cup X$$

$$Z = \cup Y$$

$$T_{WZ}(x) = T_{XY}(x) \text{ for some } X \text{ such that } x \in X$$

The (infinite) unions above are taken over all triples from $\mathcal{L}$. From the fact that $\mathcal{L}$ is linearly ordered in the sense of the relation $\leq$, it follows that:

1. T_{WZ} is well defined, i.e., its value $T_{WZ}(x)$ is independent of the choice of X such that $x \in X$.
2. T_{WZ} is an injection.

Therefore, as a result of our construction,

$$(X, Y, T_{XY}) \leq (W, Z, T_{WZ})$$

for every triple (X, Y, T_{XY}) from $\mathcal{L}$, which proves that the just constructed triple is an upper bound for $\mathcal{L}$.

Thus, by the Kuratowski-Zorn lemma, there exists a *maximal element* in $\mathcal{F}$, say a triple (X, Y, T_{XY}). We claim that either $X = A$ or $Y = B$, which proves that either $A \leq B$ or $B \leq A$. Indeed, if both X and Y were different from A and B respectively, i.e., there existed an $a \in A \backslash X$ and $b \in B \backslash Y$, then by adding element a to X, element b to Y and extending T_{XY} to $X \cup \{a\}$ by setting $T(a) = b$, we would obtain a new triple in $\mathcal{F}$ greater than (X, Y, T_{XY}). This contradicts the fact that (X, Y, T_{XY}) is a maximal element in $\mathcal{F}$. ∎

Using the linear ordering $\leq$ of cardinal numbers we can state now the Cantor theorem more precisely.

THEOREM 1.14.2
(Cantor's Theorem Reformulated)

$\#A < \#\mathcal{P}(A)$, i.e., $\#A \leq \#\mathcal{P}(A)$ and $\#A$ is different from $\#\mathcal{P}(A)$.

Proof of this theorem is left as an exercise (see Exercise 1.14.2).

We have, in particular,

$$\aleph_0 < \mathbf{c}$$

The question arises as to whether or not there exist infinite sets with cardinal numbers greater than $\aleph_0$ and smaller than $\mathbf{c}$. In other words, are there any cardinal numbers between $\aleph_0$ and $\mathbf{c}$?

It is somewhat confusing, but this question has no answer. The problem is much more general and deals with the idea of *completeness* of the axiomatic number theories (and, therefore, the foundations of mathematics as well) and it is connected with the famous result of Gödel, who showed that there may be some statements in classical number theories which cannot be assigned either "true" or "false" values. These include the problem stated above.

The *continuum hypothesis* conjectures that there does not exist a set with a cardinal number between $\aleph_0$ and $\mathbf{c}$. This has led to an occasional use of the notation $\mathbf{c} = \aleph_1$.

Exercises

1.14.1. Complete proof of Proposition 1.14.1 by showing that $\leq$ is a partial ordering of family $\mathcal{F}$.

1.14.2. Prove Theorem 1.14.2. *Hint:* Establish a one-to-one mapping showing that $\#A \leq \#\mathcal{P}(A)$ for every set A and use next Theorem 1.14.1.

1.14.3. Prove that if A is infinite, $A \times A \sim A$. *Hint:* Use the following steps:

(i) Recall that $I\!N \times I\!N \sim I\!N$ (recall Example 1.13.1).

(ii) Define a family $\mathcal{F}$ of couples (X, T_X) where X is a subset of A and $T_X: X \to X \times X$ is a bijection. Introduce a relation $\leq$ in $\mathcal{F}$ defined as

$$(X_1, T_{X_1}) \leq (X_2, T_{X_2})$$

iff $X_1 \subset X_2$ and T_{X_2} is an extension of T_{X_1}.

(iii) Prove that $\leq$ is a partial ordering of $\mathcal{F}$.

(iv) Show that family $\mathcal{F}$ with its partial ordering $\leq$ satisfies the assumptions of the Kuratowski-Zorn lemma (recall the proof of Proposition 1.14.1).

(v) Using the Kuratowski-Zorn lemma, show that $X \sim X \times X$.

Question: Why do we need the first step?

Foundations of Abstract Algebra

1.15 Operations. Abstract Systems. Isomorphism

Consider the set $\mathbb{N}$ of all natural numbers $\mathbb{N} = \{1, 2, 3, 4, \ldots\}$. If a and b are two typical elements of $\mathbb{N}$, it is clear that $a \times b \in \mathbb{N}$, $b \times a \in \mathbb{N}$, $a + b \in \mathbb{N}$, and $b + a \in \mathbb{N}$, where $\times$ and $+$ denote the familiar operations of multiplication and addition. Technically, the symbols $\times$ and $+$ describe relations between pairs of elements of $\mathbb{N}$ and another element of $\mathbb{N}$. We refer to such relations as "operations on $\mathbb{N}$" or, more specifically, as *binary operations* on $\mathbb{N}$ since pairs of elements of $\mathbb{N}$ are involved. A generalization of this property is embodied in the general concept of binary operations.

Binary Operation. A *binary operation* on a set A is a function f from a set $A \times A$ into A.

Thus a binary operation can be indicated by the usual function notation as $f: A \times A \to A$, but it is customary to use special symbols. Thus, if $b \in A$ and b is the image of $(a_1, a_2) \in A \times A$ under some binary operation, we may introduce some symbol, e.g., $\odot$, such that $a_1 \odot a_2$ denotes the image b. In other words, if $f: A \times A \to A$ and if $(a_1, a_2) \in A \times A$, $b \in A$, and b is the image of (a_1, a_2) under the mapping f, then we write

$$a_1 \odot a_2 = f\left((a_1, a_2)\right) = b$$

For obvious reasons, we say that the symbol $\odot$ describes the binary operation defined by f. The familiar symbols $+$, $-$, $\div$, $\times$ are examples of binary operations on pairs of real numbers ($\mathbb{R} \times \mathbb{R} \to \mathbb{R}$).

A binary operation $\odot$ is said to be *closed on* $E \subset A$ iff it takes elements from $E \times E$ into E, i.e., if $a, b \in E$ then $a \odot b \in E$.

The idea of binary operations on a set A can be easily generalized to k-ary operations on A. Indeed, a k-ary operation on A is a mapping from $A \times A \times \cdots \times A$ (k times) into A; it is a function whose domain is the set of the ordered k-tuples of elements of A and whose range is a subset of A.

Classification of Operations. We now cite several types of binary operations and properties of sets pertaining to binary operations of particular importance.

(i) *Commutative Operation.* A binary operation $*$ on a set is said to be *commutative* whenever

$$a * b = b * a$$

for all $a, b \in A$.

(ii) *Associative Operation.* A binary operation $*$ on a set A is said to be *associative* whenever

$$(a * b) * c = a * (b * c)$$

for all $a, b, c \in A$.

(iii) *Distributive Operations.* Let $*$ and $\circ$ denote two binary operations defined on a set A. The operation $*$ is said to be *left-distributive* with respect to $\circ$ iff

$$a * (b \circ c) = (a * b) \circ (a * c)$$

for all $a, b, c \in A$. The operation $*$ is said to be *right-distributive* with respect to $\circ$ if

$$(b \circ c) * a = (b * a) \circ (c * a)$$

for all $a, b, c \in A$. An operation $*$ that is both left- and right-distributive with respect to an operation $\circ$ is said to be simply *distributive* with respect to $\circ$.

(iv) *Identity Element.* A set A is said to have an *identity* element with respect to a binary operation $*$ on A iff there exists an element $e \in A$ with the property

$$e * a = a * e = a$$

for every $a \in A$.

(v) *Inverse Element.* Consider a set A in which an identity element e relative to a binary operation $*$ on A has been defined, and let a be an arbitrary element of A. An element $b \in A$ is called an *inverse* element of a relative to $*$ if and only if

$$a * b = e = b * a$$

Ordinarily we write a^{-1} for the inverse element of a.

(vi) *Idempotent Operation.* An operation $*$ on a set A is said to be *idempotent* whenever, for every $a \in A$, $a * a = a$.

Still other classifications of operations could be cited.

Example 1.15.1

Consider the set $A = \{a_1, a_2, a_3\}$ and suppose that on A we have a binary operation such that

$$a_1 * a_1 = a_1, \quad a_1 * a_2 = a_2, \quad a_1 * a_3 = a_3$$

$$a_2 * a_1 = a_2, \quad a_2 * a_2 = a_3, \quad a_2 * a_3 = a_1$$

$$a_3 * a_1 = a_3, \quad a_3 * a_2 = a_1, \quad a_3 * a_3 = a_2$$

We can visualize the properties of this operation more easily with the aid of the following tabular form:

$*$	a_1	a_2	a_3
a_1	a_1	a_2	a_3
a_2	a_2	a_3	a_1
a_3	a_3	a_1	a_2

Tabular forms such as this are sometimes called "Cayley squares." Clearly, the operation $*$ is commutative on A, for $a_1 * a_2 = a_2 * a_1$, $a_1 * a_3 = a_3 * a_1$, and $a_2 * a_3 = a_3 * a_2$. Finally, note that A has an identity element; namely a_1 (since $a_1 * a_1 = a_1$, $a_1 * a_2 = a_2$, and $a_1 * a_3 = a_3$). □

Example 1.15.2

On the set of integers $\mathbb{Z}$, consider an operation $*$ such that

$$a * b = a^2 b^2, \qquad a, b \in \mathbb{Z}$$

Clearly, $a^2b^2 \in Z$, so that $*$ is defined on $\mathbb{Z}$. We have the ordinary operation of addition $+$ of integers defined on $\mathbb{Z}$ also. Note that

$$a * (b + c) = a^2b^2 + 2a^2bc + a^2c^2$$

and

$$(a * b) + (a * c) = a^2b^2 + a^2c^2$$

Thus $*$ is not left-distributive with respect to $+$. $\square$

Abstract Systems. The term "system" is another primitive concept—like "set"—which is easy to grasp intuitively, but somewhat difficult to define with absolute precision. According to Webster, a system is "a regularly interacting or interdependent group of items forming a unified whole," and this definition suits our present purposes quite nicely.

Throughout this book, we deal with various kinds of *abstract mathematical systems.* These terms are used to describe any well-defined collection of mathematical objects consisting, for example, of a set together with relations and operations on the set, and a collection of postulates, definitions, and theorems describing various properties of the system.

The most primitive systems consist of only a set A and a relation R or an operation $*$ defined on the set. In such cases, we used the notation $\mathcal{S} = \{A, R\}$ or $\mathcal{S} = \{A, *\}$ to describe the system $\mathcal{S}$.

Simple systems such as this are said to have very little *structure*, which means that they contain very few component parts—e.g., only a few sets and simple operations. By adding additional components to a system (such as introducing additional sets and operations), we are said to supply a system with additional structure.

It is a fundamentally important fact that even when systems have very little structure, such as the system $\mathcal{S} = \{A, *\}$, it is possible to classify them according to whether or not they are "mathematically similar" or "mathematically equivalent." These notions are made precise by the notion of *isomorphism* between two abstract systems.

Isomorphism. Let $\mathcal{S} = \{A, *\}$ and $\mathcal{J} = \{B, \text{o}\}$ denote two abstract systems with binary operations $*$ and o being defined on A and B, respectively. Systems $\mathcal{S}$ and $\mathcal{J}$ are said to be *isomorphic* if and only if the following hold:

(i) There exists a bijective map F from A onto B.

(ii) The operations are preserved by the mapping F in the sense that if $a, b \in A$, then

$$F(a * b) = F(a) \text{ o } F(b)$$

The mapping F is referred to as an *isomorphism* or an *isomorphic* mapping of $\mathcal{S}$ onto $\mathcal{J}$.

The concept of an isomorphism provides a general way to describe the equivalence of abstract systems. If $\mathcal{S}$ and $\mathcal{J}$ are isomorphic, then we think of them as "operationally" equivalent. Indeed, literally translated, isomorphic derives from the Greek: *iso-* (same) *morphic* (form).

Example 1.15.3

Consider two algebraic systems $\mathcal{S}$ and $\mathcal{J}$ such that $\mathcal{S}$ consists of the set $A = \{1, 2, 3\}$ and a binary operation $*$ with the property defined by the table

$\mathcal{S}$

$*$	1	2	3
1	1	2	3
2	2	3	1
3	3	1	2

For example, $1 * 2 = 2, 3 * 2 = 1$, etc. Next, suppose that $\mathcal{J}$ consists of a set $B = \{x, y, z\}$ and a binary operation o defined by the table

$\mathcal{J}$

o	x	y	z
x	x	y	z
y	y	z	x
z	z	x	y

The mapping

$$f: 1 \to x, 2 \to y, 3 \to z$$

is an isomorphism of $\mathcal{S}$ onto $\mathcal{J}$, because it is one-to-one and for any $a, b \in A$ and $a_1, b_1 \in B$, if $a_1 = F(a)$ and $b_1 = F(b)$, it is clear that $F(a * b) = F(a) \text{ o } F(b) = a_1 \text{ o } b_1$. ▯

Example 1.15.4

As another example of an isomorphism, consider the system $\mathcal{S}$ consisting of the set $A = \{a_1, a_2, a_3, a_4\}$, where $a_1 = 1$, $a_2 = i = \sqrt{-1}$, $a_3 = -1$, and $a_4 = -i$ and the operation of ordinary multiplication of complex numbers. Here a_1 is an identity element since $a_1 a_i = a_i (i = 1, 2, 3, 4)$. Note also that $a_2 a_2 = a_3$, $a_4 a_3 = a_3 a_4 = a_2$, $a_4 a_4 = a_3$, etc., and $a_3^{-1} = a_3$, $a_4^{-1} = a_2, \ldots$ and so forth. Now consider the system $\mathcal{J}$ consisting of the set $B = \{\boldsymbol{b}_1, \boldsymbol{b}_2, \boldsymbol{b}_3, \boldsymbol{b}_4\}$, where $\boldsymbol{b}_i$ are 2×2 matrices,

$$\boldsymbol{b}_1 = \begin{bmatrix} 1 & 0 \\ 0 & 1 \end{bmatrix}, \quad \boldsymbol{b}_2 = \begin{bmatrix} 0 & 1 \\ -1 & 0 \end{bmatrix}, \quad \boldsymbol{b}_3 = \begin{bmatrix} -1 & 0 \\ 0 & -1 \end{bmatrix}, \quad \boldsymbol{b}_4 = \begin{bmatrix} 0 & -1 \\ 1 & 0 \end{bmatrix}$$

Using ordinary matrix multiplication as the defining operation of $\mathcal{J}$, note that $\boldsymbol{b}_1$ is an identity element since $\boldsymbol{b}_1 \boldsymbol{b}_i = \boldsymbol{b}_i$ $(i = 1, 2, 3, 4)$. We observe that $\boldsymbol{b}_2 \boldsymbol{b}_2 = \boldsymbol{b}_3$; $\boldsymbol{b}_4 \boldsymbol{b}_3 = \boldsymbol{b}_3 \boldsymbol{b}_4 = \boldsymbol{b}_2$, $\boldsymbol{b}_4 \boldsymbol{b}_4 = \boldsymbol{b}_3$, etc., and that $\boldsymbol{b}_3^{-1} = \boldsymbol{b}_3$,

$\boldsymbol{b}_4^{-1} = \boldsymbol{b}_2$, and so forth. Using the symbol $\sim$ to indicate correspondence, we can see that $a_1 \sim \boldsymbol{b}_1$, $a_2 \sim \boldsymbol{b}_2$, $a_3 \sim \boldsymbol{b}_3$, and $a_4 \sim \boldsymbol{b}_4$. This correspondence is clearly an isomorphism: it is bijective and operations in A are mapped into corresponding operations on B. □

It is clear from these examples that knowledge of an isomorphism between two systems can be very useful information. Indeed, for algebraic purposes, we can always replace a given system with any system isomorphic to it. Moreover, if we can identify the properties of any abstract system, we can immediately restate them as properties of any other system isomorphic to it.

We remark that an isomorphism of a system $\mathcal{S}$ onto itself is called an *automorphism*. We can generally interpret an automorphism as simply a rearrangement of the elements of the system.

Subsystem. Let $\mathcal{S}$ be any abstract mathematical system consisting of a set A and various operations $*, \circ, \ldots$ defined on A. Suppose there exists a subset B of A such that all operations $*, \circ, \ldots$ defined on A are *closed* on B. Then the system $\mathcal{U}$ consisting of set B and the operations induced from A is called a *subsystem* of $\mathcal{S}$.

Exercises

1.15.1. Determine the properties of the binary operations $*$ and % defined on a set $A = \{x, y, z\}$ by the tables below:

$*$	x	y	z
x	x	y	z
y	y	y	x
z	z	x	x

%	x	y	z
x	x	y	z
y	y	z	x
z	z	x	y

1.15.2. Let $*$ be a binary operation defined on the set of integers $\mathbb{Z}$ such that $a * b = c^2 b$, where $a, b \in \mathbb{Z}$. Discuss the distributive properties of $*$ with respect to addition $+$ on $\mathbb{Z}$.

1.15.3. If $*$ is a binary operation on $\mathbb{Z}$ defined by $a * b = ab$ for $a, b \in \mathbb{Z}$, is $*$ commutative? Is it associative? Is it distributive with respect to $-$?

1.15.4. Let $\mathcal{S}$ and $\mathcal{J}$ denote two systems with binary operations $*$ and #, respectively. If $\mathcal{S}$ and $\mathcal{J}$ are isomorphic, show that if

(i) The associative law holds in $\mathcal{S}$, then it holds in $\mathcal{J}$.

(ii) $\mathcal{S} = \{A, *\}$, $\mathcal{J} = \{B, \#\}$ and if $e \in A$ is an identity element in $\mathcal{S}$, then its corresponding element $f \in B$ is an identity element in $\mathcal{J}$.

1.15.5. Let $\mathcal{S}$ denote the system consisting of the set $R = \{4n : n \in I\!N\}$, where $I\!N$ is the set of natural numbers, and the operation of addition $+$. Let $\mathcal{J}$ denote the set $I\!N$ plus addition. Show that $\mathcal{S}$ and $\mathcal{J}$ are isomorphic.

1.16 Examples of Abstract Systems

We now list a number of important special abstract mathematical systems.

Groupoid. A *groupoid* is any abstract system consisting of a set on which a closed operation has been defined.

Semi-group. A *semi-group* is an associative groupoid.

Monoid. A *monoid* is a semi-group with an identity element.

Finally, we arrive at an important type of abstract system that occurs frequently in mathematical analyses:

Group. An abstract system $\mathcal{G}$ consisting of a set G and one binary operation $*$ on G is called a *group* iff the following conditions are satisfied:

(i) operation $*$ is associative;

(ii) there exists an identity element e in G;

(iii) every element a in G has its inverse a^{-1}.

Additionally, if the operation $*$ is commutative, the group is called an *Abelian* or *commutative* group.

It is understood that $*$ is closed on G; i.e., $a * b \in G$ for all a, $b \in G$.

Example 1.16.1
(Dynamics of Mechanical Systems)

The reader familiar with the dynamics of mechanical systems will appreciate this important example of semi-groups. Many dynamic systems are governed by differential equations of the form

$$\frac{d\boldsymbol{q}(t)}{dt} = \boldsymbol{A}\boldsymbol{q}(t), \qquad t > 0$$

$$\boldsymbol{q}(0) = \boldsymbol{q}_0$$

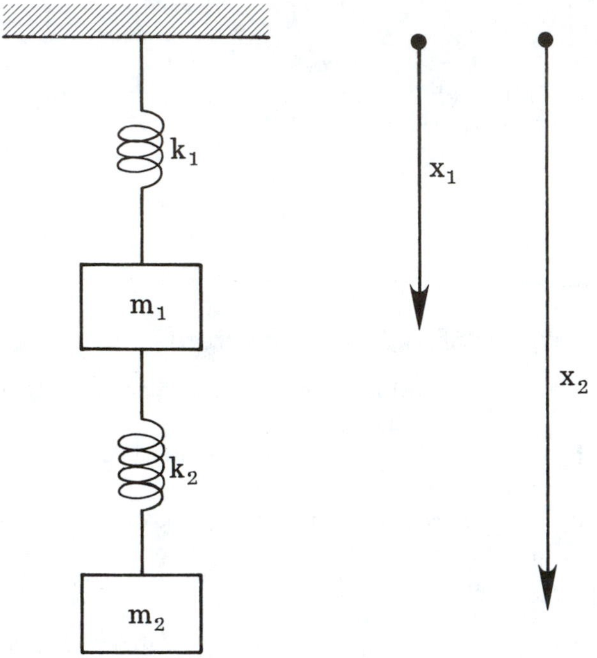

Figure 1.14
A simple system consisting of masses and linear springs.

wherein $\boldsymbol{q}(t) = \{q_1(t), q_2(t), \ldots, q_n(t)\}^T$ is an n-vector whose components are functions of time t and $\boldsymbol{A}$ is an $n \times n$ invertible matrix (those unfamiliar with these terms may wish to return to this example after reading Chapter 2). □

For example, the equations of motion of the mechanical system of masses and linear springs shown in Fig. 1.14 are

$$m_1\ddot{x}_1 + (k_1 + k_2)x_1 - k_2x_2 = 0,\ \dot{x}_1(0) = a_0,\ x_1(0) = b_0$$

$$m_2\ddot{x}_2 - k_2x_1 + k_2x_2 = 0,\ \dot{x}_2(0) = c_0,\ x_2(0) = d_0$$

where $\ddot{x}_1 = d^2x_1/dt^2$, $\dot{x}_1 = dx_1/dt$, etc. We obtain a system of first-order equations by setting $\dot{x}_1 = y_1$ and $\dot{x}_2 = y_2$. Then if

$$q_1 = y_1, \quad q_2 = y_2, \quad q_3 = x_1, \quad q_4 = x_2$$

we have

$$\boldsymbol{q} = \begin{Bmatrix} q_1 \\ q_2 \\ q_3 \\ q_4 \end{Bmatrix}, \ \dot{\boldsymbol{q}} - \boldsymbol{A}\boldsymbol{q} = \boldsymbol{0} = \begin{Bmatrix} 0 \\ 0 \\ 0 \\ 0 \end{Bmatrix}, \ \boldsymbol{q}(\boldsymbol{0}) = \boldsymbol{q_0} = \begin{Bmatrix} a_0 \\ c_0 \\ b_0 \\ d_0 \end{Bmatrix}$$

where

$$\boldsymbol{A} = -\begin{bmatrix} 0 & 0 & \left(\dfrac{k_1 + k_2}{m_1}\right) & \left(-\dfrac{k_2}{m_1}\right) \\ 0 & 0 & \left(-\dfrac{k_2}{m_2}\right) & \left(\dfrac{k_2}{m_2}\right) \\ -1 & 0 & 0 & 0 \\ 0 & -1 & 0 & 0 \end{bmatrix}$$

Solutions of these equations (when they exist) are of the form

$$\boldsymbol{q}(t) = \boldsymbol{E}(t)\boldsymbol{q}_0$$

where $\boldsymbol{E}(t)$ is an $n \times n$ matrix with the property

$$\boldsymbol{E}(t_1 + t_2) = \boldsymbol{E}(t_1) \cdot \boldsymbol{E}(t_2)$$
$$\boldsymbol{E}(t_1) \cdot (\boldsymbol{E}(t_2) \cdot \boldsymbol{E}(t_3)) = (\boldsymbol{E}(t_1) \cdot \boldsymbol{E}(t_2)) \cdot \boldsymbol{E}(t_3)$$

where $\cdot$ denotes matrix multiplication. The set of all matrices $\{\boldsymbol{E}(t)\}_{t>0}$ forms a semi-group with respect to matrix multiplication. For these reasons, the theory of semi-groups plays a fundamental role in the mathematical theory of dynamic systems. If we also admit the identity matrix I, the system becomes a monoid.

Example 1.16.2

The set of all integers forms an Abelian group with respect to the operation of addition. Clearly, the sum of two integers is an integer, addition is associative, and the identity element can be taken as 0 since the addition of 0 to any integer does not alter it. The inverse of any integer is then the negative of the integer since $a + (-a) = 0$. Groups whose basic operation is addition are sometimes called *additive* groups. $\square$

Example 1.16.3

The set of all rational numbers forms an Abelian group under addition. The identity element is again 0. □

Example 1.16.4

The four matrices of Example 1.15.4 form a finite *Abelian* group under matrix multiplication.

$$\boldsymbol{e} = \boldsymbol{b}_1 = \begin{bmatrix} 1 & 0 \\ 0 & 1 \end{bmatrix}, \quad \boldsymbol{b}_2 = \begin{bmatrix} -1 & 0 \\ 0 & -1 \end{bmatrix}$$

$$\boldsymbol{b}_3 = \begin{bmatrix} 0 & 1 \\ -1 & 0 \end{bmatrix}, \quad \boldsymbol{b}_4 = \begin{bmatrix} 0 & -1 \\ 1 & 0 \end{bmatrix}$$

Such groups are called *matrix groups*, and since there are four elements in this finite group, we say that the group is of *order* four. Note that $\boldsymbol{b}_1$ is the identity matrix and that

$$\boldsymbol{b}_2\boldsymbol{b}_3 = \boldsymbol{b}_3\boldsymbol{b}_2 = \boldsymbol{b}_4, \qquad \boldsymbol{b}_4\boldsymbol{b}_3 = \boldsymbol{b}_3\boldsymbol{b}_4 = \boldsymbol{b}_1$$

and so forth. Moreover, matrix multiplication is associative and

$$\boldsymbol{b}_1\boldsymbol{b}_1 = \boldsymbol{b}_1 = \boldsymbol{e}, \qquad \boldsymbol{b}_2\boldsymbol{b}_2 = \boldsymbol{b}_1 = \boldsymbol{e}$$

$$\boldsymbol{b}_3\boldsymbol{b}_4 = \boldsymbol{b}_1 = \boldsymbol{e}, \qquad \boldsymbol{b}_4\boldsymbol{b}_3 = \boldsymbol{b}_1 = \boldsymbol{e}$$

Group operations that characterize a finite group, such as the fourth-order group under consideration, can be arranged in a *group table* or Cayley square, as shown below, so that the group properties are immediately apparent:

*	$\boldsymbol{b}_1$	$\boldsymbol{b}_2$	$\boldsymbol{b}_3$	$\boldsymbol{b}_4$
$\boldsymbol{b}_1$	$\boldsymbol{b}_1$	$\boldsymbol{b}_2$	$\boldsymbol{b}_3$	$\boldsymbol{b}_4$
$\boldsymbol{b}_2$	$\boldsymbol{b}_2$	$\boldsymbol{b}_1$	$\boldsymbol{b}_4$	$\boldsymbol{b}_3$
$\boldsymbol{b}_3$	$\boldsymbol{b}_3$	$\boldsymbol{b}_4$	$\boldsymbol{b}_2$	$\boldsymbol{b}_1$
$\boldsymbol{b}_4$	$\boldsymbol{b}_4$	$\boldsymbol{b}_3$	$\boldsymbol{b}_1$	$\boldsymbol{b}_2$

Groups with some type of multiplication as their basic operation are referred to as *multiplicative* groups. From the preceding table, it is seen that a fourth-order group has 16 products. Clearly, an nth-order group has

n^2 products. If the group is Abelian, the group table will be symmetric. □

Example 1.16.5

The elements 1, g, g^2, g^3, where $g = \exp[2\pi i/3]$, form a group under multiplication. Here any element of the group can be expressed as a power of the group element g. Groups of this type are called *cyclic* groups. More generally,

$$1, g, g^2, g^3, \ldots, g^{n-1}$$

where n is an integer such that $g^n = 1$ is called a *cyclic group of order n*. □

Example 1.16.6
(Permutation Group)

A permutation, generally speaking, is a bijective mapping of a finite set onto itself. For a finite group of permutations of order n, however, a permutation is frequently described by designating the image of each element under a mapping of the natural numbers $1, 2, \ldots, n$ onto the elements $a_1, a_2, \ldots, a_n$. A permutation of n such elements is denoted

$$p = \begin{pmatrix} 1 & 2 & 3 & \cdots & n \\ a_1 & a_2 & a_3 & \cdots & a_n \end{pmatrix}$$

This particular permutation is called an *identity* (or *natural*) permutation, because every element a_i is mapped onto itself. For an nth-order group, there are $n!$ permutations.

Consider, for example, the case in which there are three elements and, therefore, six (3!) permutations:

$$p_1 = \begin{pmatrix} 1\,2\,3 \\ 1\,2\,3 \end{pmatrix}, \quad p_2 = \begin{pmatrix} 1\,2\,3 \\ 3\,1\,2 \end{pmatrix}$$

$$p_3 = \begin{pmatrix} 1\,2\,3 \\ 2\,3\,1 \end{pmatrix}, \quad p_4 = \begin{pmatrix} 1\,2\,3 \\ 2\,1\,3 \end{pmatrix}$$

$$p_5 = \begin{pmatrix} 1\,2\,3 \\ 3\,2\,1 \end{pmatrix}, \quad p_6 = \begin{pmatrix} 1\,2\,3 \\ 1\,3\,2 \end{pmatrix}$$

An inverse permutation is simply a reverse mapping. For example,

$$p_3^{-1} = \begin{pmatrix} 2 \ 3 \ 1 \\ 1 \ 2 \ 3 \end{pmatrix} = \begin{pmatrix} 1 \ 2 \ 3 \\ 3 \ 1 \ 2 \end{pmatrix} = p_2$$

$$p_6^{-1} = \begin{pmatrix} 1 \ 2 \ 3 \\ 1 \ 3 \ 2 \end{pmatrix} = p_6$$

$$p_2^{-1} = \begin{pmatrix} 1 \ 2 \ 3 \\ 2 \ 3 \ 1 \end{pmatrix} = p_3$$

and so forth. A product of two permutations is simply the composition of the two permutations. For example,

$$p_2 p_6 = \begin{pmatrix} 1 \ 2 \ 3 \\ 3 \ 1 \ 2 \end{pmatrix} \begin{pmatrix} 1 \ 2 \ 3 \\ 1 \ 3 \ 2 \end{pmatrix} = \begin{pmatrix} 1 \ 2 \ 3 \\ 3 \ 2 \ 1 \end{pmatrix} = p_5$$

The permutation p_6 indicates that, for example, 3 is mapped into 2. But by p_2, 2 is replaced by 1. Hence, in the product $p_2 p_6$, 3 is replaced by 1. The "paths" followed in the product are indicated by dotted lines as follows:

$$p_2 p_6 = \begin{pmatrix} 1 \ 2 \ 3 \\ 3 \ 1 \ 2 \end{pmatrix} \begin{pmatrix} 1 \ 2 \ 3 \\ 1 \ 3 \ 2 \end{pmatrix} = \begin{pmatrix} 1 \ 2 \ 3 \\ 3 \ 2 \ 1 \end{pmatrix}$$

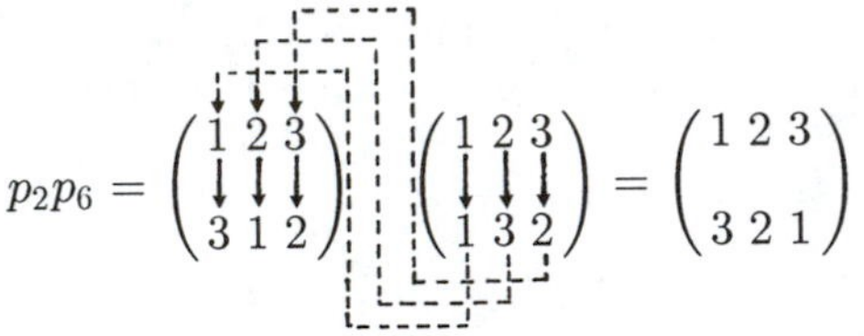

Note that $p_2 p_3 = p_1$ and $p_6 p_6 = p_1$ and that, therefore, $p_3 = p_2^{-1}$ and $p_6 = p_6^{-1}$. The group table corresponding to $p_1, p_2, \ldots, p_6$ follows.

	p_1	p_2	p_3	p_4	p_5	p_6
p_1	p_1	p_2	p_3	p_4	p_5	p_6
p_2	p_2	p_3	p_1	p_6	p_4	p_5
p_3	p_3	p_1	p_2	p_5	p_6	p_4
p_4	p_4	p_5	p_6	p_1	p_2	p_3
p_5	p_5	p_6	p_4	p_3	p_1	p_2
p_6	p_6	p_4	p_5	p_2	p_3	p_1

We observe that the table is not symmetric. Thus the permutation group is not an Abelian group.

The importance of permutation groups lies in Cayley's theorem, which states that *every finite group is isomorphic to a suitable group of permutations.*   ▯

Example 1.16.7
(Material Symmetry in Continuum Mechanics)

Group theory plays an important role in the mechanics of continuous media and in crystallography, in that it is instrumental in providing for the classification of materials according to intrinsic symmetry properties they must have.

We shall illustrate briefly the basic ideas. Suppose that relative to a (fixed) Cartesian material frame of reference, the stress tensor $\boldsymbol{\sigma}$ at particle $\boldsymbol{X}$ is related to the strain tensor γ at particle $\boldsymbol{X}$ by a relation of the form

$$\sigma = \boldsymbol{F}(\gamma) \tag{1.1}$$

where the function $\boldsymbol{F}$ is called the *response function* of the material, and the relationship itself is called a *constitutive equation* because it defines the mechanical constitution of the material (here it describes the stress produced by a given strain, and this, of course, differs for different materials).

Now, the particular form of the constitutive equation given above might be quite different had we chosen a different reference configuration of the body. Suppose that $\overline{\boldsymbol{X}}$ denotes a different labeling of the material points in the reference configuration related to the original choice by the material-coordinate transformation

$$\overline{\boldsymbol{X}} = \boldsymbol{H}\boldsymbol{X}, \qquad \det\ \boldsymbol{H} = \pm 1 \tag{1.2}$$

The linear transformation $\boldsymbol{H}$ is called a *unimodular* transformation because $|\det\ \boldsymbol{H}| = 1$. We then obtain, instead of (1.1), a constitutive equation in the transformed material coordinates,

$$\overline{\sigma} = \overline{\boldsymbol{F}}(\overline{\gamma})$$

Now, the set $\mathcal{U}$ of all unimodular transformations of the form (1.2) constitutes a group with respect to the operation of composition. Indeed, if $\boldsymbol{H}_1$, $\boldsymbol{H}_2$, and $\boldsymbol{H}_3$ are three such transformations,

(i) $\boldsymbol{H}_1(\boldsymbol{H}_2\boldsymbol{H}_3) = (\boldsymbol{H}_1\boldsymbol{H}_2)\boldsymbol{H}_3$ (associative)

(ii) $\boldsymbol{H}_1 \cdot \boldsymbol{I} = \boldsymbol{H}_1$ (identity)

(iii) $\boldsymbol{H}_1 \cdot \boldsymbol{H}_1^{-1} = \boldsymbol{I}$ (inverse)

For each material, there exists a subgroup $\mathcal{H} \subset \mathcal{U}$, called the *symmetry group* of the material, for which the form of the constitutive equation remains invariant under all transformations belonging to that group. If $\boldsymbol{H}$ is any element of the symmetry group of the material, and $\boldsymbol{H}^T$ denotes transposition, then

$$\boldsymbol{H}\boldsymbol{F}(\gamma)\boldsymbol{H}^T = \boldsymbol{F}(\boldsymbol{H}\gamma\boldsymbol{H}^T)$$

This fact is sometimes called the *principle of material symmetry.*

The group $\mathcal{O}$ of orthogonal transformations is called the *orthogonal* group, and whenever the symmetry group of a solid material in an undisturbed state is the full orthogonal group, the material is said to be *isotropic.* Otherwise it is *anisotropic.* $\square$

We now return to definitions of other algebraic systems.

Ring. An abstract system $\mathcal{R} = \{S, +, *\}$ is said to be a *ring* with respect to the binary operations $+$ and $*$ provided S contains at least two elements, and the following hold:

(i) $\{S, +\}$ is an Abelian group with identity element 0 (called *zero*).

(ii) The nonzero elements of S with the operation $*$ form a semi-group.

(iii) For every $r_1, r_2, r_3 \in S$, $*$ is distributive with respect to $+$:

$$r_1 * (r_2 + r_3) = (r_1 * r_2) + (r_1 * r_3)$$

and

$$(r_2 + r_3) * r_1 = (r_2 * r_1) + (r_3 * r_1)$$

If $\{S, *\}$ is a commutative semi-group, the ring is said to be *commutative.*

Familiar examples of rings are the sets of ordinary integers, rational numbers, real numbers, and complex numbers. These systems are rings under ordinary addition and multiplication. A ring $\mathcal{R}$ that contains a multiplicative inverse for each $a \in \mathcal{R}$ and an identity element is sometimes called a *division ring.*

Example 1.16.8

Another familiar ring is the *ring of polynomials.* Let $\mathcal{R} = \{S, +, *\}$ be a ring and define a set of functions defined on S into itself of the form

$$f(x) = a_0 + a_1 * x + \ldots + a_n * x^n$$

where $a_0, a_1, \ldots, a_n \in S$ and x^n denotes $x * \ldots * x$ (n times). Functions of this type are called *polynomials.* Defining *addition* and *multiplication* of polynomials by

$$(f+g)(x) = f(x) + g(x)$$

$$(f*g)(x) = f(x) * g(x)$$

one can show (see Exercise 1.16.10) that the set of all polynomials on S with the addition and multiplication defined above forms a *commutative ring.* $\square$

Field. An abstract system $\mathcal{F} = \{S, +, *\}$ that consists of a set containing at least two elements in which two binary operations $+$ and $*$ have been defined is a *field* if and only if the following hold:

(i) The system $\{S, +\}$ is an Abelian group with identity element 0.

(ii) The nonzero elements of S with operation $*$ form an Abelian group with identity element e.

(iii) The operation $*$ is distributive with respect to $+$.

To spell out the properties of a field in detail, we list them as follows:

(i) *Addition:* $\{S, +\}$

  (1) $a + b = b + a$.

  (2) $(a + b) + c = a + (b + c)$.

  (3) There is an identity element (zero) denoted 0 in S such that $a + 0 = a$ for every $a \in S$.

  (4) For each $a \in S$ there is an inverse element $-a$ such that $a + (-a) = 0$.

(ii) *Multiplication:* $\{S - \{0\}, *\}$

  (1) $a * b = b * a$.

  (2) $(a * b) * c = a * (b * c)$.

  (3) There is an identity element e in S such that $a * e = a$ for every $a \in S$.

  (4) For each $a \in S$ there is an inverse element a^{-1} such that $a * a^{-1} = e$.

(iii) *Distributive Property:*
For arbitrary $a, b, c \in S$, $*$ is distributive with respect to $+$.

$$(a+b)*c = (a*c)+(b*c)$$

$$a*(b+c) = (a*b)+(a*c)$$

In most of our subsequent work, we use the ordinary operations of addition and multiplication of real or complex numbers as the binary operations of fields; each $a, b \in F$ is taken to be real or complex numbers.

Note that a field is also a *commutative division ring* (though not every division ring is a field). The sets of real, rational, and complex numbers individually form fields, as does the set of all square diagonal matrices of a specified order.

Exercises

1.16.1. Let $\mathbb{Z}$ be the set of integers and let $\circ$ denote an operation on $\mathbb{Z}$ such that $a \circ b = a + b - ab$ for $a, b \in \mathbb{Z}$. Show that $\{\mathbb{Z}, \circ\}$ is a semi-group.

1.16.2. Let a, b, and c be elements of a group $\mathcal{G} = \{G, *\}$. If x is an arbitrary element of this group, prove that the equation $(a * x) * b * c = b * c$ has a unique solution $x \in G$.

1.16.3. Classify the algebraic systems formed by

(a) the irrational numbers (plus zero) under addition

(b) the rational numbers under addition

(c) the irrational numbers under multiplication.

1.16.4. Determine which of the following systems are groups with respect to the indicated operations:

(a) $\mathcal{S} = \{\{x : x \in \mathbb{Z}, x < 0\}$, addition$\}$

(b) $\mathcal{S} = \{\{x : x = 5y\ ,\ y \in \mathbb{Z}\}$, addition$\}$

(c) $\mathcal{S} = \{\{-4, -1, 4, 1\}$, multiplication$\}$

(d) $\mathcal{S} = \{\{z : z \in \mathbb{C}, |z| = 1\}$, multiplication$\}$

Here $\mathbb{Z}$ is the set of integers, and $\mathbb{C}$ is the complex number field.

1.16.5. Show that the integers $\mathbb{Z}$, the rationals $\mathbb{Q}$, and the reals $\mathbb{R}$ are rings under the operations of ordinary addition and multiplication.

1.16.6. Show that the system $\{\{a, b\}, *, \#\}$ with $*$ and $\#$ defined by

$*$	a	b
a	a	b
b	b	a

$\#$	a	b
a	a	a
b	a	b

is a ring.

1.16.7. Which of the following algebraic systems are rings?

(a) $\mathcal{S} = \{\{3x : x \in \mathbb{Z}\}, +, \cdot\}$

(b) $\mathcal{S} = \{\{x + 2 : x \in \mathbb{Z}\}, +, \cdot\}$

Here $+$ and $\cdot$ denote ordinary addition and multiplication.

1.16.8. Let $\mathcal{A} = \mathcal{P}(A) =$ the set of all subsets of a given set A. Consider the system $\mathcal{S} = \{\mathcal{A}, \otimes, \#\}$, where, if B and C are sets in $\mathcal{A}$,

$$B \otimes C = (B \cup C) - (B \cap C) \text{ and } B\#C = B \cap C$$

Show that $\mathcal{S}$ is a commutative ring.

1.16.9. Let B denote the set of ordered quadruples of real numbers of the form $(a, b, -b, a)$, $(a, b \in \mathbb{R})$. Consider the system $\mathcal{B} = \{B, \oplus, \odot\}$, where

$$(a, b, -b, a) \oplus (c, d, -d, c) = (a + c, b + d, -b - d, a + c)$$
$$(a, b, -b, a) \odot (c, d, -d, c) = (ac - bd, ad + bc, -ad - bc, ac - bd)$$

Determine if the system $\mathcal{B}$ is a field.

1.16.10. Show that the set of polynomials defined on S, where $\{S, +, *\}$ is a ring, with the operations defined in Example 1.16.8 forms a ring.

Elementary Topology in $\mathbb{R}^n$

1.17 The Real Number System

The study of properties of the real number system lies at the heart of mathematical analysis, and much of higher analysis is a direct extension or generalization of intrinsic properties of the real line. This section briefly surveys its most important features, which are are essential to understanding many of the ideas in subsequent chapters.

Our objective in this section is to give a concise review of topological properties of real line $\mathbb{R}$ or more generally the Cartesian product $\mathbb{R}^n$. By the *topological properties* we mean notions like open and closed sets, limits, continuity. We shall elaborate on all these subjects in a much more general context in Chapter 4 where we study the general theory of *topological spaces* and then the theory of *metric spaces*, of which the space $\mathbb{R}^n$ is an example.

It may seem to be somehow inefficient and redundant that we shall go over some of the notions studied in this section again in Chapter 4 in a much broader setting. This "didactic conflict" is a result of the fact that we do need the results presented in this section to develop the notion of Lebesgue integral in Chapter 3, which in turn serves as a primary tool to construct the most important examples of structures covered in Chapter 4.

We begin with a short review of the algebraic properties of the set of real numbers $\mathbb{R}$.

Real Numbers. There are many ways in which one can construct specific models for the set of real numbers. To list a few, let us mention the *Dedekind sections*, in which real numbers are identified with subsets of rational numbers, or *Cantor's representation*, in which a real number is identified with its decimal representation understood as a limit of a sequence of rational numbers. It is important to realize that in all those constructions we arrive at the same algebraic structure, or more precisely, the different models are *isomorphic* (in a specialized sense of the kind of algebraic structure we deal with). This brings us to the point at which it is not the *particular model* which is important itself but rather its *algebraic properties*, since they *fully* characterize the set of real numbers.

The properties are as follows:

(i) $\{\mathbb{R}, \cdot, +\}$ is a field.

(ii) $\leq$ is a total ordering on $\mathbb{R}$ which is order complete.
The total ordering $\leq$ on $\mathbb{R}$ is compatible with the field structure in the sense that

(iii) for $x, y \in \mathbb{R}$, if $x \leq y$, then $x + z \leq y + z$ for every $z \in \mathbb{R}$;

(iv) for $x, y \in \mathbb{R}$, if $x \geq 0$ and $y \geq 0$, then $xy \geq 0$.

Elements $x, y, z, \ldots$ of the set $\mathbb{R}$ are called *real numbers.* We generally use the symbol $\mathbb{R}$ to refer to the entire system $\{\mathbb{R}, +, \cdot, \leq\}$ and also to the field $\{\mathbb{R}, \cdot, +\}$.

Let us remember that *order complete* ordering $\leq$ on $\mathbb{R}$ distinguishes real numbers from rationals. By order complete, of course, we mean that every nonempty subset of real numbers that has an upper bound also has the least upper bound—the analogous property holds for lower bounds. The least upper bound of a set A, also called the *supremum* of the set A, will be denoted $\sup A$. We use an analogous notation for the greatest lower bound of A or the *infinum* of A denoted by $\inf A$. If A has no upper bound we frequently write that $\sup A = +\infty$. The notation is connected with the so-called *extended real line* analysis where, by the *extended real line*, we mean the set of real numbers $\mathbb{R}$ completed by the two symbols $+\infty$ and $-\infty$. Such a set can be embodied with a precise algebraic structure generalizing the algebraic structure of $\mathbb{R}$. For us, using the symbol $+\infty$ will be only an equivalent way to say that the set A has no upper bound, i.e.,

$$\sup\ A = \begin{cases} +\infty \text{ if } A \text{ has no upper bound} \\ \text{the smallest upper bound otherwise} \end{cases}$$

In the same way $\inf A = -\infty$ indicates that A has no lower bound.

Supremum and Infimum of a Real-Valued Function. If $f: X \to \mathbb{R}$ is a function defined on an arbitrary set X but taking values in the set of real numbers, its range is a subset of $\mathbb{R}$, i.e., $\mathcal{R}(f) \subset \mathbb{R}$. As a subset of real numbers, $\mathcal{R}(f)$, if bounded from above, possesses its supremum $\sup \mathcal{R}(f)$. This supremum is identified as the *supremum of function f over set X* and denoted by $\sup_{x\in X} f(x)$ or abbreviated $\sup_X f$. As stated above, $\sup_X f = +\infty$ is equivalent to the fact that $\mathcal{R}(f)$ has no upper bound.

In the same way we introduce the *infimum of f over X*, denoted $\inf_{x\in X} f(x)$ or abbreviated $\inf_X f$ and understood as the infimum of the range $\mathcal{R}(f)$.

We have the obvious inequality

$$\inf_X f \leq f(x) \leq \sup_X f$$

for every $x \in X$. If $\mathcal{R}(f)$ contains is supremum, i.e., there exists such an $x_0 \in X$ that

$$f(x_0) = \sup_X f$$

we say that function f *attains its maximum on X*. This in particular means that the *maximization problem*

$$\begin{cases} \text{Find } x_0 \in X \text{ such that} \\ f(x_0) = \sup_{x\in X} f(x) \end{cases}$$

is well posed in the sense that it has a solution. The supremum of f, $\sup_X f$ is called the *maximum of f over X*, denoted $\max_X f$ or $\max_{x\in X} f(x)$ and identified as the *greatest value* function f attains on X. Let us emphasize, however, that the use of the symbol $\max f$ is restricted *only* to the case when the maximum *exists*, i.e., f attains its supremum on X, while the use of the symbol $\sup f$ makes sense always. Replacing symbol $\sup f$ by $\max f$ without establishing the *existence of maximizers* is frequently encountered in engineering literature and leads to unnecessary confusion. The same rules apply to the notion of the *minimum of function f over set X* denoted $\min_{x \in X} f(x)$ or $\min_X f$.

Intervals. Let $a, b \in \mathbb{R}$ be fixed points in $\mathbb{R}$ such that $a < b$. The following interval notation is frequently used:

$$(a, b) = \{x: a < x < b, x \in \mathbb{R}\}$$

$$[a, b] = \{x: a \leq x \leq b, x \in \mathbb{R}\}$$

$$(a, b] = \{x: a < x \leq b, x \in \mathbb{R}\}$$

$$[a, b) = \{x: a \leq x < b, x \in \mathbb{R}\}$$

The set (a, b) is an *open* interval, $[a, b]$ is a *closed* interval, and $(a, b]$ and $[a, b)$ are *half-open* (or *half-closed*) intervals. *Infinite intervals* are of the type $(a, \infty) = \{x: a < x, x \in \mathbb{R}\}$, $[a, \infty) = \{x: a \leq x, x \in \mathbb{R}\}$, etc., while $(-\infty, \infty)$ is sometimes used to describe the entire real line. A single point is regarded as a closed interval.

Definition of $\mathbb{R}^n$. Euclidean Metric. Bounded Sets. The notion of the Cartesian product $A \times B$ of two sets A and B, discussed in Section 1.8, can be easily extended to the case of n different sets $A_i, i = 1, 2, \ldots, n$. In particular, we can speak of the *n-th power of a set A*, denoted A^n, which is understood as $A \times A \times \ldots \times A$ (n times). Thus, $\mathbb{R}^n = \mathbb{R} \times \mathbb{R} \times \ldots \times \mathbb{R}$ (n times) will consist of *n tuples* $\boldsymbol{x} = (x_1, \ldots, x_n)$ understood as finite sequences of real numbers. In the same way we define sets like $\mathbb{N}^n$, $\mathbb{Z}^n$, $\mathbb{Q}^n$ or $\mathbb{C}^n$. Set $\mathbb{R}^n$ has no longer *all* algebraic properties typical of the system $\mathbb{R}$. In general, it has no field structure and has no total ordering which would be compatible with other algebraic properties of $\mathbb{R}^n$. It does, however, share *some* very important algebraic properties with $\mathbb{R}$. Those include the notion of the *vector space* discussed in detail in the next chapter and the notion of the *Euclidean metric*. By the *Euclidean metric* $d(\boldsymbol{x}, \boldsymbol{y})$ in $\mathbb{R}^n$ we mean a nonnegative real-valued function of two variables $\boldsymbol{x}$ and $\boldsymbol{y}$ defined as follows

$$d(\boldsymbol{x},\boldsymbol{y}) = \left(\sum_{i=1}^{n}(x_i - y_i)^2\right)^{\frac{1}{2}}$$

where $\boldsymbol{x} = (x_i)_{i=1}^n$ and $\boldsymbol{y} = (y_i)_{i=1}^n$. For $n = 2(3)$, if x_i are identified as coordinates of a point $\boldsymbol{x}$ on a plane (in a space) in a Cartesian system of coordinates, $d(\boldsymbol{x},\boldsymbol{y})$ is simply the distance between points $\boldsymbol{x}$ and $\boldsymbol{y}$. The *Euclidean metric* is an example of a general notion of *metric* studied in detail in Chapter 4. The metric enables us to define precisely the notion of a *ball*. By the (open) *ball* $B(\boldsymbol{x},r)$ *centered at point* $\boldsymbol{x} \in \mathbb{R}^n$ *with radius* r we mean the collection of points $\boldsymbol{y} \in \mathbb{R}^n$ separated from the center $\boldsymbol{x}$ by a distance smaller than r. Formally,

$$B(\boldsymbol{x},r) = \{\boldsymbol{y} \in \mathbb{R}^n : d(\boldsymbol{x},\boldsymbol{y}) < r\}$$

If the strong inequality in the definition of the ball is replaced with the weak one, we talk about the *closed ball*, denoted $\overline{B}(\boldsymbol{x},r)$. By a *ball* we will always mean the *open ball* unless otherwise explicitly stated.

For $n = 1$ the open and closed balls reduce to the open or closed intervals respectively

$$B(x,r) = (x - r, x + r)$$

$$\overline{B}(x,r) = [x - r, x + r]$$

If a set $A \subset \mathbb{R}^n$ can be included in a ball $B(\boldsymbol{x},r)$ we say that it is *bounded*, otherwise set A is *unbounded*. For instance, infinite intervals in $\mathbb{R}$ are unbounded, a half plane in $\mathbb{R}^2$ is unbounded, any polygon in $\mathbb{R}^2$ is bounded, etc.

Exercises

1.17.1. Let $A,\ B \subset \mathbb{R}$ be two bounded sets of real numbers. Let $C = \{x + y : x \in A,\ y \in B\}$ (set C is called the *algebraic sum of sets A and B*). Prove that

$$\sup C = \sup A + \sup B$$

1.17.2. Let $f(x, y)$ be a real-valued function of two variables x and y defined on a set $X \times Y$, bounded from above. Define

$$g(x) = \sup_{y \in Y} f(x, y)$$

$$h(y) = \sup_{x \in X} f(x, y)$$

Prove that

$$\sup_{(x,y) \in X \times Y} f(x, y) = \sup_{x \in X} g(x) = \sup_{y \in Y} h(y)$$

In other words,

$$\sup_{(x,y) \in X \times Y} f(x, y) = \sup_{x \in X} \left(\sup_{y \in Y} f(x, y) \right) = \sup_{y \in Y} \left(\sup_{x \in X} f(x, y) \right)$$

1.17.3. Using the notation of the previous exercise, show that

$$\sup_{y \in Y} \left(\inf_{x \in X} f(x, y) \right) \leq \inf_{x \in X} \left(\sup_{y \in Y} f(x, y) \right)$$

Construct a counterexample showing that in general the equality does not hold.

1.17.4. If $|x|$ denotes the absolute value of $x \in \mathbb{R}$ defined as

$$|x| = \begin{cases} x & \text{if } x > 0 \\ -x & \text{otherwise} \end{cases}$$

prove that $|x| \leq a$ iff $-a \leq x \leq a$.

1.17.5. Prove the classical inequalities involving (including the triangle inequality) the absolute values

$$\Big| |x| - |y| \Big| \leq |x \pm y| \leq |x| + |y|$$

for every $x, y \in \mathbb{R}$.

1.17.6. Prove the *Cauchy-Schwartz inequality*

$$\left|\sum_{i=1}^{n} x_i y_i\right| \leq \left(\sum_{1}^{n} x_i^2\right)^{\frac{1}{2}} \left(\sum_{1}^{n} y_i^2\right)^{\frac{1}{2}}$$

where $x_i, y_i \in \mathbb{R}$, $i = 1, \ldots, n$.
Hint: Use the inequality

$$\sum_{i=1}^{n} (x_i \lambda + y_i)^2 \geq 0$$

for every $\lambda \in \mathbb{R}$.

1.17.7. Use the Cauchy-Schwartz inequality to prove the *triangle inequality*

$$d(\boldsymbol{x}, \boldsymbol{y}) \leq d(\boldsymbol{x}, \boldsymbol{z}) + d(\boldsymbol{z}, \boldsymbol{y})$$

for every $\boldsymbol{x}, \boldsymbol{y}, \boldsymbol{z} \in \mathbb{R}^n$.

1.18 Open and Closed Sets

We shall now examine several general properties of sets that take on special meaning when they are interpreted in connection with $\mathbb{R}^n$. While all of the subsequent ideas make some appeal to the geometrical features of the real line, they are actually much deeper, and can be easily extended to more general mathematical systems.

The so-called topology of the real line refers to notions of open sets, neighborhoods, and special classifications of points in certain subsets of $\mathbb{R}^n$. We discuss the idea of topologies and topological spaces in more detail in Chapter 4.

Neighborhoods. Let $\boldsymbol{x} = (x_1, \ldots, x_n)$ be a point in $\mathbb{R}^n$. A set $A \subset \mathbb{R}^n$ is called a *neighborhood of point* $\boldsymbol{x}$ iff there exists a ball $B(\boldsymbol{x}, \varepsilon)$, centered at $\boldsymbol{x}$, entirely contained in set A. It follows from the definition that if A is a neighborhood of $\boldsymbol{x}$ then every *superset* B of A (i.e., $A \subset B$) is also a neighborhood of $\boldsymbol{x}$.

Example 1.18.1

An open ball $B(\boldsymbol{x}, r)$ is a neighborhood of every point belonging to it.

Indeed, if $\boldsymbol{y} \in B(\boldsymbol{x}, r)$ then $d(\boldsymbol{x}, \boldsymbol{y}) < r$ and we can select ε such that

$$\varepsilon < r - d(\boldsymbol{x}, \boldsymbol{y})$$

Next, choose any $\boldsymbol{z}$ such that $d(\boldsymbol{y}, \boldsymbol{z}) < \varepsilon$. Then we have by the triangle inequality (cf. Exercise 1.17.7)

$$\begin{aligned} d(\boldsymbol{x}, \boldsymbol{z}) &\leq d(\boldsymbol{x}, \boldsymbol{y}) + d(\boldsymbol{y}, \boldsymbol{z}) \\ &\leq d(\boldsymbol{x}, \boldsymbol{y}) + \varepsilon \\ &< r \end{aligned}$$

which proves that $B(\boldsymbol{y}, \varepsilon) \subset B(\boldsymbol{x}, r)$ (cf. Fig. 1.15). ☐

Interior Points. Interior of a Set. A point $\boldsymbol{x} \in A \subset \mathbb{R}^n$ is called an *interior point of* A iff A is a neighborhood of $\boldsymbol{x}$. In other words, there exists a neighborhood N of $\boldsymbol{x}$ such that $N \subset A$ or, by the definition of neighborhood, there exists a ball $B(\boldsymbol{x}, \varepsilon)$ centered at $\boldsymbol{x}$ entirely contained in A. The collection of all interior points of a set A is called the *interior of* A and denoted by int A.

Example 1.18.2

The interior of an open ball $B(\boldsymbol{x}, r)$ coincides with the whole ball. The interior of a closed ball $\overline{B}(\boldsymbol{x}, r)$ coincides with the open ball $B(\boldsymbol{x}, r)$ centered at the same point $\boldsymbol{x}$ and of the same radius r.

Similarly, if A is, for instance, a polygon in $\mathbb{R}^2$ including its sides then its interior combines all the points of the polygon except for the points lying on the sides, etc. ☐

Open Sets. A set $A \subset \mathbb{R}^n$ is said to be *open* if int $A = A$, that is, all points of A are its interior points.

Example 1.18.3

Open balls are open. This kind of a statement usually generates a snicker when first heard, but it is by no means trivial. The first word "open" is used in context of the definition of the open ball introduced in the last section. The same word "open" used the second time refers to the notion of

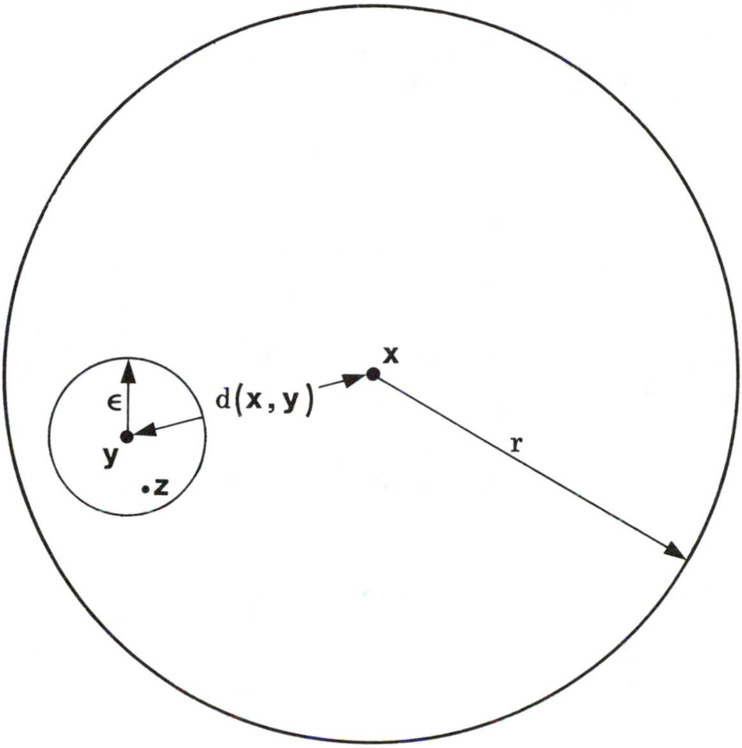

Figure 1.15
Example 1.18.1. Every open ball is a neighborhood for all its points.

the open sets just introduced. Since in this particular case the two notions coincide with each other, we do admit the repeated use of the same word. *Open* balls are indeed *open*.

By the same reasoning, *open intervals* (a, b) in $\mathbb{R}$ are *open*, too. □

Example 1.18.4

The empty set $\emptyset$ is open since it contains no points that are not interior points. Indeed, $\emptyset$ contains no points at all.

The whole set $\mathbb{R}^n$ is open (explain why). □

Properties of the open sets are summarized in the following proposition.

PROPOSITION 1.18.1

The following properties hold:

(i) *If $\mathcal{A}$ is an arbitrary family of open sets (possibly an infinite family, even not necessarily denumerable!), then the union of all sets A from $\mathcal{A}$*

$$\cup A, \quad A \in \mathcal{A}$$

is an open set.

(ii) *If $A_1, \ldots, A_n$ are open then the common part*

$$A_1 \cap A_2 \cap \ldots \cap A_n$$

is open, too.

In other words, unions of arbitrary collections of open sets are open while the common parts of only finite *families of open sets are open.*

PROOF

(i) Let $\boldsymbol{x} \in \cup A$. By the definition of the union, there exists a set $A \in \mathcal{A}$ which contains $\boldsymbol{x}$. But A is open, which means that there is a neighborhood N of $\boldsymbol{x}$ such that $N \subset A$. But $A \subset \cup A$ and therefore $N \subset \cup A$ which proves that $\boldsymbol{x}$ is an interior point of $\cup A$. Since $\boldsymbol{x}$ was an arbitrary point of $\cup A$, it proves that $\cup A$ is open.

(ii) We shall prove that the common part of two open sets is open. Then the general result follows by induction. Let A and B be two open sets. If $A \cap B = \emptyset$ the result is true since the empty set $\emptyset$ is open. Assume that there exists $\boldsymbol{x} \in A \cap B$. A is open and therefore there is a ball $B(\boldsymbol{x}, \varepsilon_1)$ contained in A. Similarly, there must be a ball $B(\boldsymbol{x}, \varepsilon_2)$ contained in B. Take $\varepsilon = \min(\varepsilon_1, \varepsilon_2)$. Obviously,

$$B(\boldsymbol{x}, \varepsilon) \subset A \cap B$$

which proves that $\boldsymbol{x}$ is an interior point of $A \cap B$ and therefore $A \cap B$ is open. ∎

Accumulation or Limit Points. Closure of a Set. Let $A \subset \mathbb{R}^n$. A point a, not necessarily in A, is called an *accumulation point* or a *limit point* of A if and only if every neighborhood of a contains at least one point of A distinct from a.

The *closure* of a set $A \subset \mathbb{R}^n$ is the set consisting of A and all of the accumulation points of A; i.e., if $\widehat{A}$ is the set of all accumulation points of a set A, then the set $\widehat{A} \cup A$ is the closure of A. The symbolism $\overline{A}$ is used to denote the closure of the set A.

Example 1.18.5

Every point $\boldsymbol{y}$ belonging to the *sphere* of the ball $B(\boldsymbol{x}, r)$, i.e., whose distance from $\boldsymbol{x}$ is exactly r, is an accumulation point of the ball. Indeed, every ball $B(\boldsymbol{y}, \varepsilon)$ (and therefore every neighborhood as well!) intersects with $B(\boldsymbol{x}, r)$ and therefore contains points from $B(\boldsymbol{x}, r)$.

Other accumulation points of $B(\boldsymbol{x}, r)$ include all the points $\boldsymbol{y}$ from the open ball, $d(\boldsymbol{x}, \boldsymbol{y}) < r$ (explain why). Thus the closure of the open ball $B(\boldsymbol{x}, r)$ coincides with the closed ball $\overline{B}(\boldsymbol{x}, r)$.

Closed Sets. If a set $A \subset \mathbb{R}^n$ coincides with its closure, $\overline{A} = A$, we say that the set A is *closed*. In other words, a set is *closed* if it contains all its accumulation points. $\square$

Example 1.18.6

Consider the set $A = \{x : x \in [0, 1) \text{ or } x = 2\} = [0, 1) \cup \{2\}$. The point 1 is an accumulation point of A, since every neighborhood of 1 contains points in A. However, 2 is *not* an accumulation point, since, for example, the neighborhood $|x - 2| < \frac{1}{2}$ contains no points in A other than 2. Clearly, A is not a closed set, since it does not contain the accumulation point 1; however, every neighborhood of 1 contains infinitely many points of A. The closure of A is the set $\overline{A} = \{x : x \in [0, 1] \text{ or } x = 2\}$. $\square$

Points such as 2 in the above example for which neighborhoods exists that contains no points of A other than itself are called *isolated points* of A.

Before we summarize the essential properties of the closed sets we shall establish a fundamental link between the notions of open and closed sets.

PROPOSITION 1.18.2

A set $A \subset \mathbb{R}^n$ is open iff its complement $A' = \mathbb{R}^n - A$ is closed.

PROOF Let $A \subset \mathbb{R}^n$ be open. We should show that $\mathbb{R}^n - A$ contains all its accumulation points. Assume instead that there exists an accumulation point $\boldsymbol{x}$ of $\mathbb{R}^n - A$ which does not belong to $\mathbb{R}^n - A$, i.e., $\boldsymbol{x} \in A$. But A is open, which means that there exists a neighborhood N of $\boldsymbol{x}$ such that $N \subset A$. But this means that N has no common points with $\mathbb{R}^n - A$ and therefore $\boldsymbol{x}$ is not an accumulation point of A', a contradiction.

Conversely, suppose that A' is closed. Again assume instead that there exists a point $\boldsymbol{x} \in A$ which is not an interior point of A. This means that *every* neighborhood N of $\boldsymbol{x}$ contains points from outside of A, i.e., from $\mathbb{R}^n - A$ and by closedness of A', the point $\boldsymbol{x}$ belongs to A', a contradiction again. ∎

This relation between open and closed sets, sometimes referred to as the *duality phenomenon*, is very useful in proving theorems in topology. As an example we shall use it to prove:

PROPOSITION 1.18.3
(Properties of Closed Sets)

(i) If $\mathcal{A}$ is an arbitrary family of closed sets, the common part of all sets A from $\mathcal{A}$

$$\cap A, \quad A \in \mathcal{A}$$

is a closed set, too.

(ii) If $A_1, \ldots, A_n$ are closed, then the union

$$A_1 \cup A_2 \cup \ldots \cup A_n$$

is closed, too.

PROOF (i) By De Morgan's laws for arbitrary unions (recall Section 1.4) we have

$$(\cap A)' = \cup A', \quad A \in \mathcal{A}$$

Now, if A are closed then by Proposition 1.18.2 their complements A' are open. According to the properties of open sets, the union $\cup A'$ is open and by the duality principle again $\cap A$ must be closed.

We prove property (ii) in exactly the same way. ∎

Example 1.18.7

There are sets in $\mathbb{R}^n$ which are neither closed nor open. For instance, if we take a closed ball $\overline{B}(\boldsymbol{x}, r)$ and remove any point $\boldsymbol{y}$ from it we get a set which is neither open nor closed (explain why).

It is perhaps more confusing that *there are* sets which are *simultaneously* closed and open. Those include the empty set $\emptyset$ and the entire $\mathbb{R}^n$ (explain why). □

Exercises

1.18.1. Let int A denote the interior of a set $A \subset \mathbb{R}^n$. Prove that the following properties hold.

(i) $\text{int}\,(\text{int}\,A) = \text{int}\,A$

(ii) $\text{int}\,(A \cup B) \supset \text{int}\,A \cup \text{int}\,B$

(iii) $\text{int}\,(A \cap B) = \text{int}\,A \cap \text{int}\,B$

(iv) If $A \subset B$ then $\text{int}\,A \subset \text{int}\,B$

1.18.2. Let $\overline{A}$ denote the closure of a set $A \subset \mathbb{R}^n$. Prove that the following properties hold.

(i) $(\overline{A})^{-} = \overline{A}$

(ii) $(\overline{A \cup B}) = \overline{A} \cup \overline{B}$

(iii) $(\overline{A \cap B}) \subset \overline{A} \cap \overline{B}$

(iv) If $A \subset B$ then $\overline{A} \subset \overline{B}$

1.18.3. Show that $\text{int}\,A = (\overline{A'})'$ where $A' = \mathbb{R}^n - A$ is the complement of set A (recall Proposition 1.18.2).

1.18.4. Construct examples showing that, in general,

$$\overline{(A \cap B)} \neq \overline{A} \cap \overline{B}$$

$$\text{int}\,(A \cup B) \neq \text{int}\,A \cup \text{int}\,B$$

1.18.5. Use the result of Exercise 1.18.3 to show that the properties listed in Exercise 1.18.2 are equivalent to those in Exercise 1.18.1.

1.18.6. Show that if $\boldsymbol{x}$ is an accumulation point of a set $A \subset \mathbb{R}^n$ then every neighborhood of $\boldsymbol{x}$ contains infinitely many points of A. Note that this in particular implies that *only infinite* sets *may* have accumulation points.

1.18.7. Show an example of an infinite set in $\mathbb{R}$ which has no accumulation points.

1.18.8. Let $A = \{x : x \in \mathbb{Q} = \text{ rationals}, 0 \le x \le 1\}$. Prove that every point x such that $0 \le x \le 1$ is an accumulation point of A, but there are no interior points of A.

1.18.9. Prove the *Bolzano-Weierstrass theorem* for sets of real numbers: if $A \subset \mathbb{R}$ is infinite and bounded, then there exists at least one point

$x \in \mathbb{R}$ that is an accumulation point of A. *Hint:* The proof of this theorem is classic and employs the *method of nested intervals:*
1) Choose $I_1 = [a_1, b_1] \supset A$ (why is this possible?) .
2) Divide I_1 into two equal intervals, and choose $I_2 = [a_2, b_2]$ so as to contain an infinity of elements of A (why is this possible?).
3) Continue this subdivision, and produce a sequence of "nested" intervals $I_1 \subset I_2 \subset I_3 \subset \cdots$ each containing infinitely many elements of A.
4) Note that if $I_n = [a_n, b_n]$, we can define $\overline{x}_n = \sup\{a_n\}$, $\overline{y}_n = \inf\{b_n\}$; show that as $n \to \infty$, $|\overline{x}_n - \overline{y}_n| \to 0$ so that $\overline{x}_n \to \overline{x}$, $\overline{y}_n \to \overline{y}$, and $\overline{x} = \overline{y}$.
5) Observe that for any $r > 0$ and sufficiently large n, $I_n \subset B(\overline{x}, r)$ and, therefore, $\overline{x}$ is an accumulation point of A (why?).

1.18.10. Most commonly, the intersection of an infinite sequence of open sets and the union of an infinite sequence of closed sets are not open or closed, respectively. Sets of this type are called sets of G_δ or F_σ *type*, i.e.,

$$A \text{ is of } G_\delta \text{ type} \quad \text{if } A = \bigcap_1^\infty A_i, \quad A_i \text{ open}$$

$$B \text{ is of } F_\sigma \text{ type} \quad \text{if } B = \bigcup_1^\infty B_i, \quad B_i \text{ closed}$$

Construct examples of a G_δ set which is not open, and an F_σ set which is not closed.

1.19 Sequences

If to every positive integer n there is assigned a number $a_n \in \mathbb{R}$, the collection $a_1, a_2, \ldots, a_n, a_{n+1}, \ldots$ is said to form a *sequence*, denoted $\{a_n\}$. For example, the rules

$$a_n = \frac{1}{n}, \quad a_n = \frac{1}{2^n}, \quad a_n = \left(\frac{n+2}{n+1}\right)^n$$

describe the sequences in $\mathbb{R}$

$$1, \frac{1}{2}, \frac{1}{3}, \ldots$$

$$\frac{1}{2}, \frac{1}{4}, \frac{1}{8}, \ldots$$

$$\frac{3}{2}, \left(\frac{4}{3}\right)^2, \left(\frac{5}{4}\right)^3, \ldots$$

In the same way, we can define a sequence $\boldsymbol{a}_n$ in $\mathbb{R}^n$. For instance, the rule

$$\boldsymbol{a}_n = \left(\frac{1}{n}, n^2\right)$$

describes a sequence of points in $\mathbb{R}^2$

$$(1,1), \left(\frac{1}{2}, 4\right), \left(\frac{1}{3}, 9\right), \ldots$$

More precisely, if A is an arbitrary set and $\mathbb{N}$ denotes the set of natural numbers then any function $s: \mathbb{N} \to A$ is called a *sequence* in A. Customarily we write s_n in place of $s(n)$. We use interchangeably the notations $\boldsymbol{a}_n$, $\{\boldsymbol{a}_n\}$ for sequences in sets $A \subset \mathbb{R}^n$.

Limit of a Sequence in $\mathbb{R}^n$. Let $\{\boldsymbol{a}_n\}$ be a sequence in $\mathbb{R}^n$. We say that $\{\boldsymbol{a}_n\}$ *has a limit* $\boldsymbol{a}$ in $\mathbb{R}^n$, or that $\{\boldsymbol{a}_n\}$ *converges to* $\boldsymbol{a}$, denoted $\boldsymbol{a}_n \to \boldsymbol{a}$, iff *for every neighborhood* $N_{\boldsymbol{a}}$ *of* $\boldsymbol{a}$ *all but a finite number of elements of sequence* $\{\boldsymbol{a}_n\}$ *lie in* $N_{\boldsymbol{a}}$.

By the definition of neighborhood we can rewrite this condition in a more practical form

> for every $\varepsilon > 0$ there exists an index $N \in \mathbb{N}$ ("$N = N(\varepsilon)$", in general, depends upon ε) such that $\boldsymbol{a}_n \in B(\boldsymbol{a}, \varepsilon)$, for every $n \geq N$

or, equivalently,

$$\forall\, \varepsilon > 0,\ \exists N \in \mathbb{N} \text{ such that } \forall\, n \geq N,\ d(\boldsymbol{a}, \boldsymbol{a}_n) < \varepsilon$$

We also use the notation

$$\lim_{n\to\infty} \boldsymbol{a}_n = \boldsymbol{a}$$

Example 1.19.1

Consider a sequence of real numbers $a_n = n^2/(1+n^2)$. Pick an arbitrary $\varepsilon > 0$. The inequality

$$\left|\frac{n^2}{1+n^2} - 1\right| < \varepsilon$$

holds iff

$$\left|-\frac{1}{n^2+1}\right| < \varepsilon$$

or, equivalently,

$$n^2 + 1 > \frac{1}{\varepsilon}$$

Choosing N equal to the integer part of $(\varepsilon^{-1} - 1)^{\frac{1}{2}}$ we see that for $n \geq N$ the original equality holds. This proves that sequence a_n converges to the limit $a = 1$. $\square$

Example 1.19.2

Consider a sequence of points in $\mathbb{R}^2$

$$\boldsymbol{a}_n = \left(\frac{1}{n}\cos(n);\ \frac{1}{n}\sin(n)\right)$$

It is easy to prove that $\boldsymbol{a}_n$ converges to the origin $(0, 0)$. $\square$

The notion of a sequence can be conveniently used to characterize the accumulation points in $\mathbb{R}^n$. We have

PROPOSITION 1.19.1

Let A be a set in $\mathbb{R}^n$. The following conditions are equivalent to each other:

(i) *$\boldsymbol{x}$ is an accumulation point of A.*

(ii) *There exists a sequence $\boldsymbol{x}_n$ of points of A, different from $\boldsymbol{x}$, such that $\boldsymbol{x}_n \rightarrow \boldsymbol{x}$.*

PROOF Implication (ii) $\rightarrow$ (i) follows directly from the definition of convergence and accumulation points. To prove that (i) implies (ii) we need to construct a sequence of points $\boldsymbol{x}_n$, different from $\boldsymbol{x}$, converging to

$\boldsymbol{x}$. Let $n \in \mathbb{N}$. Consider a ball $B(\boldsymbol{x}, \frac{1}{n})$. It follows from the definition of accumulation point that there exists a point in A, different from $\boldsymbol{x}$, which belongs to the ball. Denote it by $\boldsymbol{x}_n$. By the construction,

$$d(\boldsymbol{x}_n, \boldsymbol{x}) < \varepsilon \text{ for every } n \geq \frac{1}{\varepsilon} + 1$$

which finishes the proof. ∎

Given any sequence $\{a_1, a_2, a_3, \ldots\}$ we can form new sequences of the type $\{a_1, a_4, a_8, \ldots\}$ or $\{a_3, a_7, a_{11}\}$, etc. Such new sequences are called *subsequences* of the original sequence. More rigorously, we have:

Subsequence. Let $s: \mathbb{N} \to A$ and $t: \mathbb{N} \to A$ denote two sequences. The sequence t is a *subsequence* of s if and only if there exists a one-to-one mapping $r: \mathbb{N} \to \mathbb{N}$ such that

$$t = s \circ r$$

Example 1.19.3

Let $s(n) = 1/n$ and $t(n) = 1/n^3$. Then $s(n) = \{1, \frac{1}{2}, \frac{1}{3}, \frac{1}{4}, \ldots\}$ and $t(n) = \{1, \frac{1}{8}, \frac{1}{27}, \ldots\}$. Obviously, $t(n)$ is a subsequence of $s(n)$. To prove this, consider $r(n) = n^3$. Clearly, r is injective. □

Thus the injective map r selects particular labels $n \in \mathbb{N}$ to be identified with entries in the subsequence $t(n)$. If $\{\boldsymbol{a}_k\}$ is a sequence of $A \subset \mathbb{R}^n$, we sometimes denote subsequences as having different index labels: $\{\boldsymbol{a}_\ell\}$ or $\{\boldsymbol{a}_{k_\ell}\} \subset \{\boldsymbol{a}_k\}$.

Cluster Points. Let $\{\boldsymbol{a}_k\} \in \mathbb{R}^n$ be a sequence of points in $\mathbb{R}^n$. If $\{\boldsymbol{a}_k\}$ has a subsequence, say $\boldsymbol{a}_l$, converging to a point $\boldsymbol{a} \in \mathbb{R}^n$, we call $\boldsymbol{a}$ a *cluster point* of the sequence $\{\boldsymbol{a}_k\}$. A sequence may have infinitely many cluster points, a *convergent* sequence has only one—its limit.

Example 1.19.4

Consider a sequence of real numbers

$$a_k = (-1)^k \left(1 + \frac{1}{k}\right)$$

Obviously, subsequence a_{2k} (even indices) converges to 1 while the subsequence a_{2k-1} converges to -1. Thus a_k has two cluster points: $+1$ and -1. □

Limits Superior and Inferior. In the case of sequences of real numbers, the notion of cluster points allows us to generalize the concept of the limit. A sequence $a_n \in \mathbb{R}$ is *bounded* whenever a ball (interval) $B(x,r)$ exists that contains all of the entries a_n of the sequence. Let a_n be a bounded sequence of real numbers and let A denote the set of its cluster points. According to the Bolzano-Weierstrass theorem for sequences (see Exercise 1.19.2) A is not empty. We refer to its *supremum* as the *limit superior* of a_n, denoted $\limsup_{n\to\infty} a_n$, and to its *infimum* as the *limit inferior* of a_n, denoted $\liminf_{n\to\infty} a_n$. Symbols $+\infty$ and $-\infty$ are used accordingly if A has no upper or lower bounds (recall Section 1.17).

Thus, for the sequence from the previous example, its limit superior is 1 while its limit inferior is -1.

Notions of limit superior and limit inferior may be conveniently characterized in many different ways. The following proposition summarizes some of them for the limit inferior. Analogous results hold for limit superior.

PROPOSITION 1.19.2

Let a_n be a sequence of real numbers and A denote the set of its cluster points. Let a denote the limit inferior *of a_n, i.e.,*

(i) $a = \inf A$.
The following characterization holds:

(ii) $a = \sup_N \inf_{n \geq N} \{a_n\} = \lim_{N\to\infty} \inf_{n\geq N} \{a_n\}$.
Moreover, when a is finite then

(iii) $a = \min A$.

In other words, in the case of a finite limit inferior the $\liminf_{n\to\infty} a_n$ can be characterized as the smallest *cluster point of A.*

PROOF Equality in (ii) follows from the fact that the sequence

$$b_N = \inf_{n\geq N}\{a_n\}$$

is increasing (the infimum is taken over smaller and smaller sets), and therefore the limit of b_N coincides with the supremum of the set $\{b_N\}$ (see Exercise 1.19.1).

Case 1. $\inf A = -\infty$. By definition, there exists a subsequence a_{n_k} converging (diverging) to $-\infty$, which implies that a_n is not bounded below. Consequently, $b_N \equiv -\infty$ and the limit in (ii) is equal to $-\infty$ as well.

Case 2. $\inf A = +\infty$. There is only one cluster point, equal to the limit of the sequence, equal to $+\infty$. The definition of convergence (divergence) to $+\infty$ reads as follows:

$$\forall\ M \exists N = N(M) \text{ such that } \forall\ n \geq N\ a_n \geq M$$

As a result,

$$\sup_N b_N \geq \sup_M b_{N(M)} \geq \sup_M M = +\infty$$

Case 3. $\inf A$ is finite. We first show that (iii) holds, i.e., that there exists a subsequence a_{n_k} of a_n such that $a_{n_k} \to a$. We will construct the subsequence recursively. First of all, note that eliminating a *finite number* of elements from a sequence does not change its cluster points (why?). Now, it follows from the definition of infimum that for every k there exists a cluster point c_k such that

$$c_k < a + \frac{1}{2k}$$

In turn, it follows from the definition of the cluster points that we can find an element $a_{n_k} = a_{n(k)}$ such that

$$|a_{n_k} - c_k| < \frac{1}{2k}$$

Since the extraction of a finite number of points from the sequence does not change its cluster points we can assume that $n(k)$ is different from $n(1), \ldots, n(k-1)$. As a result of the construction, $a_{n(k)}$ is a well-defined subsequence of a_n and

$$|a - a_{n(k)}| < \frac{1}{k}$$

which proves that $a_{n(k)} \to a$.

Now let a_{n_k} be a subsequence converging to a. Then, obviously,

$$b_{n_k} = \inf_{n \geq n_k} \{a_n\} \leq a_{n_k}$$

and passing to the limit with k we get

$$\lim_{N\to\infty} \inf_{n \geq N} \{a_n\} = \lim_{n\to\infty} b_N = \lim_{k\to\infty} b_{n_k} \leq \lim_{k\to\infty} a_{n_k} = a$$

To prove the inverse equality we need to show that the point defined in (ii) is a cluster point. We proceed recursively again. For an arbitrary k we select an element $a_{n(k)}$ such that $n(k)$ is different from $n(1), \ldots, n(k-1)$ and such that

$$a_{n(k)} \leq \inf_{n \geq n(k)} \{a_n\} + \frac{1}{k}$$

Consequently,

$$\lim_{k\to\infty} a_{n(k)} = \lim_{k\to\infty} \left\{ \inf_{n\geq n(k)} \{a_n\} + \frac{1}{k} \right\}$$

$$= \lim_{k\to\infty} \inf_{n\geq n(k)} \{a_n\}$$

$$= \lim_{N\to\infty} \inf_{n\geq N} \{a_n\}$$

which finishes the proof. ∎

Let us emphasize that the equivalence of conditions (i) and (iii) allows us in particular to speak about the limit inferior as the *smallest* cluster point. Similarly, the limit superior can be understood as the *largest* cluster point. This identification of limit superior and limit inferior with the regular limits of certain subsequences allows us to use the notions in calculus in the very same way as the notion of the limit. Typical properties are summarized in the following proposition.

PROPOSITION 1.19.3

Let a_n and b_n be two sequences of real numbers. Then the following properties hold:

(i) *If $a_n \leq b_n$, for every n, then*

$$\liminf a_n \leq \liminf b_n \text{ and } \limsup a_n \leq \limsup b_n$$

(ii) $\liminf a_n + \liminf b_n \leq \liminf(a_n + b_n)$.

(iii) $\limsup(a_n + b_n) \leq \limsup a_n + \limsup b_n$.

PROOF

(i) Pick a subsequence b_{n_k} such that

$$b = \lim_{k\to\infty} b_{n_k} = \lim_{n\to\infty} \inf b_n$$

Since the weak inequality is conserved in the limit, convergent *subsequences* of the corresponding subsequence a_{n_k} have limits less or equal than b. But this proves that all cluster points of a_{n_k} are less or equal than b and, consequently, the smallest cluster point of a_n is less or equal than b, too.

Analogously, we prove that the limit superior satisfies the weak inequality, too.

(ii) We have the obvious inequality

$$\inf_{n\geq N}\{a_n\} + \inf_{n\geq N}\{b_n\} \leq a_n + b_n$$

for every $n \geq N$.

Taking the infimum on the right-hand side we get

$$\inf_{n\geq N}\{a_n\} + \inf_{n\geq N}\{b_n\} \leq \inf_{n\geq N}\{a_n + b_n\}$$

Finally, passing to the limit with $n \to \infty$ and using condition (iii) from Proposition 1.19.2 we get the result required.

The proof of property (iii) is identical. ∎

Exercises

1.19.1. A sequence is said to be *monotone increasing* if $a_{n+1} \geq a_n$ for all n; it is *monotone decreasing* if $a_{n+1} \leq a_n$ for all n. Further, a sequence is said to be *bounded above* if a number b exists such that $a_n \leq b$ for all n; a sequence is *bounded below* if a number a exists such that $a \leq a_n$ for all n.
Prove that every monotone increasing (decreasing) and bounded above (below) sequence in $\mathbb{R}$ is convergent.

1.19.2. Prove the *Bolzano-Weierstrass theorem for sequences:* every bounded sequence in $\mathbb{R}$ has a convergent subsequence. *Hint:* Let A be the set of values of the sequence. Consider two cases:

(i) A is finite,

(ii) A is infinite.

Use the Bolzano-Weierstrass theorem for sets (Exercise 1.18.9) in the second case.

1.19.3. Establish the convergence or divergence of the sequences $\{x_n\}$, where

$$(a)\ x_n = \frac{n^2}{1+n^2} \quad (b)\ x_n = \sin(n)$$

$$(c)\ x_n = \frac{3n^2+2}{1+3n^2} \quad (d)\ x_n = \frac{(-1)^n n^2}{1+n^2}$$

1.19.4. Let $x_1 \in \mathbb{R}$ be > 1 and let $x_2 = 2 - 1/x_1, \ldots, x_{n+1} = 2 - 1/x_n$. Show that this sequence is monotone and bounded and determine its limit.

1.19.5. Prove that if $\{x_n\}$ and $\{y_n\}$ are sequences converging to x and y, respectively, and that if $x_n \le y_n \ \forall\ n$, then $x \le y$.

1.19.6. Let $\boldsymbol{x}^k = (x_1^k, \ldots, x_n^k)$ be a sequence of points in $\mathbb{R}^n$, $k = 1, 2, \ldots$. Prove that

$$\lim_{k\to\infty} \boldsymbol{x}^k = \boldsymbol{x} \ \text{ iff } \lim_{k\to\infty} x_i^k = x_i, \ \text{ for every } i = 1, \ldots, n$$

where $\boldsymbol{x} = (x_1, \ldots, x_n)$.

1.19.7. Let $\boldsymbol{x}_k$ be a sequence in $\mathbb{R}^n$. Prove that $\boldsymbol{x}$ is a cluster point of $\boldsymbol{x}_k$ iff every neighborhood of $\boldsymbol{x}$ contains infinitely many members of $\boldsymbol{x}_k$.

1.19.8. Let $\boldsymbol{x}^k = (x_1^k, x_2^k)$ be a sequence in $\mathbb{R}^2$ given by the formula

$$x_i^k = (-1)^{k+i}\frac{k+1}{k}, \ i = 1, 2, \ k \in \mathbb{N}$$

Determine the cluster points of the sequence.

1.19.9. Calculate lim inf and lim sup of the following sequence in $\mathbb{R}$:

$$a_n = \begin{cases} n/(n+3) & \text{for } n = 3k \\ n^2/(n+3) & \text{for } n = 3k+1 \\ n^2/(n+3)^2 & \text{for } n = 3k+2 \end{cases}$$

where $k \in \mathbb{N}$.

1.19.10. Formulate and prove a theorem analogous to Proposition 1.19.2 for limit superior.

1.20 Limits and Continuity

We now examine the fundamental concepts of limits and continuity of functions $f: \mathbb{R}^n \to \mathbb{R}^m$ defined on $\mathbb{R}^n$. In real analysis, the concept of continuity of a function follows immediately from that of limit of a function.

Limit of a Function. Let $f: A \subset \mathbb{R}^n \to \mathbb{R}^m$ denote a function defined on a set $A \subset \mathbb{R}^n$ and let $\boldsymbol{x}_0$ be an accumulation point of A. Then f is said to have a limit $\boldsymbol{a}$ at the point $\boldsymbol{x}_0$ if, for every $\varepsilon > 0$, there is another number $\delta > 0$ such that whenever $d(\boldsymbol{x}, \boldsymbol{x}_0) < \delta$, $d(f(\boldsymbol{x}), \boldsymbol{a}) < \varepsilon$.

The idea is illustrated in Fig. 1.16a. If x is sufficiently near x_0, $f(x)$ can be made as near to a as is desired. Figure 1.16b shows a case in which $f(x)$ is *discontinuous* at x_0. Clearly, if we pick an $\varepsilon > 0$, there exists no interval in the domain of $f(x)$ for which, say, $|f(x) - a_1| < \varepsilon$ whenever $|x - x_0| < \delta$. If we choose $x < x_0$, then $|f(x) - a_1| < \varepsilon$ whenever $|x - x_0| < \delta$; or, if $x > x_0$ then $|f(x) - a_2| < \varepsilon$ whenever $|x - x_0| < \delta$. Then a_1 is called the *left limit* of $f(x)$ at x_0 and a_2 is called the *right limit* of $f(x)$ at x_0. The function $f(x)$ has a limit a at x_0 iff $a_1 = a_2 = a$, and we write

$$\lim_{x \to x_0} f(x) = a$$

Continuity (The Limit Definition). A function $f: A \to \mathbb{R}$ on a set $A \subset \mathbb{R}^n$ is *continuous* at the accumulation point $\boldsymbol{x}_0 \in A$ if and only if

(i) $f(\boldsymbol{x}_0)$ exists

(ii) $\lim_{\boldsymbol{x} \to \boldsymbol{x}_0} f(\boldsymbol{x}) = f(\boldsymbol{x}_0)$

If $\boldsymbol{x}_0$ is not an accumulation point of A, we only require (i) for continuity. Note that this in particular implies that function f is always continuous at isolated points of A.

The definition of continuity can be rewritten without referring to the notion of a limit.

Continuity ($\varepsilon - \delta$ Definition, Cauchy). A function $f: \mathbb{R}^n \supset A \to \mathbb{R}^m$ is continuous at a point $\boldsymbol{x}_0 \in A$ (this automatically means that $f(\boldsymbol{x}_0)$ exists) iff for every $\varepsilon > 0$ there exists a $\delta > 0$ such that

$$d(f(\boldsymbol{x}_0), f(\boldsymbol{x})) < \varepsilon \text{ whenever } d(\boldsymbol{x}_0, \boldsymbol{x}) < \delta, \boldsymbol{x} \in A$$

Observe that the two metrics here may be different, with the first being the Euclidean metric on $\mathbb{R}^n$ and the second the Euclidean metric on $\mathbb{R}^m$, with the possibility that $m \neq n$.

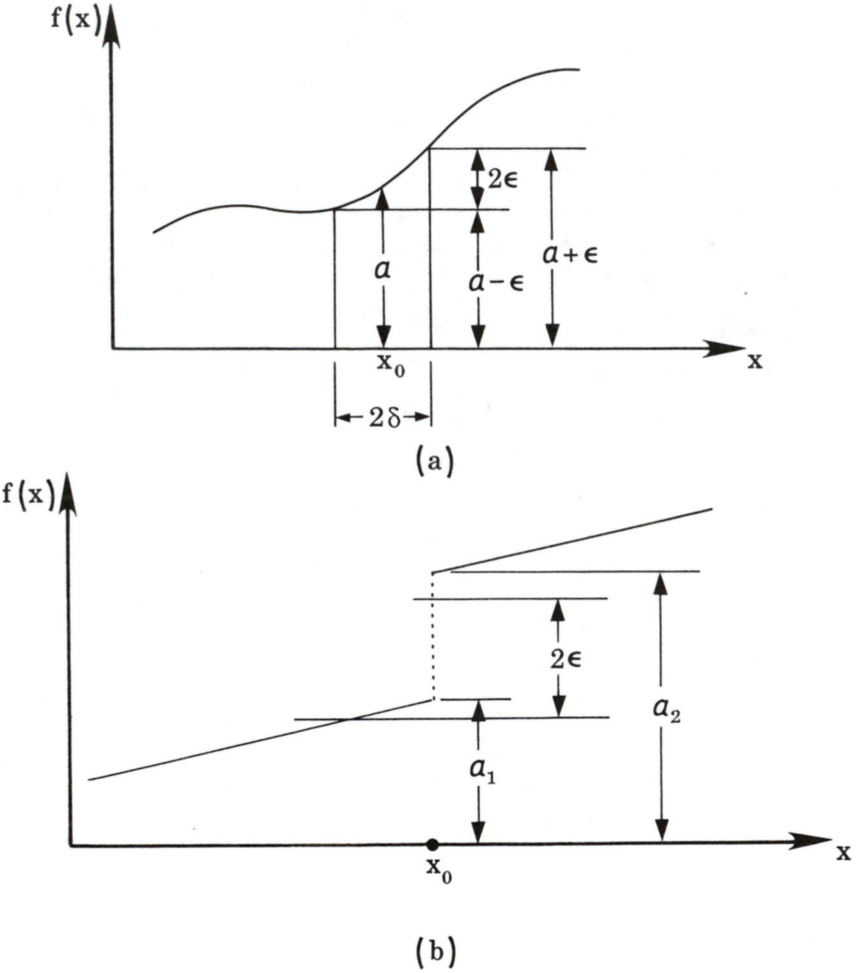

Figure 1.16
Continuous and discontinuous functions.

The essence of the notion of continuity is actually tied up in the concept of neighborhoods rather than limits or $\varepsilon - \delta$ arguments. Using the notion of neighborhood we can rewrite the definition of continuity in the following, equivalent way: A function $f: \mathbb{R}^n \supset A \to \mathbb{R}^m$ is continuous at a point $\boldsymbol{x}_0 \in A$ iff *for every neighborhood* N *of* $f(\boldsymbol{x}_0)$ *there exists a neighborhood* M *of* $\boldsymbol{x}_0$ *such that*

$$f(\boldsymbol{x}) \in N \text{ whenever } \boldsymbol{x} \in M \cap A$$

or simply

$$f(M) \subset N$$

Using the notion of sequences we can introduce the notion of continuity yet in another way.

Sequential Continuity (Heine). A function $f: \mathbb{R}^n \supset A \to \mathbb{R}^m$ is said to be *sequentially continuous at a point* $\boldsymbol{x}_0 \in A$ iff for every sequence $\boldsymbol{x}_n \in A$ converging to $\boldsymbol{x}_0$, sequence $f(\boldsymbol{x}_n)$ converges to $f(\boldsymbol{x}_0)$, i.e.,

$$\boldsymbol{x}_n \in A, \ \boldsymbol{x}_n \to \boldsymbol{x}_0 \text{ implies } f(\boldsymbol{x}_n) \to f(\boldsymbol{x}_0)$$

It turns out that the two notions are equivalent to each other.

PROPOSITION 1.20.1

A function $f: \mathbb{R}^n \supset A \to \mathbb{R}^m$ is continuous at a point $\boldsymbol{x}_0 \in A$ iff it is sequentially continuous at $\boldsymbol{x}_0$.

PROOF We show first that continuity implies the sequential continuity. Let $\boldsymbol{x}_k$ be an arbitrary sequence converging to $\boldsymbol{x}_0$. To prove that $f(\boldsymbol{x}_k)$ converges to $f(\boldsymbol{x}_0)$ one has to show that for every $\varepsilon > 0$ there exists an index N such that $d(f(\boldsymbol{x}_0), f(\boldsymbol{x}_k)) < \varepsilon$ whenever $k \geq N$. By continuity of f at $\boldsymbol{x}_0$ follows that there is a $\delta > 0$ such that $d(\boldsymbol{x}_0, \boldsymbol{x}) < \delta$ implies $d(f(\boldsymbol{x}_0), f(\boldsymbol{x})) < \varepsilon$. Since $\boldsymbol{x}_k$ converges to $\boldsymbol{x}_0$, there exists an N such that $d(\boldsymbol{x}_0, \boldsymbol{x}_k) < \delta$ for $k \geq N$, which in turn implies that $d(f(\boldsymbol{x}_0), f(\boldsymbol{x}_k)) < \varepsilon$ for $k \geq N$. Thus f is sequentially continuous at $\boldsymbol{x}_0$.

Assume now that f is sequentially continuous at $\boldsymbol{x}_0$, but it is not continuous at $\boldsymbol{x}_0$. Negating the condition for continuity we get that *there exists* $\varepsilon > 0$ *such that for every* $\delta > 0$ *there exists an* $\boldsymbol{x} \in A$ *such that*

$$d(\boldsymbol{x}_0, \boldsymbol{x}) < \delta \text{ but } d(f(\boldsymbol{x}_0), \boldsymbol{x}) \geq \varepsilon$$

Select $\delta = \frac{1}{k}$ and define a sequence $\boldsymbol{x}_k$ of points satisfying the condition above. By the construction, $\boldsymbol{x}_k$ converges to $\boldsymbol{x}_0$, but $f(\boldsymbol{x}_k)$ does not converge to $f(\boldsymbol{x}_0)$. This proves that f is *not* sequentially continuous at $\boldsymbol{x}_0$, a contradiction. ∎

Globally Continuous Functions. Let $f: \mathbb{R}^n \supset A \to \mathbb{R}^m$ be a function. We say that f is *globally continuous on* A, or shortly, f *is continuous on* A iff f is continuous at every point of A.

We have the following fundamental characterization of globally continuous functions.

PROPOSITION 1.20.2

Let $f: \mathbb{R}^n \to \mathbb{R}^m$ be a function. The following conditions are equivalent to each other:

(i) f *is globally continuous in* $\mathbb{R}^n$.

(ii) *For every open set* $G \subset \mathbb{R}^m$, *the inverse image* $f^{-1}(G)$ *by* f *is open in* $\mathbb{R}^n$.

(iii) *For every closed set* $H \subset \mathbb{R}^m$, *the inverse image* $f^{-1}(H)$ *by* f *is closed in* $\mathbb{R}^n$.

The proof of this result is left as an exercise (see Exercise 1.20.1). We shall prove this theorem in Chapter 4 in the context of general topological spaces.

The notion of continuous functions is fundamental in real analysis. Continuous functions satisfy many important properties which distinguish them from other classes of functions. One of them is the ultimate link between continuity of functions and the so-called compact sets.

Compact Set. A set $K \subset \mathbb{R}^n$ is said to be *compact* iff K *is bounded and closed.*

Thus, for instance, every closed ball is compact, but a half plane in $\mathbb{R}^2$ is not. Compact sets can be conveniently characterized using sequences. It turns out that if A is compact then every sequence of points $\boldsymbol{x}_k$ from A has a subsequence converging to a point $\boldsymbol{x}_0$ in A. We also say that A is then *sequentially compact.*

To show that compact sets do indeed have this property, pick an arbitrary sequence $\boldsymbol{x}_k$ in A. The sequence of real numbers $x_{k,1}$ identified as the first components of points $\boldsymbol{x}_k$ is certainly bounded and, therefore, by the Bolzano-Weierstrass theorem, has a convergent subsequence to an element, say $x_1 \in \mathbb{R}$. Proceeding in the same way, we can extract a subsequence from this subsequence such that also the second components converge to a number, say $x_2 \in \mathbb{R}$ and so on until a subsequence $\boldsymbol{x}_i$ is selected such that each of n components converge to a number $x_n \in \mathbb{R}$. But this means (recall

Exercise 1.19.6) that the subsequence $\boldsymbol{x}_i$ converges to $\boldsymbol{x} = (x_1, \ldots, x_n) \in \mathbb{R}$. By closedness of A, $\boldsymbol{x}$ must be in A, which proves the result.

This simple result allows us to establish the famous *Weierstrass theorem* for functions continuous on compact sets in $\mathbb{R}^n$.

THEOREM 1.20.1

Let K be a compact set in $\mathbb{R}^n$ and f a continuous function defined on K taking on values in $\mathbb{R}$. Then f attains both its supremum and infimum on K. In other words, there exist points $\boldsymbol{x}_{\min}$ and $\boldsymbol{x}_{\max}$ such that

$$f(\boldsymbol{x}_{\min}) = \inf_K f \ , \ f(\boldsymbol{x}_{\max}) = \sup_K f$$

PROOF We shall prove that f attains its maximum on K. The proof for the minimum is analogous. We first note that the number $a = \sup\limits_K f$ is an accumulation point of the range of f, $\mathcal{R}(f)$. If it were not, a ball $B(a, \varepsilon_1)$ could be found containing upper bounds of $\mathcal{R}(f) \leq a$ but no points in $\mathcal{R}(f)$, and this would contradict the fact that a is the least upper bound. Thus, there exists a sequence $\boldsymbol{x}_k \in K$ such that $f(\boldsymbol{x}_k) \to \sup\limits_K f$ (this may include the case when $\sup\limits_K f = +\infty$, i.e., the sequence $f(\boldsymbol{x}_k)$ is divergent to infinity). Since K is compact and, therefore, sequentially compact as well, there exists a subsequence, say $\boldsymbol{x}_i \in K$, converging to, say, $\boldsymbol{x}_0 \in K$. But f is continuous and therefore $f(\boldsymbol{x}_i) \to f(\boldsymbol{x}_0)$, which proves that

1. $\sup\limits_K f$ is finite;
2. $f(\boldsymbol{x}_0) = \sup\limits_K f$.

∎

The notion of compact sets will be studied in a much more general setting in Chapter 4.

Exercises

1.20.1. Prove Proposition 1.20.2.

1.20.2. Let $g \circ f$ denote the composition of a function $f: \mathbb{R}^n \supset A \to \mathbb{R}^m$ and $g: \mathbb{R}^m \supset B \to \mathbb{R}^k$ where the range of f is contained in B so the composition makes sense. Prove that if f is continuous at $\boldsymbol{x}_0 \in A$, g is continuous at $f(\boldsymbol{x}_0) \in B$, then $g \circ f$ is continuous at $\boldsymbol{x}_0$.

1.20.3. Let $f, g: \mathbb{R}^n \to \mathbb{R}^m$ be two continuous functions defined on the same set $A \subset \mathbb{R}^n$. Prove that the linear combinations of f, g defined as

$$(\alpha f + \beta g)(\boldsymbol{x}) = \alpha f(\boldsymbol{x}) + \beta g(\boldsymbol{x})$$

is also continuous.

1.20.4. Prove the *Weierstrass intermediate value theorem:*
Let f be a function from $\mathbb{R}$ into $\mathbb{R}$ that is continuous on a closed interval $[a, b]$. Then $f(x)$ assumes every value between $f(a)$ and $f(b)$.

1.20.5. Let $f: \mathbb{R}^n \supset A \to B \subset \mathbb{R}^m$ be a bijective continuous map. Prove that the inverse function $f^{-1}: \mathbb{R}^m \supset B \to A \subset \mathbb{R}^n$ is also continuous.

1.20.6. Determine a point $a_0 \in (0, 1)$ at which the following function is continuous:

$$f(x) = \begin{cases} 1 - x, & x \text{ is rational} \\ x, & x \text{ is irrational} \end{cases}$$

1.20.7. Show that

$$f(x) = \begin{cases} x \sin \dfrac{1}{x}, & x \neq 0 \\ 0, & x = 0 \end{cases}$$

is continuous on all of $\mathbb{R}$.

1.20.8. Let $\overline{I} = [0, 1]$. Give examples of

(i) a function $f: \overline{I} \to \mathbb{R}$ that is not bounded

(ii) a function $f: \overline{I} \to \mathbb{R}$ that does not achieve its supremum $\sup \{f(x): x \in \overline{I}\}$ or its infimum $\inf \{f(x): x \in \overline{I}\}$

(iii) a bounded continuous function $f: \mathbb{R} \to \mathbb{R}$ that does not achieve its supremum or infimum on $\mathbb{R}$

Elements of Differential and Integral Calculus

1.21 Derivatives and Integrals of Functions of One Variable

For completeness, we now give a brief glimpse of the concepts of differentiation and Riemann integration of real-valued functions of one real variable. A more elaborate account of these ideas would carry us too far afield; and some of the more basic theorems are dealt with as exercises.

Derivative of a Function at a Point. Let a be an accumulation point of a set $A \subset \mathbb{R}$ and let f be a function defined from A into $\mathbb{R}$. A real number K is said to be the *derivative* of f at a if, for every $\epsilon > 0$, there is a number $\delta(\epsilon) > 0$ such that if $x \in A$ and $0 < |x - a| < \delta$, then

$$\left| \frac{f(x) - f(a)}{x - a} - K \right| < \epsilon$$

When a number such as K exists, we write $K = f'(a)$.

Alternatively, $f'(a)$ is defined as the limit

$$\lim_{x \to a} \frac{f(x) - f(a)}{x - a} = f'(a)$$

If we use the classical notations

$$\Delta f(a) = f(a + \Delta x) - f(a)$$

then also

$$f'(a) = \lim_{\Delta x \to 0} \frac{\Delta f(a)}{\Delta x}$$

which is the basis for the classical Leibnitz notation

$$f'(a) = \frac{df}{dx}(a)$$

If $f'(a)$ exists, we say that the function f is *differentiable* at a. If f is differentiable at every point $x \in A$, we say f is *differentiable on* A.

PROPOSITION 1.21.1

If a function f is differentiable at a point $a \in A \subset \mathbb{R}$ then f is continuous at a.

PROOF To show f is continuous at a, we must show that $\lim_{x\to a} f(x)$ exists and equals $f(a)$.

Pick $\epsilon = 1$, and select $\delta = \delta(1)$ such that

$$\left| \frac{f(x) - f(a)}{x - a} - f'(a) \right| < 1$$

$\forall\, x$ satisfying $0 < |x - a| < \delta(1)$. Using the triangle inequality, we have

$$\begin{aligned} |f(x) - f(a)| &= |f(x) - f(a) + (x - a)f'(a) - (x - a)f'(a)| \\ &\leq |f(x) - f(a) - (x - a)f'(a)| + |x - a||f'(a)| \\ &\leq |x - a| + |x - a||f'(a)| \end{aligned}$$

Clearly, $|f(x) - f(a)|$ can be made less than ϵ if we pick $|x - a| < \min\{\delta, \epsilon/(1 + |f'(a)|)\}$. Hence $\lim_{x\to a} f(x) = f(a)$, which was to be proved. ∎

Example 1.21.1

The converse of Proposition 1.21.1 is not true. In fact, the function

$$f(x) = \begin{cases} 1 + x, & x \leq 0 \\ 1 - 2x, & x > 0 \end{cases}$$

is continuous at $x = 0$, but not differentiable there. Indeed, $\lim_{x\to 0^+} [(f(x) - f(0))/(x - 0)] = 1$, whereas $\lim_{x\to 0^-} [(f(x) - f(0))/(x - 0)] = -2$.

If $f: \mathbb{R} \to \mathbb{R}$ is differentiable at every $a \in A$, a function prescribing for every $a \in A$ the derivative of f at a, denoted f', is called the *derivative function of f* or simply the *derivative of f*.

Relative (Local) Extrema of a Function. A function f on a set $A \subset \mathbb{R}$ is said to have a *relative* or *local maximum (minimum)* at a point c in A if there exists a neighborhood N of c such that $f(x) \leq f(c) (f(x) \geq f(c))$, for every $x \in N \cap A$.

The following theorem brings the fundamental characterization of local extrema for differentiable functions. □

THEOREM 1.21.1

Let $f: \mathbb{R} \supset A \to \mathbb{R}$ be differentiable at a point c in the interior of the set A. Suppose f has a relative maximum at c.

Then $f'(c) = 0$.

PROOF Suppose the conditions of the theorem hold, but $f'(c) > 0$. Since f is differentiable at c, it is continuous at c by virtue of Proposition 1.21.1. Thus, pick $\epsilon > 0$ and let $\delta(\epsilon)$ be such that $0 < |x - c| < \delta$ implies that $|(f(x) - f(c))/(x - c) - f'(c)| < \epsilon$. Then, if $c < x < c + \delta$, $-\epsilon(x - c) < f(x) - f(c) - f'(c)(x - c)$, we may choose ϵ so that

$$0 < (f'(c) - \epsilon)(x - c) < f(x) - f(c)$$

That is, $f(x) > f(c)$. This contradicts the hypothesis that f has a relative maximum at c. Hence $f'(c)$ cannot be > 0. A similar argument leads to the conclusion that $f'(c)$ also cannot be < 0. Hence $f'(c) = 0$. ∎

An analogous result can be proved for the local minimum.

THEOREM 1.21.2
(Rolle's Theorem)

Let f be a function continuous on the closed interval $[a, b] \subset \mathbb{R}$ and let f be differentiable on the open interval (a, b). Suppose $f(a) = f(b) = 0$. Then there exists a point $c \in (a, b)$ such that $f'(c) = 0$.

PROOF If $f = 0 \ \forall \ x \in (a, b)$, we can take as point c any point in (a, b). Thus suppose f does not vanish identically on (a, b). Then f (or $-f$) assumes some positive values on (a, b). We recall from the Weierstrass theorem (Theorem 1.20.1) that a continuous function defined on a compact set attains its supremum at some point c on this set, and, since $f(a) = f(b) = 0 (\neq f(c))$, c must satisfy $a < c < b$. Now, $f'(c)$ exists, by hypothesis, so $f'(c)$ must be zero by Theorem 1.21.1. ∎

This brings us to one of the most fundamental theorems of calculus.

THEOREM 1.21.3
(The Mean-Value Theorem)

Let f be continuous on $[a,b] \subset \mathbb{R}$ and let f have a derivative everywhere in (a,b). Then there exists a point $c \in (a,b)$ such that

$$f(b) - f(a) = (b-a)f'(c)$$

PROOF Let $g(x)$ be defined by

$$g(x) = f(x) - f(a) - \frac{f(b)-f(a)}{b-a}\,(x-a)$$

Clearly, $g(x)$ is continuous on $[a,b]$ and has a derivative on (a,b) and $g(a) = g(b) = 0$. From Rolle's theorem (Theorem 1.21.2), there is a point c in (a,b) such that $g'(c) = 0$. The assertion of the theorem follows immediately from this fact. ∎

We now pass on to a review of the concept of integration.

Partitions. A *partition* P of an interval $I = [a,b]$ is a finite collection of nonoverlapping intervals whose union is I, generally described by specifying a finite set of numbers; e.g.,

$$a = x_0 \le x_1 \le x_2 \le \cdots \le x_n = b$$

For example, if

$$I_k = [x_{k-1}, x_k], \quad 1 \le k \le n$$

then P is given by

$$I = \bigcup_{k=1}^{n} I_k$$

The quantity

$$\rho(P) = \max_k |x_k - x_{k-1}|$$

is known as the *radius of partition* P.

Riemann Sums and Integrals. Let P be a partition of an interval $I = [a,b] \subset \mathbb{R}$ and let f be a function defined on I. The real number

$$R(P,f) = \sum_{k=1}^{n} f(\xi_k)(x_k - x_{k-1})$$

where $x_{k-1} \le \xi_k \le x_k, 1 \le k \le n$, is called the *Riemann sum* of f corresponding to the partition $P = (x_0, x_1, \ldots, x_n)$ and the choice of intermediate points ξ_k. The function f is said to be *Riemann integrable on I* if for every sequence of partitions P_n converging to zero in the sense that $\rho(P_n) \to 0$, with an arbitrary choice of the intermediate points ξ_k, the corresponding sequence of Riemann sums converges to a common value J.

The number J is called the *Riemann integral* of f over $[a, b]$ and is denoted

$$J = \int_a^b f(x)dx = \int_a^b f dx$$

The function f is called the *integrand* of J.

Example 1.21.2

Let $f(x) = 1$ if x is rational and let $f(x) = 0$ if f is irrational. It is easily verified that the limit of Riemann sum depends upon the choices of ξ_k. The function f is, therefore, *not* Riemann integrable.

Necessary and sufficient conditions for a function f to be Riemann integrable will be studied in Chapter 3. It will turn out that, in particular, every function continuous everywhere except for a finite number of points is integrable in the Riemann sense. Obviously, the function just considered does not satisfy this assumption. □

THEOREM 1.21.4
(The Mean-Value Theorem of Integral Calculus)

Let f be a continuous function on $[a, b]$. Then there exists a point $c \in [a, b]$ such that

$$\int_a^b f(x)dx = f(c)(b-a)$$

PROOF If f is constant then the result is trivial. Suppose that f is not constant. By the Weierstrass theorem f attains both minimum and maximum in $[a, b]$, say at points c_1 and c_2, respectively.

It follows that the function $g(x)$ defined by

$$g(x) = f(x)(b-a) - \int_a^b f(s)ds$$

takes on a negative value at c_1 and a positive value at c_2 (why?). By the Weierstrass intermediate value theorem (Exercise 1.20.4) g must assume every intermediate value between c_1 and c_2, in particular, the zero value, which proves the theorem. ∎

THEOREM 1.21.5
(The First Fundamental Theorem of Integral Calculus)

Let f be continuous on $[a, b]$. Then the function $F(x)$ defined by

$$F(x) = \int_a^x f(s)ds$$

is differentiable in $[a, b]$ and $F'(x) = f(x)$.

PROOF Let $x \in [a, b]$ and Δx be such that $x + \Delta x \in [a, b]$, too. By virtue of the mean-value theorem of integral calculus there exists a point c between x and $x + \Delta x$ such that

$$F(x + \Delta x) - F(x) = \int_a^{x+\Delta x} f(s)ds - \int_a^x f(s)ds$$

$$= \int_x^{x+\Delta x} f(s)ds = f(c)[(x + \Delta x) - x]$$

$$= f(c)\Delta x$$

Since f is continuous and $c \to x$ when $\Delta x \to 0$, we have

$$\lim_{\Delta x \to 0} \frac{F(x + \Delta x) - F(x)}{\Delta x} = \lim_{\Delta x \to 0} f(c) = f(x)$$

which ends the proof. ∎

A function $F(x)$ such that $F'(x) = f(x)$ is called a *primitive function of* f. It follows immediately that a primitive function can be determined only up to an additive constant.

As an immediate corollary of Theorem 1.21.5 we get

THEOREM 1.21.6
(The Second Fundamental Theorem of Integral Calculus)

Let f be continuous on $[a, b]$ and F denote a primitive function of f. Then

$$\int_a^b f(x)dx = F(b) - F(a)$$

PROOF Let $\widehat{F}$ be the primitive function of the function defined in Theorem 1.21.5. Then the equality follows by the definition of $\widehat{F}$. If F is an arbitrary primitive function of f, then F differs from $\widehat{F}$ by a constant, say c.

$$F(x) = \widehat{F}(x) + c$$

This implies that

$$F(b) - F(a) = \widehat{F}(b) - \widehat{F}(a)$$

which finishes the proof. ∎

Exercises

1.21.1. Prove Theorem 1.21.1 for the case in which f assumes a relative minimum on (a, b) (i.e., let there exist a point $c \in (a, b)$ and a $\delta > 0$ such that $f(x) \geq f(c)$ whenever $|x - c| < \delta$ and show that $f'(c) = 0$).

1.21.2. The derivative of the derivative of a function f is, of course, the *second derivative* of f and is denoted f''. Similarly, $(f'')' = f'''$, etc. Let n be a positive integer and suppose f and its derivatives $f', f'', \ldots, f^{(n)}$ are defined and are continuous on $[a, b]$ and that $f^{(n+1)}$ exists in (a, b). Let c and x belong to $[a, b]$. Prove *Taylor's formula* for f; i.e., show that there exists a number ξ satisfying $c < \xi < x$ such that

$$f(x) = f(c) + \frac{1}{1!}f'(c)(x - c) + \frac{1}{2!}f''(c)(x - c)^2$$

$$+ \cdots + \frac{1}{n!}f^{(n)}(c)(x - c)^n + \frac{1}{(n + 1)!}f^{(n+1)}(\xi)(x - c)^{n+1}$$

1.21.3. Let f be continuous on $[a, b]$ and differentiable on (a, b). Prove the following:

(i) If $f'(x) = 0$ on (a, b), then $f(x) =$ constant on (a, b).

(ii) If $f'(x) = g'(x)$ on (a, b), then $f(x) - g(x) =$ constant.

(iii) If $f'(x) < 0 \ \forall \ x \in (a, b)$ and if $x_1 < x_2 \in (a, b)$, then $f(x_1) > f(x_2)$.

(iv) If $|f'(x)| < M < \infty$ on (a, b), then

$$|f(x_1) - f(x_2)| \leq M|x_1 - x_2|, \ \forall \ x_1, x_2 \in (a, b)$$

1.21.4. Let f and g be continuous on $[a, b]$ and differentiable on (a, b). Prove that there exists a point $c \in (a, b)$ such that $f'(c)(g(b) - g(a)) = g'(c)(f(b) - f(a))$. This result is sometimes called the *Cauchy mean-value theorem.* *Hint:* Consider the function $h(x) = (g(b) - g(a))(f(x) - f(a)) - (g(x) - g(a))(f(b) - f(a))$.

1.21.5. Prove L'Hôspital's rule: If $f(x)$ and $g(x)$ are differentiable on (a, b), with $g'(x) \neq 0$ $\forall \boldsymbol{x} \in (a, b)$ and if $f(c) = g(c) = 0$ and the limit $K = \lim_{x \to c} f'(x)/g'(x)$ exists, then $\lim_{x \to c} f(x)/g(x) = K$. *Hint:* Use the results of Exercise 1.21.4.

1.21.6. Let f_1 and f_2 be Riemann integrable on $I = [a, b]$. Show that for any real numbers α and β, $\alpha f_1 + \beta f_2$ is integrable, and

$$\int_a^b (\alpha f_1 + \beta f_2) dx = \alpha \int_a^b f_1 dx + \beta \int_a^b f_2 dx$$

1.21.7. Let f and g be Riemann integrable on $[a, b]$ and suppose that F and G are functions with the property that $F'(x) = f(x)$ and $G'(x) = g(x)$ $\forall$ $x \in [a, b]$. Prove the *integration by parts formula:*

$$\int_a^b F(x) g(x) dx = F(b)G(b) - F(a)G(a) - \int_a^b f(x) G(x) dx$$

1.21.8. Prove that if f is Riemann integrable on $[a, c]$ and $[c, b]$, then it is also Riemann integrable on $[a, b]$ and

$$\int_a^b f dx = \int_a^c f dx + \int_c^b f dx, \quad a < c < b$$

1.21.9. Let f be a Riemann integrable function defined on $[a, b]$ and let $x(u)$ denote a C^1 bijective map from an interval $[c, d]$ onto $[a, b]$. Use the definition of the Riemann integral to show that

$$\int_a^b f(x) dx = \int_c^d f(x(u)) |\frac{dx}{du}| du$$

1.22 Multidimensional Calculus

Directional and Partial Derivatives of a Function. Let $\boldsymbol{f}: \mathbb{R}^n \to \mathbb{R}^m$ be a function defined on a set $A \subset \mathbb{R}^n$. Equivalently, $\boldsymbol{f}$ can be identified as a composite function $\boldsymbol{f} = (f_1, \ldots, f_m)$ where each of the component functions f_i is a real-valued function defined on A. Let $\boldsymbol{x}$ be an accumulation point of set A and $\boldsymbol{u}$ denote a *unit vector* in $\mathbb{R}^n$, i.e., a point $\boldsymbol{u} = (u_1, \ldots u_n) \in \mathbb{R}^n$, such that

$$u_1^2 + u_2^2 + \ldots + u_n^2 = 1$$

The limit

$$\lim_{\varepsilon \to 0, \varepsilon > 0} \frac{f_j(\boldsymbol{x} + \varepsilon \boldsymbol{u}) - f_j(\boldsymbol{x})}{\varepsilon}$$

if it exists, is called the *directional derivative of the j-th component function f_j at $\boldsymbol{x}$ in the direction $\boldsymbol{u}$* and denoted by

$$D^{\boldsymbol{u}} f^j(\boldsymbol{x})$$

The *directional derivative of* $\boldsymbol{f}$ at $\boldsymbol{x}$ *in the direction* $\boldsymbol{u}$ is defined as

$$\boldsymbol{D}^{\boldsymbol{u}} \boldsymbol{f}(\boldsymbol{x}) = (D^{\boldsymbol{u}} f_1(\boldsymbol{x}), \ldots, D^{\boldsymbol{u}} f_m(\boldsymbol{x}))$$

The derivative of the function of a single variable t,

$$t \to f_j(x_1, \ldots, t^{(i)}, \ldots, x_n)$$

if it exists, is called the *i-th partial derivative of the j-th component function* f_i at $\boldsymbol{x}$ and denoted by

$$\frac{\partial f_j}{\partial x_i}(\boldsymbol{x})$$

The composite function

$$\frac{\partial \boldsymbol{f}}{\partial x_i} = \left(\frac{\partial f_1}{\partial x_i}, \ldots, \frac{\partial f_m}{\partial x_i} \right)$$

is identified as the partial derivative of $\boldsymbol{f}$ with respect to the i-th coordinate.

It follows from the definitions that partial derivatives, if they exist, coincide with the directional derivatives in the direction of the respective "axis of coordinates," i.e.,

$$\frac{\partial \boldsymbol{f}}{\partial x_i} = \boldsymbol{D}^{\boldsymbol{e}_i} \boldsymbol{f}$$

where $\boldsymbol{e}_i = (0, \ldots, 1 \text{ on the } i\text{-th place}, \ \ldots 0)$.

As in the case of functions of one variable, *functions* prescribing at every point $\boldsymbol{x}$ a particular partial or directional derivatives at this point are called the *partial* or *directional derivatives (functions) of* $\boldsymbol{f}$. This allows us to introduce higher-order derivatives understood as derivatives of derivatives.

In this book we shall adopt the multiindex notation for the partial derivatives of higher order of the following form:

$$\boldsymbol{D}^{\boldsymbol{\alpha}} \boldsymbol{f} = \frac{\partial^{|\boldsymbol{\alpha}|} \boldsymbol{f}}{\partial \boldsymbol{x}^{\boldsymbol{\alpha}}}$$

where $\boldsymbol{\alpha} = (\alpha_1, \ldots, \alpha_n)$ is the multiindex and where we accept the following symbols

$$\begin{aligned} |\boldsymbol{\alpha}| \quad &= \alpha_1 + \alpha_2 + \ldots + \alpha_n \\ \partial \boldsymbol{x}^{\boldsymbol{\alpha}} &= \underbrace{\partial x_1 \ldots \partial x_1}_{\alpha_1} \underbrace{\partial x_2, \ldots \partial x_2}_{\alpha_2} \ldots \underbrace{\partial x_n \ldots \partial x_n}_{\alpha_n} \end{aligned}$$

The number $|\alpha|$ is called the *order* of the derivative.

Example 1.22.1

Note the fact that the existence of partial derivatives at a point *does not* imply the continuity at the point. Take, for instance, the function $f: \mathbb{R}^2 \to \mathbb{R}$ defined by

$$f(x_1, x_2) = \begin{cases} 1 & \text{if } x_1, x_2 = 0 \\ 0 & \text{otherwise} \end{cases}$$

Function f has both partial derivatives at (0,0) (equal 0), but it is certainly discontinuous at $\mathbf{0} = (0, 0)$. □

Later on in this book we return to the notion of *differentiability of functions* of many variables. If a function f is differentiable at a point then it is automatically continuous at the point and, in particular, possesses all partial and directional derivatives at it. For functions of one variable the notion of differentiability reduces to the existence of the derivative.

Class of C^k Functions. Let $\boldsymbol{f}: \mathbb{R}^n \to \mathbb{R}^m$ be a function defined on an *open set* $\Omega \subset \mathbb{R}^n$. We say that f *is of class* $C^k(\Omega)$ if all partial derivatives of order less than or equal to k exist and are continuous on Ω. The symbols $C^0(\Omega)$ or $C(\Omega)$ is reserved for functions which are just continuous on Ω.

Riemann Integrals in $\mathbb{R}^n$. The notion of the Riemann integral can be generalized to scalar-valued functions in $\mathbb{R}^n$. If (a_i, b_i) $i = 1, \ldots, n$ denotes an open interval in $\mathbb{R}$, the Cartesian product

$$\sigma = (a_1, b_1) \times \ldots \times (a_n, b_n) \subset \mathbb{R}^n$$

is called (*open*) *cube* in $\mathbb{R}^n$.

Assume for simplicity that we are given a function $f: \mathbb{R}^n \to \mathbb{R}$ defined on a cube $E \subset \mathbb{R}^n$. By a *partition* P of E we understand a finite family of pairwise disjoint cubes $\sigma \subset E$ such that

$$E \subset \cup \overline{\sigma}, \ \sigma \in P$$

where $\overline{\sigma}$ denotes the closure of σ,

$$\overline{\sigma} = [a_1 b_1] \times \ldots \times [a_n, b_n]$$

If a single *cube radius* is defined as

$$r(\sigma) = \left(\sum_i^n (b_i - a_i)^2 \right)^{\frac{1}{2}}$$

the *radius of the partition* is defined as

$$r(P) = \max_{\sigma \in P} r(\sigma)$$

Choosing an arbitrary (intermediate) point ξ_σ from every cube $\sigma \in P$ we define the *Riemann sum* as

$$R = R(P, \boldsymbol{\xi}) = \sum_{\sigma \in P} f(\boldsymbol{\xi}_\sigma) m(\sigma)$$

where $m(\sigma)$ denotes the measure (area, volume) of the cube σ defined

$$m(\sigma) = (b_1 - a_1) \ldots (b_n - a_n)$$

The function f is said to be *Riemann integrable* on E iff for *every sequence* P_k of partitions such that

$$r(P_k) \to 0$$

and an *arbitrary* choice of the intermediate point $\boldsymbol{\xi}_\sigma$ the corresponding sequence of Riemann sums converges to a common value J. The number J is called again the *Riemann integral of f over E* and is denoted

$$J = \int_E f dE = \int_E f(\boldsymbol{x}) d\boldsymbol{x} = \int_E f(x_1, \ldots, x_n) dx_1 \ldots dx_n$$

It is possible to extend the notion for more general sets E including all open sets in $\mathbb{R}^n$.

We shall return to this and related subjects in Chapter 3.

Exercises

1.22.1. Let $\boldsymbol{f}: \mathbb{R}^n \to \mathbb{R}^m$ be a function defined on a set $E \subset \mathbb{R}^n$ and $\boldsymbol{x}$ be an internal point of E. Suppose that $\boldsymbol{f}$ has all its partial derivatives at all $\boldsymbol{x} \in E$ and that a directional derivative $\boldsymbol{D}^{\boldsymbol{u}}\boldsymbol{f}(\boldsymbol{x})$ exists, where $\boldsymbol{u} = (u_1, \dots, u_n)$ is a unit vector in $\mathbb{R}^n$. Show that

$$\boldsymbol{D}^{\boldsymbol{u}}\boldsymbol{f}(\boldsymbol{x}) = \sum_{i=1}^{n} \frac{\partial \boldsymbol{f}}{\partial x_i}(\boldsymbol{x}) u_i$$

1.22.2. Let $\boldsymbol{z} = \boldsymbol{z}(t)$ be a one-to-one function from $[a, b]$ into $\mathbb{R}^2$. The image c of $\boldsymbol{z}$ in $\mathbb{R}^2$ is identified as a *curve* in $\mathbb{R}^2$ and function $\boldsymbol{z}$ as its parametrization. Assume that $\boldsymbol{z}$ is a class C^1.

Let $f: \mathbb{R}^2 \to \mathbb{R}$ be now a continuous function on $\mathbb{R}^2$. Show that the integral

$$J = \int_a^b f(\boldsymbol{z}(t)) \sqrt{\left(\frac{dz_1}{dt}\right)^2 + \left(\frac{dz_2}{dt}\right)^2}\, dt$$

exists. Use the result from Exercise 1.21.9 to show that J is independent of the parametrization of curve c, i.e., if $\hat{\boldsymbol{z}}(T)$ is another injective function defined on a different interval $[\bar{a}, \bar{b}] \subset \mathbb{R}$, but with the *same* image c in $\mathbb{R}^2$ as $\boldsymbol{z}(t)$, then the corresponding integral $\widehat{J}$ is equal to J.

The number J depends thus on the curve c only and is known as the *line integral* of f along c, denoted by

$$\int_c f \, dc$$

1.22.3. Let Ω be an open set in $\mathbb{R}^2$ with a boundary $\partial\Omega$ which is a (closed) C^1 curve in $\mathbb{R}^2$. Let $f, g: \mathbb{R}^2 \to \mathbb{R}$ be two C^1 functions defined on a set $\Omega_1 \subset \overline{\Omega}$ (functions f and g are in particular continuous along the boundary $\partial\Omega$).

Prove the *elementary Green's formula* (also known as the multidimensional version of the integration by parts formula)

$$\int_{\Omega} f \frac{\partial g}{\partial x_i} d\Omega = - \int_{\Omega} \frac{\partial f}{\partial x_i} g d\Omega + \int_{\partial\Omega} f g n_i d(\partial\Omega), \; i = 1, 2$$

where $\boldsymbol{n} = (n_1, n_2)$ is the outward normal unit vector to the boundary $\partial\Omega$.

Hint: Assume that the integral over Ω can be calculated as the iterated integral (see Fig. 1.17)

$$\int_a^b \left(\int_{c(x_1)}^{d(x_1)} \frac{\partial g}{\partial x_i} dx_2 \right) dx_1$$

and use the integration by parts formula for functions of one variable.

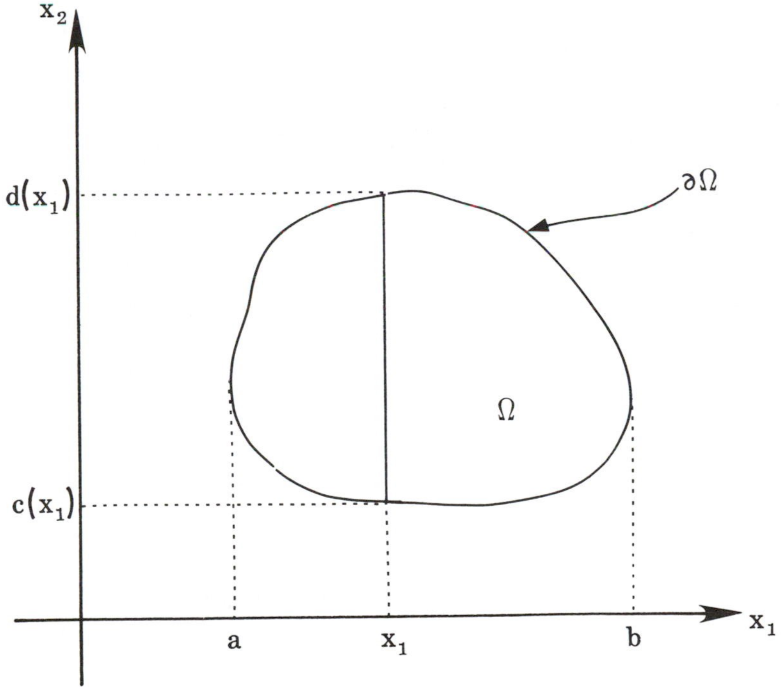

Figure 1.17
Exercise 1.22.3. Notation for the iterated integral.

2

Linear Algebra

Vector Spaces—The Basic Concepts

2.1 Concept of a Vector Space

An important abstract mathematical system that embodies a generalization of the familiar concept of a vector is the vector space. We first cite a formal definition of an abstract vector space and then proceed to identify the two most significant cases: real and complex spaces.

Definition of an (Abstract) Vector Space. An abstract mathematical system $\{X, +, \mathbb{F}, +, \times, *\}$ consisting of sets X, $\mathbb{F}$, and operations $+$, $+$ $\times$, $*$ is an (abstract) *vector space* iff

1. $\{X, +\}$ is an Abelian group with identity $\mathbf{0}$
2. $\{\mathbb{F}, +, \times\}$ is a field, with identities 0 with respect to $+$ and 1 with respect to $\times$
3. $* : \mathbb{F} \times X \to X$ is a binary operation satisfying the conditions

(i) $\alpha * (\boldsymbol{x} + \boldsymbol{y}) = \alpha * \boldsymbol{x} + \alpha * \boldsymbol{y}$

(ii) $(\alpha + \beta) * \boldsymbol{x} = \alpha * \boldsymbol{x} + \beta * \boldsymbol{x}$

(iii) $\alpha * (\beta * \boldsymbol{x}) = (\alpha \times \beta) * \boldsymbol{x}$

(iv) $1 * \boldsymbol{x} = \boldsymbol{x}$

$\forall\, \alpha, \beta \in \mathbb{F}$ and $\boldsymbol{x}, \boldsymbol{y} \in X$.

We refer to such a system as a *vector space* X *over the field* $\mathbb{F}$ (using again X for both the underlying set and the entire abstract system and using $\mathbb{F}$ for both the field and its underlying set).

The elements $\boldsymbol{x}, \boldsymbol{y}, \boldsymbol{z}, \ldots \in X$ are called *vectors* and the operation $+$ on X is *vector addition.*

The elements $\alpha, \beta, \gamma, \ldots \in \mathbb{F}$ are called *scalars* and the operation $*$,

$$\mathbb{F} \times X \ni (\alpha, \boldsymbol{x}) \to \alpha * \boldsymbol{x} \in X$$

is *scalar multiplication* of vectors.

Since no confusion is likely between addition $+$ of vectors and addition $+$ of scalars we shall henceforth use the simpler notation

$$\boldsymbol{x} + \boldsymbol{y} \in X, \qquad \text{i.e., } + \to +$$

Since $\{X, +\}$ is an Abelian group, the operation of vector addition has the following properties:

(i) $\boldsymbol{x} + (\boldsymbol{y} + \boldsymbol{z}) = (\boldsymbol{x} + \boldsymbol{y}) + z$ associative law

(ii) There exists a "zero" element $\mathbf{0} \in \boldsymbol{X}$ such that $\boldsymbol{x} + \mathbf{0} = \mathbf{0} + \boldsymbol{x} = \boldsymbol{x}$

(iii) For every $\boldsymbol{x} \in X$ there exists an inverse element $-\boldsymbol{x} \in X$ such that $\boldsymbol{x} + (-\boldsymbol{x}) = (-\boldsymbol{x}) + \boldsymbol{x} = \mathbf{0}$

(iv) $\boldsymbol{x} + \boldsymbol{y} = \boldsymbol{y} + \boldsymbol{x}$ commutative law

One can easily verify that:

1. $\mathbf{0} = 0 * \cdot \boldsymbol{x}$, i.e., vector $\mathbf{0}$ may be constructed by multiplying the 0–scalar by any vector $\boldsymbol{x}$.

2. $-\boldsymbol{x} = (-1) * \boldsymbol{x}$, i.e., multiplying vector $\boldsymbol{x}$ by a scalar (-1) opposite to identity element 1 (opposite with respect to scalar addition) one obtains the vector $-\boldsymbol{x}$ opposite to $\boldsymbol{x}$ with respect to vector addition. (Beware of the fact that now both vector and scalar additions are denoted by the same symbol $+$, it being left to the context in which they are used as to exactly which of the two operations we have in mind.)

Throughout this text we shall confine our attention to the two most common types of vector spaces: real spaces over the field $\mathbb{R}$ and complex vector spaces over the field $\mathbb{C}$. For the sake of consiseness we drop also the notations for both multiplications, writing compactly $\alpha\beta$ instead of $\alpha \times \beta$ and $\alpha\boldsymbol{x}$ in place of $\alpha * \boldsymbol{x}$. Thus, e.g., axiom *(iii)* can be rewritten in the form

$$\alpha(\beta\boldsymbol{x}) = (\alpha\beta)\boldsymbol{x}$$

Frequently, we also write $\boldsymbol{x} - \boldsymbol{y}$ in place of $\boldsymbol{x} + (-\boldsymbol{y})$ and call it *vector subtraction*. In general we shall use capital Latin letters from the end of the alphabet to denote vector spaces–U, V, W, X, Y, etc.–and lowercase Latin letters a, b, c, x, y, etc., for scalars.

To fix the idea, let us consider a number of examples.

Example 2.1.1

The most well-known example of a real vector space involves the case in which V is the set of all n-tuples of real numbers, n being a fixed positive integer.

$$V = \mathbb{R}^n = \mathbb{R} \times \ldots \times \mathbb{R} \ (n \text{ times })$$

Thus an element $\boldsymbol{a} \in V$ represents a n-tuple $(a_1, \ldots, a_n)$. Given two vectors $\boldsymbol{a}$ and $\boldsymbol{b}$ and a scalar α, we define vector addition and multiplication by a scalar in the following way:

$$\boldsymbol{a} + \boldsymbol{b} = (a_1, \cdots, a_n) + (b_1, \cdots, b_n) \stackrel{\text{def}}{=} (a_1 + b_1, \ldots, a_n + b_n)$$

$$\alpha \boldsymbol{a} \quad = \alpha(a_1, \ldots, a_n) \stackrel{\text{def}}{=} (\alpha a_1, \ldots, \alpha a_n)$$

One can easily verify that all the axioms are satisfied. □

Example 2.1.2

Most commonly, the term "vector" is associated with a pair of points or, equivalently, a directed segment of line in a plane or "space." We define the two operations in the usual way. The vector addition is constructed using the ancient parallelogram law or in the tip-to-tail fashion as it is shown in Figs. 2.1 and 2.2.

The zero vector is a line segment of zero length and the inverse of a segment $\boldsymbol{a}$, denoted by $-\boldsymbol{a}$, is a segment of equal length, but of opposite direction.

Multiplication of a vector $\boldsymbol{a}$ by a scalar α changes its length by a factor $|\alpha|$ (modulus of α) and reverses its direction if α is negative, as indicated in Fig. 2.3. For example, $-2\boldsymbol{a}$ is a line segment twice the length of $\boldsymbol{a}$, but in a direction opposite to $\boldsymbol{a}$.

Again, it is easy to see that all the axioms are satisfied.

REMARK 2.1.1. In mechanics we distinguish between a fixed vector with a specified point of application, a free vector with no point of applica-

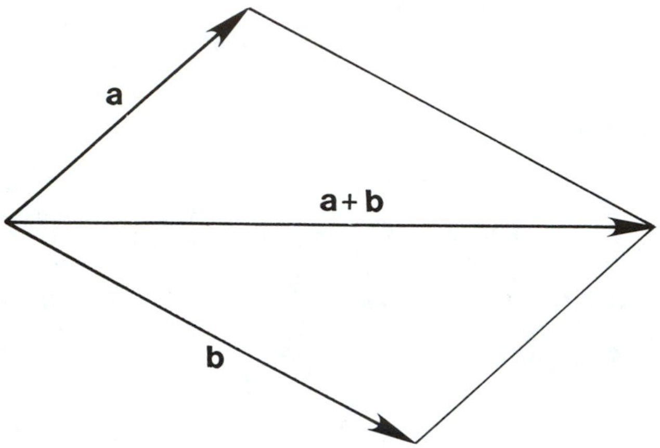

Figure 2.1
Vector addition by the parallelogram law.

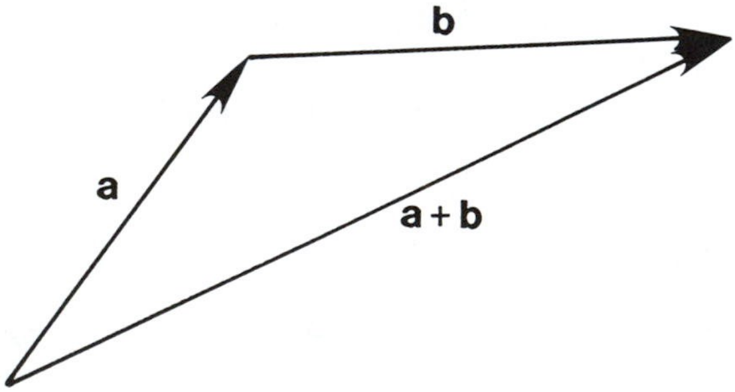

Figure 2.2
Vector addition in the tip-to-tail fashion.

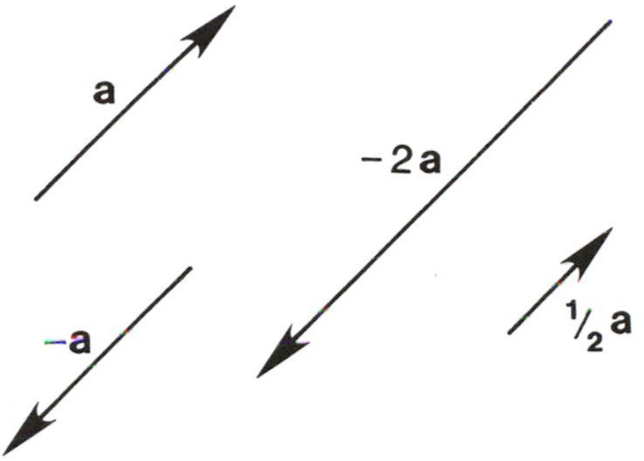

Figure 2.3
Multiplication by a scalar.

tion and even sliding vectors with specified lines of action only. Obviously, in our example we deal with free vectors, which more precisely can be identified with families (classes of equivalence) of fixed vectors possessing the same lines of action, directions and magnitudes. ■

□

Example 2.1.3

It is a common opinion that every object which can be identified with a directed segment of line ("an arrow"), i.e., characterized by a magnitude, line of action and direction, must be necessarily "a vector." The following example of a "vector" of finite rotation contradicts this point of view.

Recall that in three-dimensional kinematics of a rigid body rotating about a point, rotation about an axis passing through the point can be identified with a "vector" directed along the line of rotation with a direction specified by the right-hand rule and magnitude equal to the angle of rotation. According to Euler's theorem on rigid rotations, a composition of two rotations yields a new rotation with a corresponding "vector" which can be considered as a natural candidate for the sum of the two vectors corresponding to the rotations considered. Such "vectors," however, do not obey the commutative law (iv), and hence they cannot be identified as vector quantities.

To show this, consider the two finite rotations $\boldsymbol{\theta}_1 + \boldsymbol{\theta}_2$ applied to the block in Fig. 2.4a. Each rotation has a magnitude of 90° and a direction defined

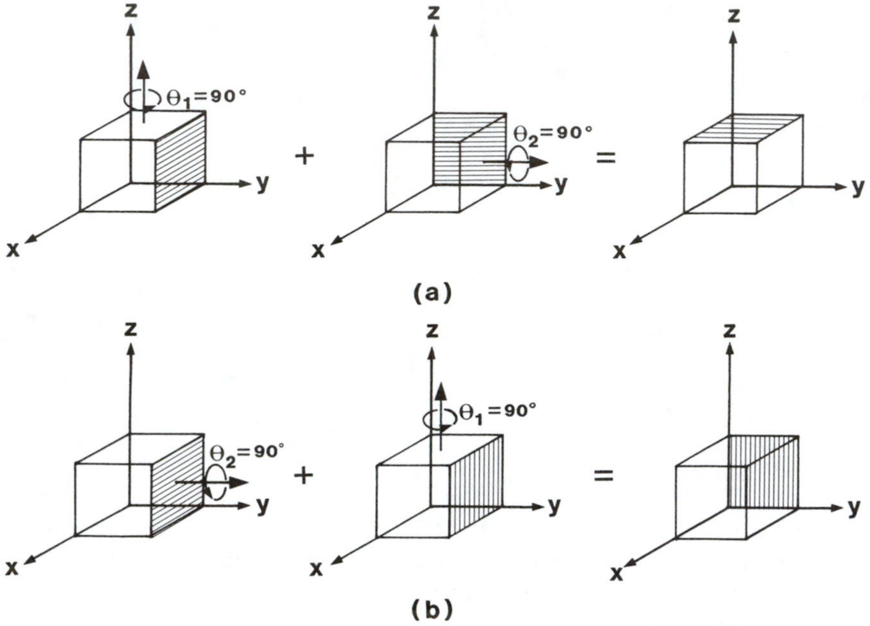

Figure 2.4
Composition of finite rotations.

by the right-hand rule, as indicated by the black arrowhead. The resultant orientation of the block is shown at the right. When these two rotations are applied in the reverse order $\boldsymbol{\theta}_2 + \boldsymbol{\theta}_1$ as shown in Fig. 2.4b, the resultant position of the block is not the same as it is in Fig. 2.4a. Consequently, finite rotations do not form a vector space. □

REMARK 2.1.2. If smaller, yet finite, rotations had been used to illustrate the example, e.g., 5° instead of 90°, the resultant orientation of the block after each combination of rotations would also be different; however, in this case only by a small amount. In the limit, both orientations are the same and for that reason we can speak about infinitesimal rotations, angular velocities or virtual angular displacements as vectors. ■

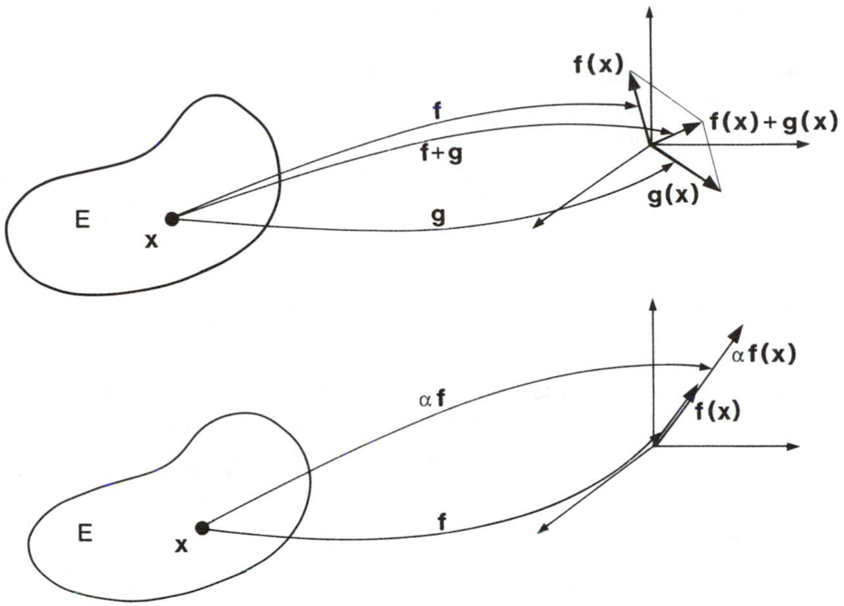

Figure 2.5
Vector addition and multipication by a scalar in function spaces.

Example 2.1.4

Let V be a vector space and E an arbitrary set. Recall that by V^E we denote the set of all functions defined on E with values in V. The family (set) V^E can be embodied with the vector structure provided the two operations are defined as follows:
vector (function) addition

$$(f+g)(x) \stackrel{\text{def}}{=} f(x)+g(x)$$

multiplication of a vector (function) by a scalar

$$(\alpha f)(x) \stackrel{\text{def}}{=} \alpha f(x)$$

As usual, the same symbol "+" is used to indicate the sum of two functions f and g (the left-hand side of definition) and the sum of their values at x on the right-hand side. The concept of the algebraic operations is illustrated in Fig. 2.5.

The reader is encouraged to check the axioms. We emphasize that in this example we have defined operations in V^E using operations on vector space V.

The function vector space V^E is the most general example of function (vector) spaces. Usually we are more specific with assumptions on set E and space V and frequently we incorporate in the definition some extra assumptions on the regularity of functions considered. If V is chosen as the space of real (complex) numbers we speak of real- (complex-) valued function. If $V = \mathbb{R}^n(\mathbb{C}^n)$, we speak about vector- (complex vector-) valued functions. The following is a very preliminary list of function (vector) spaces. In all of the definitions $\Omega \subset \mathbb{R}^n$ denotes a domain (an open, connected set) in $\mathbb{R}^n$.

$C^k(\Omega)$ = space of all real- or complex-valued functions defined on Ω of class k, meaning functions with derivatives of order k which are continuous functions with domain Ω; thus

k = 0 denotes continuous functions;

k = $1, 2, \ldots$ denotes functions differentiable up to k-th

order with k-th order derivatives continuous;

k = ∞ means that all derivatives of any arbitrary order exist;

k = ω shall denote analytic functions.

$\boldsymbol{C}^k(\Omega)$ = denotes space of real (complex) vector-valued functions (usually in $\mathbb{R}^k$, $k = 2, 3$ in applications) with the derivatives of order k continuous on Ω.

REMARK 2.1.3. The fact that both $C^k(\Omega)$ or $\boldsymbol{C}^k(\Omega)$ are vector spaces is not an immediate consequence of Example 2.1.4. Imposing some extra conditions on functions considered, one has to check that the two operations, vector addition and multiplication, by a scalar are closed with respect to these conditions. In other words, one has to verify that the sum of two C^k functions (which *a priori* is only a function on Ω and belongs to the space $\mathbb{R}^\Omega$) is C^k itself, i.e., falls into the category of functions considered. The same concerns multiplication by a scalar: a product of a scalar and a C^k function is a C^k function itself. ∎

REMARK 2.1.4. It makes only a little sense to speak about C^k classes for complex-valued functions of complex variable. It is a well known fact that (complex) differentiability implies analyticity (complex analytic functions are called holomorfic). Thus, for complex functions $C^k, k = 1, 2, \ldots, \infty, \omega$ means the same class of functions. ∎

□

It is often desirable, especially in the study of partial differential equations, to speak of boundary values of functions defined on the set Ω. Since set Ω is open, the boundary points do not belong to Ω and therefore functions defined on Ω are not necessarily specified at these points. An attempt to define functions directly on the closed set $\overline{\Omega}$ in general fails since the notion of differentiability only makes sense for open sets.

To overcome this technical difficulty, we introduce spaces $C^k(\overline{\Omega}), k = 0, 1 \ldots, \infty$. A function f belongs to the space $C^k(\overline{\Omega})$ if there exists an open set Ω_1, (depending on f) and an extension f_1, such that

1. $\overline{\Omega} \subset \Omega_1$
2. $f_1 \in C^k(\Omega_1)$
3. $f_1|_\Omega = f$

According to the definition, a function f from $C^k(\overline{\Omega})$ can be extended to a function f_1 defined on a larger set containing, particularly, boundary $\partial\Omega$, and values of the extension f_1 can be identified as values on f on the boundary $\partial\Omega$. One can easily verify (see Exercises 2.1.7. and 2.1.8) that $C^k(\overline{\Omega})$ is a vector space.

Exercises

2.1.1. Let V be an abstract vector space over a field $\mathbb{F}$. Denote by 0 and 1 inverse elements with respect to addition and multiplication of scalars, resp. Let -1 be an element opposite to 1 (with respect to scalar addition). Prove the identities

$$(i)\ \mathbf{0} = 0 \cdot \boldsymbol{x} \qquad \text{for every } \boldsymbol{x} \in V$$
$$(ii)\ -\boldsymbol{x} = (-1) \cdot \boldsymbol{x} \qquad \text{for every } \boldsymbol{x} \in V$$

where $\mathbf{0}$ is a zero-vector (neutral element with respect to vector addition) and $-\boldsymbol{x}$ denotes the opposite vector to $\boldsymbol{x}$.

2.1.2. Let $\mathbb{C}$ denote the field of complex numbers. Prove that $\mathbb{C}^n$ is a vector space with operations defined in the analogous way as in Example 2.1.1.

2.1.3. Prove Euler's theorem on rigid rotations. Consider a rigid body fixed at a point A in an initial configuration Ω. Suppose the body is carried from the configuration Ω to a new configuration Ω_1, by a rotation about an axis 1_1, and next from Ω_1, to a configuration Ω_2 by a rotation about another axis 1_2. Show that there exists a unique axis 1 and a corresponding rotation carrying the rigid body from the initial configuration Ω to the final one Ω_2 directly. Consult any textbook on rigid body dynamics, if necessary.

2.1.4. Construct an example showing that the sum of two finite rotation "vectors" does not need to lie in a plane generated by those "vectors."

2.1.5. Let $\mathcal{P}^k(\Omega)$ denote a set of all real- or complex-valued polynomials defined on a set $\Omega \subset \mathbb{R}^n(\mathbb{C}^n)$. Show that $\mathcal{P}^k(\Omega)$ with customary defined operations (see Example 2.1.4) is a vector space.

2.1.6. Let $G_k(\Omega)$ denote a set of all polynomials of order greater or equal to k. Is $G_k(\Omega)$ a vector space? Why?

2.1.7. The extension f_1 in the definition of a function f from class $C^k(\overline{\Omega})$ does not have to be unique. The boundary values of f_1 however do not depend on a particular extension. Explain why.

2.1.8. Show that $C^k(\overline{\Omega}), k = 0, 1, \ldots, \infty$, is a vector space.

2.2 Subspaces

In most of our studies of vector spaces, we are not concerned with the entire space only, but also with certain subsystems called subspaces.

Linear Subspace. Let V be a vector space. A nonempty subset W of V, say $W \subset V$, is called a (linear) subspace of V if W with operations restricted from V is a vector space (satisfies axioms of the vector space definition) itself.

PROPOSITION 2.2.1

A nonempty subset $W \subset V$ is a linear subspace of V if and only if it is closed with respect to both operations: vector addition and multiplication by a scalar, i.e.,

$$\boldsymbol{u}, \boldsymbol{v}, \in W \Rightarrow \boldsymbol{u} + \boldsymbol{v} \in W$$

$$\alpha \in \mathbb{R}(\mathbb{C}), \boldsymbol{u} \in W \Rightarrow \alpha \boldsymbol{u} \in W$$

PROOF Denote by $+$ and $\cdot$ the operations in V. If $W = \{W; +; \cdot\}$ is a vector space then it must be closed with respect to both operations from the definition of vector space. Conversely, if W is closed with respect to the operations then it makes sense to speak about sums $\boldsymbol{u}+\boldsymbol{v}$ and products $\alpha\boldsymbol{u}$ as elements of W, and all axioms which are satisfied in V are automatically satisfied in W. ∎

Example 2.2.1

Consider a subset W_c of $\mathbb{R}^n$ of the form

$$W_c = \{\boldsymbol{x} = (x_1, \ldots, x_n) \in \mathbb{R}^n : \sum_{i=1}^{n} \alpha_i x_i = c\}$$

where $\alpha_i \in \mathbb{R}$.

Let $\boldsymbol{x}, \boldsymbol{y} \in W_c$. It follows particularly that

$$\sum \alpha_i (x_i + y_i) = \sum \alpha_i x_i + \sum \alpha_i y_i = 2c$$

Thus W_c is closed with respect to vector addition if and only if $2c = c$, i.e., $c = 0$. The same holds for the multiplication by a scalar. Concluding, the set W is a linear subspace of $\mathbb{R}^n$ if and only if $c = 0$.

Another way to see why c must be equal to zero is to observe that W_c as a vector space must contain zero vector $\mathbf{0} = (0, \ldots, 0)$. Substituting zero coordinates to the definition of W we get immediately that $c = 0$.

REMARK 2.2.1. For $c \neq 0$, W_c can be interpreted as a linear subspace (corresponding to $c = 0$) translated in $\mathbb{R}^n$ by a vector. Such a subset is called a *linear manifold.* ∎

□

Example 2.2.2

Each of the function spaces defined before on the domain $\Omega \subset \mathbb{R}^n$ can be identified as a linear subspace of $\mathbb{R}^\Omega$ (recall Remark 2.1.3). □

Example 2.2.3

One of the fundamental concepts in the variational theory of value problems in mechanics is the notion of the space (set) of all kinematically admissible displacements. Consider, for example, a membrane occupying a domain $\Omega \subset \mathbb{R}^2$ with a boundary $\Gamma = \partial\Omega$ consisting of two disjoint parts Γ_u and Γ_t. Recall the classical formulation of the boundary value problem.

Find $u = u(x, y)$, such that

$$-\Delta u = f \quad \text{in } \Omega$$

$$u = u_0 \quad \text{on } \Gamma_u$$

$$\frac{\partial u}{\partial n} = g \quad \text{on } \Gamma_t$$

In the above, Δ denotes the Laplacian operator ($\Delta = \boldsymbol{\nabla} \cdot \boldsymbol{\nabla} = \partial^2/\partial x^2 + \partial^2/\partial y^2$), $\frac{\partial u}{\partial n}$ the normal derivative of u ($\boldsymbol{n}$ is an outward normal unit to Γ_t), functions f and g specify a given load of the membrane, inside of Ω and on Γ_t, respectively, and u_0 is a given displacement of the membrane along part Γ_u. We call the first boundary condition the *essential* or *kinematic boundary condition* since it is expressed directly in the displacement u, while the second one is called the *natural* or *static boundary condition.* The *set of all kinematically admissible displacements* is defined as

$$W = \{u \in C^k(\overline{\Omega}) : u = u_0 \text{ on } \Gamma_u\}$$

Obviously, W is a subset of the vector space $C^k(\overline{\Omega})$. The *regularity* of the solution u is, in this example, characterized by the order k of the space

$C^k(\overline{\Omega})$ and this order always depends upon the regularity of the domain Ω, of u_0, and of the force data f and g.

In a manner exactly the same as in Example 2.2.1., we prove that W is a linear subspace of $C(\overline{\Omega})$ if and only if function u_0 is identically equal to zero. In such a case we speak of the *space* of all kinematically admissible displacements. $\square$

Given two subspaces of a vector space V, we can define their algebraic sum and intersection.

Algebraic Sum of Subspaces. Let $X, Y \subset V$ denote two subspaces of the vector space V. The set of all vectors of the form

$$\boldsymbol{z} = \boldsymbol{x} + \boldsymbol{y},$$

where $\boldsymbol{x} \in X$ and $\boldsymbol{y} \in Y$ is also a vector subspace of V, denoted $X + Y$, and called the algebraic sum of X and Y.

The algebraic sum should not be confused with the union of subspaces $X, Y (X \cup Y)$. The first one possesses a linear structure while the second one is merely a subset of V.

Intersection of Subspaces. Contrary to the set operation of the union of sets, the usual intersection operation preserves the linear structure and the intersection $X \cap Y$ is a linear subspace of V. Note that $X \cap Y$ is never empty since it must contain at least the zero vector.

Algebraic Sum of a Vector and a Subspace. Let $\boldsymbol{x} \in V$ and Y be a subspace of the linear space V. The set

$$\boldsymbol{x} + Y \stackrel{\text{def}}{=} \{\boldsymbol{x} + \boldsymbol{y} : \boldsymbol{y} \in Y\}$$

is called the algebraic sum of vector $\boldsymbol{x}$ and subspace Y. The concepts of algebraic sums and intersection are illustrated in Fig. 2.6.

Example 2.2.4

Consider again the set $W_c \subset \mathbb{R}^n$ defined in Example 2.2.1. Let $c = 0$ and $\boldsymbol{x}$ denote an arbitrary element of W_c. One can easily prove that

$$W_c = \boldsymbol{x} + W_0$$

(recall Remark 2.2.1). $\square$

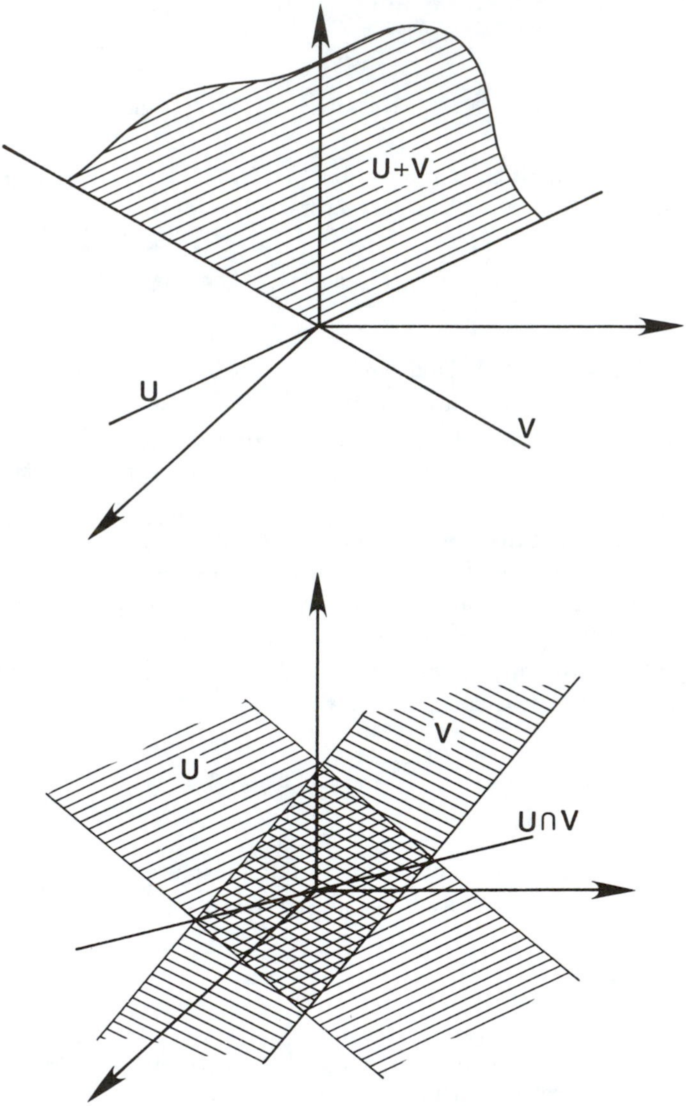

Figure 2.6

Algebraic sum and common part of subspaces in $\mathbb{R}^3$.

Example 2.2.5

Let W be the set of all kinematically admissible displacements from Example 2.2.3 and let W_0 denote its counterpart for $u_0 = 0$. Finally, let us suppose that function u_0 can be extended to a function denoted by the same symbol, but defined on the entire $\overline{\Omega}$. One can see that

$$W = u_0 + W_0$$

□

When the common part of two subspaces X and Y consists of zero vector only, their algebraic sum gets a new name.

Direct Sums. Complements. Let X, Y denote two subspaces of a vector space V such that $X \cap Y = \{\mathbf{0}\}$. In such a case their algebraic sum $X + Y$ is denoted by $X \oplus Y$ and is called the *direct sum* of X and Y. In other words,

$$X + Y = X \oplus Y \text{ if and only if } X \cap Y = \{\mathbf{0}\}$$

If there exist two subspaces X and Y such that the entire space V is the direct sum of X and Y, Y is called a *complement* of X; conversely, X is a complement of Y.

THEOREM 2.2.1

A linear space V is a direct sum of its two subspaces X and Y, $V = X \oplus Y$, if and only if every vector $\boldsymbol{v} \in V$ has a unique representation

$$\boldsymbol{v} = \boldsymbol{x} + \boldsymbol{y}$$

for some $\boldsymbol{x} \in X$ and $\boldsymbol{y} \in Y$.

PROOF If $V = X \oplus Y$, then $V = X + Y$ and every $\boldsymbol{v} \in V$ can be expressed as $\boldsymbol{x} + \boldsymbol{y}$ for appropriate choices of vectors $\boldsymbol{x} \in X, \boldsymbol{y} \in Y$. If, in addition, $\boldsymbol{v} = \hat{\boldsymbol{x}} + \hat{\boldsymbol{y}}$, where $\hat{\boldsymbol{x}} \in X$, $\hat{\boldsymbol{y}} \in Y$, then

$$\boldsymbol{x} + \boldsymbol{y} = \hat{\boldsymbol{x}} + \hat{\boldsymbol{y}}$$

or

$$\boldsymbol{x} - \hat{\boldsymbol{x}} = \hat{\boldsymbol{y}} - \boldsymbol{y}$$

But $\boldsymbol{x} - \hat{\boldsymbol{x}} \in X$ and $\hat{\boldsymbol{y}} - \boldsymbol{y} \in Y$. Thus both $\boldsymbol{x} - \hat{\boldsymbol{x}}$ and $\hat{\boldsymbol{y}} - \boldsymbol{y}$ belong to both X and Y. However, $X \cap Y = \{\mathbf{0}\}$, which implies that both $\boldsymbol{x} - \hat{\boldsymbol{x}} = \mathbf{0}$ and $\hat{\boldsymbol{y}} - \boldsymbol{y} = \mathbf{0}$. Hence $\boldsymbol{x} = \hat{\boldsymbol{x}}$ and $\hat{\boldsymbol{y}} = \boldsymbol{y}$ and $\boldsymbol{v}$ has a unique representation as the sum $\boldsymbol{x} + \boldsymbol{y}$.

Conversely, assume that the representation is unique and take a vector $\boldsymbol{w} \in X \cap Y$. Then

$$\boldsymbol{w} = \boldsymbol{w} + \mathbf{0} = \mathbf{0} + \boldsymbol{w}$$

where in the first sum $\boldsymbol{w} \in X, \mathbf{0} \in \mathbf{Y}$ and in the second sum $\mathbf{0} \in \mathbf{X}, \boldsymbol{w} \in \mathbf{Y}$. Since the representation is unique we must conclude $\boldsymbol{w} = 0$ and we have $X \cap Y = \{\mathbf{0}\}$. ∎

Example 2.2.6

Take $V = C(\overline{\Omega})$ for some bounded domain $\Omega \subset \mathbb{R}^n$ and take

$$X = \{u \in C(\overline{\Omega}) : \int_\Omega u d\Omega = 0\}$$

$$Y = \{u \in C(\overline{\Omega}) : u = \text{const}\}$$

Then $V = X \oplus Y$. Indeed, if $u \in V$, then u can always be represented as

$$u = v + w$$

where w is a constant function, $w = \text{meas}(\Omega)^{-1} \int_\Omega u d\Omega$, and $v = u - w$ belongs to X. Obviously $w \in X \cap Y$ implies $w = 0$. The subspace of constant functions is therefore the complement of the subspace consisting of functions whose mean value on Ω is equal to zero and vice versa. □

2.3 Equivalence Relations and Quotient Spaces

Recall that a relation R in a set V has been called an equivalence relation whenever R satisfies three axioms:

(i)	xRx	(reflexivity)
(ii)	$xRy \Rightarrow yRx$	(symmetricity)
(iii)	$xRy, yRz \Rightarrow xRz$	(transitivity)

Let V be now a vector space. The simplest example of an equivalence relation in V is constructed by taking a subspace $M \subset V$ and defining the relation R_M by

$$\boldsymbol{x} R_M \boldsymbol{y} \overset{\text{def}}{\Leftrightarrow} \boldsymbol{x} - \boldsymbol{y} \in M$$

It is easily verified that the three conditions are satisfied. Consequently, we can use the notion of an equivalence class $[\boldsymbol{x}]$ consisting of all elements equivalent to $\boldsymbol{x}$. In other words,

$$[\boldsymbol{x}] = \{\boldsymbol{y} \in V : \boldsymbol{y} - \boldsymbol{x} \in M\}$$

which is equivalent to

$$[\boldsymbol{x}] = \boldsymbol{x} + M$$

Thus equivalent class $[\boldsymbol{x}]$ can be identified as an affine subspace "parallel" to M, passing through vector $\boldsymbol{x}$. Subspace M, particularly, can be identified as an equivalence class of the zero vector $\mathbf{0}$.

It is easily verified that the quotient set V/R_M is a vector space under the operations of vector addition and multiplication by a scalar as follows:

$$[\boldsymbol{x}] + [\boldsymbol{y}] \overset{def}{=} [\boldsymbol{x} + \boldsymbol{y}]$$
$$\alpha[\boldsymbol{x}] \overset{\text{def}}{=} [\alpha \boldsymbol{x}]$$

The quotient space V/R_M is denoted V/M and referred to as the quotient space of V modulo M. The concept of equivalence class $[\boldsymbol{x}] = \boldsymbol{x} + M$ and quotient space V/M is illustrated in Fig. 3.2.1.

Example 2.3.1

Consider in the space $\mathbb{R}^2$ a subspace

$$M = \{\boldsymbol{x} = (x_1, x_2) \in \mathbb{R}^2 : \alpha_1 x_1 + \alpha_2 x_2 = 0\}$$

(recall Example 2.2.1). M can be identified as a straight line passing through the origin.

Let $\boldsymbol{y} = (y_1, y_2)$ denote an arbitrary vector. The equivalence class

$$\begin{aligned}[\boldsymbol{y}] &= \{\boldsymbol{x} \in \mathbb{R}^2 : \boldsymbol{x} - \boldsymbol{y} \in M\} \\ &= \{\boldsymbol{x} : \alpha_1(x_1 - y_1) + \alpha_2(x_2 - y_2) = 0\} \\ &= \{\boldsymbol{x} : \alpha_1 x_1 + \alpha_2 x_2 = c\}, \text{ where } c = \alpha_1 y_1 + \alpha_2 y_2\end{aligned}$$

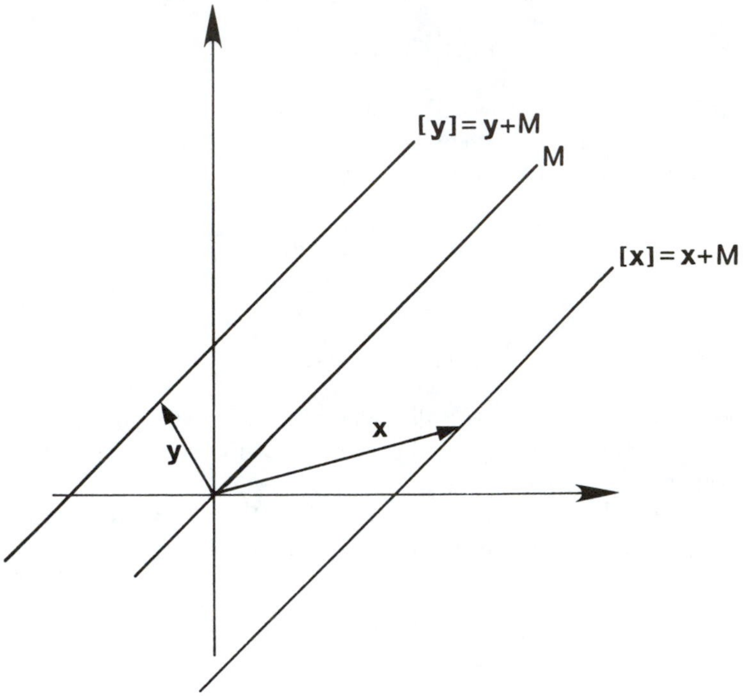

Figure 2.7

Equivalence relation in $\mathbb{R}^2$.

is identified as a straight line parallel to M and passing through point $\boldsymbol{y}$ (see Fig. 2.7).

The quotient space $\mathbb{R}^2/M$ consists of all such lines parallel to M. □

Example 2.3.2

Consider again the membrane problem (recall Example 2.2.3) and suppose that $\Gamma_u = \emptyset$, i.e., no part of the membrane boundary is supported. A solution to the corresponding boundary-value problem, called Neumann problem, if it exists, is certainly not unique. Given a solution $u(x, y)$, we easily see that $u + c$, for any $c \in \mathbb{R}$, must be a solution as well.

Identifying the space W of kinematically admissible displacements with the entire space $C^\infty(\overline{\Omega})$, we introduce the following subspace of (infinitesimal, linearized) rigid body motions:

$$M = \{u(x, y) \, : \, u = \text{ const in } \Omega\}$$

It turns out that the quotient space W/M is a natural candidate space for a solution to the Neumann problem. Two deflections u and v belong to the same class of equivalence if they differ by a constant function. We say that the solution of such a membrane problem is determined up to a (linearized or infinitesimal) rigid body motion. □

Example 2.3.3

In continuum mechanics a deformable body is identified with a set Ω satisfying usually some extra regularity assumptions. The body Ω can occupy different *configurations* in the space $\mathbb{R}^n (n = 2, 3)$ which may be identified as (open) sets Ω_τ in $\mathbb{R}^n$ or, more precisely, by the transformations which map Ω onto Ω_τ:

$$\tau : \Omega \to \Omega_\tau \subset \mathbb{R}^n$$

If two configurations are considered, say Ω_{τ_1}, and Ω_{τ_2} the composition

$$\tau_2 \circ \tau_1^{-1} : \Omega_{\tau_1} \ni \boldsymbol{X} \to \tau_2 \circ \tau_1^{-1}(\boldsymbol{X}) \in \Omega_{\tau_2}$$

is called a (relative) *deformation* of configuration τ_2 with respect to τ_1. One has, of course, to assume that both τ_1 and τ_2 are invertible. Moreover, assuming that the composition $\tau_2 \circ \tau_1^{-1}$ is C^1, we introduce the so-called (relative) *deformation gradient* as

$$\boldsymbol{F} = \boldsymbol{\nabla}\chi(\boldsymbol{X}, t) \quad , F^k_n = x^k_{,i} = \frac{\partial \chi^k}{\partial X^L}$$

where $\boldsymbol{x} = \chi = \tau_2 \circ \tau_1^{-1}$. The composition $\boldsymbol{C} = \boldsymbol{F}^T \circ \boldsymbol{F}$ is called the *right Cauchy-Green tensor.* In the Cartesian system of coordinates in $\mathbb{R}^n$ it takes the form

$$c_{ij} = x^k_{ij} x^k_{ij}$$

provided the standard summation convention is being used.

Sometimes in place of the deformation it is more convenient to consider the displacement vector $\boldsymbol{u}$

$$\boldsymbol{u}(\boldsymbol{X}) = \boldsymbol{x}(\boldsymbol{X}) - \boldsymbol{X}$$

and the relative *Green strain tensor* defined as

$$2\boldsymbol{E} = \boldsymbol{C} - \mathbf{1}, \;\; E_{ij} = C_{ij} - \delta_{ij}$$

with δ_{ij} being the usual Kronecker symbol. From the definition of $\boldsymbol{E}$ and $\boldsymbol{C}$ the simple formula follows:

$$E_{ij} = \frac{1}{2}(u_{i,j} + u_{j,i} + u_{k,i}u_{k,j})$$

We shall define the following relation R in the set S of all configurations τ.

$$\tau_1 R \tau_2 \overset{\text{def}}{\Leftrightarrow} \boldsymbol{E} = \mathbf{0}$$

Physically, this means that the body is carried from the configuration τ_1 to τ_2 by a rigid body motion. We shall prove that R is an equivalence relation.

Toward this goal let us make a few observations first. First of all, $\boldsymbol{E} = \mathbf{0}$ if and only if $\boldsymbol{C} = \mathbf{1}$. Secondly, if τ_1, τ_2, τ_3 denote three configurations and ${}^{\alpha}_{\beta}\boldsymbol{F}$ denotes the relative deformation gradient of configuration τ_α with respect to configuration τ_β, then from the chain rule of differentiation it follows that

$${}^{3}_{1}\boldsymbol{F} = {}^{3}_{2}\boldsymbol{F}\,{}^{2}_{1}\boldsymbol{F}$$

Recall now that an equivalence relation must be reflexive, symmetric and transitive. Relation R is obviously reflexive since the relative deformation gradient of a configuration with respect to itself equals $\mathbf{1}$. Next, from the identity

$${}^{3}_{1}\boldsymbol{C} = \left({}^{3}_{1}\boldsymbol{F}\right)^T {}^{3}_{1}\boldsymbol{F} = \left({}^{3}_{2}\boldsymbol{F}\,{}^{2}_{1}\boldsymbol{F}\right)^T {}^{3}_{2}\boldsymbol{F}\,{}^{2}_{1}\boldsymbol{F} = \left({}^{2}_{1}\boldsymbol{F}\right)^T {}^{3}_{2}\boldsymbol{C}\,{}^{2}_{1}\boldsymbol{F}$$

it follows that relation R is transitive. Indeed, if $\tau_2 R \tau_3$ then ${}^{3}_{2}\boldsymbol{C} = \mathbf{1}$ and ${}^{3}_{1}\boldsymbol{C} = \left({}^{2}_{1}\boldsymbol{F}\right)^T {}^{2}_{1}\boldsymbol{F} = {}^{2}_{1}\boldsymbol{C}$. Consequently, if also $\tau_1 R \tau_2$ then ${}^{2}_{1}\boldsymbol{C} = \mathbf{1}$ and, finally, ${}^{3}_{1}\boldsymbol{C} = \mathbf{1}$, which means that $\tau_1 R \tau_3$.

Finally, let $\tau_1 R \tau_2$. Let ${}^{2}_{1}\boldsymbol{F} = \boldsymbol{F}$. Obviously, ${}^{1}_{2}\boldsymbol{F} = \boldsymbol{F}^{-1}$ and we have $\boldsymbol{F}\boldsymbol{F}^{-1} = \mathbf{1}$, which implies that

$$\left(\boldsymbol{F}\boldsymbol{F}^{-1}\right)^T \left(\boldsymbol{F}\boldsymbol{F}^{-1}\right) = \left(\boldsymbol{F}^{-1}\right)^T \boldsymbol{F}^T\boldsymbol{F}\boldsymbol{F}^{-1} = \left(\boldsymbol{F}^{-1}\right)^T {}^{2}_{1}\boldsymbol{C}\boldsymbol{F}^{-1} = \mathbf{1}$$

Since $\tau_1 R \tau_2$ then ${}^{2}_{1}\boldsymbol{C} = \mathbf{1}$ and $\left(\boldsymbol{F}^{-1}\right)^T \boldsymbol{F}^{-1} = \mathbf{1}$, which proves that $\tau_2 R \tau_1$. Thus R is reflexive and therefore R is an equivalence relation.

An equivalence class in this relation can be interpreted as a set of all configurations which are "connected" by rigid body motions. Thus the quotient set S/R consists of all configurations "up to a rigid body motion." □

Example 2.3.4

To understand better the notion of the linearized rigid motion let us return now to Example 2.3.3 of the equivalence relation R in the class of configurations τ. Although the configurations are vector-valued functions defined on Ω, they do not form a vector space, the reason being, for instance, that the only candidate for zero-vector, the zero-function, cannot be identified as a configuration (is not invertible!).

To formulate the problem in terms of vector spaces we shall introduce a reference configuration $\tau_R: \Omega \to \Omega_R$ and consider displacements from that configuration to a given one instead of configurations themselves. If $\boldsymbol{u}$ and $\boldsymbol{v}$ are two displacements from Ω_R to Ω_1 and Ω_2, respectively, then $\boldsymbol{v} - \boldsymbol{u}$ prescribes the displacement vector from Ω_1 to Ω_2 (cf. Fig. 2.8). In a manner identical to the one in Example 2.3.3, we introduce in the space of displacements defined on Ω_R the equivalence relation: We say that displacement $\boldsymbol{u}$ is related to displacement $\boldsymbol{v}$, $\boldsymbol{u}R\boldsymbol{v}$, if the Green strain tensor corresponding to the displacement $\boldsymbol{v} - \boldsymbol{u}$ vanishes. For the same reasons as before, the relation satisfies the three axioms of equivalence relations.

A natural question arises: can the introduced relation be induced by a subspace M? The answer is no and there are many ways to verify this. One of them is to notice that if R had been introduced by a subspace M then the equivalence class of zero displacement would have to coincide with M and particularly would have to possess the structure of a vector space. This is, however, not true. To see this, take two displacements $\boldsymbol{u}$ and $\boldsymbol{v}$ describing rigid body motions, i.e., such that $\boldsymbol{E}(\boldsymbol{u}) = \boldsymbol{0}$ and $\boldsymbol{E}(\boldsymbol{v}) = \boldsymbol{0}$ and check whether $\boldsymbol{E}(\boldsymbol{u} + \boldsymbol{v}) = \boldsymbol{0}$. Due to nonlinearity of $\boldsymbol{E}$ with respect to $\boldsymbol{u}$ the answer is negative.

The situation changes if we use the linearized geometrical relations, i.e., we replace the Green strain tensor $\boldsymbol{E}$ with the infinitesimal strain tensor

$$\varepsilon_{ij} = \frac{1}{2}\left(u_{i,j} + u_{j,i}\right)$$

We leave the reader to check that the equivalence relation generated by the infinitesimal strain tensor is induced by the subspace of linearized (infinitestimal) rigid body motions (review also Exercises 2.3.2, 2.3.3). The concept of two equivalence relations is illustrated in Fig. 2.9. □

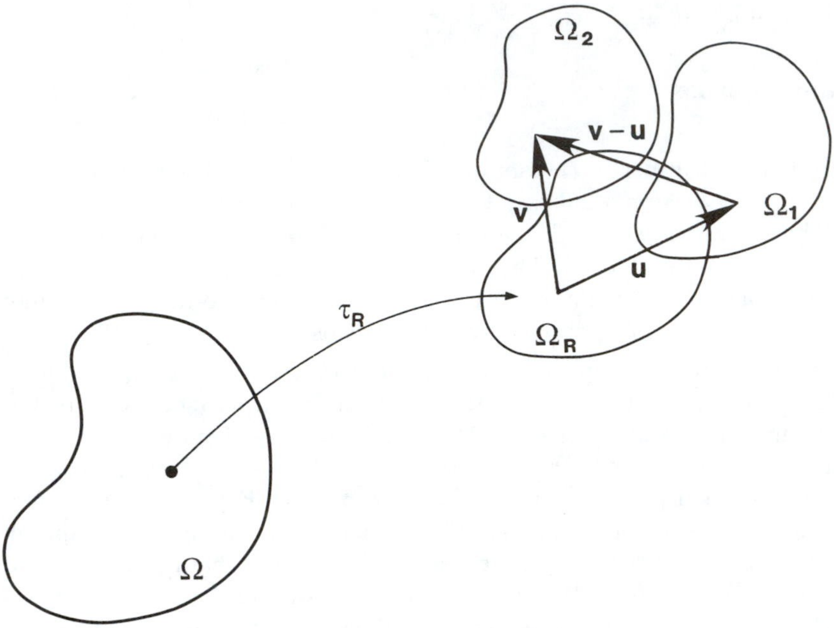

Figure 2.8
Reference configuration and concept of displacements.

Exercises

2.3.1. Prove that the operations in the quotient space V/M are well defined, i.e., the equivalence classes $[\boldsymbol{x}+\boldsymbol{y}]$ and $[\alpha\boldsymbol{x}]$ do not depend upon the choice of elements $\boldsymbol{x} \in [\boldsymbol{x}]$ and $\boldsymbol{y} \in [\boldsymbol{y}]$.

2.3.2. Let M be a subspace of a real space V and R_M the corresponding equivalence relation. Together with three equivalence axioms (i)–(iii), relation R_M satisfies two extra conditions:

$$\text{(iv) } \boldsymbol{x}R\boldsymbol{y}, \boldsymbol{u}R\boldsymbol{v} \Leftrightarrow (\boldsymbol{x}+\boldsymbol{u})R(\boldsymbol{y}+\boldsymbol{v})$$

$$\text{(v) } \boldsymbol{x}R\boldsymbol{y} \Leftrightarrow (\alpha\boldsymbol{x})R(\alpha\boldsymbol{y}) \qquad \forall\alpha \in \mathbb{R}$$

We say that R_M is *consistent* with linear structure on V. Let R be an arbitrary relation satisfying conditions (i)–(v), i.e., an equivalence

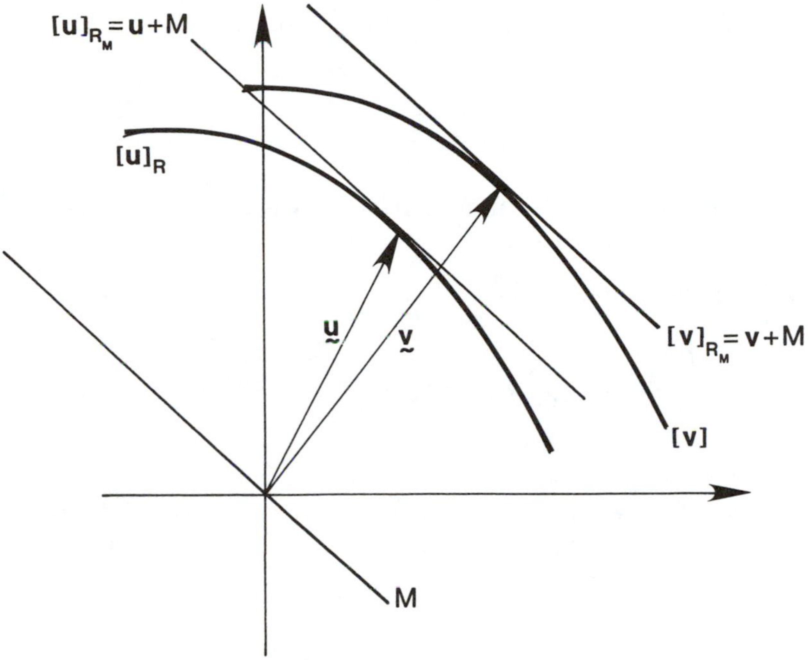

Figure 2.9
Concept of two equivalence relations in the space of displacements.

relation consistent with linear structure on V. Show that there exists a unique subspace M of V such that $R = R_M$, i.e., R is generated by the subspace M.

2.3.3. Another way to see the difference between two equivalence relations discussed in Example 2.3.3. is to discuss the equations of rigid body motions. For the sake of simplicity let us consider the two-dimensional case.

(i) Prove that $\boldsymbol{E}(\boldsymbol{u}) = \boldsymbol{0}$ if and only if $\boldsymbol{u}$ takes the form

$$u_1 = c_1 + \cos\theta(x_1 - d_1) + \sin\theta(x_2 - d_2) - x_1$$

$$u_2 = c_2 - \sin\theta(x_1 - d_1) + \cos\theta(x_2 - d_2) - x_2$$

where θ is the angle of rotation.

(ii) Prove that $\varepsilon_{ij}(\boldsymbol{u}) = 0$ if and only if $\boldsymbol{u}$ has the following form:

$$u_1 = a_1 + \theta x_2$$

$$u_2 = a_2 - \theta x_1$$

One can see that for small values of angle θ ($\cos\theta \approx 1, \sin\theta \approx \theta$) the second set of equations can be obtained by linearizing the first.

2.4 Linear Dependence and Independence. Hamel Basis. Dimension.

Linear Combination. Given k vectors $\boldsymbol{x}_1, \ldots, \boldsymbol{x}_k$ and k scalars $\alpha_1, \ldots, \alpha_k$, the vector

$$\sum_{i=1}^{k} \alpha_i \boldsymbol{x}_i = \alpha_1 \boldsymbol{x}_1 + \ldots + \alpha_k \boldsymbol{x}_k$$

is called a *linear combination* of the vectors $\boldsymbol{x}_1, \ldots, \boldsymbol{x}_k$.

Linear Dependence. We say that a vector $\boldsymbol{x}$ is *linearly dependent* on vectors $\boldsymbol{x}_1, \ldots, \boldsymbol{x}_k$ if there exists a linear combination of $\boldsymbol{x}_i$ equal to $\boldsymbol{x}$, i.e.,

$$\boldsymbol{x} = \alpha_1 \boldsymbol{x}_1 + \ldots + \alpha_k \boldsymbol{x}_k$$

Vectors $\boldsymbol{x}_1, \ldots, \boldsymbol{x}_k$ are called *linearly independent* if none of them is linearly dependent upon the remaining ones. If not, they are called *linearly dependent.*

PROPOSITION 2.4.1

The following conditions are equivalent:

(i) $\boldsymbol{x}_1, \ldots, \boldsymbol{x}_k$ *are linearly independent.*

(ii) $\displaystyle\sum_{i=1}^{k} \alpha_i \boldsymbol{x}_i = \boldsymbol{0} \Leftrightarrow \alpha_1 = \ldots = \alpha_k = 0.$

PROOF

(i) $\rightarrow$ (ii). Suppose, to the contrary, that $\sum \alpha_i \boldsymbol{x}_i = 0$ and that there exists $\alpha_l \neq 0$. It follows that

$$\alpha_l \boldsymbol{x}_l = \sum_{i \neq l} -\alpha_i \boldsymbol{x}_i$$

and, consequently,

$$\boldsymbol{x}_l = \sum_{i \neq l} -\frac{\alpha_i}{\alpha_l}\boldsymbol{x}_i$$

which proves that $\boldsymbol{x}_l$ linearly depends on the remaining $\boldsymbol{x}_i$.

(ii) $\Rightarrow$ (i). Suppose, to the contrary, again that there is a vector $\boldsymbol{x}_l$ such that

$$\boldsymbol{x}_l = \sum_{i \neq l} \beta_i \boldsymbol{x}_i$$

Taking

$$\alpha_i = \begin{cases} \beta_i & i \neq l \\ -1 & i = l \end{cases}$$

we easily construct the combination $\sum \alpha_i \boldsymbol{x}_i = \boldsymbol{0}$ with not all coefficients equal to zero. ∎

COROLLARY 2.4.1

(i) None of a set of linearly independent vectors $\boldsymbol{x}_1, \ldots, \boldsymbol{x}_k$ is equal to zero.

(ii) Any subset of linearly independent vectors is linearly independent.

Example 2.4.1

Consider the classical free vectors (recall Example 2.1.2) in three-dimensional space. Three vectors are linearly dependent if and only if they are coplanar. □

Example 2.4.2

Consider the space $\mathbb{R}^n$ and a set of vectors

$$\boldsymbol{e}_i = (0, \ldots, 1_{(i)}, \ldots, 0) \qquad i = 1, \ldots, n$$

Obviously, $\sum \alpha_i \boldsymbol{e}_i = (\alpha_1, \ldots, \alpha_n)$ and therefore if $\sum \alpha_i \boldsymbol{e}_i = \boldsymbol{0}$ then all α_i must be zero. By Proposition 2.4.1 vectors $\boldsymbol{e}_i$ are linearly independent. □

Example 2.4.3

Let $\Omega \subset I\!R$ denote an open interval and $\mathcal{P}^k(\Omega)$ the space of polynomials up to the k-th order defined on Ω. It is easy to see that the set of monomials $1, x, x^2, \ldots, x^k$ is linearly independent in $\mathcal{P}^k(\Omega)$. □

So far we have talked about linear dependence or independence of finite sets of vectors only. It is possible to extend this concept also to *infinite* sets.

Linear Independence in Infinite Sets. Let V be a vector space and P an arbitrary subset of V. We say that P is *linearly independent* if every finite subset of P is linearly independent.

Example 2.4.4

Let V denote a set of infinite sequences of real numbers $\boldsymbol{x} = \{x_i\}_1^\infty = (x_1, \ldots); x_i \in I\!R, i = 1, 2, \ldots$ with a property that in every such sequence only a finite number of elements is different from zero. Formally, we can write:

$$V = \{\boldsymbol{x} = \{x_i\}_1^\infty, x_i \in I\!R, i = 1, 2 \ldots : \exists\, k = k(\boldsymbol{x}) \colon x_i = 0 \ \forall i > k\}$$

Since infinite sequences are nothing else than real-valued functions defined on the set of positive integers $I\!N$, V can be embodied with natural operations from $I\!R^{I\!N}$. One can easily verify that V is closed with respect to the operations and therefore V is a vector space (a subspace of $I\!R^{I\!N}$).

Consider the infinite set of vectors

$$B = \{\boldsymbol{e}_i, \quad i = 1, 2 \ldots\} \text{ where}$$

$$\boldsymbol{e}_i = \{0, \ldots, 1_{(i)}, \ldots\}\ , \ i = 1, 2, \ldots$$

If A is a finite subset of B then there must be an integer n such that all vectors from A possess zero components on places with indices greater than n. Consequently,

$$\sum_A \alpha_i \boldsymbol{e}_i = \sum_{i=1}^n \beta_i \boldsymbol{e}_i, \text{ where } \beta_i = \begin{cases} \alpha_i, \text{ if } \boldsymbol{e}_i \in A \\ 0 \text{ otherwise} \end{cases}$$

Consequently, if

$$\sum_A \alpha_i \boldsymbol{e}_i = (\beta_1, \ldots, \beta_n, 0, \ldots) = \boldsymbol{0}$$

then all coefficients β_i, and therefore α_i, must vanish too. This proves that set B is linearly independent. □

Hamel Basis. A linearly independent subset B of a vector space V is called a *Hamel basis* on V if it is maximal, i.e., no linearly independent subset S of V exists such that B is a proper subset of S.

THEOREM 2.4.1

Let V be a vector space. $B \subset V$ is a Hamel basis of V if and only if every vector $\boldsymbol{x} \in V$ has a unique representation in the form of the linear combination

$$\boldsymbol{x} = \alpha_1 \boldsymbol{a}_1 + \ldots + \alpha_n \boldsymbol{a}_n \qquad (*)$$

where vectors $\boldsymbol{a}_i$ belong to set B and α_i are scalars.

PROOF Let B be a basis. Every element of B has a trivial representation of the form $(*)$. Suppose that $\boldsymbol{x} \notin B$. Then $B \cup \{\boldsymbol{x}\}$ is linearly dependent and there must be a finite subset of $B \cup \{\boldsymbol{x}\}$ (containing $\boldsymbol{x}$; explain why) which is linearly dependent, i.e., there exists a nonzero linear combination of the form

$$\alpha \boldsymbol{x} + \alpha_1 \boldsymbol{a}_1 + \ldots + \alpha_n \boldsymbol{a}_n = \boldsymbol{0} \quad , \quad \boldsymbol{a}_i \in B$$

Since $\boldsymbol{a}_1, \ldots, \boldsymbol{a}_n$ are linearly independent, we conclude that $\alpha \neq 0$ and therefore $\boldsymbol{x}$ can be represented in the form

$$\boldsymbol{x} = -\frac{\alpha_1}{\alpha} \boldsymbol{a}_1 + \ldots + \frac{\alpha_n}{\alpha} \boldsymbol{a}_n$$

Suppose now that two representations exist, say

$$\boldsymbol{x} = \sum \alpha_i \boldsymbol{a}_i = \sum \alpha_i' \boldsymbol{a}_i$$

Completing both combinations by zero elements we can assume that both combinations contain the same vectors $\boldsymbol{a}_i$ and, consequently,

$$\sum (\alpha_i - \alpha_k') \boldsymbol{a}_i = \boldsymbol{0}$$

By linear independence of vectors $\boldsymbol{a}_i$, all coefficients $\alpha_i - \alpha_i'$ are zero, i.e., $\alpha_i = \alpha_i'$, which proves that the representation is unique.

Suppose now that every vector has a unique representation of the form (∗). Consider an arbitrary linear combination of the form

$$\alpha_1 \boldsymbol{a}_1 \ldots + \alpha_n \boldsymbol{a}_n = \mathbf{0} \;\; \boldsymbol{a}_1, \cdots, \boldsymbol{a}_n \in B$$

Since the zero-vector $\mathbf{0}$ already has a trivial representation

$$\mathbf{0} = 0\boldsymbol{a}_1 + \ldots + 0\boldsymbol{a}_n$$

and the representation is unique, it follows that all α_i are zero and therefore B is linearly independent. B is also maximal since any element from outside B has a representation (*), which implies that $B \cup \{\boldsymbol{x}\}$ is linearly dependent. ∎

REMARK 2.4.1. The scalars $\alpha_1 \ldots \alpha_n$ are sometimes called the components of vector $\boldsymbol{x}$ relative to basis B. ∎

Span. A set of vectors $P = \{\boldsymbol{a}_1 \ldots, \boldsymbol{a}_k \ldots\}$ of a vector space V is said to *span* V if every vector $\boldsymbol{x} \in V$ can be expressed as a linear combination of vectors from P,

$$\boldsymbol{x} = \sum \alpha_i \boldsymbol{a}_i, \quad \boldsymbol{a}_i \in P$$

Making use of Theorem 2.4.1 one can easily prove

PROPOSITION 2.4.2

Let B be a subset of a vector space V. The following conditions are equivalent:

(i) B is a Hamel basis of V.

(ii) B is linearly independent and spans V.

We hasten to point out that the Hamel basis is merely one of many kinds of bases encountered in studying various mathematical systems. It portrays a purely algebraic property of vector spaces and is intimately connected with the linear algebraic properties of such spaces. In studying topological properties in Chapter 4, we again encounter bases of certain spaces, but there we are interested in topological properties, and the structure of topological bases is quite different from that of the bases considered here. The term basis (or base) means roughly what we might expect it to: a basis for communication. Once a basis is established and perfectly understood by all interested parties, we may proceed to describe the properties of the system under investigation relative to that basis. A reasonable mathematical system always has reasonable properties that are often useful to know. The particular form in which these properties manifest themselves may well depend upon what basis we choose to study them.

We emphasize that even in the context of vector spaces the notion of the basis is not unique. In the case of infinite-dimensional spaces, which we discuss later in this section, a purely algebraic structure turns out to be insufficient for our purposes and a topological one must be added. This leads to a new definition of the basis in certain infinite-dimensional spaces, which we describe later. Contrary to infinite-dimensional spaces, useful properties of finite-dimensional spaces can be studied within the pure algebraic structure and the notion of basis is practically unique. The examples illustrate the concept of Hamel basis.

So long as it is well understood that our aims in this part of our study are purely albegraic, we can drop the adjective Hamel and simply refer to sets of vectors as possible bases of vector spaces. When the context requires, we shall be specific about the types of bases.

Example 2.4.5

Consider free vectors in a plane. Let $\boldsymbol{a}_1, \boldsymbol{a}_2$ denote two arbitrary, but not collinear, vectors. Pick an arbitrary vector $\boldsymbol{b}$ and project it along the line of action of $\boldsymbol{a}_1$ in the direction of $\boldsymbol{a}_2$ (cf. Fig. 2.4.1). Denote the projection by $\boldsymbol{b}_1$ and the corresponding projection along $\boldsymbol{a}_2$ by $\boldsymbol{b}_2$. Obviously, $\boldsymbol{b} = \boldsymbol{b}_1 + \boldsymbol{b}_2$. Vectors $\boldsymbol{a}_1$ and $\boldsymbol{b}_1$ are collinear and therefore there must exist a scalar β_1 such that $\boldsymbol{b}_1 = \beta_1 \boldsymbol{a}_1$. Similarly, there exists a scalar β_2 such that $\boldsymbol{b}_2 = \beta_2 \boldsymbol{a}_2$ and, consequently,

$$\boldsymbol{b} = \boldsymbol{b}_1 + \boldsymbol{b}_2 = \beta_1 \boldsymbol{a}_1 + \beta_2 \boldsymbol{a}_2$$

Thus vectors $\boldsymbol{a}_1, \boldsymbol{a}_2$ span the entire space. Since none of them can be represented as a product of a number and the other vector (they are not collinear), they are also linearly independent. Concluding, any two noncollinear vectors form a basis for free vectors in a plane. The coefficients β_1 and β_2 are components of $\boldsymbol{b}$ with respect to that basis.

Similarly, we show that any three noncoplanar vectors form a basis for free vectors in a space.

□

Example 2.4.6

Consider the space $\mathbb{R}^n$ and set of vectors $\boldsymbol{e}_i$ considered in Example 2.4.2. Any vector $\boldsymbol{x} = (x_1, x_2 \ldots x_n) \in \mathbb{R}^n$ can be represented in the form

$$\boldsymbol{x} = \sum_{i=1}^{n} x_i \boldsymbol{e}_i$$

Thus vectors $\boldsymbol{e}_i, i = 1, \ldots, n$ span the entire $\mathbb{R}^n$, and since they are linearly independent, they form a basis. Such a basis is called a canonical basis in $\mathbb{R}^n$. $\square$

Example 2.4.7

Monomials from Example 2.4.3 form a basis for the space $\mathcal{P}^k(\Omega)$. $\square$

Example 2.4.8

Let V be a vector space defined in Example 2.4.4. The set $B = \{\boldsymbol{e}_i, i = 1, 2, \ldots\}$ is linearly independent and simultaneously spans the space V (explain why), so B is a Hamel basis for V. $\square$

Example 2.4.9

Consider the vector space whose elements are again infinite sequences of real numbers $\boldsymbol{x} = \{x_i\}_1^\infty$, but this time such that $\sum_1^\infty x_i^2 < +\infty$. We will encounter this space many times in subsequent chapters; it is a special vector space, endowed with a norm, and is usually referred to as ℓ^2. We are interested only in algebraic properties now. Obviously, the space V from the previous example is a subspace of ℓ^2. The set B which has been a Hamel basis for V is not a basis for ℓ^2 any more. It is still linearly independent, but it does not *span* the entire space ℓ^2. It spans V, which is only a proper subspace of ℓ^2.

It is true that all elements in ℓ^2 can be written in the form

$$\boldsymbol{x} = \sum_{i=1}^{\infty} \alpha_i \boldsymbol{e}_i$$

but such a sum makes no sense in spaces with purely algebraic structure:

an *infinite* series requires that we specify a mode of *convergence*, and convergence is a topological concept, not an algebraic one.

We can overcome this difficulty by adding topological structure to ℓ^2, and we do just that in Chapter 4. There we endow ℓ^2 with a norm,

$$||\boldsymbol{x}||_{\ell^2} = \left(\sum_{k=1}^{\infty} |x_k|^2\right)^{\frac{1}{2}}$$

which allows us to describe not only the "length" of $\boldsymbol{x}$, but also the "distance" between vectors $\boldsymbol{x}$ and $\boldsymbol{y}$ in ℓ^2. In this particular setting we define as a *basis* (not a Hamel basis) any countable infinite set of linearly independent vectors $\{\boldsymbol{x}_i\}_{i=1}^{\infty}$, such that $\forall\ \boldsymbol{x} \in \ell^2$,

$$\forall\ \varepsilon > 0\ \exists N = N(\varepsilon) : ||\boldsymbol{x} - \sum_{k=1}^{\ell} \alpha_k \boldsymbol{a}_k|| < \varepsilon\ \forall\ \ell > N$$

We will prove that the set $B = \{\boldsymbol{e}_i\}_1^{\infty}$ considered in this sense is a basis for ℓ^2.

Besides its practical meaning, the notion of the Hamel basis allows us to define the fundamental concept of the dimension of a vector space and, as a consequence, distinguish between finite- and infinite-dimensional spaces. To do it, however, we need to prove the two following fundamental theorems. □

THEOREM 2.4.2

Let V be a vector space, B a (Hamel) basis and $P \subset V$ an arbitrary linearly independent set. Then

$$\#P \leq \#B$$

Before we proceed with the proof let us make an important corollary.

COROLLARY 2.4.2

Every two bases in a vector space have the same number of elements or, more precisely, the same cardinal number. Indeed, if B_1 and B_2 denote two bases, then B_2 is linearly independent and according to the theorem, $\#B_2 \leq \#B_1$. Conversely, $\#B_1 \leq \#B_1$ and the equality holds.

Dimension. The cardinal number of any basis of a vector space V is called the *dimension* of the space and denoted $\dim V$.

If $\dim V = n < +\infty$, the space is called a finite-dimensional (n-dimensional) space; if not, we then speak of infinite-dimensional vector spaces. Although several properties are the same for both cases, the differences are very significant. The theory which deals with finite-dimensional spaces is customarily called "linear algebra," while the term "functional analysis" is reserved for the case of infinite-dimensional spaces, the name coming from function spaces which furnish the most common example of spaces of infinite dimension.

By this time a careful reader would have noticed that we have skipped over a very important detail. In everything we have said so far based on the concept of a basis, we have been implicity assuming that such a basis *exists* in fact in every vector space. Except for a few cases where we can construct a basis explicitly, this is not a trivial assertion and has to be proved.

THEOREM 2.4.3

Every linearly independent set A in a vector space X can be extended to a (Hamel) basis. In particular, every nontrivial vector space ($X \neq \{\mathbf{0}\}$), possesses a basis.

Proofs of Theorems 2.4.2 and 2.4.3 are pretty technical and can be skipped during the first reading. The fundamental tool in both cases is the Kuratowski-Zorn lemma.

PROOF ***of Theorem 2.4.3.***

Let $\mathcal{U}$ be a class of all linearly independent sets containing set A. $\mathcal{U}$ is nonempty since it contains A. We shall introduce a partial ordering in the family $\mathcal{U}$ in the following way.

$$B_1 \leq B_2 \overset{\text{def}}{\Leftrightarrow} B_1 \subset B_2$$

Now let $\mathcal{B}$ denote a linearly ordered set in family $\mathcal{U}$. We shall construct an upper bound for $\mathcal{B}$. Toward this goal define set B_0 as the union of all sets from the family

$$B_0 = \bigcup_{B \in \mathcal{B}} B$$

Obviously, $A \subset B_0$. B_0 is linearly independent since every finite subset must be contained in a certain B and all B's are linearly independent.

Thus, according to the Kuratowski-Zorn lemma, there exists a maximal element in $\mathcal{U}$ which is nothing else than a (Hamel) basis in X.

To prove the second assertion it suffices to take as A a subset consisting of one, single nonzero vector. $\blacksquare$

PROOF ***of Theorem 2.4.2.***

When B is finite the proof is standard and does not require the use of the Kuratowski-Zorn lemma. Obviously, the proof we present holds for both finite- and infinite-dimensional cases. We shall show that there exists an injection (one-to-one mapping) from P to B.
Denote by $\mathcal{F}$ a class of all injections satisfying the following conditions:

(i) $P \cap B \subset \operatorname{dom} f \subset P, \quad \operatorname{im} f \subset B.$

(ii) The set $(P - \operatorname{dom} f) \cup \operatorname{im} f$ is linearly independent.

$\mathcal{F}$ is nonempty (explain why) and can be ordered by the following partial ordering in $\mathcal{F}$:

$$f \leq g \overset{\text{def}}{\Leftrightarrow} \operatorname{dom} f \subset \operatorname{dom} g \text{ and } g|\operatorname{dom} f = f$$

Let $\mathcal{G}$ be a linearly ordered set in the class $\mathcal{F}$. The union of functions $f \in \mathcal{G}$, denoted F, is a well-defined injection satisfying condition (i). Let $A = A_1 \cup A_2, A_1 \subset P - \operatorname{dom} F$, $A_2 \subset \operatorname{im} F$. It must be an f from $\mathcal{F}$ such that $A_2 \subset \operatorname{im} f$. Obviously, $A_1 \subset P - \operatorname{dom}(\cup f) \subset P - \operatorname{dom} f$, and, therefore, according to condition (ii), A must be linearly independent.

Thus, F is an upper bound for the family $\mathcal{G}$ and according to the Kuratowski-Zorn lemma, there exists a maximal element f in the class $\mathcal{F}$. It is sufficient to show that $\operatorname{dom} f = P$.

Suppose to the contrary that $P - \operatorname{dom} f \neq \emptyset$. It implies that also $\operatorname{im} f \neq B$. Indeed, if it were $\operatorname{im} f = B$ then from the fact that B is a basis and that the set

$$\operatorname{im} f \cup (P - \operatorname{dom} f) = B \cup (P - \operatorname{dom} f)$$

is linearly independent, it would follow that $P - \operatorname{dom} f \subset B$ and, consequently, $P - \operatorname{dom} f \subset P \cap B \subset \operatorname{dom} f$, which is impossible.

So pick a vector $\boldsymbol{v}_0 \in B - \operatorname{im} f$. Two cases may exist. Either $\boldsymbol{v}_0$ is a linear combination of elements from $(P - \operatorname{dom} f) \cup \operatorname{im} f$, or not. In the second case a union of f and $\{(\boldsymbol{u}_0.\boldsymbol{v}_0)\}$, where $\boldsymbol{u}_0$ is an arbitrary element from $P - \operatorname{dom} f$, denoted f_1 belongs to family $\mathcal{F}$. Indeed, f_1 satisfies trivially condition (i) and the set

$$(P - \operatorname{dom} f_1) \cup \operatorname{im} f_1, = \{P - (\operatorname{dom} f \cup \{\boldsymbol{u}_0\})\} \cup \operatorname{im} f \cup \{\boldsymbol{v}_0\}$$

is linearly independent, so f_1 is a proper extension of f and belongs to $\mathcal{F}$, which contradicts the fact that f is maximal.

Consider the first case. Vector $\boldsymbol{v}_0$ can be represented in the form

$$\boldsymbol{v}_0 = \lambda_0 \boldsymbol{u}_0 + \ldots + \lambda_n \boldsymbol{u}_n + \mu_0 \boldsymbol{w}_0 + \ldots + \mu_m \boldsymbol{w}_m$$

where $\boldsymbol{u}_0 \ldots \boldsymbol{u}_n \in P - \operatorname{dom} f$, $\boldsymbol{w}_0 \ldots \boldsymbol{w}_m \in \operatorname{im} f$. One of the numbers $\lambda_0, \lambda_1, \ldots, \lambda_n$, say λ_0, must be different from zero since in the other case set B would be linearly dependent. Consider again the extension $f_1 = f \cup \{(\boldsymbol{u}_0, \boldsymbol{v}_0)\}$. If $(P - \operatorname{dom} f_1) \cup \operatorname{im} f_1$ were linearly dependent then $\boldsymbol{v}_0$ would be a linear combination of elements from $(P - \operatorname{dom} f) \cup \operatorname{im} f - \{\boldsymbol{u}_0\}$, which is impossible since $\lambda_0 \neq 0$. So, again f has the proper extension f_1 in the family $\mathcal{F}$ and therefore cannot be maximal. ∎

Construction of a Complement. One of the immediate consequences of Theorems 2.4.2 and 2.4.3 is a possibility of constructing a complement Y to an arbitrary subspace X of a vector space V. Toward this goal pick an arbitrary basis B for X (which, according to Theorem 2.4.3, exists). According to Theorem 2.4.3, basis B can be extended to a basis C for the whole V. The complement space Y is generated by vectors from $C - B$. Indeed, $X + Y = V$ and $X \cap Y = \{\mathbf{0}\}$ due to the linear independence of C. Except for the trivial case when $X = V$, subspace X possesses many (infinitely many, in fact) complements Y.

We conclude this section with a summary of the classification of the spaces which we have used in this chapter according to dimension.

Example 2.4.10

1. Free vectors in a plane form a two-dimensional subspace of a three-dimensional vector space;

2. $\dim \mathbb{R}^n = n, \dim \mathbb{C}^n = n$;

3. $\dim \mathcal{P}^k(\Omega) = k + 1$ if Ω is an interval in $\mathbb{R}$;

4. spaces V from Example 2.4.4 and ℓ^2 from Example 2.4.9 are infinite dimensional.

□

Linear Transformations

2.5 Linear Transformations—The Fundamental Facts

Each notion of an algebraic structure is accompanied by specific operations, functions which reflect the basic features of the considered structures. The linear transformation plays such a role for vector spaces.

Linear Transformation. Let V and W be two vector spaces, both over the same field $\mathbb{F}$. A *linear transformation* $T : V \to W$ is a mapping of V into W such that the following hold:

$$(i)\ \ T(\boldsymbol{x}+\boldsymbol{y}) = T(\boldsymbol{x}) + T(\boldsymbol{y}) \text{ for every } \boldsymbol{x}, \boldsymbol{y} \in V.$$

$$(ii)\ T(\alpha\boldsymbol{x}) = \alpha T(\boldsymbol{x}) \text{ for every } \boldsymbol{x} \in V, \text{ scalar } \alpha \in \mathbb{F}.$$

We say that T is *additive and homogeneous.* One can combine properties (i) and (ii) into the more concise definition: the transformation $T : V \to W$ is linear if and only if

$$T(\alpha\boldsymbol{x} + \beta\boldsymbol{y}) = \alpha T(\boldsymbol{x}) + \beta T(\boldsymbol{y})$$

We have two simple observations:

1. One can easily generalize this law for combinations of more than two vectors

$$T(\alpha_1\boldsymbol{x} + \ldots + \alpha_n\boldsymbol{x}_n) = \alpha_1 T(\boldsymbol{x}_1) + \ldots + \alpha_n T(\boldsymbol{x}_n)$$

2. If T is linear then an image of zero vector in V must be a necessary zero vector in W, since

$$T(\mathbf{0}) = T(\boldsymbol{x} + (-1)\boldsymbol{x}) = T(\boldsymbol{x}) - T(\boldsymbol{x}) = \mathbf{0}$$

The term *transformation* is synonymous with *function, map* or *mapping.* One should emphasize, however, that the linear transformation is defined on the *whole* space V; its domain of definition coincides with the entire V. The term *linear operator* is frequently reserved for linear functions which are defined in general only on a *subspace* of V. Its use is basically restricted to infinite-dimensional spaces.

Example 2.5.1

A function can be additive and not homogeneous. For example, let $z = x + iy \left(i = \sqrt{-1}\right)$ denote a complex number, $z \in \mathbb{C}$, and let $T : \mathbb{C} \to \mathbb{C}$ be a complex *conjugation*; i.e.,

$$T(z) = \overline{z} = x - iy$$

Then

$$T(z_1 + z_2) = x_1 + x_2 - i(y_1 + y_2) = T(z_1) + T(z_2)$$

However, if $a = \alpha + i\beta$ is a complex scalar,

$$T(az) = \alpha x - \beta y - i(\alpha y + \beta x) \neq aT(z) = \alpha x + \beta y + i(\beta x - \alpha y)$$

Hence complex conjugation is not a linear transformation. □

Example 2.5.2

There are many examples of functions that are homogeneous and not additive. For example, consider the map $T : \mathbb{R}^2 \to \mathbb{R}$ given by

$$T((x_1, x_2)) = \frac{x_1^2}{x_2}$$

Clearly,

$$T((x_1, x_2) + (y_1, y_2)) = \frac{(x_1 + y_1)^2}{x_2 + y_2} \neq \frac{x_1^2}{x_2} + \frac{y_1^2}{y_2}$$

However,

$$T(\alpha(x_1, x_2)) = \frac{(\alpha x_1)^2}{\alpha x_2} = \alpha \frac{x_1^2}{x^2} = \alpha T((x_1, x_2))$$

□

Example 2.5.3

One of the most common examples of a linear transformation is that associated with the integration of real-valued functions. Let $V = C(\overline{\Omega})$. Define $T : V \to V$

$$Tf(\boldsymbol{x}) = \int_{\Omega} K(\boldsymbol{x}, \boldsymbol{y}) f(\boldsymbol{y}) dy$$

Function $K(\boldsymbol{x}, \boldsymbol{y})$ is called the *kernel* of the *integral transformation* T and it usually carries some regularity assumptions to assure the existence of the integral. If, for instance, K is continuous and bounded then the integral exists and it can be understood as the classical Riemann integral.

One easily verifies that transformation T is linear. $\square$

Example 2.5.4

Another common example is associated with the operation of differentiation. Let $f(x)$ be a real-valued function of real variable x. Then T is defined as follows:

$$Tf(x) = f'(x)$$

Clearly,

$$(f + g)' = f' + g'$$

$$(\alpha f)' \quad = \alpha f'$$

where α is a real number. Thus T qualifies as a linear function. As we have mentioned before, an important issue is the domain of definition of T. If we assume for V the space of (continuously) differentiable functions $C^1(0,1)$ then T is defined on the whole V and we will use the term *transformation*. If, however, we choose for V, for instance, the space of continuous functions $C(0,1)$, then the derivative makes only sense for a subspace of V and we would use rather the term *operator*. $\square$

REMARK 2.5.1. In many cases, especially in the context of linear operators, we shall simplify the notion writing Tu in place of $T(u)$ (cf. Examples 2.5.3 and 2.5.4). In general this rule is reserved for linear transformations or operators only. $\blacksquare$

Example 2.5.5

Let $u : \Omega \rightarrow I\!R$ be a real-valued function. We denote by $\boldsymbol{f}$ the operator defined on the set of real functions on Ω by the formula:

$$\boldsymbol{f}(u)(\boldsymbol{x}) = f(\boldsymbol{x}, u(\boldsymbol{x}))$$

where $f(\boldsymbol{x}, t)$ is a certain function. This operator, nonlinear in general, is called the Nemytskii operator and plays a fundamental role in the study of a broad class of nonlinear integral equations. If function f is linear in t, i.e., $f(\boldsymbol{x},t) = g(\boldsymbol{x})t$, where g is a function of variable $\boldsymbol{x}$ only, then operator f becomes linear. For a precise definition, of course, one has to specify more precisely the domain of f involving usually some regularity assumptions on functions u. □

The General Form of Linear Transformation in Finite-Dimensional Spaces. Let V and W be two finite-dimensional spaces, $\dim V = n, \dim W = m$, and let $T : V \to W$ denote an arbitrary linear transformation. Let $\boldsymbol{e}_1, \ldots, \boldsymbol{e}_n$ and $\boldsymbol{f}_1, \ldots, \boldsymbol{f}_m$ denote two arbitrary bases for V and W, respectively. Every vector $\boldsymbol{v} \in V$ can be represented in the form

$$\boldsymbol{v} = v_1\boldsymbol{e}_1 + \ldots + v_n\boldsymbol{e}_n$$

where $v_i, i = 1, \ldots, n$ are the components of $\boldsymbol{v}$ with respect to basis $\boldsymbol{e}_i$. Since T is linear, we have

$$T(\boldsymbol{v}) = v_1 T(\boldsymbol{e}_1) + \ldots + v_n T(\boldsymbol{e}_n)$$

Each of vectors $T(\boldsymbol{e}_j)$ belongs to space W and therefore has its own representation with respect to basis $\boldsymbol{f}_i$. Denoting components of $T(\boldsymbol{e}_j)$ with respect to basis $\boldsymbol{f}_i$ by T_{ij}, i.e.,

$$T(\boldsymbol{e}_j) = T_{1j}\boldsymbol{f}_1 + \ldots + T_{mj}\boldsymbol{f}_m = \sum_{i=1}^{m} T_{ij}\boldsymbol{f}_i$$

we have

$$T(\boldsymbol{v}) = \sum_{j=1}^{n} v_j T(\boldsymbol{e}_j) = \sum_{j=1}^{n} v_j \sum_{i=1}^{m} T_{ij}\boldsymbol{f}_i = \sum_{i=1}^{m}\sum_{j=1}^{n} T_{ij} v_j \boldsymbol{f}_i$$

Thus values of T are *uniquely determined* by matrix T_{ij}. If this matrix is known then in order to calculate components of $T(\boldsymbol{v})$ one has to *multiply matrix* T_{ij} by *vector* of *components* v_j. In other words, if $\boldsymbol{w} = T(\boldsymbol{v})$ and $w_i, i = 1, \ldots, m$ stand for the components of $\boldsymbol{w}$ with respect to basis $\boldsymbol{f}_i, i = 1, \ldots, m$, then

$$w_i = \sum_{j=1}^{n} T_{ij} v_j$$

Writing T_{ij} in the form of a two-dimensional array

$$\begin{bmatrix} T_{11} & T_{12} & T_{13} \dots T_{1n} \\ \\ T_{21} & T_{22} & \dots T_{2n} \\ \\ \\ \\ T_{m1} & T_{m2} & \dots T_{mn} \end{bmatrix}$$

we associate the first index i with the row number, while j indicates the column number. Therefore, in order to *multiply* matrix T_{ij} by vector v_j one has to multiply rows of T_{ij} by vector v_j. According to our notation, a j-th column of matrix T_{ij} can be interpreted as components of the image of vector $\boldsymbol{e}_j, T(\boldsymbol{e}_j)$, with respect to basis $\boldsymbol{f}_i$. Array T_{ij} is called the *matrix representation* of linear transformation T with respect to bases $\boldsymbol{e}_j$ and $\boldsymbol{f}_i$. Conversely, if T can be represented in such a form, then T is linear. We will return to this important issue in Section 2.8.

REMARK 2.5.2. From the proof of Theorem 2.5.1 follows one of the most fundamental properties of linear transformations. A linear transformation T is uniquely determined through its values $T(\boldsymbol{e}_1), T(\boldsymbol{e}_2) \dots$ for a certain basis $\boldsymbol{e}_1, \boldsymbol{e}_2, \dots$. Let us emphasize that this assertion holds for finite- and infinite-dimensional spaces as well. The practical consequence of this observation is that a linear transformation may be defined by setting its values on an arbitrary basis. ∎

Example 2.5.6

Let us find the matrix representation of a rotation T in a plane with respect to a basis of two perpendicular unit vectors, $\boldsymbol{e}_1$ and $\boldsymbol{e}_2$ (see Fig. 2.10). Let θ denote angle of rotation. One can easily see that

$$T\boldsymbol{e}_1 = \cos\theta \boldsymbol{e}_1 + \sin\theta \boldsymbol{e}_2$$

$$T\boldsymbol{e}_2 = -\sin\theta \boldsymbol{e}_1 + \cos\theta \boldsymbol{e}_2$$

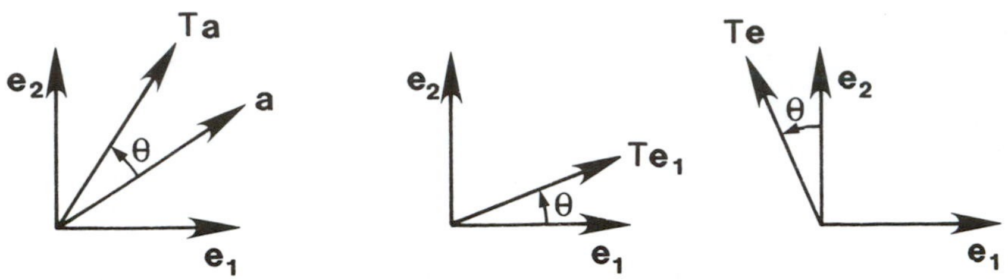

Figure 2.10
Rotation in a plane.

Thus the matrix representation takes the form:

$$\begin{pmatrix} \cos\theta \ \sin\theta \\ \\ -\sin\theta \ \cos\theta \end{pmatrix}$$

□

Range Space and Null Space of a Linear Transformation. Let $T : V \to W$ be an arbitrary linear transformation from a vector space V into a vector space W. Together with the *range* of T we consider the *kernel* of T, defined as a subset of V consisting of all elements in V that are mapped into the zero vector $\mathbf{0}$ of W. Formally,

$$\mathrm{KerT} = \{\boldsymbol{x} \in \mathrm{V} : \ \mathrm{T}(\boldsymbol{x}) = \mathbf{0}\}$$

One can easily check that both kernel and range of T form *linear subspaces* of V and W, respectively. We call them the *null space* and the *range space*, denoted $\mathcal{N}(T)$ and $\mathcal{R}(T)$, respectively. If no confusion occurs we will supress the letter T and write shortly $\mathcal{N}$ and $\mathcal{R}$.

Rank and Nullity. The rank $r(T)$ of a linear transformation T is the dimension of its range space $\mathcal{R}(T)$

$$r(T) = \dim \mathcal{R}(T)$$

The nullity $n(T)$ is the dimension of its null space $\mathcal{N}(T)$

$$n(T) = \dim \mathcal{N}(T)$$

Monomorphism, Epimorphism, Isomorphism. Injective or surjective functions in the context of linear transformations acquire new names. An injective (one-to-one) linear transformation is called a *monomorphism* or a *nonsingular transformation*; a surjective (onto) transformation is called an *epimorphism*. Finally, a bijective linear transformation carries the name of *isomorphism*. We have the following simple observation.

PROPOSITION 2.5.1

Let $T : V \to W$ be a linear transformation. Then T is nonsingular (monomorphism) if and only if $\mathcal{N}(T) = \{\mathbf{0}\}$.

PROOF Let $\boldsymbol{x} \in \mathcal{N}(T)$, i.e., $T(\boldsymbol{x}) = \mathbf{0}$. If T is one-to-one, $\boldsymbol{x}$ must be equal to $\mathbf{0}$ since already $T(\mathbf{0}) = \mathbf{0}$. Conversely, let $\mathcal{N}(T) = \{\mathbf{0}\}$ and suppose that $T(\boldsymbol{x}) = T(\boldsymbol{y})$. Then

$$T(\boldsymbol{x} - \boldsymbol{y}) = T(\boldsymbol{x}) - T(\boldsymbol{y}) = \mathbf{0}$$

which implies that $\boldsymbol{x} - \boldsymbol{y} \in \mathcal{N}(T)$ and, consequently, $\boldsymbol{x} - \boldsymbol{y} = \mathbf{0}$ or $\boldsymbol{x} = \boldsymbol{y}$. ∎

Before we proceed with the next examples we shall prove a fundamental equality relating rank and nullity of a linear transformation T defined on a finite-dimensional space.

THEOREM 2.5.1

Let V be a finite-dimensional vector space and $T : V \to W$ denote a linear transformation from V into another vector space W. Then

$$\dim V = \dim \mathcal{N}(T) + \dim \mathcal{R}(T)$$

i.e., the sum of rank and nullity of linear transformation T equals the dimension of space V.

PROOF Denote $n = \dim V$ and let $\boldsymbol{e}_1 \ldots \boldsymbol{e}_k$ be an arbitrary basis of the null space. According to Theorem 2.4.2, the basis $\boldsymbol{e}_1, \ldots, \boldsymbol{e}_k$ can be extended to a basis $\boldsymbol{e}_1, \ldots, \boldsymbol{e}_k, \boldsymbol{e}_{k+1}, \ldots, \boldsymbol{e}_n$ for the whole V with vectors $\boldsymbol{e}_{k+1}, \ldots, \boldsymbol{e}_n$ forming a basis for a complement of $\mathcal{N}(T)$ in V. We claim that vectors $T(\boldsymbol{e}_{k+1}), \ldots, T(\boldsymbol{e}_n)$ are linearly independent and that they span the range space $\mathcal{R}(T)$. To prove the second assertion pick an arbitrary vector $\boldsymbol{w} = T(\boldsymbol{v})$. Representing vector $\boldsymbol{v}$ in basis $\boldsymbol{e}_i$, we get

$$\begin{aligned}\boldsymbol{w} &= T(v_1\boldsymbol{e}_1 + \ldots + \boldsymbol{v}_k\boldsymbol{e}_k + v_{k+1}\boldsymbol{e}_{k+1} + \ldots + v_n\boldsymbol{e}_n)\\ &= v_1T(\boldsymbol{e}_1) + \ldots + v_kT(\boldsymbol{e}_k) + v_{k+1}T(\boldsymbol{e}_{k+1}) + \ldots + v_nT(\boldsymbol{e}_n)\\ &= v_{k+1}T(\boldsymbol{e}_{k+1}) + \ldots + v_nT(\boldsymbol{e}_n)\end{aligned}$$

since the first k vectors vanish. Thus $T(\boldsymbol{e}_{k+1})$, . . . , $T(\boldsymbol{e}_n)$ span $\mathcal{R}(T)$. Consider now an arbitrary linear combination with coefficients $\alpha_{k+1}, \ldots, \alpha_n$ such that

$$\alpha_{k+1}T(\boldsymbol{e}_{k+1}) + \ldots + \alpha_nT(\boldsymbol{e}_n) = \boldsymbol{0}$$

But T is linear, which means that

$$T(\alpha_{k+1}\boldsymbol{e}_{k+1} + \ldots + \alpha_n\boldsymbol{e}_n) = \alpha_{k+1}T(\boldsymbol{e}_{k+1}) + \ldots + \alpha_nT(\boldsymbol{e}_n) = \boldsymbol{0}$$

and, consequently,

$$\alpha_{k+1}\boldsymbol{e}_{k+1} + \ldots + \alpha_n\boldsymbol{e}_n \in \mathcal{N}(T)$$

The only vector, however, which belongs simultaneously to $\mathcal{N}(T)$ and its complement is the zero vector and therefore

$$\alpha_{k+1}\boldsymbol{e}_{k+1} + \ldots + \alpha_n\boldsymbol{e}_n = \boldsymbol{0}$$

which, since $\boldsymbol{e}_{k+1}, \ldots, \boldsymbol{e}_n$ are linearly independent, implies that $\alpha_{k+1} = \ldots = \alpha_k = 0$, from which in turn follows that $T(\boldsymbol{e}_{k+1}), \ldots, T(\boldsymbol{e}_n)$ are linearly independent as well.

Thus vectors $T(\boldsymbol{e}_{k+1}), \ldots, T(\boldsymbol{e}_n)$ form a basis for the range space $\mathcal{R}(T)$ and, consequently, $\dim \mathcal{R}(T) = n - k$, which proves the theorem. ∎

Theorem 2.5.1 has several simple, but important, consequences which we shall summarize in the following proposition.

PROPOSITION 2.5.2

Let V and W be two finite-dimensional spaces and $T : V \to W$ denote an arbitrary linear transformation. Then the following holds:

(i) *If* $\dim V = n$ *then*
$$T \text{ is a monomorphism if and only if } \operatorname{rank} T = n$$

(ii) *If* $\dim W = m$ *then*
$$T \text{ is an epimorphism if and only if } \operatorname{rank} T = m$$

(iii) *If* $\dim V = \dim W$ *then*
$$T \text{ is an isomorphism if and only if } \operatorname{rank} T = n$$

In particular, in the third case T is an isomorphism if and only if it is a monomorphism or epimorphism only.

Example 2.5.7

Let $\Omega \subset \mathbb{R}^2$ be a domain and $\mathcal{P}_k(\Omega)$ be a space of all polynomials defined on Ω of order less than or equal to k. One can check that $\dim \mathcal{P}^k(\Omega) = (k+1)(k+2)/2$. Let $\Delta = \left(\frac{\partial^2}{\partial x^2} + \frac{\partial^2}{\partial y^2}\right)$ be the Laplacian operator. Obviously, Δ is linear and maps $\mathcal{P}^k(\Omega)$ into itself. Since the null space $\mathcal{N}(\Delta)$ is generated by monomials $(1, x, y, xy)$ and therefore $\dim \mathcal{N} = 4$, according to Theorem 2.5.1, $\dim \mathcal{R}(\Delta) = (k+1)(k+2)/2 - 4$. $\square$

Example 2.5.8

Let V be a finite-dimensional space, $\dim V = n$ and let M be a subspace of V, $\dim M = m < n$. Let V/M be the quotient space. Introduce the mapping

$$\iota : V \ni x \to [x] \in V/M$$

Obviously, ι is linear and its null space coincides with subspace M. Since ι is also surjective, we have $\dim V/M = \dim V - \dim M = n - m$ $\square$

2.6 Isomorphic Vector Spaces

One of the most fundamental concepts in abstract algebra is the idea of isomorphic spaces. If two algebraic structures are isomorphic (in a proper sense corresponding to the kind of structure considered) all algebraic properties of one structure are carried by the isomorphism to the second one and the two structures are *indistinguishable.* We shall frequently speak in this book about different isomorphic structures in the context of topological and algebraic properties.

Isomorphic Vector Spaces. Two vector spaces, X and Y, are said to be *isomorphic* if there exists an isomorphism $\iota : X \to Y$, from space X onto space Y.

To get used to this fundamental notion we will first study a series of examples.

Example 2.6.1

Let V be a finite-dimensional (real) space, $\dim V = n$ and let $\boldsymbol{a}_1, \dots, \boldsymbol{a}_n$ denote an arbitrary basis for V. Consider now the space $\mathbb{R}^n$ with the canonical basis $\boldsymbol{e}_i = (0, \dots, \underset{(i)}{1}, \dots, 0)$ and define a linear transformation ι by setting

$$\iota(\boldsymbol{e}_i) \stackrel{\text{def}}{=} \boldsymbol{a}_i, \quad i = 1, \dots, n$$

Consequently, if $\boldsymbol{x} = (x_1 \dots x_n) \in \mathbb{R}^n$, then for $\boldsymbol{v} = \iota(\boldsymbol{x})$

$$\boldsymbol{v} = \iota(\boldsymbol{x}) = \iota\left(\sum_1^n x_i \boldsymbol{e}_i\right) = \sum_1^n x_i \iota(\boldsymbol{e}_i) = \sum_1^n x_i \boldsymbol{a}_i$$

Thus $x_i, i = 1, \dots, n$ can be identified as components of vector $\boldsymbol{v}$ with respect to basis $\boldsymbol{a}_i$. The two spaces V and $\mathbb{R}^n$ are of the same dimension, the map ι is obviously surjective, and, therefore, according to Proposition 2.5.2 (iii), ι is an isomorphism as well.

Thus we have proved a very important assertion. *Every finite-dimensional (real) space V is isomorphic to $\mathbb{R}^n$*, where $n = \dim V$. Similarly, complex spaces are isomorphic to $\mathbb{C}^n$. The $\mathbb{R}^n$, frequently called the *model space* carries all linear properties of finite-dimensional vector spaces and for this reason many authors of text books on linear algebra of finite-dimensional spaces restrict themselves to spaces $\mathbb{R}^n$. $\square$

Example 2.6.2

Consider the space of free vectors in a plane. By choosing any two noncollinear vectors $\boldsymbol{a}_1, \boldsymbol{a}_2$ we can set an isomorphism from $\mathbb{R}^2$ into the space of free vectors. An image of a pair of two numbers(x_1, x_2) is identified with a vector $\boldsymbol{x}$ whose components with respect to $\boldsymbol{a}_1$ and $\boldsymbol{a}_2$ are equal to x_1 and x_2, respectively, (see Fig. 2.11). $\square$

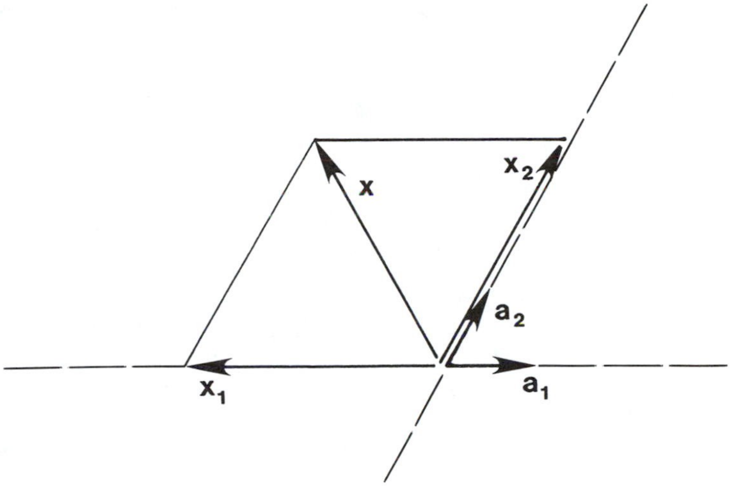

Figure 2.11
A system of coordinates in the space of free vectors.

Example 2.6.3

Let (a, b) be an interval in $\mathbb{R}$ and let $\mathcal{P}^k(a, b)$ denote the space of polynomials on (a, b) of order less than or equal to k. Since monomials $1, x, \ldots, x^k$ form a basis in $\mathcal{P}^k$, the space $\mathcal{P}^k$ is isomorphic to $\mathbb{R}^{k+1}$. An image of a vector $\lambda = (\lambda_0, \ldots, \lambda^k) \in \mathbb{R}^{k+1}$ is identified with a polynomial of the form

$$\lambda_0 + \lambda_1 x + \lambda_2 x^2 + \ldots + \lambda_k x^k$$

□

Example 2.6.4

Let V be a vector space and X its subspace. Denote by Y a complement of X and consider the quotient space V/X. Define a mapping $\iota : Y \to V/X$ as follows:

$$i : Y \ni \boldsymbol{y} \to [\boldsymbol{y}] = \boldsymbol{y} + X \in V/X$$

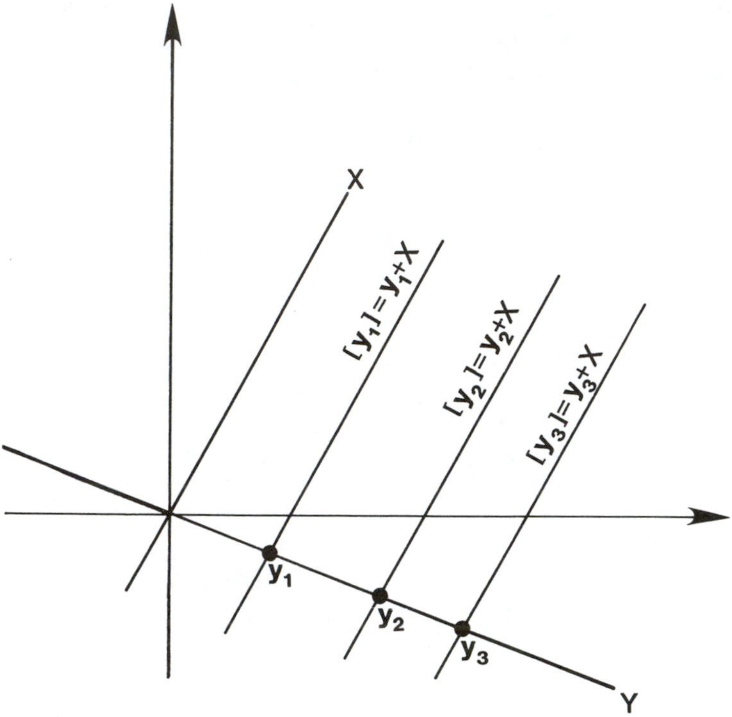

Figure 2.12
Concept of isomorphism between V/X and a complement $Y(X \oplus Y = V)$.

The map ι is trivially linear, is injective, since the only common element for X and Y is zero vector, and is surjective, since $V = X + Y$. Thus, the quotient space V/X is isomorphic to an arbitrary complement of X. The concept of isomorphism ι in context of $\mathbb{R}^2$ is illustrated in Fig. 2.12. □

Cartesian Products of Vector Spaces. Some of the isomorphisms are so natural that we hardly distinguish between the corresponding isomorphic vector spaces. For example, let X and Y be two vector spaces. One can easily verify that the Cartesian product $X \times Y$ is a vector space with the following operations:

$$(\boldsymbol{x}_1, \boldsymbol{y}_1) + (\boldsymbol{x}_2, \boldsymbol{y}_2) \stackrel{\text{def}}{=} (\boldsymbol{x}_1 + \boldsymbol{x}_2, \boldsymbol{y}_1 + \boldsymbol{y}_2)$$

$$\alpha(\boldsymbol{x}, \boldsymbol{y}) \stackrel{\text{def}}{=} (\alpha\boldsymbol{x}, \alpha\boldsymbol{y})$$

Consequently, one can consider the space of functions defined on a set Ω with values in $X \times Y$, the space $(X \times Y)^\Omega$. Similarly, one can consider first spaces of function X^Ω and Y^Ω and next their Cartesian product $X^\Omega \times Y^\Omega$. Spaces $(X \times Y)^\Omega$ and $X^\Omega \times Y^\Omega$ are different, but they are related by the natural isomorphism

$$\begin{aligned}\iota \ : \ (\Omega \ni x \to (u(x), v(x)) \in X \times Y) \to \\ (\Omega \ni x \to u(x) \in X; \Omega \ni x \to v(x) \in Y)\end{aligned}$$

The concept can be easily generalized for more than two vector spaces.

Example 2.6.5

Together with the membrane problem discussed in Example 2.2.3, one of the most common examples throughout this book will be that of boundary-value problems in linear elasticity. Consider, once again, a domain $\Omega \subset \mathbb{R}^n (n = 2 \text{ or } 3)$ with the boundary $\Gamma = \partial\Omega$ consisting of two disjoint parts Γ_u and Γ_t (see Fig. 2.13). The classical formulation of the problem is as follows:

Find $\boldsymbol{u} = \boldsymbol{u}(\boldsymbol{x})$, such that

$$\begin{aligned}-\text{div}\ \boldsymbol{\sigma}(\boldsymbol{u}) = \boldsymbol{X} \quad &\text{in } \Omega \\ \boldsymbol{u} = \boldsymbol{u}_0 \ &\text{on } \Gamma_u \\ \boldsymbol{t}(\boldsymbol{u}) = \boldsymbol{g} \ &\text{on } \Gamma_t\end{aligned}$$

where

$\boldsymbol{u}(\boldsymbol{x})$	is a displacement of point $\boldsymbol{x} \in \Omega$
$\boldsymbol{X}$	denotes body forces
$\boldsymbol{u}$ and $\boldsymbol{g}$	are prescribed displacements and tractions only
$\boldsymbol{\sigma}$	is the stress tensor (we shall consider precisely the notion of tensors later in this chapter) of the form

$$\sigma_{ij} = E_{ijkl}\varepsilon_{kl}$$

where E_{ijkl} denotes the tensor of elasticities and the infinitesimal strain tensor ε_{kl} is given by the formula

$$\varepsilon_{kl} = \frac{1}{2}(u_{k,l} + u_{l,k})$$

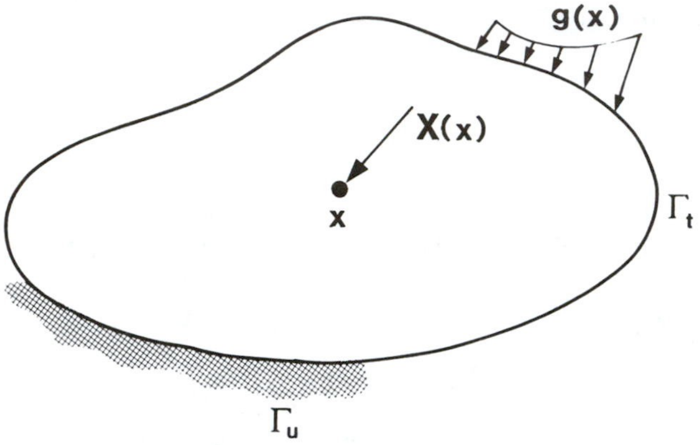

Figure 2.13
The classical problem of linear elasticity.

Finally, $\boldsymbol{t}(\boldsymbol{u})$ denotes the stress vector associated with displacement $\boldsymbol{u}$ by the formula

$$t_i = \sigma_{ij} n_j$$

where $n_j, j = 1, \ldots, n$ denote the components of the outward normal unit to $\partial\Omega$. In both formulas for σ_{ij} and t_i the standard summation convention has been used: repeated indices are summed throughout their ranges $1, 2, \cdots, n$.

One of the first steps toward a variational formulation of this problem is the definition of the set of kinematically admissible displacements

$$\boldsymbol{V} = \{\boldsymbol{u}(\boldsymbol{x}) : \boldsymbol{u} = \boldsymbol{u}_0 \text{ on } \Gamma_u\}$$

At this point we point out only two possible models for the displacement fields, corresponding to the discussion preceding this example. For every point $\boldsymbol{x} \in \Omega$ displacement $\boldsymbol{u}(\boldsymbol{x})$ is a vector in $\mathbb{R}^n$ and therefore $\boldsymbol{u}$ can be identified as a function on Ω with values in $\mathbb{R}^n$. With customary regularity assumptions, we shall write

$$\boldsymbol{u} \in \boldsymbol{C}(\overline{\Omega}) \subset (\mathbb{R}^n)^{\Omega}$$

where $\boldsymbol{C}(\overline{\Omega})$ denotes the space of all vector-valued functions continuous in the closure $\overline{\Omega}$ (recall Section 2.1). At the same time, we consider each of the components of $\boldsymbol{u}$ separately, regarding them as real functions defined on $\overline{\Omega}$, i.e., $u_i \in C(\overline{\Omega}) \subset \mathbb{R}^{\Omega}$ and, consequently, $\boldsymbol{u}$ may be considered as an element of the Cartesian product $C(\overline{\Omega}) \times \ldots \times C(\overline{\Omega})$ (n times). The two spaces $(\mathbb{R}^{\Omega})^n$ and $(\mathbb{R}^n)^{\Omega}$ are, however, isomorphic and in practice we do not distinguish between two constructions, writing simply

$$\boldsymbol{u} \in \boldsymbol{C}(\overline{\Omega})$$

□

We shall return to the notion of isomorphic spaces and their examples many times in this book, beginning already in Section 2.8.

2.7 More About Linear Transformations

In this section we intend to complete the fundamental facts about linear transformations.

Composition of Linear Transformations. Let U, V and W be vector spaces and let $T: U \to V$ and $S: V \to W$ denote two linear transformations from U and V into V and W, respectively. It follows from the definition of linear transformation that the *composition* (called also the *product*) of transformations $ST: U \to W$ defined by

$$ST(u) = S(T(u))$$

is also linear.

Let us note that only in the case when the three spaces coincide, i.e., $U = V = W$, does it makes sense to speak about both compositions ST and TS simultaneously. In general

$$ST \neq TS$$

i.e., composition of linear transformations is generally not commutative.

Inverse of a Linear Transformation. Let V and W be two vector spaces and $T: V \to W$ be an isomorphism, i.e., a bijective linear transformation. Then the inverse function $T^{-1}: W \to V$ is also linear. Indeed, let $\boldsymbol{w}_1$ and $\boldsymbol{w}_2$ denote two arbitrary vectors in W. T is bijective, so there exists vectors $\boldsymbol{v}_1$ and $\boldsymbol{v}_2$ such that $T(\boldsymbol{v}_1) = \boldsymbol{w}_1, T(\boldsymbol{v}_2) = \boldsymbol{w}_2$. We have

$$T^{-1}(\alpha_1\boldsymbol{w}_1 + \alpha_2\boldsymbol{w}_2) = T^{-1}(\alpha_1 T(\boldsymbol{v}_1) + \alpha_2 T(\boldsymbol{v}_2))$$

$$= T^{-1}(T(\alpha_1\boldsymbol{v}_1 + \alpha_2\boldsymbol{v}_2)) = \alpha_1\boldsymbol{v}_1 + \alpha_2\boldsymbol{v}_2$$

$$= \alpha_1 T^{-1}(\boldsymbol{w}_1) + \alpha_2 T^{-1}(\boldsymbol{w}_2)$$

so T^{-1} is linear.

Projection. We are familiar with the concept of a projection from elementary notions of geometry. For example, the projection of a directed line segment on a plane can be roughly visualized as the "shadow" it casts on the plane. For example, a film is used with a movie projector to project a three-dimensional image on a two-dimensional screen. In much the same way, we speak here of functions that project one linear space onto another or, more specifically, onto a subspace of possibly lower dimension. What is the essential feature of such projections? In the case of the shadow produced as the projection of a line, the shadow is obviously the image of itself; in other words, if P is a projection, and $P(v)$ is the image of a vector v and P, then the image of this image under P is precisely $P(v)$.

We make these concepts precise by introducing formally the following definition: A linear transformation P on a vector space V into itself is a *projection* if and only if

$$P^2 = P \circ P = P$$

i.e., if $P(\boldsymbol{v}) = \boldsymbol{w}$, then $P(\boldsymbol{w}) = P(P(\boldsymbol{v})) = P^2(\boldsymbol{v}) = \boldsymbol{w}$.

The following proposition shows that the definition does, in fact, imply properties of projections that are consistent with our intuitive ideas of projections.

PROPOSITION 2.7.1

The following conditions are equivalent:

(i) $T: V \to V$ *is a projection.*

(ii) *There exist subspaces* X *and* Y *such that* $V = X \oplus Y$, *and* $T(\boldsymbol{v}) = \boldsymbol{x}$, *where* $\boldsymbol{v} = \boldsymbol{x} + \boldsymbol{y}, \boldsymbol{x} \in X, \boldsymbol{y} \in Y$ *is the unique decomposition of* $\boldsymbol{v}$.

PROOF

(ii) $\to$ (i). Let $\boldsymbol{v} = \boldsymbol{x} + \boldsymbol{y}$ by the unique decomposition of a vector $\boldsymbol{v}$. Simultaneously, $\boldsymbol{x} = \boldsymbol{x} + \boldsymbol{0}$ is the unique decomposition of vector $\boldsymbol{x}$. We have

$$T^2(\boldsymbol{v}) = T(T(\boldsymbol{v})) = T(\boldsymbol{x}) = \boldsymbol{x} = T(\boldsymbol{v}), \text{ i.e., } T^2 = T$$

(i) $\rightarrow$ (ii). Define $X = \mathcal{R}(T), Y = \mathcal{N}(T)$. From the decompostion

$$\boldsymbol{v} = T(\boldsymbol{v}) + \boldsymbol{v} - T(\boldsymbol{v})$$

and the fact that $T(\boldsymbol{v} - T(\boldsymbol{v})) = T(\boldsymbol{v}) - T^2(\boldsymbol{v}) = \mathbf{0}$ follows that $V = X + Y$.

Suppose now that $\boldsymbol{v} \in \mathcal{R}(T) \cap \mathcal{N}(T)$. This implies that there exists $\boldsymbol{w} \in V$ such that $T(\boldsymbol{w}) \in \mathcal{N}(T)$, i.e., $T(T(\boldsymbol{w})) = \mathbf{0}$. But $T(T(\boldsymbol{w})) = T^2(\boldsymbol{w}) = T(\boldsymbol{w}) = \boldsymbol{v}$, so $\boldsymbol{v} = \mathbf{0}$, which proves the assertion. ∎

COROLLARY 2.7.1

Let X be an arbitrary subspace of V. There exists a (not unique) projection T such that $X = \mathcal{R}(T)$.

Example 2.7.1

Let V be the space of free vectors in a plane. Let X and Y denote two arbitary, different straight lines which can be identified with two one-dimensional subspaces of V. Obviously, $V = X \oplus Y$. Pick an arbitrary vector $\boldsymbol{v}$ and denote by $T(\boldsymbol{v})$ the classical projection along line X in the direction Y (recall Fig. 2.14). Obviously, $T^2 = T$. □

Example 2.7.2

Let $V = C(\overline{\Omega})$ for some bounded domain Ω in $\mathbb{R}^n$. Define $T: f \rightarrow Tf$ where Tf is a constant function given by the formula

$$Tf = \text{meas}(\Omega)^{-1} \int_{\Omega} f(x)dx$$

Obviously $T^2 = T$ and therefore T is a projection. For the interpretation of $\mathcal{N}(T)$ and $\mathcal{R}(T)$ see Example 2.2.6. □

Linear Transformations on Quotient Spaces. Let V, W be vector spaces and $T: V \rightarrow W$ denote an arbitrary linear transformation. Suppose

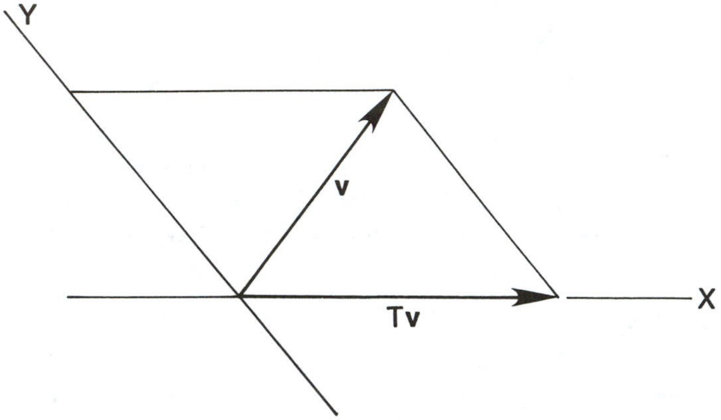

Figure 2.14
Example of a projection.

M is a linear subspace of V such that $M \subset \mathcal{N}(T)$. Define

$$\overline{T}: V/M \to W, \quad \overline{T}([\boldsymbol{v}]) = T(\boldsymbol{v})$$

Transformation $\overline{T}$ is well defined and linear. Indeed, let $\boldsymbol{v}, \boldsymbol{w} \in [\boldsymbol{v}]$. It implies that $\boldsymbol{v} - \boldsymbol{w} \in M$ and therefore $T(\boldsymbol{v}) - T(\boldsymbol{w}) = T(\boldsymbol{v} - \boldsymbol{w}) = \boldsymbol{0}$, so the definition of $\overline{T}$ is independent of the choice of $\boldsymbol{w} \in [\boldsymbol{v}]$. Linearity of $\overline{T}$ follows immediately from linearity of T.

COROLLARY 2.7.2

In the most common case we choose $M = \mathcal{N}(T)$. Then $\overline{T}$ becomes a monomorphism from V/M into Y.

Example 2.7.3

Let $V = X \oplus Y$ and let P denote the projection onto Y in the direction of X, i.e., $P\boldsymbol{v} = \boldsymbol{y}$, where $\boldsymbol{v} = \boldsymbol{x} + \boldsymbol{y}$ is the unique decomposition of vector $\boldsymbol{v}$. Obviously, P is surjective and $\mathcal{N}(T)$ coincides with X. Thus the quotient transformation $\overline{P}: V/X \to Y$ becomes an isomorphism and the two spaces

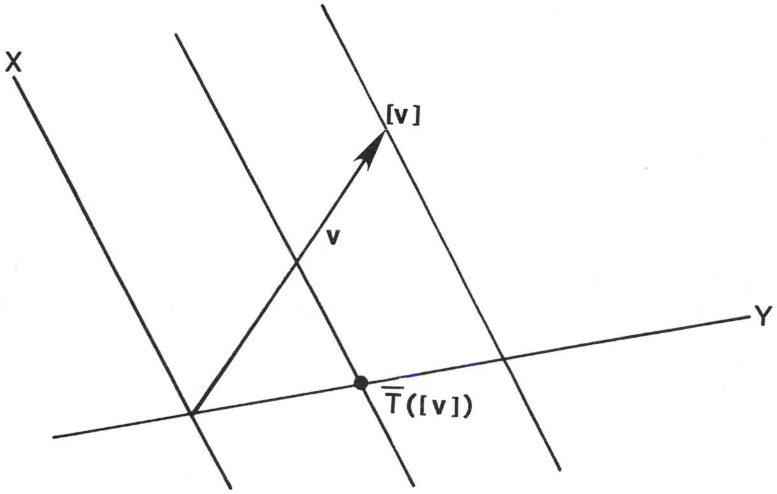

Figure 2.15
Identification of quotient space V/X with a complement Y through the quotient projection $\overline{T}$.

V/X and Y—the complement of X, are isomorphic. In other words, every equivalence class $[\boldsymbol{x}]$ can be identified with a common "point" of $[\boldsymbol{x}]$ and Y. The concept is illustrated in Fig. 2.15. $\square$

Example 2.7.4

Consider again the membrane problem (recall Examples 2.2.3 and 2.3.2) with $\Gamma_u = \emptyset$. Let $W = C^\infty(\overline{\Omega})$ be the space of all kinematically admissible displacements. Define the operator

$$L: C^\infty(\overline{\Omega}) \to C^\infty(\overline{\Omega}) \times C^\infty(\partial\Omega)$$

$$L(u) = \left(-\Delta u, \frac{\partial u}{\partial n}\right)$$

Obviously, the space of infinitesimal rigid body motions

$$M = \{u \in C^\infty(\overline{\Omega}) \,:\, u \doteq \text{ const in } \Omega\}$$

forms a subspace of kernel of operator L, i.e., $M \subset \mathcal{N}(L)$. Thus the quotient operator $\overline{L}$ is a well-defined, linear operator on the quotient space W/M. $\square$

Example 2.7.5

In much the same manner as in Example 2.7.4, we can define the quotient elasticity operator in the case of pure traction boundary conditions (Neumann problem). If $\Gamma_u = \emptyset$ then the space of kinematically admissible displacements V can be identified with the whole space $\boldsymbol{C}^\infty(\overline{\Omega})$. We define on V the operator

$$L: V \to \boldsymbol{C}^\infty(\overline{\Omega}) \times \boldsymbol{C}^\infty(\partial\Omega)$$

$$L\boldsymbol{u} = (-\mathrm{div}\boldsymbol{\sigma}(\boldsymbol{u}), \boldsymbol{t}(\boldsymbol{u}))$$

(see Example 2.6.5 for definitions of stress tensor $\boldsymbol{\sigma}$ and stress vector $\boldsymbol{t}$). Recall the definition of the space of infinitesimal rigid body motions (cf. Example 2.3.3).

$$M = \{\boldsymbol{u} \in \boldsymbol{C}^\infty(\overline{\Omega}) : \varepsilon_{ij}(\boldsymbol{u}) = 0\}$$

Obviously, $M \subset \mathcal{N}(L)$ and therefore the quotient operator $\overline{L}$ is well defined on the quotient space V/M. $\square$

The Space $L(X, Y)$ of Linear Transformations. We have already learned that for any set Ω, the set of functions defined on Ω with values in a vector space Y, Y^Ω forms a vector space. In a very particular case we can choose for Ω a vector space X and restrict ourselves to linear transformations only. A linear combination of linear transformations is linear as well, so the set of linear transformations from X to Y forms a linear subspace of Y^X. We denote this space by $L(X, Y)$ or shortly $L(X)$ if $X = Y$. In the case of $X = Y$ a new operation can be defined on $L(X)$ — the composition of transformation ST. With this extra operation the vector space $L(X)$ satisfied axioms of an algebraic structure called *linear algebra.*

Definition of Linear Algebra. A vector space V over the field $\mathbb{F}$ is called a linear algebra if to vector addition and multiplication by a scalar a new operation $\circ: V \times V \to V$ can be added such that the following axioms hold.

(i) $(\boldsymbol{x} \circ \boldsymbol{y}) \circ \boldsymbol{z} = \boldsymbol{x} \circ (\boldsymbol{y} \circ \boldsymbol{z})$ associative law

(ii) $(\boldsymbol{x} + \boldsymbol{y}) \circ \boldsymbol{z} = \boldsymbol{x} \circ \boldsymbol{z} + \boldsymbol{y} \circ \boldsymbol{z}$

(iii) $\boldsymbol{z} \circ (\boldsymbol{x} + \boldsymbol{y}) = \boldsymbol{z} \circ \boldsymbol{x} + \boldsymbol{z} \circ \boldsymbol{y}$ distributive laws

(iv) $(\alpha \boldsymbol{x}) \circ \boldsymbol{y} = \alpha(\boldsymbol{x} \circ y) = \boldsymbol{x} \circ (\alpha \boldsymbol{y})$

The first three axioms together with the axioms imposed on the vector addition + (cf. Section 2.1) indicate that with respect to operations + and $\circ$ V is a ring. Thus, roughly speaking, V is a linear algebra (or briefly an algebra) if V is simultaneously a vector space and a ring and the two structures are consistent in the sense that the condition (iv) holds.

Let us check now that the space $L(X)$ satisfies the axioms of linear algebra. Indeed, conditions (i) and (ii) and the first of equalities in (iv) hold for arbitrary functions, not necessarily linear. In other words, the composition of functions $f \circ g$ is always associative and behaves linearly with respect to the "external" function f. This follows directly from the definition of the composition of functions. To the contrary, to satisfy axioms (iii) and the second equality in (iv), we need linearity of f; more precisely, the composition $f \circ g$ is linear with respect to g if the function f is linear. Indeed, we have

$$(f \circ (\alpha_1 g_1 + \alpha_2 g_2))(\boldsymbol{x}) = f(\alpha_1 g_1(\boldsymbol{x}) + \alpha_2 g_2(\boldsymbol{x}))$$

$$= \alpha_1 f(g_1(\boldsymbol{x})) + \alpha_2 f(g_2(\boldsymbol{x}))$$

$$= (\alpha_1 (f \circ g_1) + \alpha_2 (f \circ g_2))(\boldsymbol{x})$$

if and only if function f is linear. Thus $L(X)$ is an algebra.

Let us finally note that conditions (i)–(iv) and the product of linear transformations itself make sense in a more general context of different vector spaces. For example, if X, Y and Z denote three different vector spaces and $f, g \in L(Y, Z)$ and $h \in L(X, Y)$ then

$$(f + g) \circ h = f \circ h + g \circ h$$

Of course, in the case of different spaces we cannot speak about the structure of linear algebra.

The algebraic theory of linear algebras is a separate subject in abstract algebra. We shall not study this concept further, restricting ourselves to the single example of the space $L(X)$. The main goal of introducing this definition is a better understanding of the next section which deals with matrices.

Exercises

2.7.1. Let V be a vector space and id_V the identity transformation on V.

Prove that a linear transformation $T: V \to V$ is a projection if and only if $\mathrm{id}_V - T$ is a projection.

2.8 Linear Transformations and Matrices

Most readers are probably familiar with the concept of matrix multiplication and other operations on matrices. In the most common treatment of this subject, especially in engineering literature, matrices are treated as tables or columns of objects on which certain simple algebraic operations can be defined. In this section we shall show the intimate relation between the algebra of matrices and that of linear transformations.

We have already discussed the concept of *isomorphic vector spaces*: two spaces X and Y are *isomorphic* if there exists an isomorphism, i.e., a linear and bijective transformation ι, from X into Y. So far the two spaces X and Y with their linear structures were given *a priori* and the bijection ι, when defined, had to be checked for linearity. One of the most fundamental concepts in abstract algebra is to *transfer* an algebraic structure from an *algebraic object* X onto another *set* Y through a bijection ι which becomes automatically an isomorphism. More precisely, let V be a vector space, W an arbitrary *set* and suppose that there exists a bijection ι from V onto W, i.e., we have one-to-one correspondence of *vectors* from V with *elements* of W.

We shall introduce operations in W in the following way:

$$w_1 + w_2 \stackrel{\text{def}}{=} \iota(\iota^{-1}(w_1) + \iota^{-1}(w_2))$$

$$\alpha w \stackrel{\text{def}}{=} \iota(\alpha\iota^{-1}(w))$$

In other words, in order to add two elements w_1 and w_2 in W we have to find their counterparts $v_1 = \iota^{-1}(w_1)$ and $v_2 = \iota^{-1}(w_2)$ in V first, then add v_1 to v_2 and next find the image of $v_1 + v_2$ through bijection ι. The concept is illustrated in Fig. 2.16. We interpret the multiplication by a scalar in the same way.

We leave for the reader the lengthy, but trivial, verification that W with such defined operations satisfies the axioms of a vector space. Moreover, it *follows from the definition of operations* in W that ι is *linear*. Thus V and W become two isomorphic vector spaces. If, additionally, V is a linear algebra we can transfer in the same way the multiplication from V defining

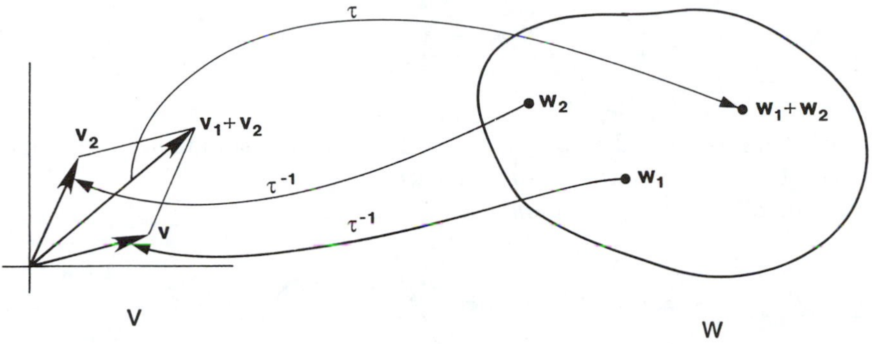

Figure 2.16
Transfer of vector addition through a bijection ι.

the multiplication in W by

$$w_1 \circ w_2 \stackrel{\text{def}}{=} \iota(\iota^{-1}(w_1) \circ \iota^{-1}(w_2))$$

Then W becomes a linear algebra too, and ι is an isomorphism of two algebras V and W.

For the rest of this section we shall assume that we deal with only finite-dimensional spaces. Studying linear transformations in Section 2.5, we have found that there exists a one-to-one correspondence between linear transformations and matrices. More precisely, if X and Y are two vector spaces, $\dim X = n, \dim Y = m$ and $(\boldsymbol{e}_1, \ldots, \boldsymbol{e}_n), (\boldsymbol{f}_1, \ldots, \boldsymbol{f}_m)$ denote bases in X and Y, respectively, then a transformation $T: X \to Y$ is linear if and only if T is of the form

$$T(\boldsymbol{x}) = \sum_{i=1}^{m} \left(\sum_{j=1}^{n} T_{ij} x_j \right) \boldsymbol{f}_i$$

i.e., if $\boldsymbol{y} = T(\boldsymbol{x})$ is a value of T for $\boldsymbol{x}$ and y_i denote the components of $\boldsymbol{y}$ with respect to basis $\boldsymbol{f}_i$, then

$$y_i = \sum_{j=1}^{n} T_{ij} x_j, \qquad i = 1 \ldots m$$

Recall that the j-th column of matrix T_{ij} indicate components of $T(\boldsymbol{e}_j)$

with respect to basis $\boldsymbol{f}_i$, i.e.,

$$T(\boldsymbol{e}_j) = \sum_{i=1}^{m} T_{ij}\boldsymbol{f}_i$$

We shall now use the bijection between linear transformations and matrices to define the operations on matrices.

Multiplication by a Scalar. Obviously,

$$(\alpha T)(\boldsymbol{e}_j) = \alpha(T(\boldsymbol{e}_j)) = \alpha \sum_{i=1}^{m} T_{ij}\boldsymbol{f}_i = \sum_{i=1}^{m}(\alpha T_{ij})\boldsymbol{f}_i$$

and therefore we define the product of a scalar α and matrix T_{ij} as a new matrix which is obtained by multiplying elements of T_{ij} by scalar α.

Matrix Addition. Similarly,

$$(T+R)(\boldsymbol{e}_j) = T(\boldsymbol{e}_j) + R(\boldsymbol{e}_j) = \sum_{i=1}^{m} T_{ij}\boldsymbol{f}_i + \sum_{i=1}^{m} R_{ij}\boldsymbol{f}_i = \sum_{i=1}^{m}(T_{ij}+R_{ij})\boldsymbol{f}_i$$

and, consequently, we add two matrices element by element.

Matrix Multiplication. Suppose we are given a third vector space Z with a basis $(\boldsymbol{g}_1 \ldots \boldsymbol{g}_l)$ and two linear transformations $T: X \to Y, R: Y \to Z$ with representations T_{ij} and R_{ki}, respectively. Let us denote $S = R \circ T$ and try to calculate the corresponding representation S_{kj}. We have

$$\sum_{k=1}^{\ell} S_{kj}\boldsymbol{g}_k = S(\boldsymbol{e}_j) = R(T(\boldsymbol{e}_j)) = R\left(\sum_{i=1}^{m} T_{ij}\boldsymbol{f}_i\right) = \sum_{i=1}^{m} T_{ij}R(\boldsymbol{f}_i)$$

$$= \sum_{i=1}^{m} T_{ij} \sum_{k=1}^{\ell} R_{ki}\boldsymbol{g}_k \quad = \sum_{k=1}^{\ell}\left(\sum_{i=1}^{m} R_{ki}T_{ij}\right)\boldsymbol{g}_k$$

and therefore by a direct comparison of both sides we get the product formula for matrices:

$$S_{kj} = \sum_{i=1}^{m} R_{ki}T_{ij}$$

Thus in order to multiply matrix T_{ij} by matrix R_{ki} we need to multiply rows of R_{ki} by columns of T_{ij}. The well-known formula gets its natural explanation.

According to our construction, the set of matrices $m \times n$ with operations defined above forms a vector space and in the case of square matrices ($m = n$) has a structure of linear algebra. The two spaces (algebras): the space (algebra) of linear transformations $L(X,Y)(L(X,X))$ and the space of matrices $m \times n$ (square matrices), become isomorphic. All notions and facts concerning transformations may be transferred to matrices, and, consequently, everything that is known for matrices may be reinterpreted in terms of corresponding transformations. We shall return to this one-to-one correspondence many times. For the beginning let us record a few fundamental facts.

Noncommutativity of Product of Transformations. It is easy to construct two square matrices A and B such that

$$\sum_{k=1}^{n} A_{ik}B_{kj} \neq \sum_{k=1}^{n} B_{ik}A_{kj}$$

(cf. Exercise 2.11.3) and therefore the multiplication of matrices is generally noncommutative. Consequently, product of transformations does not commute as well.

Rank of Matrix. Let $T: X \to Y$ be a linear transformation and T_{ij} the corresponding matrix. We define by the *rank of matrix* T_{ij}, the rank of corresponding transformation T, i.e., the dimension of the image space $\mathcal{R}(T)$. We have the following simple observation.

PROPOSITION 2.8.1

Rank of a matrix $T_{ij}, i = 1, \ldots, m, j = 1, \ldots, n$ is equal to the maximal number of linearly independent column vectors (treated as vectors in $\mathbb{R}^m$).

PROOF Obviously, rank of the corresponding transformation T equals the maximal number of linearly independent vectors $T(\boldsymbol{e}_j)$, where $\boldsymbol{e}_j$ denotes a basis in vector space X. Since the column vectors are precisely the components of $T(\boldsymbol{e}_j)$ with respect to basis $\boldsymbol{f}_i$ and $\mathbb{R}^m$ is isomorphic with Y through any basis ($\boldsymbol{f}_i$ in particular), the number of linearly independent column vectors must be precisely equal to the number of linearly independent vectors $T(\boldsymbol{e}_j), j = 1, \ldots, n$. ∎

Inverse of a Matrix. Let $\dim X = \dim Y = n$ and let T be an isomorphism, i.e., a one-to-one linear transformation from X to Y. Let T_{ij} be a representation of transformation T with respect to bases $\boldsymbol{e}_j$ and $\boldsymbol{f}_i$. Since T is invertible, we may speak about a representation T^{-1}_{ji} of its inverse T^{-1}. Matrix T^{-1}_{ji} is called the *inverse matrix* of matrix T_{ij}. According to the product formula for matrix multiplication, we have, equivalently,

$$\sum_{k=1}^{n} T_{ik} T_{kj}^{-1} = \sum_{k=1}^{n} T_{jk}^{-1} T_{ki} = \delta_{ij}$$

which follows from the definition of the inverse transformation

$$TT^{-1} = \text{id}_Y, \quad T^{-1}T = \text{id}_X$$

and the fact that the matrix representation for the identity transformation (in every space, with respect to any basis) can be visualized as the Kronecker symbol.

2.9 Solvability of Linear Equations

One of the fundamental problems following the concept of the linear transformation is that of solvability of linear equations. Suppose we are given spaces X and Y and a linear transformation $T: X \to Y$. For a given vector $\boldsymbol{y} \in Y$ we may ask two fundamental questions:

(i) Does an element $\boldsymbol{x} \in X$ exist, such that

$$T\boldsymbol{x} = \boldsymbol{y}$$

(ii) Is such an element unique?

The above equation is called a linear equation for $\boldsymbol{x}$ and the two questions deal with problems of *existence* and *uniqueness* of solutions for one specific "right-hand" $\boldsymbol{y}$ or for every $\boldsymbol{y} \in Y$. Let us record some simple observations:

(i) If T is an isomorphism, then there exists a unique solution $\boldsymbol{x} = T^{-1}\boldsymbol{y}$ for an arbitrary vector $\boldsymbol{y}$.

(ii) For a given $\boldsymbol{y}$ there exists a solution $\boldsymbol{x}$ if and only if $\boldsymbol{y} \in \mathcal{R}(T)$.

(iii) A solution $\boldsymbol{x}$ is unique if and only if T is injective or, equivalently, $\mathcal{N}(T) = \{\boldsymbol{0}\}$.

The trivial observations gain their important interpretation in context of finite-dimensional spaces. More precisely, if $\boldsymbol{e}_j, j = 1, \ldots, n, \boldsymbol{f}_i, i = 1, \ldots, m$

denote bases in X and Y, respectively, T_{ij} is the matrix representation of T, the linear equation $T\boldsymbol{x} = \boldsymbol{y}$ is equivalent to the system of m linear algebraic equations of n unknowns in the form

$$\begin{cases} T_{11}x_1 + T_{12}x_2 + \ldots + T_{1n}x_n = y_1 \\ . \\ . \\ . \\ T_{m1}x_1 + T_{m2}x_2 + \ldots + T_{mn}x_n = y_m \end{cases}$$

Let us discuss some particular cases:

1. *Number of equations equals number of unknowns,* $m = n$. If matrix T_{ij} is nonsingular, i.e., transformation T is an isomorphism (which is equivalent to saying that $\det T_{ij} \neq 0$) the system of equations possesses a unique solution for an arbitrary right-hand side vector $\boldsymbol{y}_i, i = 1, \ldots, m$. In particular, for a homogeneous system of equations (zero right-hand side) the only solution is the trivial, zero vector.
If matrix T_{ij} is singular then $\mathcal{N}(T) \neq \{\boldsymbol{0}\}$ and a solution, if it exists, is *never unique*. Since $\dim X = n = \dim \mathcal{N}(T) + \dim \mathcal{R}(T)$, $\dim \mathcal{R}(T) < n$ and, consequently, $\mathcal{R}(T) \not\subseteq Y$, which implies that the system of equations has a solution only for some right-hand side vectors $\boldsymbol{y}$.
A necessary and sufficient condition for existence of solutions may be formulated using the notion of the rank of matrix. Toward this goal let us note that the range space $\mathcal{R}(T)$ is generated by vectors $T(\boldsymbol{e}_j), j = 1, \ldots, n$ and, consequently, the system has a solution if and only if vector $\boldsymbol{y}$ belongs to $\mathcal{R}(T)$ or, equivalently, the dimension of a space generated by both vectors $T(\boldsymbol{e}_j), j = 1, \ldots, n$ and the vector $\boldsymbol{y}$ equals dimension of $\mathcal{R}(T)$, i.e., the rank of T. This is equivalent to saying that rank of matrix T_{ij} must be equal to the rank of the so-called *augmented matrix*, i.e., matrix T_{ij} with added vector y_i:
$$\begin{pmatrix} T_{11}, \ldots, T_{1n}, y_1 \\ \\ T_{m1}, \ldots, T_{mn}, y_m \end{pmatrix}$$

2. *Number of equations is smaller than number of unknowns.* From the fundamental identity $n = \dim \mathcal{N}(T) + \dim \mathcal{R}(T)$ it follows that such a system must be $\dim \mathcal{N}(T) > 0$ and therefore such a system *never* has a unique solution. Again, ranks of the matrix T_{ij} and its

augmented counterpart can be compared to determine whether a solution exists.

3. *Number of equations is bigger than number of unknowns.* Again, it follows from the fundamental identity that the range space must be a proper subspace of Y, i.e., a solution exists only for some right-hand sides y_i. For a verification we may once again compare the ranks of matrix T_{ij} and T_{ij} augmented.

Exercises

2.9.1. **Equivalent and Similar Matrices** Given matrices $\boldsymbol{A}$ and $\boldsymbol{B}$, when nonsingular matrices $\boldsymbol{P}$ and $\boldsymbol{Q}$ exist such that

$$\boldsymbol{B} = \boldsymbol{P}^{-1}\boldsymbol{A}\boldsymbol{Q}$$

we say that the *matrices* $\boldsymbol{A}$ and $\boldsymbol{B}$ are *equivalent.* If $\boldsymbol{B} = \boldsymbol{P}^{-1}\boldsymbol{A}\boldsymbol{P}$, we say $\boldsymbol{A}$ and $\boldsymbol{B}$ are *similar.*

Let $\boldsymbol{A}$ and $\boldsymbol{B}$ be similar $n \times n$ matrices. Prove that $\det \boldsymbol{A} = \det \boldsymbol{B}$, $r(\boldsymbol{A}) = r(\boldsymbol{B})$, $n(\boldsymbol{A}) = n(\boldsymbol{B})$.

2.9.2. Let S and T be similar linear transformations on n-dimensional vector spaces. Prove that the ranks $r(S)$ and $r(T)$ are the same.

2.9.3. Let T_1 and T_2 be two different linear transformations from an n-dimensional linear vector space $\mathcal{U}$ into itself. Prove that T_1 and T_2 are represented relative to two different bases by the *same* matrix if and only if there exists a nonsingular transformation $\mathcal{Q}$ on $\mathcal{U}$ such that $T_2 = \mathcal{Q}^{-1}T_1\mathcal{Q}$.

2.9.4. Let T be a linear transformation represented by the matrix

$$\boldsymbol{A} = \begin{bmatrix} 1 & -1 & 4 \\ 0 & 3 & 2 \end{bmatrix}$$

relative to bases $\{\boldsymbol{a}_1, \boldsymbol{a}_2\}$ of $\mathbb{R}^2$ and $\{\boldsymbol{b}_1, \boldsymbol{b}_2, \boldsymbol{b}_3\}$ of $\mathbb{R}^3$. Compute the matrix representing T relative to the new bases:

$$\begin{aligned} \boldsymbol{\alpha}_1 &= 4\boldsymbol{a}_1 - \boldsymbol{a}_2 & \boldsymbol{\beta}_1 &= 2\boldsymbol{b}_1 - \boldsymbol{b}_2 + \boldsymbol{b}_3 \\ \boldsymbol{\alpha}_2 &= \boldsymbol{a}_1 + \boldsymbol{a}_2 & \boldsymbol{\beta}_2 &= \boldsymbol{b}_1 - \boldsymbol{b}_3 \\ & & \boldsymbol{\beta}_3 &= \boldsymbol{b}_1 + 2\boldsymbol{b}_2 \end{aligned}$$

2.9.5. Let $\boldsymbol{A}$ be an $n \times n$ matrix. Show that transformations which

(a) interchange rows or columns of $\boldsymbol{A}$

(b) multiply any row or column of $\boldsymbol{A}$ by a scalar $\neq 0$

(c) add any multiple of a row or column to a parallel row or column

produce a matrix with the same rank as $\boldsymbol{A}$.

2.9.6. Let $\mathcal{V}^3$ be the space of ordered triples such that $(v_1, v_2, v_3) \in \mathcal{V}^3$. Consider the two bases

$$\begin{aligned} \boldsymbol{v}_1 &= (1,0,0), \quad \boldsymbol{v}_2 = (0,1,0), \quad \boldsymbol{v}_3 = (0,0,1) \\ \bar{\boldsymbol{v}}_1 &= (1,-1,0), \bar{\boldsymbol{v}}_2 = (0,2,-1), \bar{\boldsymbol{v}}_3 = (0,1,2) \end{aligned}$$

If $\boldsymbol{A}$ is a matrix relative to $\boldsymbol{v}_1, \boldsymbol{v}_2, \boldsymbol{v}_3$, and $\widetilde{\boldsymbol{A}}$ is a similar matrix relative to $\bar{\boldsymbol{v}}_1, \bar{\boldsymbol{v}}_2, \bar{\boldsymbol{v}}_3$, construct $\widetilde{\boldsymbol{A}}$ if

$$\boldsymbol{A} = \begin{bmatrix} 1 & -1 & 0 \\ 3 & 2 & 1 \\ 1 & 0 & 1 \end{bmatrix}$$

2.9.7. Let $\{\boldsymbol{e}_1, \boldsymbol{e}_2\}$ and $\{\boldsymbol{i}_1, \boldsymbol{i}_2\}$ be two bases for $\mathcal{U} = \mathbb{R}^2 = \mathbb{R} \times \mathbb{R}$, where $\boldsymbol{e}_1 = (-1,2), \boldsymbol{e}_2 = (0,3)$, and $\boldsymbol{i}_1 = (1,0), \boldsymbol{i}_2 = (0,1)$. Let $T: \mathcal{U} \to \mathcal{U}$ be given by $T(x,y) = (3x - 4y, x + y)$. Find the matrices for T for each choice of basis and show that these matrices are similar. Then, if $\boldsymbol{u} = (5,2)$, find the column vector $\boldsymbol{v} = T(\boldsymbol{u})$ for each choice of basis.

Algebraic Duals

2.10 The Algebraic Dual Space, Dual Basis

Let V denote a vector space over a field $\mathbb{F}$. A mapping f from V into $\mathbb{F}$ is called a *functional* on V. If f is additionally linear ($\mathbb{F}$ is obviously a vector space over itself) we speak of a *linear functional* on V.

The Algebraic Dual Space. We recall that if W is a vector space and A an arbitrary set then the space of functions defined on A with values in

W, denoted W^A, forms a new vector space. Since $\mathbb{F}$ is a vector space over itself, the space of linear functionals on $V, L(V, \mathbb{F})$ is a vector space, too. This vector space is denoted V^* and is called the *algebraic dual* of V.

Example 2.10.1

A familiar example of a linear functional is found in connection with the space of all real-valued and continuous functions in the closure of a bounded domain $\Omega \subset \mathbb{R}^n$, the space $C(\overline{\Omega})$. Since each such a function is Riemann integrable, it makes sense to define a linear transformation T in the form

$$T(f) = \int_\Omega g(x) f(x) dx$$

where g is a given function from $C(\overline{\Omega})$. Clearly, T is a linear functional on $C(\overline{\Omega})$. $\square$

Example 2.10.2

Consider again the space $C(\overline{\Omega})$. Pick an arbitrary point $\boldsymbol{x}_\circ \in \Omega$ and define a functional

$$\delta_{\boldsymbol{x}_\circ}(f) = f(\boldsymbol{x}_\circ)$$

This functional, called *Dirac's functional* or shortly *Dirac's delta* at point $\boldsymbol{x}_\circ$, plays a fundamental role in the theory of distributions.

More generally, given a finite sequence of points $\boldsymbol{x}_j \in \Omega, j = 1, \ldots, m$, we can define a linear functional on $C(\overline{\Omega})$ in the form

$$T(f) = \sum_{j=1}^{m} \alpha_j f(\boldsymbol{x}_j)$$

Obviously, T is a linear combination of the corresponding Dirac deltas at points $\boldsymbol{x}_j$. $\square$

Example 2.10.3

The Dirac functional can be "applied" to a function or to its derivatives as well. Consider, for instance, the space $C^k(\overline{\Omega})$, a point $\boldsymbol{x}_\circ \in \overline{\Omega}$ and a multiindex $\boldsymbol{\alpha} \in \mathbb{Z}^n$, $|\boldsymbol{\alpha}| = k$. The following is a linear functional on $C^k(\overline{\Omega})$:

$$T(f) = D^{\boldsymbol{\alpha}} f(\boldsymbol{x}_\circ) \quad \square$$

The General Form of a Linear Functional in Finite-Dimensional Spaces. Let V be a finite-dimensional space, $\dim V = n$, with a basis $(\boldsymbol{e}_1, \ldots, \boldsymbol{e}_n)$ and consider an arbitrary linear functional $f \in V^*$. We have

$$f(\boldsymbol{v}) = f\left(\sum_{i=1}^{n} v_i \boldsymbol{e}_i\right) = \sum_{i=1}^{n} v_i f(\boldsymbol{e}_i) = \sum_{i=1}^{n} f_i v_i$$

where $f_i \stackrel{\text{def}}{=} f(\boldsymbol{e}_i)$. Conversely, any functional of this form is linear and therefore the formula above constitutes the general form of a linear functional in finite-dimensional space V. The whole information about the functional is stored in the sequence of n numbers, f_i being equal to the values of f on arbitrary basis $\boldsymbol{e}_i$.

As a particular choice for every $j = 1, \ldots, n$ we can define a linear functional $\boldsymbol{e}_j^*$ by setting

$$\boldsymbol{e}_j^*(\boldsymbol{e}_i) = \delta_{ij}$$

Consequently,

$$\boldsymbol{e}_j^*(\boldsymbol{v}) = \boldsymbol{e}_j^*\left(\sum_{i=1}^{n} v_i \boldsymbol{e}_i\right) = \sum_{i=1}^{n} v_i \boldsymbol{e}_j^*(\boldsymbol{e}_i) = \sum_{1}^{n} v_i \delta_{ij} = v_j$$

and $\boldsymbol{e}_j^*$ can be interpreted as a functional prescribing for a vector $\boldsymbol{v}$ its j-th component with respect to basis $\boldsymbol{e}_i$.

We have the following simple observation.

PROPOSITION 2.10.1

Functionals $\boldsymbol{e}_j^*$ form a basis in dual space V^*.

PROOF Indeed, every functional $f \in V^*$ can be represented in the form

$$f(\boldsymbol{v}) = \sum_{1}^{n} v_i f_i = \sum_{1}^{n} \boldsymbol{e}_i^*(\boldsymbol{v}) f_i$$

i.e.,

$$f = \sum_{1}^{n} f_i \boldsymbol{e}_i^*$$

and therefore $\boldsymbol{e}_i^*$ span V. To prove the linear independence consider a linear combination

$$\sum_1^n \alpha_j \boldsymbol{e}_j^* = \boldsymbol{0}$$

We have

$$0 = \left(\sum_1^n \alpha_j \boldsymbol{e}_j^* \right) (\boldsymbol{e}_i) = \sum_1^n \alpha_j \boldsymbol{e}_j^*(\boldsymbol{e}_i) = \sum_1^n \alpha_j \delta_{ij} = \alpha_i$$

for every $i = 1, \ldots, n$, which proves that $\boldsymbol{e}_j^*$ are linearly independent. ∎

Basis $\{\boldsymbol{e}_j^*\}_{j=1}^n$ is called the *dual (reciprocal, biorthogonal)* basis to basis $\{\boldsymbol{e}_i\}_{i=1}^n$. In particular, Proposition 2.10.1 implies that $\dim V^* = \dim V = n$. As two finite-dimensional spaces of the same dimension, V and its dual V^* are isomorphic and each linear functional could be identified with a specific element in V. Such an identification, though possible, is not unique (it depends on a particular choice of a basis). By choosing to distinguish V^* from V we will have a chance to uncover a variety of interesting properties.

Bilinear Functionals. Given two vector spaces, V and W, we may consider a functional a of two variables

$$a : V \times W \to \mathbb{F}(\mathbb{R} \text{ or } \mathbb{C}), a(v, w) \in \mathbb{F}$$

Functional a is called bilinear if it is linear with respect to each of the variables separately. In the case when $V = W$ we speak of bilinear functionals defined on V. The notion can be easily generalized to the case of multilinear (m-linear) functionals defined on $V_1 \times \ldots \times V_m$ or in the case $V_1 = V_2 = \ldots = V_m$, simply on V.

Example 2.10.4

Consider again the space $C^1(\overline{\Omega})$. The following are examples of bilinear functionals.

$$a(u, v) = \int_\Omega \sum_{i,j=1}^n a_{ij}(\boldsymbol{x}) D^i u(\boldsymbol{x}) D^j v(\boldsymbol{x}) d\boldsymbol{x}$$

$$+ \int_\Omega \sum_{i=1}^n b_i(\boldsymbol{x}) D^i u(\boldsymbol{x}) d\boldsymbol{x} + \int_\Omega c(\boldsymbol{x}) u(\boldsymbol{x}) v(\boldsymbol{x}) d\boldsymbol{x}$$

$$a(u, v) = u(\boldsymbol{x}_\circ) v(\boldsymbol{x}_\circ)$$

□

The General Form of a Bilinear Functional in Finite-Dimensional Spaces. Let V and W be two finite-dimensional spaces, $\dim V = n$, $\dim W = m$ and $(\boldsymbol{e}_1, \ldots, \boldsymbol{e}_n)$ and $(\boldsymbol{g}_1, \ldots, \boldsymbol{g}_m)$ denote two bases in V and W, respectively. Let $a: V \times W \to \mathbb{R}$ denote an arbitrary bilinear functional. Representing vectors $\boldsymbol{v} \in V$ and $\boldsymbol{w} \in W$ in the bases, we get

$$a(\boldsymbol{v}, \boldsymbol{w}) = a\left(\sum_1^n v_i \boldsymbol{e}_i, \boldsymbol{w}\right) = \sum_1^n v_i a(\boldsymbol{e}_i, \boldsymbol{w})$$

$$= \sum_1^n v_i a\left(\boldsymbol{e}_i, \sum_1^m w_j \boldsymbol{g}_j\right) = \sum_1^n v_i \sum_1^m w_j a(\boldsymbol{e}_i, \boldsymbol{g}_j)$$

Conversely, setting arbitrary values of a functional a on bases $\boldsymbol{e}_i, \boldsymbol{g}_j$ we easily check that the functional of the form above is bilinear. Thus, introducing notation

$$a_{ij} = a(\boldsymbol{e}_i, \boldsymbol{g}_j)$$

we get the *representation formula for bilinear functionals in finite-dimensional spaces:*

$$a(\boldsymbol{v}, \boldsymbol{w}) = \sum_{i=1}^n \sum_{j=1}^n a_{ij} v_j w_j$$

Space of Bilinear Functionals $M(\boldsymbol{X}, \boldsymbol{Y})$. Given two vector spaces X and Y, we denote the set of all bilinear functionals defined on $X \times Y$ by $M(X, Y)$. Obviously, a linear combination of bilinear functionals is bilinear as well and therefore $M(X, Y)$ is a vector space.

Duality Pairing. Given a vector space V and its dual V^*, we define the *duality pairing* as the functional

$$V^* \times V \ni (f, \boldsymbol{v}) \to \ < f, \boldsymbol{v} > \ \stackrel{\text{def}}{=} f(\boldsymbol{v}) \in \mathbb{R}(\mathbb{C})$$

It follows from the definition of function addition that the duality pairing is linear with respect to the first variable and from linearity of functionals f that it is linear with respect to the second variable. Thus the duality pairing is a bilinear functional on $V^* \times V$. Two easy properties follow from the definition.

PROPOSITION 2.10.2

The following properties hold:

(i) $< f, \boldsymbol{v} > \ = \ 0 \quad \forall \quad \boldsymbol{v} \in V$ implies $f = \boldsymbol{0}$.

(ii) $< f, \boldsymbol{v} > \ = \ 0 \quad \forall \quad f \in V^*$ implies $\boldsymbol{v} = \boldsymbol{0}$.

PROOF

(i) follows trivially from the definition of duality pairing. To prove (ii) assume, to the contrary, that there exists a $\boldsymbol{v} \neq \boldsymbol{0}$ such that $< f, \boldsymbol{v} > = \boldsymbol{0}$ for every $f \in V^*$. Consider a direct sum

$$V = \mathbb{R}v \oplus W$$

where W is a complement of one-dimensional subspace $\mathbb{R}v$. Setting $f(v) = 1, f \equiv 0$ on W, we extend f by linearity

$$f(\alpha v + w) = \alpha f(v) + f(w) = \alpha$$

and therefore f is a well-defined linear functional on V. Obviously, $f(\boldsymbol{v}) \neq 0$, which contradicts the assumption. ∎

Orthogonal Complement. Let U be subspace of a vector space V. Due to bilinearity of duality pairing the set

$$\{\boldsymbol{v}^* \in V^* : < \boldsymbol{v}^*, \boldsymbol{v} > = 0 \text{ for every } \boldsymbol{v} \in U\}$$

is a linear subspace of V^*. We call it the orthogonal complement of U and denote it by $U^\perp$. Thus

$$< \boldsymbol{v}^*, \boldsymbol{v} > = 0 \text{ for every } \boldsymbol{v}^* \in v^\perp, \boldsymbol{v} \in U$$

Similarly, if W is a subspace of V^*, the set (space) of all such vectors $\boldsymbol{v} \in V$ that

$$< \boldsymbol{v}^*, \boldsymbol{v} > = 0 \text{ for every } \boldsymbol{v}^* \in W$$

denoted $W^\perp$, is called the orthogonal complement of W. In other words, if $W \subset V^*$ is the orthogonal complement of $U \subset V$ then U is the orthogonal complement of W.

We leave as an exercise proof of the following proposition.

PROPOSITION 2.10.3

Let U be a subspace of a finite-dimensional vector space $V, \dim V \doteq n$. Then the following hold:

(i) $\dim U^\perp = n - \dim U$.

(ii) *If $V = U \oplus W$ then $V^* = W^\perp \oplus U^\perp$.*

Bidual Space. Having defined the dual space V^*, we are tempted to proceed in the same manner and define the (*algebraic*) *bidual* as the dual of the dual space, i.e.,

$$V^{**} \stackrel{\text{def}}{=} (V^*)^*$$

Proceeding in the same way, we could introduce the "three stars," "four stars," etc., spaces. This approach, though theoretically possible, has (fortunately!) only a little sense in the case of finite-dimensional spaces, since it turns out that the bidual space V^{**} is in a natural way isomorphic with the space V.

PROPOSITION 2.10.4

The following map is an isomorphism between a finite-dimensional vector space V and its bidual V.

$$\iota: V \ni \boldsymbol{v} \to \{V^* \ni \boldsymbol{v}^* \to < \boldsymbol{v}^*, \boldsymbol{v} > \in \mathbb{R}(\mathbb{C})\} \in V^{**}$$

PROOF Due to linearity of v^* the functional

$$V^* \ni v^* \to < \boldsymbol{v}^*, \boldsymbol{v} > \in \mathbb{R}(\mathbb{C})$$

(called the *evaluation* at $\boldsymbol{v}$) is linear and therefore map ι is well defined. Checking for linearity, we have

$$\{\boldsymbol{v}^* \to < \boldsymbol{v}^*, \alpha_1 \boldsymbol{v}_1 + \alpha_2 \boldsymbol{v}_2 >\} = \alpha_1 \{\boldsymbol{v}^* \to < \boldsymbol{v}^*, v_1 >\} + \alpha_2 \{\boldsymbol{v}^* \to < \boldsymbol{v}^*, \boldsymbol{v}_2 >\}.$$

Map ι is also injective since $< \boldsymbol{v}^*, \boldsymbol{v} > = 0$ for every $\boldsymbol{v}^* \in V^*$ implies that (Proposition 1.10.2) $\boldsymbol{v} = 0$. Since all the spaces are of the same dimension this implies that ι is also surjective, which ends the proof. ∎

Thus, according to Proposition 2.10.4 in the case of a finite-dimensional space V, we identify bidual V^{**} with the original space V, "tridual" V^{***} with dual V^*, etc. The two spaces V and its dual V^* (the "lone star space") with the duality pairing $< \boldsymbol{v}^*, \boldsymbol{v} >$ are treated symmetrically.

2.11 Transpose of a Linear Transformation

Transpose of a Linear Transformation. Let V and W be two vector spaces over the same field $\mathbb{F}$ and T denote an arbitrary linear transformation from V into W. Denoting by W^* and V^* algebraic duals to W and V, respectively, we introduce a new transformation T^T from the algebraic dual W^* into algebraic dual V^* defined as follows:

$$T^T: W^* \to V^*, \quad T^T(\boldsymbol{w}^*) = \boldsymbol{w}^* \circ T$$

Transformation T^T is well defined, i.e., composition $\boldsymbol{w}^* \circ T$ defines (due to linearity of both T and functional $\boldsymbol{w}^*$) a linear functional on V. T^T is also linear. Transformation T^T is called the *transpose* of the linear transformation T. Using the duality pairing notation, we may express the definition of the transpose in the equivalent way:

$$\langle T^T \boldsymbol{w}^*, \boldsymbol{v} \rangle = \langle \boldsymbol{w}^*, T\boldsymbol{v} \rangle \quad \forall\, \boldsymbol{v} \in V, \boldsymbol{w}^* \in W$$

Let us note that the transpose T^T acts in the opposite direction to T; it maps dual W^* into dual V^*. We may illustrate this by the following simple diagram:

$$V \xrightarrow{T} W$$

$$V^* \xleftarrow{T^T} W^*$$

A number of algebraic properties of transpose transformations are easily proved:

PROPOSITION 2.11.1

(i) *Let T and $S \in L(V,W)$. Then*

$$(\alpha T + \beta S)^T = \alpha T^T + \beta S^T$$

(ii) *If $T \in L(U,V)$ and $S \in L(V,W)$, then*

$$(S \circ T)^T = T^T \circ S^T$$

(iii) *If id_V denotes the identity transformation on V, then*

$$(\mathrm{id}_V)^T = \mathrm{id}_{V^*}$$

where id_{V^*} is the identity transformation on V^*.

(iv) *Let $T \in L(V, W)$ be an isomorphism, i.e., T^{-1} exists. Then $(T^T)^{-1}$ exists too, and*

$$(T^T)^{-1} = (T^{-1})^T$$

PROOF The first three assertions follow immediately from the definition of the transpose. For instance, to prove the second one, we need to notice that

$$(S \circ T)^T(\boldsymbol{w}^*) = \boldsymbol{w}^* \circ S \circ T = (\boldsymbol{w}^* \circ S) \circ T = (S^T \boldsymbol{w}^*) \circ T$$

$$= T^T(S^T \boldsymbol{w}^*) = (T^T \circ S^T)(\boldsymbol{w}^*)$$

From the second and third statements it follows that

$$\mathrm{id}_{V^*} = (\mathrm{id}_V)^T = (T^{-1} \circ T)^T = T^T \circ (T^{-1})^T$$

and, similarly,

$$\mathrm{id}_{W^*} = (T^{-1})^T \circ T^T$$

which proves that $(T^T)^{-1}$ exists and is equal to $(T^{-1})^T$. ∎

In the case of finite-dimensional spaces we have the following:

PROPOSITION 2.11.2

Let $T \in L(X, Y)$ and assume that Y is finite-dimensional. Then

$$\mathrm{rank}\ T = \mathrm{rank}\ T^T$$

PROOF One can always choose such a basis in Y, say $\boldsymbol{b}_1, \dots, \boldsymbol{b}_k, \boldsymbol{b}_{k+1}, \dots, \boldsymbol{b}_m$, that the first k-elements constitute a basis for the range space $\mathcal{R}(T)$. Next, let us consider the null space of the transpose T^T. We have

$$\mathcal{N}(T^T) = \{\boldsymbol{y}^* \in Y^* : T^T(\boldsymbol{y}^*) = \boldsymbol{0}\} = \{\boldsymbol{y}^* \in Y^* : \boldsymbol{y}^* \circ T = \boldsymbol{0}\}$$

We shall prove that $\dim \mathcal{N}(T^T) = m - k$. Toward this goal consider a mapping

$$\iota \colon \mathcal{N}(T^T) \ni \boldsymbol{y}^* \longrightarrow (\boldsymbol{y}^*(\boldsymbol{b}_{k+1}), \dots, \boldsymbol{y}^*(\boldsymbol{b}_m)) \in \mathbb{R}^{m-k}$$

Every functional $\boldsymbol{y}^* \in \mathcal{N}(T^T)$ vanishes on $\mathcal{R}(T)$ and therefore is uniquely determined by its values on the vectors $\boldsymbol{b}_{k+1}, \dots, \boldsymbol{b}_m$. This proves that the mapping ι is an injection. Moreover, by setting arbitrary values of a functional $\boldsymbol{y}^*$ on vectors $\boldsymbol{b}_{k+1}, \dots, \boldsymbol{b}_m$ and zero value on the first k vectors $\boldsymbol{b}_1, \dots, \boldsymbol{b}_k$, we define a functional from $\mathcal{N}(T^T)$ and therefore ι is also a surjection. Thus ι is an isomorphism and, consequently,

$$\dim \mathcal{N}(T^T) = m - k$$

Since, simultaneously,

$$m = \dim \mathcal{N}(T^T) + \dim \mathcal{R}(T^T)$$

by comparison,

$$\dim \mathcal{R}(T^T) = k = \dim \mathcal{R}(T)$$

which finishes the proof. ∎

Matrix Representation of a Transpose in Finite Dimensional Spaces. Let X and Y be finite-dimensional spaces with corresponding bases $\boldsymbol{a}_1, \dots, \boldsymbol{a}_n$ and $\boldsymbol{b}_1, \dots, \boldsymbol{b}_m$, respectively. Let $T \in L(X,Y)$ and let T_{ij} denote the corresponding matrix representation for T, i.e.,

$$T_{ij} = \langle \boldsymbol{b}_i^*, T(\boldsymbol{a}_j) \rangle$$

A natural question arises: what is a matrix representation of the transpose T^T? The matrix representation depends obviously on the choice of bases. The natural choice for X^* and Y^* are the dual bases $\boldsymbol{a}_j^*$ and $\boldsymbol{b}_i^*$. Since every basis can be identified with its bidual, we have

$$T_{ji}^* = \left\langle \boldsymbol{a}_j^{**}, T^T(\boldsymbol{b}_i^*) \right\rangle = \left\langle T^T(\boldsymbol{b}_i^*), \boldsymbol{a}_j \right\rangle = \langle \boldsymbol{b}_i^*, T(\boldsymbol{a}_j) \rangle = T_{ij}$$

Thus, as we might have expected from previous comments and nomenclature itself, the matrix representation of transpose T^T (in dual bases) is obtained by interchanging rows and columns of the matrix representing T. If A is the matrix corresponding to T, the one corresponding to T^T is the transpose matrix, denoted A^T. As usual, all properties of transformations can be reinterpreted in terms of matrices.

PROPOSITION 2.11.3

(i) Let $\boldsymbol{A}, \boldsymbol{B} \in Matr(n,m)$. Then $(\alpha \boldsymbol{A} + \beta \boldsymbol{B})^T = \alpha \boldsymbol{A}^T + \beta \boldsymbol{B}^T$.

(ii) *Let* $\boldsymbol{A} \in Matr(k,n), \boldsymbol{B} \in Matr(n,m)$. *Then* $(\boldsymbol{BA})^T = \boldsymbol{A}^T\boldsymbol{B}^T$.

(iii) *Let* $\mathbf{1}$ *be the identity matrix,* $\mathbf{1} \in Matr\ (n,n)$. *Then*

$$\mathbf{1}^{\mathbf{T}} = \mathbf{1}.$$

(iv) *Let* $\boldsymbol{A} \in Matr(n,n)$ *and suppose that* $\boldsymbol{A}^{-1} \in Matr(n,n)$ *exists. Then*

$$(\boldsymbol{A}^{-1})^T = (\boldsymbol{A}^T)^{-1}$$

(v) *Let* $\boldsymbol{A} \in Matr(n,m)$. *Then rank* $\boldsymbol{A}^T$ *= rank* $\boldsymbol{A}$.

Exercises

2.11.1. Prove Proposition 2.11.3.

2.11.2. Using the properties of linear transformations, develop the rules of *matrix algebra* (here $i = 1, 2, \ldots, m$; $j = 1, 2, \ldots, n$):

(a) *Equality of Matrices.* $\boldsymbol{A} = \boldsymbol{B}$ iff $\boldsymbol{A}$ and $\boldsymbol{B}$ are of the same order and $A_{ij} = B_{ij}$.

(b) *Matrix Addition.* $\boldsymbol{A} \pm \boldsymbol{B} = \boldsymbol{C}$ iff $\boldsymbol{A}$, $\boldsymbol{B}$, and $\boldsymbol{C}$ are of the same order and if $A_{ij} \pm B_{ij} = C_{ij}$.

(c) *Scalar Multiplication.* If $p, q \in \mathbb{R}$,

$$q(p\boldsymbol{A}) = (qp)\boldsymbol{A};\ (q+P)\boldsymbol{A} = q\boldsymbol{A} + p\boldsymbol{A};\ q(\boldsymbol{A}+\boldsymbol{B}) = q\boldsymbol{A} + q\boldsymbol{B}$$

(d) *Matrix Multiplication.* Two matrices $\boldsymbol{A}$ and $\boldsymbol{B}$ are conformable for multiplication in the order $\boldsymbol{AB}$ iff the number of columns r of $\boldsymbol{A}$ equals the number of rows of $\boldsymbol{B}$. If $\boldsymbol{A}$ is of order $m \times r$ and $\boldsymbol{B}$ is of order $r \times n$, then $\boldsymbol{AB} = \boldsymbol{C}$ is of order $m \times n$ and

$$C_{ij} = \sum_{k=1}^{r} A_{ik}B_{kj}$$

Show that matrix multiplication describes a composition of two linear transformations: $A\colon \mathcal{U}^m \to \mathcal{W}^r$, $B\colon \mathcal{W}^r \to \mathcal{V}^n$, $C = A \circ b :$ $\mathcal{U}^m \to \mathcal{V}^n$.

2.11.3. Construct an example of square matrices $\boldsymbol{A}$ and $\boldsymbol{B}$ such that

(a) $\boldsymbol{AB} \neq \boldsymbol{BA}$

(b) $\boldsymbol{AB} = \mathbf{0}$, but neither $\boldsymbol{A} = \mathbf{0}$ nor $\boldsymbol{B} = \mathbf{0}$

(c) $\boldsymbol{AB} = \boldsymbol{AC}$, but $\boldsymbol{B} \neq \boldsymbol{C}$

2.11.4. If $\boldsymbol{A} = [A_{ij}]$ is an $m \times n$ rectangular matrix and its *transpose* $\boldsymbol{A}^T$ is the $n \times m$ matrix, $\boldsymbol{A}^T_{n\times m} = [A_{ji}]$. Prove that

$$
\begin{aligned}
(\boldsymbol{A}^T)^T &= \boldsymbol{A} \\
(\boldsymbol{A}+\boldsymbol{B})^T &= \boldsymbol{A}^T + \boldsymbol{B}^T \\
(\boldsymbol{ABC}\cdots\boldsymbol{XYZ})^T &= \boldsymbol{Z}^T\boldsymbol{Y}^T\boldsymbol{X}^T\cdots\boldsymbol{C}^T\boldsymbol{B}^T\boldsymbol{A}^T \\
(q\boldsymbol{A})^T &= q\boldsymbol{A}^T
\end{aligned}
$$

2.11.5. In this exercise, we develop a classical formula for the inverse of a square matrix. Let $\boldsymbol{A} = [a_{ij}]$ be a matrix of order n. We define the *cofactor* A_{ij} of the element a_{ij} of the i-th column of $\boldsymbol{A}$ as the determinant of the matrix obtained by deleting the i-th row and j-th column of $\boldsymbol{A}$, multiplied by $(-1)^{i+j}$:

$A_{ij} = $ cofactor a_{ij}

$$
-(-1)^{i+j}\begin{vmatrix}
A_{11} & a_{12} & & a_{1,j-1}a_{1,j+1} & & a_{1n} \\
 & & & \cdots & & \\
a_{i-1,1} & a_{i-1,2} & \cdots & a_{i-1,j-1}a_{i-1,j+1} & \cdots & a_{i-1,n} \\
a_{i+1,1} & a_{i+1,2} & \cdots & a_{i+1,j-1}a_{i+1,j+1} & \cdots & a_{i+1,n} \\
 & & & \cdots & & \\
a_{n1} & a_{n2} & \cdots & a_{n,j-1}a_{n,j+1} & \cdots & a_{nn}
\end{vmatrix}
$$

(a) Show that

$$
\delta_{ij} \det \boldsymbol{A} = \sum_{k=1}^{n} a_{ik}A_{kj}, \quad 1 \le i,j \le n
$$

where δ_{ij} is the Kronecker delta ($\delta_{ij} = 1$ if $i = j$, $\delta_{ij} = 0$ if $i \neq j$).

Hint: Look up the definition of the determinant of a square matrix.

(b) Using the result in (a), conclude that

$$
\boldsymbol{A}^{-1} = \frac{1}{\det \boldsymbol{A}}[A_{ij}]^T
$$

(c) Use (b) to compute the inverse of $\boldsymbol{A} = \begin{bmatrix} 1 & 2 & 2 \\ 1 & -1 & 0 \\ 2 & 1 & 3 \end{bmatrix}$ and verify your answer by showing that

$$A^{-1}A = AA^{-1} = I$$

2.11.6. Consider the matrices

$$A = \begin{bmatrix} 1 & 0 & 4 & 1 \\ 2 & -1 & 3 & 0 \end{bmatrix}, \quad B = \begin{bmatrix} -1 & 4 \\ 12 & 0 \\ 0 & 1 \end{bmatrix}, \quad C = [1, -1, 4, -3]$$

$$D = \begin{bmatrix} 2 \\ 3 \end{bmatrix}, \quad E = \begin{bmatrix} 1 & 0 & 2 & 3 \\ -1 & 4 & 0 & 1 \\ 1 & 0 & 2 & 4 \\ 0 & 1 & -1 & 2 \end{bmatrix}$$

If possible, compute the following:

(a) $AA^T + 4D^TD + E^T$ (b) $C^TC + E - E^2$
(c) B^TD (d) $B^TBD - D$
(e) $EC - A^TA$ (f) $A^TDC(E - 2I)$

2.11.7. Let $\mathcal{V}^4$ denote the vector space of ordered quadruples of real numbers. Do the following vectors provide a basis for $\mathcal{V}^4$?

$$a = (1, 0, -1, 1), \ b = (0, 1, 0, 22),$$
$$c = (3, 3, -3, 9), \ d = (0, 0, 0, 1)$$

2.11.8. Evaluate the determinant of the matrix

$$A = \begin{bmatrix} 1 & -1 & 0 & 4 \\ 1 & 0 & 2 & 1 \\ 4 & 7 & 1 & -1 \\ 1 & 0 & 1 & 2 \end{bmatrix}$$

2.11.9. Invert the following matrices (see Exercise 2.11.5).

$$A = \begin{bmatrix} 1 & -1 \\ 1 & 2 \end{bmatrix}, \quad B = \begin{bmatrix} 4 & 2 & 1 \\ 2 & 4 & 2 \\ 1 & 2 & 2 \end{bmatrix}$$

2.11.10. Prove that if A is symmetric and nonsingular, so also is A^{-1}.

2.11.11. Prove that if $\boldsymbol{A}, \boldsymbol{B}, \boldsymbol{C}$ and $\boldsymbol{D}$ are nonsingular matrices of the same order,

$$(\boldsymbol{ABCD})^{-1} = \boldsymbol{D}^{-1}\boldsymbol{C}^{-1}\boldsymbol{B}^{-1}\boldsymbol{A}^{-1}$$

2.11.12. Consider the linear problem:

$$\boldsymbol{Tx} = \boldsymbol{y}$$

where

$$\boldsymbol{T} = \begin{bmatrix} 0 & 1 & 3 & -2 \\ 2 & 1 & -4 & 3 \\ 2 & 3 & 2 & -1 \end{bmatrix}, \quad \boldsymbol{y} = \begin{bmatrix} 1 \\ 5 \\ 7 \end{bmatrix}$$

(i) Determine the rank of $\boldsymbol{T}$.

(ii) Determine the null space of $\boldsymbol{T}$.

(iii) Obtain a particular solution and the general solution.

(iv) Determine the range space of $\boldsymbol{T}$.

2.11.13. Construct examples of linear systems of equations having (1) no solutions, (2) infinitely many solutions, (3) if possible, unique solutions for the following cases:

(a) 3 equations, 4 unknowns
(b) 3 equations, 3 unknowns

2.11.14. Determine the rank of the following matrices:

$$\boldsymbol{T} = \begin{bmatrix} 2 & 1 & 4 & 7 \\ 0 & 1 & 2 & 1 \\ 2 & 2 & 6 & 8 \\ 4 & 4 & 14 & 10 \end{bmatrix}, \quad \boldsymbol{T}_2 = \begin{bmatrix} 1 & 2 & 1 & 3 & 4 & 4 \\ 2 & 0 & 3 & 2 & 1 & 5 \\ 1 & 1 & 1 & 2 & 1 & 3 \end{bmatrix}, \quad \boldsymbol{T}_3 = \begin{bmatrix} 2 & -1 & 1 \\ 2 & 0 & 1 \\ 0 & 1 & 1 \end{bmatrix}$$

2.11.15. Solve, if possible, the following systems:

$$\text{(i)}\quad \begin{aligned} 4x_1 \qquad\quad + 3x_3 - x_4 + 2x_5 &= 2 \\ x_1 - x_2 + x_3 - x_4 + x_5 &= 1 \\ x_1 + x_2 + x_3 - x_4 + x_5 &= 1 \\ x_1 + 2x_2 + x_3 \qquad + x_5 &= 0 \end{aligned}$$

$$
\begin{aligned}
\text{(ii)} \quad -4x_1 - 8x_2 + 5x_3 &= 1 \\
2x_1 - 2x_2 + 3x_3 &= 2 \\
5x_1 + x_2 + 2x_3 &= 4
\end{aligned}
$$

$$
\begin{aligned}
\text{(iii)} \quad 2x_1 + 3x_2 + 4x_3 + 3x_4 &= 0 \\
x_1 + 2x_2 + 3x_3 + 2x_4 &= 0 \\
x_1 + x_2 + x_3 + x_4 &= 0
\end{aligned}
$$

2.12 Tensor Products. Covariant and Contravariant Tensors

Let A and B be two arbitrary sets. Given two functionals f and g defined on A and B, respectively, we can define a new functional on the Cartesian product $A \times B$, called the product of f and g, as

$$A \times B \ni (x, y) \to f(x)g(y) \in \mathbb{R}(\mathbb{C})$$

In the case of vector spaces and linear functionals this simple construction leads to some very important algebraic results.

Tensor Product of Linear Functionals. Given two vector spaces X and Y with their duals X^*, Y^*, we define the tensor product of two functions as

$$(\boldsymbol{x}^* \otimes \boldsymbol{y}^*)(\boldsymbol{x}, \boldsymbol{y}) = \boldsymbol{x}^*(\boldsymbol{x})\boldsymbol{y}^*(\boldsymbol{y}) \qquad \text{for } \boldsymbol{x} \in X, \boldsymbol{y} \in Y$$

It is easy to see that the tensor product $\boldsymbol{x}^* \otimes \boldsymbol{y}^*$ is a bilinear functional on $X \times Y$ and therefore the tensor product can be considered as an operation from the Cartesian product $X \times Y$ to the space of bilinear functionals $M(X, Y)$:

$$\otimes : X^* \times Y^* \ni (\boldsymbol{x}, \boldsymbol{y}) \to \boldsymbol{x}^* \otimes \boldsymbol{y}^* \in M(X, Y)$$

PROPOSITION 2.12.1

Let X and Y be two vector spaces. The following properties hold:

(i) The tensor product operation

$$\otimes : X^* \times Y^* \to M(X, Y) \cdot$$

is a bilinear map from $X^* \times Y^*$ into $M(X,Y)$.

(ii) If, additionally, X and Y are finite dimensional and $\mathbf{e}_i, i = 1,\ldots,n$ and $\boldsymbol{g}_j, j = 1,\ldots,m$ denote two bases for X and Y, respectively, with $\boldsymbol{e}_i^*, \boldsymbol{g}_j^*$ their dual bases, then the set $\boldsymbol{e}_i^* \otimes \boldsymbol{g}_j^*$ forms a basis for $M(X,Y)$.

PROOF

(i) follows directly from the definition.

(ii) According to the representation formula for bilinear functionals in finite-dimensional spaces, we have for $a \in M(X,Y)$

$$\begin{aligned} a(x,y) &= \sum_{i=1}^{n}\sum_{j=1}^{m} a_{ij}x_i y_j \\ &= \sum_{i=1}^{n}\sum_{j=1}^{m} a_{ij}\boldsymbol{e}_i^*(\boldsymbol{x})\boldsymbol{g}_j^*(\boldsymbol{y}) \\ &= \sum_{i=1}^{n}\sum_{j=1}^{m} a_{ij}(\boldsymbol{e}_i^* \otimes g_j^*)(\boldsymbol{x},\boldsymbol{y}) \end{aligned}$$

and therefore

$$a = \sum_{i=1}^{n}\sum_{j=1}^{m} a_{ij}\boldsymbol{e}_i^* \otimes \boldsymbol{g}_j^*$$

which means that $\boldsymbol{e}_j^* \otimes \boldsymbol{g}_j^*$ spans the space $M(X,Y)$. To prove linear independence assume that

$$\sum_1^n\sum_1^m a_{ij}\boldsymbol{e}_i^* \otimes \boldsymbol{g}_j^* = \mathbf{0}$$

Taking consecutively pairs $\boldsymbol{e}_k, \boldsymbol{g}_l$, we get

$$\begin{aligned} 0 &= \sum_1^n\sum_1^m a_{ij}(\boldsymbol{e}_i^* \otimes \boldsymbol{g}_j^*)(\boldsymbol{e}_K, \boldsymbol{g}_l) \\ &= \sum_1^n\sum_1^m a_{ij}\boldsymbol{e}_i^*(\boldsymbol{e}_k)\boldsymbol{g}_j^*(\boldsymbol{g}_l) \\ &= \sum_1^n\sum_1^m a_{ij}\delta_{ik}\delta_{jl} = a_{kl} \end{aligned}$$

which ends the proof. ■

REMARK 2.12.1. It follows from Proposition 2.12.1 that

$$\dim M(X,Y) = \dim X \dim Y$$

■

Tensor Product of Finite-Dimensional Vector Spaces. The algebraic properties of space $M(X,Y)$ give rise to the definition of an abstract tensor product of two finite-dimensional vector spaces. We say that a vector space Z is a *tensor product* of finite-dimensional spaces X and Y, denoted $Z = X \otimes Y$, provided the following conditions hold:

(i) There exists a bilinear map

$$\otimes : X \times Y \ni (\boldsymbol{x}, \boldsymbol{y}) \to \boldsymbol{x} \otimes \boldsymbol{y} \in X \otimes Y$$

($x \otimes y$ is called the *tensor product of vectors* x *and* y).

(ii) If $\boldsymbol{e}_j, i = 1, \ldots, n$ and $\boldsymbol{g}_j, j = 1, \ldots, m$ are two bases for X and Y respectively, then $\boldsymbol{e}_i \otimes \boldsymbol{g}_j$ is a basis for $X \otimes Y$.

Obviously, according to Proposition 2.12.1, the tensor product for duals X^* and Y^* can be identified as $M(X,Y)$. But what can we say for arbitrary vector spaces X and Y? Does their tensor product exist? And if the answer is yes, is it unique? The following proposition answers these questions.

PROPOSITION 2.12.2

Let X and Y be two finite-dimensional spaces. Then the following spaces satisfy the axioms of tensor product $X \otimes Y$.

(i) $M(X^*, Y^*)$ *— the space of bilinear functionals on* $X^* \times Y^*$. *The tensor product of two vectors* $\boldsymbol{x} \otimes \boldsymbol{y}$ *is identified as*

$$\boldsymbol{x} \otimes \boldsymbol{y} \colon X^* \times Y^* \ni (\boldsymbol{x}^*, \boldsymbol{y}^*) \to \boldsymbol{x}^*(\boldsymbol{x})\boldsymbol{y}^*(\boldsymbol{y}) \in \mathbb{R}(\mathbb{C})$$

(ii) $M^*(X,Y)$ *— the dual to the space* $M(X,Y)$ *of bilinear functionals on* $X \times Y$.

$$\boldsymbol{x} \otimes \boldsymbol{y} \colon M(X,Y) \ni f \to f(\boldsymbol{x}, \boldsymbol{y}) \in \mathbb{R}(\mathbb{C})$$

(iii) $L(X^*,Y)$ — *the space of linear transformations from dual* X^* *to space* Y.

$$\boldsymbol{x} \otimes \boldsymbol{y}: X^* \ni \boldsymbol{x}^*, \rightarrow \langle \boldsymbol{x}^*, \boldsymbol{x} \rangle \, \boldsymbol{y} \in Y$$

PROOF follows directly from the definition and is left as an exercise.

So, as we can see, the problem is not one of existence, but rather with the uniqueness of the tensor product $X \otimes Y$. Indeed, there are many *models* of the tensor product; but, fortunately, one can show that all of them are *isomorphic.* In other words, no matter which of the models we discuss, as long as we deal with (linear) algebraic properties only, all the models yield the same results. ∎

Covariant and Contravariant Tensors. One can easily generalize the definition of tensor product to more than two finite-dimensional vector spaces. In particular, we can consider a tensor product of a space X with itself and its dual.

Let X be a finite-dimensional vector space with a basis $\boldsymbol{e}_1, \ldots, \boldsymbol{e}_n$. Elements of the tensor product

$$\underbrace{X \otimes \ldots \otimes X}_{p \text{ times}} \otimes \underbrace{X^* \otimes \ldots \otimes X^*}_{q \text{ times}}$$

are called tensors of order (p, q). A tensor of order $(p, 0)$ is called a *contravariant tensor of order* p, a tensor of order $(0, q)$ is a *convariant tensor of order* q, and if $p > 0, q > 0$, the tensor is referred to as a *mixed tensor.*

Let T be a tensor of order (p, q). Using the summation convention, we can write

$$\boldsymbol{T} = T^{i_1,\ldots,i_p}_{j_1,\ldots,j_q} \boldsymbol{e}_{i_1} \otimes \ldots \otimes \boldsymbol{e}_{i_p} \otimes \boldsymbol{e}^*_{j_1} \otimes \ldots \otimes \boldsymbol{e}^*_{j_q}$$

where $\boldsymbol{e}^*_j$ is the dual basis. The quantities $T^{i_1,\ldots,i_p}_{j_1,\ldots,j_q}$ are called components of tensor T with respect to basis $\boldsymbol{e}_1, \ldots \boldsymbol{e}_n$.

Obviously, according to the definition just stated, vectors $\boldsymbol{x} \in X$ are identified with contravariant tensors of order 1 while functionals from the dual space are identified with covariant tensors of order 1.

Let $\boldsymbol{e}_k, k = 1, \ldots, n$ be a basis in X and $\boldsymbol{x} \in X$ denote an arbitrary vector. Using the notation for tensors, we can write

$$\boldsymbol{x} = x^k \boldsymbol{e}_k$$

where x^k are components of vector $\boldsymbol{x}$ with respect to basis $\boldsymbol{e}_k$. One of the fundamental issues in linear algebra, especially in applications, is to ask about a transformation formula for vector components when the basis is changed. Toward establishing such a formula, consider a new basis $\overline{e}_j, j = 1, \ldots n$ with a corresponding representation for $\boldsymbol{x}$ in the form

$$\boldsymbol{x} = \overline{x}^j \boldsymbol{e}_j$$

Thus we have the identity

$$\overline{x}^j \ \overline{\mathbf{e}}_j = x^k \boldsymbol{e}_k$$

Applying to both sides the dual basis functional $\overline{\mathbf{e}}^*$, we get

$$\overline{x}^i = \overline{x}^j \delta_{ij} = \overline{x}^j \left\langle \overline{\mathbf{e}}_i^*, \overline{e}^j \overline{\mathbf{e}}_j \right\rangle = \left\langle \overline{\mathbf{e}}_i^*, \overline{x}^j \overline{\mathbf{e}}_j \right\rangle = \left\langle \overline{\mathbf{e}}_i^*, x^k \boldsymbol{e}_k \right\rangle = \left\langle \overline{\mathbf{e}}_i^*, \boldsymbol{e}_k \right\rangle x^k$$

Introducing the matrix

$$\alpha_{ik} = \left\langle \overline{\mathbf{e}}_i^*, \boldsymbol{e}_k \right\rangle$$

we obtain the transformation formula for vector components in the form

$$\overline{x}^i = \alpha^{i\cdot}_{\cdot k} x^k$$

The matrix $\boldsymbol{\alpha} = [\alpha^{i}_{\cdot k}]\alpha^{i}_{\cdot k}$ is called the *transformation matrix* from basis $\boldsymbol{e}_k$ to basis $\overline{\mathbf{e}}_j$. Thus in order to calculate the new components of vector $\boldsymbol{x}$, we must multiply the transformation matrix by old components of the same vector.

From the formula for the transformation matrix it follows easily that

$$\boldsymbol{e}_k = \alpha^{i\cdot}_{\cdot k} \overline{\mathbf{e}}_i$$

Indeed, applying to both sides functional $\overline{\mathbf{e}}_j^*$, we check that

$$\left\langle \overline{\mathbf{e}}_j^*, \boldsymbol{e}_k \right\rangle = \alpha^{i\cdot}_{\cdot k} \left\langle \overline{\mathbf{e}}_j^*, \overline{\mathbf{e}}_i \right\rangle = \alpha^{i\cdot}_{\cdot k} \delta^{j\cdot}_{\cdot i} = \alpha^{j\cdot}_{\cdot k}.$$

Having found the transformation formula for vectors, we may seek a corresponding formula for linear functionals, elements from the dual space X^*. Using tensor notation, we have for an arbitrary functional $f \in X^*$

$$\overline{f}_j \overline{\mathbf{e}}_j^* = f = f_k \boldsymbol{e}_k^*$$

Applying both sides to vector $\boldsymbol{e}_i$, we get

$$\overline{f}_i = \overline{f}_j \delta_{ij} = \overline{f}_j \left\langle \overline{\mathbf{e}}_j^*, \overline{e}_i \right\rangle = \left\langle \overline{f}_j \overline{\mathbf{e}}_j^*, \overline{e_i} \right\rangle = \left\langle f_k \boldsymbol{e}_k^*, \overline{\mathbf{e}}_i \right\rangle = \left\langle \boldsymbol{e}_k^*, \overline{\mathbf{e}}_i \right\rangle f_k = \beta^{k\cdot}_{\cdot i} f_k$$

where

$$\beta^k_{\cdot i} = \langle \boldsymbol{e}^*_k, \overline{\boldsymbol{e}}_i \rangle$$

is the transformation matrix from the *new* basis $\overline{\boldsymbol{e}}_i$ to the old basis $\boldsymbol{e}_k$. Note that this time in order to obtain the new components of the functional f we have to multiply the *transpose* of matrix $\boldsymbol{\beta}$ by old components of f. From the formula for the matrix $\boldsymbol{\beta}$ it follows that

$$\boldsymbol{e}^*_i = \beta^l_j \overline{\boldsymbol{e}}^*_j$$

Indeed, applying both sides to vector $\overline{\boldsymbol{e}}_i$, we check that

$$\langle \boldsymbol{e}^*_l, \overline{\boldsymbol{e}}_i \rangle = \beta^{l\cdot}_{\cdot j} \langle \overline{\boldsymbol{e}}^*_j, \overline{\boldsymbol{e}}_i \rangle = \beta^{l\cdot}_{\cdot i}$$

Finally, from the definition of transformation matrices it follows that matrix $\boldsymbol{\beta}$ is the inverse matrix of matrix $\boldsymbol{\alpha}$. Indeed, from

$$\boldsymbol{e}_k = \alpha^{i\cdot}_{\cdot k} \overline{\boldsymbol{e}}_i = \alpha^{i\cdot}_{\cdot k} \beta^{l\cdot}_{\cdot i} \boldsymbol{e}_l$$

it follows that

$$\beta^{l\cdot}_{\cdot i} \alpha^{i\cdot}_{\cdot k} = \delta^{l\cdot}_{\cdot k}$$

which proves the assertion.

We conclude this section with the statement of a general transformation formula for tensors of an arbitrary order (p, q). For simplicity let us restrict ourselves, for instance, to a tensor of order $(2, 1)$.

Again, let $\overline{\boldsymbol{e}}_i$ and $\boldsymbol{e}_k$ be a new and old basis and $\overline{\boldsymbol{e}}^*_i, \boldsymbol{e}^*_k$ denote their duals. For a tensor T of order $(2, 1)$ we have

$$\boldsymbol{T} = \overline{T}^{ij\cdot}_{\cdot\cdot k} \overline{\boldsymbol{e}}_j \otimes \overline{\boldsymbol{e}}_j \otimes \overline{\boldsymbol{e}}^*_k$$

and, simultaneously,

$$\boldsymbol{T} = T^{lm\cdot}_{\cdot\cdot n} \boldsymbol{e}_l \otimes \boldsymbol{e}_m \otimes \overline{\boldsymbol{e}}^*_n$$

From the transformation formulas for vectors $\boldsymbol{e}_l, \overline{\boldsymbol{e}}^*_n$ and the properties of tensor product it follows that

$$\boldsymbol{e}_l \otimes \boldsymbol{e}_m \otimes \overline{\boldsymbol{e}}^*_n = \alpha^{i\cdot}_{\cdot l} \alpha^{j\cdot}_{\cdot m} \beta^{n\cdot}_{\cdot k} \overline{\boldsymbol{e}}_i \otimes \overline{\boldsymbol{e}}_j \otimes \overline{\boldsymbol{e}}^*_k$$

or, after substitution into the second formula for $\boldsymbol{T}$,

$$\boldsymbol{T} = \alpha^{i\cdot}_{\cdot \iota} \alpha^{j\cdot}_{\cdot m} \beta^{n\cdot}_{\cdot k} T^{\iota m\cdot}_{\cdot\cdot n}\ \overline{\boldsymbol{e}}_i \otimes \overline{\boldsymbol{e}}_j \otimes \overline{\boldsymbol{e}}^*_k$$

Finally, comparing both formulas for $\boldsymbol{T}$, we get the transformation formula for components of tensor $\boldsymbol{T}$,

$$\overline{T}^{ij\cdot}_{\cdot\cdot k} = \alpha^{i}_{\cdot l}\alpha^{j}_{\cdot m}\beta^{n}_{\cdot k}T^{lm\cdot}_{\cdot\cdot n}$$

Note the difference between the multiplication of contra- and covariant indices of tensor T.

Euclidean Spaces

2.13 Scalar (Inner) Product. Representation Theorem in Finite-Dimensional Spaces

In this section we shall deal with a generalization of the "dot product" or inner product of two vectors.

Scalar (Inner) Product. Let V be a complex vector space. A complex-valued function from $V \times V$ into $\mathbb{C}$ that associates with each pair $\boldsymbol{u}, \boldsymbol{v}$ of vectors in V a scalar, denoted $(\boldsymbol{u}, \boldsymbol{v})_V$ or shortly $(\boldsymbol{u}, \boldsymbol{v})$ if no confusion occurs, is called a *scalar (inner) product* on V if and only if

(i) $(\alpha_1\boldsymbol{u}_1 + \alpha_2\boldsymbol{u}_2, \boldsymbol{v}) = \alpha_1(\boldsymbol{u}_1, \boldsymbol{v}) + \alpha_2(\boldsymbol{u}_2, \boldsymbol{v})$, i.e., $(\boldsymbol{u}, \boldsymbol{v})$ is linear with respect to $\boldsymbol{u}$

(ii) $(\boldsymbol{u}, \boldsymbol{v}) = \overline{(\boldsymbol{v}, \boldsymbol{u})}$, where $\overline{(\boldsymbol{v}, \boldsymbol{u})}$ denotes the complex conjugate of $(\boldsymbol{v}, \boldsymbol{v})$ (antisymmetry)

(iii) $(\boldsymbol{u}, \boldsymbol{u})$ is positively defined, i.e., $(\boldsymbol{u}, \boldsymbol{u}) \geq 0$ and $(\boldsymbol{u}, \boldsymbol{u}) = 0$ implies $\boldsymbol{u} = \boldsymbol{0}$

Let us note that due to antisymmetry $(\boldsymbol{u}, \boldsymbol{u})$ is a real number and therefore it makes sense to speak about positive definiteness. The first two conditions imply that $(\boldsymbol{u}, \boldsymbol{v})$ is *antilinear* with respect to the second variable:

$$(\boldsymbol{u}, \beta_1\boldsymbol{v}_1 + \beta_2\boldsymbol{v}_2) = \overline{\beta}_1(\boldsymbol{u}, \boldsymbol{v}_1) + \overline{\beta}_2(\boldsymbol{u}, \boldsymbol{v}_2)$$

In most of the developments to follow, we shall deal with real vector spaces only. Then property (ii) becomes one of symmetry

(ii) $(\boldsymbol{u}, \boldsymbol{v}) = (\boldsymbol{v}, \boldsymbol{u})$

and the inner product becomes a bilinear functional.

Inner Product (Pre-Hilbert, Unitary) Spaces. A vector space V on which an inner product has been defined, is called an *inner product space.* Sometimes the names *pre-Hilbert* or *unitary spaces* are also used for such spaces.

Orthogonal Vectors. Two vectors $\boldsymbol{u}$ and $\boldsymbol{v}$ of an inner product space V are said to be orthogonal if

$$(\boldsymbol{u}, \boldsymbol{v}) = 0$$

A number of elementary properties of an inner product follow immediately from the definition. We shall begin with the Cauchy-Schwarz inequality.

PROPOSITION 2.13.1

(The Cauchy-Schwarz inequality)

Let u and v be arbitrary vectors of an inner product space. Then

$$|(\boldsymbol{u}, \boldsymbol{v})| \leq (\boldsymbol{u}, \boldsymbol{u})^{\frac{1}{2}} (\boldsymbol{v}, \boldsymbol{v})^{\frac{1}{2}}$$

PROOF If $\boldsymbol{v} = \boldsymbol{0}$, the inequality is obviously satisfied. Suppose $\boldsymbol{v} \neq \boldsymbol{0}$. Then for an arbitrary scalar $\alpha \in \mathbb{C}(\mathbb{R})$,

$$0 \leq (\boldsymbol{u} - \alpha\boldsymbol{v}, \boldsymbol{u} - \alpha\boldsymbol{v}) = (\boldsymbol{u}, \boldsymbol{u}) - \alpha(\boldsymbol{v}, \boldsymbol{u}) - \overline{\alpha}(\boldsymbol{u}, \boldsymbol{v}) + \alpha\overline{\alpha}(\boldsymbol{v}, \boldsymbol{v})$$

Take $\alpha = \overline{(\boldsymbol{v}, \boldsymbol{u})}/(\boldsymbol{v}, \boldsymbol{v})$. Then $\overline{\alpha} = \overline{(\boldsymbol{u}, \boldsymbol{v})}/(\boldsymbol{v}, \boldsymbol{v})$ and

$$(\boldsymbol{u}, \boldsymbol{u}) - \frac{|(\boldsymbol{v}, \boldsymbol{u})|^2}{(\boldsymbol{v}, \boldsymbol{v})} - \frac{|(\boldsymbol{u}, \boldsymbol{v})|^2}{(\boldsymbol{v}, \boldsymbol{v})} + \frac{|(\boldsymbol{u}, \boldsymbol{v})|^2}{(\boldsymbol{v}, \boldsymbol{v})} \geq 0$$

or

$$(\boldsymbol{u}, \boldsymbol{u})(\boldsymbol{v}, \boldsymbol{v}) - |(\boldsymbol{u}, \boldsymbol{v})|^2 \geq 0$$

from which the assertion follows. ∎

The Cauchy-Schwarz inequality is a useful tool in many proofs in analysis. For brevity, we shall follow common practice and refer to it as simply the Schwarz inequality.

Example 2.13.1

Let $V = \mathbb{C}^n$. The following is an inner product on V:

$$(\boldsymbol{v}, \boldsymbol{w}) = ((v_1, \ldots, v_n), (w_1, \ldots, w_n)) = \sum_{i=1}^{n} v_i \overline{w}_i$$

In the case of $V = \mathbb{R}^n$ the same definition can be modified to yield

$$(\boldsymbol{v}, \boldsymbol{w}) = \sum_{1}^{n} v_i w_i$$

resulting in the classical formula for a dot product of two vectors in a Cartesian system of coordinates. $\square$

The inner product is by no means unique! Within one vector space it can be introduced in many different ways. The simplest situation is observed in the case of finite-dimensional spaces. Restricting ourselves to real spaces only, recall that given a basis $\boldsymbol{e}_1, \ldots, \boldsymbol{e}_n$ in V, the most general formula for a bilinear functional a on V (cf. Section 2.10) is

$$a(\boldsymbol{v}, \boldsymbol{v}) = \sum_{i,j=1}^{n} a_{ij} v_i v_j$$

where $a_{ij} = a(\boldsymbol{e}_i, \boldsymbol{e}_j)$. Symmetry and positive definiteness of a are equivalent to symmetry and positive definiteness of matrix a_{ij}. Thus, setting an arbitrary symmetric and positive definite matrix a_{ij}, we may introduce a corresponding inner product on V. The classical formula from Example 2.13.1 corresponds to the choice of canonical basis in $\mathbb{R}^n$ and matrix $a_{ij} = \delta_{ij}$, where δ_{ij} denotes the Kronecker delta.

Throughout the rest of this chapter we shall restrict ourselves to finite-dimensional spaces.

Suppose we are given an inner product finite dimensional space V. Choosing a vector $\boldsymbol{u} \in V$, we may define a corresponding linear functional $\boldsymbol{u}^*$ on V by

$$\langle \boldsymbol{u}^*, \boldsymbol{v} \rangle = (\boldsymbol{v}, \boldsymbol{u})$$

The mapping $R : V \ni \boldsymbol{u} \to \boldsymbol{u}^* \in V$ is linear for real vector spaces and antilinear for complex vector spaces. It is also injective because $(\boldsymbol{v}, \boldsymbol{u}) = 0$ for every $\boldsymbol{v} \in V$ implies $\boldsymbol{u} = \boldsymbol{0}$ (pick $\boldsymbol{v} = \boldsymbol{u}$ and make use of positive definitness of the inner product). Since both V and its dual V^* are of the same dimension, in the case of real spaces, the map R is an isomorphism and the two spaces are isomorphic. We summarize these observations in the following theorem.

THEOREM 2.13.1
(Representation Theorem for Finite-Dimensional Spaces)

Let V be a finite-dimensional vector space with inner product $(\cdot,\cdot)$. Then for every linear functional $\boldsymbol{u}^* \in V^*$ there exists a unique vector $\boldsymbol{u} \in V$ such that

$$\langle \boldsymbol{u}^*, \boldsymbol{v} \rangle = (\boldsymbol{v}, \boldsymbol{u}) \qquad \forall\, \boldsymbol{v} \in V$$

The map

$$R : V \ni \boldsymbol{u} \longrightarrow \boldsymbol{u}^* \in V^*$$

called the Riesz map, establishing the one-to-one correspondence between elements of V and V^* is linear for real and antilinear for complex vector spaces.

PROOF It remains to show only the surjectivity of R in the case of a complex space V. Let $\boldsymbol{u}^*$ be an arbitrary linear functional on V. Representing $\boldsymbol{u}^*$ in the form

$$\begin{aligned} \langle \boldsymbol{u}^*, \boldsymbol{v} \rangle &= Re\langle \boldsymbol{u}^*, \boldsymbol{v} \rangle + i\; Im\langle \boldsymbol{u}^*, \boldsymbol{v} \rangle \\ &= f(\boldsymbol{v}) + ig(\boldsymbol{v}) \end{aligned}$$

we easily verify that both f and g are linear functionals in the real sense. (Every complex vector space is automatically a real vector space, if we restrict ourselves to real scalars only.) It follows also from the *complex* homogeneity of $\boldsymbol{u}^*$ that

$$\begin{aligned} f\langle i\boldsymbol{v} \rangle + ig(i\boldsymbol{v}) &= \langle \boldsymbol{u}^*, i\boldsymbol{v} \rangle = i\langle \boldsymbol{u}^*, \boldsymbol{v} \rangle \\ &= i\,(f(\boldsymbol{v}) + ig(\boldsymbol{v})) \\ &= -g(\boldsymbol{v}) + if(\boldsymbol{v}) \end{aligned}$$

which implies that f and g are not independent of each other. In fact,

$$g(\boldsymbol{v}) = -f(i\boldsymbol{v}) \qquad \forall\, \boldsymbol{v} \in V$$

Decomposing in the same way the scalar product

$$(\boldsymbol{v}, \boldsymbol{u}) = Re(\boldsymbol{v}, \boldsymbol{u}) + i\ Im(\boldsymbol{v}, \boldsymbol{u})$$

we can easily verify that

1. $Re(\boldsymbol{v}, \boldsymbol{u})$ is a scalar product in the real sense on V
2. $Im(\boldsymbol{v}, \boldsymbol{u}) = -Re(i\boldsymbol{v}, \boldsymbol{u})$

In particular, it follows from condition (ii) for inner products that the imaginary part $Im(\boldsymbol{v}, \boldsymbol{u})$ is *antisymmetric*, i.e.,

$$Im(\boldsymbol{v}, \boldsymbol{u}) = -Im(\boldsymbol{u}, \boldsymbol{v})$$

Applying now the representation theorem for real spaces, we conclude that there exists a vector $\boldsymbol{u}$ such that

$$Re(\boldsymbol{v}, \boldsymbol{u}) = f(\boldsymbol{v}) \qquad \forall\, \boldsymbol{v} \in V$$

But making use of the relations between the real and imaginary parts of both functional $\boldsymbol{u}^*$ and inner product $(\boldsymbol{v}, \boldsymbol{u})$, we have

$$Im(\boldsymbol{v}, \boldsymbol{u}) = -Re(i\boldsymbol{v}, \boldsymbol{u}) = -f(i\boldsymbol{v}) = g(\boldsymbol{v})$$

and, consequently,

$$(\boldsymbol{v}, \boldsymbol{u}) = Re(\boldsymbol{v}, \boldsymbol{u}) + i\ Im(\boldsymbol{v}, \boldsymbol{u}) = f(\boldsymbol{v}) + ig(\boldsymbol{v}) = \langle \boldsymbol{v}^*, \boldsymbol{v} \rangle$$

which finishes the proof. ∎

2.14 Basis and Cobasis. Adjoint of a Transformation. Contra- and Covariant Components of Tensors

The representation theorem with the Riesz map allows us, in the case of a finite-dimensional inner product space V, to identify the dual V^* with the original space V. Consequently, every notion which has been defined for dual space V^* can now be reinterpreted in the context of the inner product space.

Throughout this section V will denote a *finite-dimensional vector* space with an inner product $(\cdot,\cdot)$ and the corresponding Riesz map

$$R: V \ni \boldsymbol{u} \to \boldsymbol{u}^* = (\cdot, \boldsymbol{u}) \in V^*$$

Cobasis. Let $\mathbf{e}_i, i = 1, \ldots, n$ be a basis and $\mathbf{e}_j^*, j = 1, \ldots, n$ its dual basis. Consider vectors

$$\mathbf{e}^j = R^{-1}\mathbf{e}_j^*$$

According to the definition of the Riesz map, we have

$$(\mathbf{e}_i, \mathbf{e}^j) = (\mathbf{e}_i, R^{-1}\mathbf{e}_j^*) = \langle \mathbf{e}_j^*, \mathbf{e}_i \rangle = \delta_{ij}$$

We have the following:

PROPOSITION 2.14.1

For a given basis $\mathbf{e}_i, i = 1, \ldots, n$ there exists a unique basis $\mathbf{e}^j$ (called cobasis*) such that*

$$(\mathbf{e}_i, \mathbf{e}^j) = \delta_{ij}$$

PROOF For a real space V the assertion follows from the fact that R is an isomorphism. For complex vector spaces the proof follows precisely the lines of the proof of the representation theorem and is left as an exercise. ∎

Orthogonal Complements. Let U be a subspace of V and $U^\perp$ denote its orthogonal complement in V. The inverse image of $U^\perp$ by the Riesz map

$$R^{-1}(U^\perp)$$

denoted by the same symbol $U^\perp$ will also be called the *orthogonal complement* (in V) of subspace U. Let $\boldsymbol{u} \in U, \boldsymbol{v} \in U^\perp$. We have

$$(\boldsymbol{u}, \boldsymbol{v}) = \langle R\boldsymbol{v}, \boldsymbol{u} \rangle = 0$$

Thus the orthogonal complement can be expressed in the form

$$U^\perp = \{\boldsymbol{v} \in \boldsymbol{V} : (\boldsymbol{u}, \boldsymbol{v}) = 0 \text{ for every } \boldsymbol{u} \in U\}$$

Adjoint Transformations. Let W denote another finite-dimensional space with an inner product $(\cdot,\cdot)_W$ and the corresponding Riesz map R_W.

Recalling the diagram defining the transpose of a transformation, we complete it by the Riesz maps

$$\begin{array}{ccc} V & \xrightarrow{T} & W \\ \downarrow R_V & & \downarrow R_W \\ V^* & \xleftarrow{T^T} & W^* \end{array}$$

and set a definition of the adjoint transformation T^* as the composition

$$T^* = R_V^{-1} \circ T^T \circ R_W$$

It follows from the definition that T^* is a well-defined linear transformation from W^* to V^* (also in the complex case). We have

$$\begin{aligned}(\boldsymbol{v}, T^*\boldsymbol{w})_V = (\boldsymbol{v}, (R_V^{-1} \circ T^T \circ R_W)\boldsymbol{w})_V &= \langle (T^T \circ R_W)\boldsymbol{w}, \boldsymbol{v} \rangle \\ &= \langle R_W\boldsymbol{w}, T\boldsymbol{v} \rangle = (T\boldsymbol{v}, \boldsymbol{w})_W\end{aligned}$$

which proves the following:

PROPOSITION 2.14.2

Let $T \in L(V, W)$. There exists a unique adjoint transformation $T^* \in L(W, V)$ such that

$$(\boldsymbol{v}, T^*\boldsymbol{w})_V = (T\boldsymbol{v}, \boldsymbol{w})_W \qquad \textit{for every } \boldsymbol{v} \in V, \boldsymbol{w} \in W$$

Reinterpreting all the properties of the transpose of a transformation in terms of the adjoints, we get the following:

PROPOSITION 2.14.3

Let U, V, W be finite-dimensional spaces with inner products. Then the following properties hold:

(i) $T, S \in L(V, W) \quad \Rightarrow \quad (\alpha T + \beta S)^* = \overline{\alpha} T^* + \overline{\beta} S^*$
where $\overline{\alpha}, \overline{\beta}$ are the complex conjugates of α and β if spaces are complex.

(ii) *If* $T \in L(U, V)$ *and* $S \in L(V, W)$*, then*

$$(S \circ T)^* = T^* \circ S^*$$

(iii) $(\mathrm{id}_V)^* = \mathrm{id}_V$.

(iv) *Let* $T \in L(V, W)$ *be an isomorphism. Then*

$$(T^*)^{-1} = (T^{-1})^*$$

(v) *rank* T^* = *rank* T.

(vi) *Let* $T \in L(V, W)$ *and* T_{ij} *denote its matrix representation with respect to bases* $\boldsymbol{a}_1, \ldots, \boldsymbol{a}_n \in V$ *and* $\boldsymbol{b}_1, \ldots, \boldsymbol{b}_m \in W$*, i.e.,*

$$T_{ij} = < \boldsymbol{b}_i^*, T(\boldsymbol{a}_j) > = (T(\boldsymbol{a}_j), \boldsymbol{b}^i)$$

Then the transpose matrix $(T_{ij})^{\perp} = T_{ji}$ *is the matrix representation for the adjoint* T^* *with respect to cobases* $\boldsymbol{a}^1, \ldots, \boldsymbol{a}^n \in V$ *and* $\boldsymbol{b}^1, \ldots, \boldsymbol{b}^m \in W$*, i.e.,*

$$T_{ji} = (T^*)_{ij} =< \boldsymbol{a}^{i*}, T^*(\boldsymbol{b}^j) >= (T^*(\boldsymbol{b}^j), \boldsymbol{a}_i)$$

PROOF follows directly from the results collected in Section 2.11 and the definition of the adjoint. ∎

Contravariant and Covariant Components of Tensors. Once we have decided to identify a space X with its dual X^*, all the tensor products

$$\underbrace{X \otimes \ldots \otimes X}_{p \text{ times}} \otimes \underbrace{X^* \otimes \ldots \otimes X^*}_{q \text{ times}}$$

such that $p + q = k$ for a fixed k, are isomorphic and we simply speak of tensors of order k. Thus, for example, for tensors of order 2 we identify the following spaces:

$$X \otimes X, \quad X \otimes X^*, \quad X^* \otimes X^*$$

We do, however, distinguish between different components of tensors. To explain this notion, consider, for instance, a tensor T of second order. Given a basis $\boldsymbol{e}_i$ and its cobasis $\boldsymbol{e}^j$, we can represent T in three different ways:

$$\boldsymbol{T} = T^{ij}_{..}\boldsymbol{e}_i \otimes \boldsymbol{e}_j$$

$$\boldsymbol{T} = T^{i\cdot}_{\cdot j}\boldsymbol{e}_i \otimes \boldsymbol{e}^j$$

$$\boldsymbol{T} = T^{..}_{ij}\boldsymbol{e}^i \otimes \boldsymbol{e}^j$$

Matrices $T^{ij}_{..}, T^{i\cdot}_{\cdot j}$ and $T^{..}_{ij}$ are called contravariant components of tensor T. It is easy to see that different representation formulas correspond to different, but isomorphic, definitions of tensor product (cf. Section 2.12).

Let $\overline{\boldsymbol{e}}_i$ denote now a new basis and $\overline{\boldsymbol{e}}^k$ its cobasis. Following Section 2.12, we define the transformation matrix from the old basis to the new basis $\overline{\boldsymbol{e}}_k$ as

$$\alpha_{ik} \stackrel{\text{df}}{=} (\overline{\boldsymbol{e}}^i, \boldsymbol{e}_k)$$

and its inverse—the transformation matrix from the new basis to the old one—as

$$\beta_{ik} \stackrel{\text{df}}{=} (\boldsymbol{e}^i, \overline{\boldsymbol{e}}_k)$$

Let T be, for instance, a tensor of third order and $T^{ij\cdot}_{..k}$ denote its mixed components. The following transformation formula follows directly from the formula derived in Section 2.12.

$$\overline{T}^{ij\cdot}_{..k} = \alpha^{i\cdot}_{\cdot l}\alpha^{j\cdot}_{\cdot m}\beta^{n\cdot}_{\cdot k}T^{lm\cdot}_{..n}$$

Orthonormal Bases. A basis $\boldsymbol{e}_i, i = 1, \ldots, n$ is called *orthonormal* if

$$(\boldsymbol{e}_i, \boldsymbol{e}_j) = \delta_{ij}$$

i.e., vectors $\boldsymbol{e}_i$ are orthogonal to each other in the sence of the inner product and normalized, i.e., $(\boldsymbol{e}_i, \boldsymbol{e}_i) = 1$. (Later on, we will interpret $||\boldsymbol{x}|| = (\boldsymbol{x}, \boldsymbol{x})^{\frac{1}{2}}$ as a *norm*, a notion corresponding in elementary algebra to the notion of the length of a vector.)

As an immediate consequence of the definition we observe that for an orthonormal basis, the basis and its cobasis coincide with each other. Consequently, we cannot distinguish between different components of tensors. They are all the same! The transformation matrix falls into the category of so-called orthonormal matrices for which

$$\boldsymbol{\alpha}^{-1} = \boldsymbol{\alpha}^T$$

i.e., the inverse matrix of matrix α coincides with its transpose (cf. Exercise 2.1.4). Thus, the inverse transformation matrix satisfies the equality

$$\beta^{n\cdot}_{\cdot k} = \alpha^{k\cdot}_{\cdot n}$$

and the transformation formula for tensors, for instance, of the third order gets the form

$$\overline{T}_{ijk} = \alpha_{il}\alpha_{jm}\alpha_{nk}T_{lmn}$$

In the case of the orthonormal bases, since we do not distinguish between different components, all the indices are placed on the same level.

3

Lebesgue Measure and Integration

Lebesgue Measure

3.1 Elementary Abstract Measure Theory

We shall begin our study of Lebesgue measure and integration theory from some fundamental, general notions.

The concept of measure of a set arises from the problem of generalizing the notion of "size" of sets in $\mathbb{R}$ and $\mathbb{R}^n$ and extending such notions to arbitrary sets. Thus, the measure of a set $A = (a, b) \subset \mathbb{R}$ is merely its length, the measure of a set $A \subset \mathbb{R}^2$ is its area, and of a set $A \subset \mathbb{R}^3$, its volume. In more general situations, the idea of the size of a set is less clear. Measure theory is the mathematical theory concerned with these generalizations and is an indispensable part of functional analysis. The benefits of generalizing ideas of size of sets are substantial, and include the development of a rich and powerful theory of integration that extends and generalizes elementary Riemann integration outlined in Chapter 1. Now, we find that the basic mathematical properties of sizes of geometrical objects, such as area and volume, are shared by other types of sets of interest, such as sets of random events and the probability of events taking place. Our plan here is to give a brief introduction to this collection of ideas, which includes the ideas of Lebesgue measure and integration essential in understanding fundamental examples of metric and normed spaces dealt with in subsequent chapters. We begin with the concept of σ-algebra.

σ-Algebra of (Measurable) Sets. Suppose we are given a set X. A nonempty class $S \subset \mathcal{P}(X)$ is called a *σ-algebra* of sets if the following conditions hold:

(i) $A \in S \Rightarrow A' \in S$.

(ii) $A_i \in S, i = 1, 2, \ldots \quad \bigcup_1^{\infty} A_i \in S.$

Numerous other definitions of similar algebraic structures exists. The letter "σ" corresponds to the countable unions in the second condition. If only finite unions are considered, one talks about an *algebra* of sets without the symbol "σ".

Some fundamental corollaries follow immediately from the definition.

PROPOSITION 3.1.1

Let $S \subset \mathcal{P}(X)$ be a σ-algebra. The following properties hold:

(i) $A_1, \ldots, A_n \in S \Longrightarrow \bigcup_1^n A_i \in S.$

(ii) $\emptyset, X \in S.$

(iii) $A_1, A_2, \ldots \in S \Longrightarrow \bigcap_1^{\infty} A_i \in S.$

(iv) $A_1, \ldots, A_n \in S \Longrightarrow \bigcap_1^n A_i \in S.$

(v) $A, B \in S \Longrightarrow A - B \in S.$

PROOF (i) follows from the second axiom and the observation that $A_1 \cup \ldots \cup A_n = A_1 \cup \ldots \cup A_n \cup A_n \cup \ldots$. To prove (ii) pick a set $A \in S$ (S is nonempty). According to the first axiom $A' \in S$ and therefore from (i) it follows that $X = A \cup A'$ belongs to S. But the empty set $\emptyset$ as the complement of the whole S must belong therefore to S as well. The third and the fourth assertions follow from De Morgan's law

$$\left(\bigcap_1^{\infty} A_i\right)' = \bigcup_1^{\infty} A_i'$$

and the last one from the formula

$$A - B = A \cap B'$$

∎

It is a matter of a direct check of the axioms to prove the following.

PROPOSITION 3.1.2

Let $S_\iota \subset \mathcal{P}(X)$ denote a family of σ-algebras. Then the common part $\bigcap_\iota S_\iota$ is a σ-algebra as well.

Example 3.1.1

Two trivial examples of σ-algebras are the family consisting of the space X and the empty set $\emptyset$ only: $S = \{X, \emptyset\}$ and the entire family of all subsets of X : $S = \mathcal{P}(X)$. □

Example 3.1.2

One of the most significant and revealing examples of σ-algebras in measure theory is provided by the concept of the probability of a random event. We denote by X the *set of elementary events*. For example, X might be the set of possible spots on a die: ·, ··, ···, ∷, ⁙, ⠿. Giving each possible event i a label e_i, we can write $X = \{e_1, e_2, e_3, e_4, e_5, e_6\}$. Thus, e_4 is the event that when throwing a die, the face with four spots will arise. □

The subsets of a σ-algebra $S \subset \mathcal{P}(X)$ are identified as *random events*. In our case simply $S = \mathcal{P}(X)$ and the axioms for a σ-algebra of sets are trivially satisfied. For example, $A = (e_2, e_4, e_6)$ is the random event that an even face will apear when casting a die, $B = (e_1, e_2, e_3, e_4, e_5)$ is the event that face e_6 will not appear. The event $\emptyset \in S$ is the *impossible event* and $X \in S$ is the *sure event*.

The set of random events provides an important example of a σ-algebra to which all of measure theory applies, but which is not based on extensions of elementary measurements of length, etc.

Definition of a σ-Algebra Generated by a Family of Sets. Let $K \subset \mathcal{P}(X)$ be an arbitrary family of sets. We will denote by $S(K)$ the smallest σ-algebra containing K. Such a σ-algebra always exists since according to Proposition 3.1.2, it can be constructed as the common part of all σ-algebras containing K

$$S(K) = \bigcap \{S \text{ σ-algebra } : S \supset K\}$$

Note that the family of σ-algebras containing K is non-empty since it contains $\mathcal{P}(X)$ (cf. Example 3.1.1.).

Borel Sets. A nontrivial example of a σ-algebra is furnished by the σ-algebra of so-called *Borel sets* generated by the family K of all open sets in $\mathbb{R}^n$. We shall denote this σ-algebra by $\mathcal{B}(\mathbb{R}^n)$ or simply $\mathcal{B}$. Since closed sets are complements of open sets, the family of Borel sets contains both open and closed sets. Moreover, it contains also sets of G_δ- and F_σ-type as countable intersections or unions of open and closed sets, respectively (see Section 4.1).

The algebraic structure of a σ-algebra can be transferred through a mapping from one space onto another.

PROPOSITION 3.1.3

Let $f: X \to Y$ be a mapping prescribed on the whole X, i.e., dom $f = X$. Let $S \subset \mathcal{P}(X)$ be a σ-algebra of sets. Then the family

$$R \stackrel{def}{=} \{E \in \mathcal{P}(Y) : \ f^{-1}(E) \in S\}$$

is a σ-algebra in Y.

PROOF $X = f^{-1}(Y) \in S$ and therefore $Y \in R$, which proves that R is nonempty. The first axiom follows from the fact that

$$f^{-1}(E') = (f^{-1}(E))'$$

and the second from the identity

$$f^{-1}\left(\bigcup_i A_i\right) = \bigcup_i f^{-1}(A_i)$$

∎

COROLLARY 3.1.1

Let $f: X \to Y$ be a bijection and $S \subset \mathcal{P}(X)$ a σ-algebra. Then the following hold:

(i) $f(S) \stackrel{\text{def}}{=} \{f(A) : \ A \in S\}$ is a σ-algebra in Y.

(ii) If K generates S in X then $f(K)$ generates $f(S)$ in Y.

We shall prove now two fundamental properties of Borel sets in conjunction with continuous functions.

PROPOSITION 3.1.4

Let $f : \mathbb{R}^n \to \mathbb{R}^m$ be a continuous function. Then

(i) $B \in \mathcal{B}(\mathbb{R}^m)$ *implies* $f^{-1}(B) \in \mathcal{B}(\mathbb{R}^n)$.
If, additionally, f is bijective then

(ii) $f(\mathcal{B}(\mathbb{R}^n)) = \mathcal{B}(\mathbb{R}^n)$.

PROOF Consider the σ-algebra (Proposition 3.1.3)

$$R = \{E \in \mathcal{P}(\mathbb{R}^m) : f^{-1}(E) \in \mathcal{B}(\mathbb{R}^n)\}$$

Since the inverse image of an open set through a continuous function is open, R contains the open sets in $\mathbb{R}^m$ and therefore the whole σ-algebra of Borel sets as the smallest σ-algebra containing the open sets. The second assertion follows immediately from Corollary 3.1.1. ∎

We shall conclude our considerations concerning the Borel sets with the following important result.

PROPOSITION 3.1.5

Let $E \in \mathcal{B}(\mathbb{R}^n)$, $F \in \mathcal{B}(\mathbb{R}^m)$. Then $E \times F \in \mathcal{B}(\mathbb{R}^{n+m})$. In other words, the Cartesian product of two Borel sets is a Borel set.

PROOF
Step 1. Pick an open set $G \subset \mathbb{R}^n$ and consider the family

$$\{F \subset \mathbb{R}^m : G \times F \in \mathcal{B}(\mathbb{R}^n \times \mathbb{R}^m)\}$$

One can easily prove that the family is a σ-algebra. Since it contains open sets (the Cartesian product of two open sets is open) it must contain the whole σ-algebra or Borel sets. In conclusion, the Cartesian products of open and Borel sets are Borel.

Step 2. Pick a Borel set $F \subset \mathbb{R}^m$ and consider the family

$$\{E \subset \mathbb{R}^n : E \times F \in \mathcal{B}(\mathbb{R}^n \times \mathbb{R}^m)\}$$

Once again, one can prove that the family is a σ-algebra and according to Step 1, it contains open sets. Thus it must contain all Borel sets as well, which ends the proof. ∎

Definition of a Measure. The second fundamental notion we shall discuss in this section is the notion of an abstract measure. Suppose we are given a set (space) X and a σ-algebra of sets $S \subset \mathcal{P}(X)$.

A nonnegative scalar-valued function $\mu: S \to [0, +\infty]$ is called a *measure* provided the following two conditions hold:

(i) $\mu \not\equiv +\infty$.

(ii) $\mu\left(\bigcup_1^\infty E_i\right) = \sum_1^\infty \mu(E_i) \stackrel{\text{def}}{=} \lim_{N\to\infty} \sum_{i=1}^N \mu(E_i)$
for $E_i \in S$ pairwise disjoint.

Both axioms are intuitively clear. The first one excludes the trivial case of measure identically equal to $+\infty$; the second assures that the notion of measure (thinking again of such ideas as length, area, volume, etc.) of a countable union of pairwise disjoint sets is equal to the (infinite) sum of their measures. If a measure μ is defined on S, the number $\mu(A)$ associated with a set $A \subset S$ is called the *measure* of A and A is said to be *measurable.* Notice that the second condition corresponds to the second axiom in the definition of σ-algebra. According to it, the infinite sum $\bigcup_1^\infty A_i$ is measurable and it makes sense to speak of its measure.

Surprisingly many results follow from the definition. We shall summarize them in the following proposition.

PROPOSITION 3.1.6

Let $\mu: S \to [0, +\infty]$ be a measure. Then the following conditions hold:

(i) $\mu(\emptyset) = 0$.

(ii) $\mu\left(\bigcup_1^n E_i\right) = \sum_1^n \mu(E_i)$, *$E_i \in S$ pairwise disjoint.*

(iii) $E, F \in S,\ E \subset F \Longrightarrow \mu(E) \le \mu(F)$.

(iv) $E_i \in S, i = 1, 2, \ldots \Longrightarrow \mu\left(\bigcup_1^\infty E_i\right) \le \sum_1^\infty \mu(E_i)$.

(v) $E_i \in S, i = 1, 2, \ldots\ E_1 \subset E_2 \subset \ldots \Longrightarrow \mu\left(\bigcup_1^\infty E_i\right) = \lim_{n\to\infty} \mu(E_n)$.

(vi) $E_i \in S, i = 1,2,\ldots\ E_1 \supset E_2 \supset \ldots,\ \exists m : \mu(E_m) < \infty \Longrightarrow \mu\left(\bigcap_1^\infty E_i\right) = \lim_{n\to\infty} \mu(E_n)$.

PROOF

(i) Since measure is not identically equal $+\infty$, there exists a set $E \in S$ such that $\mu(E) < \infty$. One has

$$\mu(E) = \mu(E \cup \emptyset \cup \emptyset \cup \ldots) = \mu(E) + \mu(\emptyset) + \mu(\emptyset) + \ldots$$

from which it follows that $\mu(\emptyset) = 0$.

(ii) Let $E_i \in S, i = 1, \ldots, n$ be pairwise disjoint sets. Completing sequence E to an infinite family by empty sets, we get

$$\mu\left(\bigcup_1^n E_i\right) = \mu\left(E_1 \cup \ldots E_n \cup \emptyset \cup \emptyset \cup \ldots\right) = \sum_1^n \mu(E_i)$$

(iii) Follows immediately from the decomposition $F = E \cup (F - E)$ and nonnegativeness of measure.

(iv) One has

$$\bigcup_1^\infty E_i = E_1 \cup (E_2 - E_1) \cup \ldots \left(E_k - \bigcup_1^{k-1} E_i\right) \cup \ldots$$

and therefore

$$\mu\left(\bigcup_1^\infty E_i\right) = \sum_{k=1}^\infty \mu\left(E_k - \bigcup_1^{k-1} E_i\right) \le \sum_1^\infty \mu(E_k)$$

in accordance with (iii).

(v) One has

$$E_n = E_1 \cup \ldots \cup E_n = E_1 \cup (E_2 - E_1) \cup \ldots \cup (E_n - E_{n-1})$$

and, consequently

$$\mu(E_n) = \mu(E_1) + \mu\left(E_2 - E_1\right) + \ldots + \mu\left(E_n - E_{n-1}\right)$$

which implies that

$$\lim_{n\to\infty} \mu(E_n) = \mu(E_1) + \sum_2^\infty \mu\left(E_i - E_{i-1}\right)$$

the last sum being equal to

$$\mu\left(E_1 \cup \bigcup_{2}^{\infty}(E_i - E_{i-1})\right) = \mu\left(\bigcup_{1}^{\infty} E_i\right).$$

(vi) Taking advantage of condition (v), we have

$$\begin{aligned}\mu\left(E_m - \bigcap_{1}^{\infty} E_i\right) &= \mu\left(\bigcup_{1}^{\infty}(E_m - E_i)\right) = \lim_{n\to\infty}\mu\left(E_m - E_n\right)\\ &= \lim_{n\to\infty}(\mu(E_m) - \mu(E_n))\\ &= \mu(E_m) - \lim_{n\to\infty}\mu(E_n)\end{aligned}$$

On the other side $\bigcap_{1}^{\infty} E_i \subset E_m$, so

$$\mu\left(E_m - \bigcap_{1}^{\infty} E_i\right) = \mu(E_m) - \mu\left(\bigcap_{1}^{\infty} E_i\right)$$

Making use of the fact that $\mu(E_m) < +\infty$, we end the proof. ∎

REMARK 3.1.1. The example of the family of sets in $\mathbb{R}$, $E_i = (i, +\infty)$ and the measure $\mu((a,b)) = (b-a)$ shows the necessity of the assumption $\mu(E_m) < +\infty$ in assertion (vi). ∎

REMARK 3.1.2. By definition, measures may take on the $+\infty$ value. The set of real numbers $\mathbb{R}$ enlarged with symbols $+\infty$ and $-\infty$ is called the *extended set of real numbers* and denoted by $\overline{\mathbb{R}}$

$$\overline{\mathbb{R}} = \mathbb{R} \cup \{-\infty\} \cup \{+\infty\}$$

Many of the algebraic properties of $\mathbb{R}$ are extended to $\overline{\mathbb{R}}$. In particular, by definition,

$$\infty + c = c + \infty = \infty \quad \forall c \in \mathbb{R}$$

$$-\infty + c = c - \infty = -\infty \quad \forall c \in \mathbb{R}$$

$$\pm\infty \cdot c = \begin{cases} \pm\infty \text{ for } c > 0 \\ \mp\infty \text{ for } c < 0 \end{cases}$$

By definition, also,

$$0 \cdot \pm\infty = 0$$

This implies particularly that for any constant $0 \leq \alpha < +\infty$ and measure μ, product $\alpha\mu$ is a measure, too (prescribed on the same σ-algebra).

Symbols $\infty + (-\infty)$, $-\infty + \infty$ remain undetermined! ∎

Substitution of a Function for a Measure. Let $f: Z \to X$ be a bijection, $S \subset \mathcal{P}(X)$ a σ-algebra and $\mu: S \to [0, +\infty]$ a measure. One can easily check that the following is a measure on $f^{-1}(S)$ (cf. Corollary 3.1.1):

$$(\mu \circ f)(E) \stackrel{\text{def}}{=} \mu(f(E))$$

Two immediate identities follow.

COROLLARY 3.1.2

(i) $\mu \circ (f \circ g) = (\mu \circ f) \circ g$.

(ii) $\alpha(\mu \circ f) = (\alpha\mu) \circ f$.

Measure Space. A triple (X, S, μ) consisting of a set (space) X, a σ-algebra $S \subset \mathcal{P}(X)$ of (measurable) sets and a measure $\mu: S \to [0, +\infty]$ is called a *measure space*.

We conclude this section with a number of simple examples.

Example 3.1.3

Let $S = \{\emptyset, X\}$ and $\mu \equiv 0$. Then the triple (X, S, μ) is an example of a trivial measure space. □

Example 3.1.4

A natural example of a measure space is furnished by the case of a finite set $X, \#X < +\infty$. Assuming $S = \mathcal{P}(X)$, one defines $\mu(A) = \#A$. One can easily check that the triple (X, S, μ) satisfies conditions of measure space. $\square$

Example 3.1.5

Returning to Example 3.1.2, let S denote the σ-algebra of *random events* of a certain type of elementary events X. A measure $p : S \to [0, \infty]$ is called a *probability* iff

$$0 \leq p(A) \leq 1 \quad \forall A \in S$$

with $p(X) = 1$ (the sure event has probability 1.0).

According to Proposition 3.1.6(i), the impossible event has zero probability: $p(\emptyset) = 0$. $\square$

Exercises

3.1.1. Prove Proposition 3.1.2.

3.1.2. Prove Corollary 3.1.1.

3.1.3. Prove the details from the proof of Proposition 3.1.5.

3.1.4. Let X be a set, $S \subset \mathcal{P}(X)$ any nontrivial σ-algebra of sets and y an element of X. Define

$$\mu(A) \overset{\text{def}}{=} \begin{cases} 1 & \text{if } y \in A \\ 0 & \text{otherwise} \end{cases}$$

Prove that μ is a measure.

3.2 Construction of Lebesgue Measure in $\mathbb{R}^n$

Though many interesting examples of measure spaces are possible, we will focus our attention almost exclusively on the most important case—the concept of Lebesgue measure and Lebesgue measurable sets. The present section is devoted to one of many possible constructions of it. The two notions: the Lebesgue measure and Lebesgue measurable sets will be constructed simultaneously.

Partition of $\mathbb{R}^n$. For a given positive integer k, we will consider the following *partition* of the real line

$$\zeta_k = \left\{ \left[\frac{\nu}{2^k}; \frac{\nu+1}{2^k} \right) \; : \; \nu \in \mathbb{Z} \right\}$$

and the corresponding partition of $\mathbb{R}^n$

$$\zeta_k^n = \left\{ \sigma = \left[\frac{\nu_1}{2^k}; \frac{\nu_1+1}{2^k} \right) \times \cdots \times \left[\frac{\nu_n}{2^k}; \frac{\nu_n+1}{2^k} \right) \; : \; \boldsymbol{\nu} = (\nu_1, \ldots, \nu_n) \in \mathbb{Z}^n \right\}$$

So the whole $\mathbb{R}^n$ has been partitioned into half-open, half-closed *cubes* σ of the same size. The diagonal length

$$\delta_k = 2^{-k}\sqrt{n}$$

will be called the *radius of the partition.*

Partition of an Open Set. Let $G \subset \mathbb{R}^n$ be an open set. Given a positive integer k we define a *partition of the open set* G as the family of all cubes belonging to the partition of $\mathbb{R}^n$ whose closures are contained in G.

$$\zeta_k(G) = \{\sigma \in \zeta_k^n \; : \; \overline{\sigma} \subset G\}$$

The union of cubes belonging to $\zeta_k(G)$ will be denoted by

$$S_k(G) = \bigcup \{\sigma \in \zeta_k(G)\}$$

The concept of the partition of an open set G is illustrated in Fig. 3.1.

Two immediate corollaries follow:

COROLLARY 3.2.1

(i) The sequence $S_k(G)$ is increasing, i.e.,

$$S_k(G) \subset S_{k+1}(G)$$

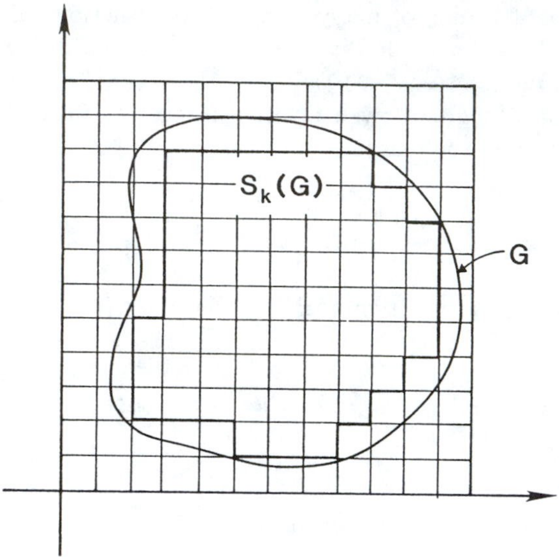

Figure 3.1
Concept of the partition of an open set G.

(ii) The set G is a union of its partitions

$$G = \bigcup_{1}^{\infty} S_k(G)$$

PROOF To prove the second assertion pick an arbitrary point $\boldsymbol{x} \in G$. Since G is open, there exists a ball $B(\boldsymbol{x}, r)$ such that $B(\boldsymbol{x}, r) \subset G$. Consider now a partition $\zeta_k(G)$ with k sufficiently small so $\delta_k < r$ and let σ denote a cube which contains $\boldsymbol{x}$. It must be $\overline{\sigma} \subset B(\boldsymbol{x}, r)$ and therefore $\overline{\sigma} \subset G$, which proves that $x \in S_k(G)$ for the chosen k. Thus G is contained in the right-hand side which is trivially contained in G and therefore the two sets are equal. ∎

"Measure" of Open Sets. With every open set $G \subset \mathbb{R}^n$ we associate a positive number $m(G)$ defined as follows:

$$m(G) = \lim_{k \to \infty} \frac{1}{2^{kn}} \# \zeta_k(G)$$

where $\#\zeta_k(G)$ denotes the number of cubes in the family $\zeta_k(G)$. Note that, with the natural assumption that the measure (volume) of a single cube is 2^{-kn}, the right-hand side of the above definition prescribes as a natural measure of $S_k(G)$, an approximation of G itself. Considering finer and finer partitions, we make this approximation better and better, which corresponds to the limit on the right-hand side. Note also that according to Corollary 3.2.1, the limit (of the increasing sequence of real numbers) always exists and is finite when G is bounded.

Quotation marks around the word "measure" are intended to emphasize that the family of open sets does not form a σ-algebra and therefore we cannot speak about a measure in a strict, mathematical sense, yet.

Some other simple observations following directly from the definition will be summarized in the following proposition.

PROPOSITION 3.2.1

Let G and H be open sets. The following properties hold:

(i) $G \subset H \Longrightarrow m(G) \le m(H)$.

(ii) $G \cap H = \emptyset \Longrightarrow m(G \cup H) = m(G) + m(H)$.

(iii) $m(G \times H) = m(G) \cdot m(H)$.
Moreover, for open intervals $(a_i, b_i) \subset \mathbb{R}$,

(iv) $m((a_1, b_1) \times \ldots \times (a_n, b_n)) = (b_1 - a_1) \ldots (b_n - a_n)$.

PROOF

(i) Follows from the fact that $S_k(G) \subset S_k(H)$.

(ii) Every cube contained with its closure in the union $G \cup H$ must be entirely contained either in G or in H and therefore $S_k(G \cup H) = S_k(G) \cup S_k(H)$. (This intuitively simple property follows from the fact that two points, one from G, another from H, cannot be connected by a broken line entirely contained in $G \cup H$ [the notion of connectivity].) Passing to the limit, we get the result required.

(iii) This follows from the fact that

$$\zeta_k(G \times H) = \{\sigma = \sigma_1 \times \sigma_2 \,:\, \sigma_1 \in \zeta_k(G), \quad \sigma_2 \in \zeta_k(H)\}$$

To prove (iv) one needs to use (iii) and prove that

$$m((a, b)) = b - a$$

We leave the proof of this result as a simple exercise. ∎

Before we prove the next property of function m we will need to consider the following simple lemma.

LEMMA 3.2.1

Let G be an open set. Then

$$m(G) = \sup\{m(H) : H \text{ open}, \overline{H} \text{ compact } \subset G\}$$

PROOF Choose $H_k = \text{int}\, S_k(G)$. Two cases are possible. Either $m(H_k) < +\infty$ for every k or $m(H_k) = +\infty$ for k sufficiently large. In the first case all H_k must be bounded and therefore $\overline{H}_k$ are compact, contained in G and according to Proposition 3.2.1 (iv), $m(H_k) = 2^{-kn}\#\zeta_k(G)$, which ends the proof. In the second case $m(G) = +\infty$ and one can consider intersections $H_{kj} = H_k \cap B(\mathbf{0}, j)$ in place of H_k. Obviously, H_{kj} satisfy all the necessary conditions and $m(H_{kj}) \to +\infty$. ∎

Having Lemma 3.2.1 in hand, we can prove the following simple, but technical, proposition.

PROPOSITION 3.2.2

The following hold:

(i) $m\left(\bigcup_1^\infty G_i\right) \leq \sum_1^\infty m(G_i)$, G_i *open.*

(ii) $\inf\{m(G - F) : F \textit{ closed} \subset G\} = 0$ *for* G *open.*

PROOF
Part (i). Step 1. We shall prove first that for two open sets G and H

$$m(G \cup H) \leq m(G) + m(H)$$

Toward this goal pick an open set D such that its closure $\overline{D}$ is compact and contained in $G \cup H$. Sets $\overline{D} - G$ and $\overline{D} - H$ are compact and disjoint and therefore they must be separated by a positive distance $\rho > 0$. One can easily see that for k such that $\rho > \delta_k$ one has

$$\zeta_k(D) \subset \zeta_k(G) \cup \zeta_k(H)$$

which implies that

$$\#\zeta_k(D) \leq \#\zeta_k(G) + \#\zeta_k(H)$$

and, consequently,

$$m(D) \leq m(G) + m(H)$$

Making use of Lemma 3.2.1. and taking the supremum over all open sets D with $\overline{D}$ compact contained in $G \cup H$, we arrive at the result required.

Step 2. By induction we have immediately

$$m\left(\bigcup_1^n G_i\right) \leq \sum_1^n m(G_i), \quad G_i \text{ open}$$

Finally, consider a sequence of open sets $G_i, i = 1, 2, \ldots$ and an arbitrary open set D such that

$$\overline{D} \text{ compact and } \overline{D} \subset \bigcup_1^\infty G_i$$

We will prove in Chapter 4 that *every open covering of a compact set contains a finite subcovering*, which implies that there exists an integer n such that

$$\overline{D} \subset \bigcup_1^n G_i$$

Consequently,

$$m(D) \leq m\left(\bigcup_1^n G_i\right) \leq \sum_1^n m(G_i) \leq \sum_1^\infty m(G_i)$$

Taking supremum over sets D, we end the proof of Part (i).

Part (ii). Case 1. Assume additionally that G is bounded. Pick an $\varepsilon > 0$. According to Lemma 3.2.1, there exists an open set H such that

$$\overline{H} \text{ compact } \subset G \text{ and } m(G) - \varepsilon \leq m(H)$$

Certainly,

$$H \cap (G - \overline{H}) = \emptyset \quad \text{and} \quad H \cup (G - \overline{H}) \subset G$$

and therefore

$$m(H) + m\left(G - \overline{H}\right) \leq m(G)$$

and, consequently,

$$m\left(G - \overline{H}\right) \leq m(G) - m(H) \leq \varepsilon$$

The choice $F = \overline{H}$ ends the proof of this case. ∎

Case 2. G arbitrary. Consider again the increasing family of balls $B_n = B(\mathbf{0}, n)$. Next pick an $\varepsilon > 0$. For every n set $G \cap B_n$ is bounded and according to Case 1 of this proof, there exists a closed set F_n such that

$$m\left(G \cap B_n - F_n\right) < \frac{\varepsilon}{2^n}$$

A simple calculation shows that

$$G - \bigcap_{n=1}^{\infty} (B_n' \cup F_n) = \bigcup_{1}^{\infty} (G \cap B - F_n)$$

and therefore

$$m\left(G - \bigcap_{1}^{\infty} (B_n' \cup F_n)\right) \leq \sum_{1}^{\infty} \frac{\varepsilon}{2^n} = \varepsilon$$

The choice

$$F = \bigcap_{1}^{\infty} (B_n' \cup F_n)$$

ends the proof.

The Lebesgue Measure—a Prototype. We shall extend now the "measure" function $m(G)$ to arbitrary sets. Given an arbitrary set $E \subset \mathbb{R}^n$ we define

$$m^*(E) \overset{\text{def}}{=} \inf\{m(G) : G \text{ open }, E \subset G\}$$

COROLLARY 3.2.2

(i) $m^*(G) = m(G)$ for open sets G.

(ii) $E \subset F$ implies that $m^*(E) \leq m^*(F)$.

(iii) $m^*\left(\bigcup_1^\infty E_i\right) \leq \sum_1^\infty m^*(E_i)$.

PROOF Conditions (i) and (ii) follow directly from the definition. To prove (iii), pick an $\varepsilon > 0$. According to the definition, for every i there exists an open set G_i such that

$$m(G_i) \leq m^*(E_i) + \frac{\varepsilon}{2^i}\,, \; E_i \subset G_i$$

Thus for the open set

$$G = \bigcup_1^\infty G_i$$

we have

$$\bigcup_1^\infty E_i \subset G \text{ and } m^*\left(\bigcup_1^\infty E_i\right) \leq m(G) \leq \sum_1^\infty m(G_i) \leq \sum_1^\infty m^*(E_i) + \varepsilon$$

Taking infimum over $\varepsilon > 0$ finishes the proof. ∎

Lebesgue Measurable Sets and Lebesgue Measure. Though function m^* has been assigned to every set E, only some of them will satisfy the axioms of σ-algebra.

PROPOSITION 3.2.3

The following three families of sets coincide with each other

(i) $\{E \subset \mathbb{R}^n : \inf_{E \subset G \text{ open}} m^*(G - E) = 0\}$.

(ii) $\{E \subset \mathbb{R}^n : \inf_{F \text{ closed} \subset E \subset G \text{ open}} m^*(G - F) = 0\}$.

(iii) $\{E \subset \mathbb{R}^n : \inf_{F \text{ closed } \subset E} m^*(E - F) = 0\}$.

PROOF

(i) $\subset$ (ii). Pick an $\varepsilon > 0$. There exists a G open such that

$$m^*(G - E) < \frac{\varepsilon}{4}$$

According to the definition of m^* there is an open set H such that

$$G - E \subset H \quad \text{and} \quad m(H) < \frac{\varepsilon}{2}$$

Making use of Proposition 3.2.2 (ii), we can find a closed set $F \subset G$ such that

$$m\,(G - F) < \frac{\varepsilon}{2}$$

Obviously, set $F - H$ is closed and

$$F - H = F \cap H' \subset F \cap \Big((G \cap E')'\Big) = F \cap (G' \cup E) = F \cap E \subset E$$

Finally, for G open and $F - H$ closed, we have

$$G - (F - H) = G \cap (F' \cup H) = (G - F) \cup (G \cap H) \subset (G - F) \cup H$$

and according to Proposition 3.2.2 (i),

$$m\,(G - (F - H)) \leq m\,(G - F) + m(H) < \varepsilon$$

(iii) $\supset$ (ii). Pick an $\varepsilon > 0$. There exists an F closed such that

$$m^*\,(E - F) < \frac{\varepsilon}{4}$$

According to the definition of m^*, there is an open set H such that

$$E - F \subset H \text{ and } m(H) < \frac{\varepsilon}{2}$$

Consider now the closed set F. According to 3.2.2 (ii), for the open set F' there exists an open set G (equivalently, G' is closed) such that

$$G' \subset F' \text{ (equivalently } F \subset G\text{) and}$$

$$m\,(G - F) = m\,(F' - G') < \tfrac{\varepsilon}{2}$$

Finally, for $G \cup H$ open and F closed, we have

$$m\,((G \cup H) - F) = m\,((G - F) \cup (H - F)) \leq m\,(G - F) + m(H) < \varepsilon$$

Inclusions (ii) $\subset$ (iii) and (iii) $\subset$ (ii) follow directly from the definition of m^*. ∎

The family defined in the three different, but equivalent, ways in Proposition 3.2.3 is called the family of *Lebesgue measurable sets* and denoted $\mathcal{L}(\mathbb{R}^n)$ or compactly $\mathcal{L}$. Intuitively, a Lebesgue measurable set is approximated from "inside" by closed and from "outside" by open sets. We shall prove now two fundamental facts: first that the family $\mathcal{L}$ is a σ-algebra containing Borel sets and, second, that m^* restricted to $\mathcal{L}$ satisfies axioms of a measure.

THEOREM 3.2.1

The following hold:

(i) $\mathcal{L}$ *is a* σ*-algebra,* $\mathcal{B} \subset \mathcal{L}$.

(ii) $m \stackrel{\text{def}}{=} m^*|_{\mathcal{L}}$ *is a measure.*

PROOF

(i) *Step 1.* Let $E \in \mathcal{L}$. F closed $\subset E \subset G$ open implies G' closed $\subset E' \subset F'$ open. Moreover $G - F = F' - G'$ and therefore, according to Proposition 3.2.3 (ii), $E' \in \mathcal{L}$.

Step 2. Assume $E_i \in \mathcal{L}$, $i = 1, 2, \ldots$. Pick an $\varepsilon > 0$. According to the first definition of $\mathcal{L}$, there exist open sets G_i such that

$$E_i \subset G_i \quad \text{and} \quad m^*(G_i - E_i) < \frac{\varepsilon}{2^i}$$

Obviously, $G = \bigcup_1^\infty G_i \supset \bigcup_1^\infty E_i$ is open and

$$m^*\left(\bigcup_1^\infty G_i - \bigcup_1^\infty E_i\right) \le m^*\left(\bigcup_1^\infty (G_i - E_i)\right) \le \sum_1^\infty m^*(G_i - E_i) \le \varepsilon$$

In conclusion, $\mathcal{L}$ is a σ-algebra. Since it contains open sets it must contain all Borel sets as well.

(ii) Obviously, $m \not\equiv +\infty$. To prove additivity we shall show first that for E_1 and E_2 disjoint

$$m^*(E_1 \cup E_2) \ge m^*(E_1) + m^*(E_2)$$

Toward this goal pick an open set $G \supset E_1 \cup E_2$ and a constant $\varepsilon > 0$. According to the definition of $\mathcal{L}$, there exist F_i closed $\subset E_i$ such that $m^*(E_i - F_i) < \frac{\varepsilon}{2}$. One can show that there exist two open sets G_i, both contained in G and separating F_i, i.e.,

$$F_i \subset G_i \subset G \ , \quad G_i \text{ disjoint}$$

It follows that

$$m^*(F_1) + m^*(F_2) \leq m(G_1) + m(G_2) = m(G_1 \cup G_2) \leq m(G)$$

and, consequently,

$$m^*(E_1) + m^*(E_2) \leq m(G) + \varepsilon$$

Taking infimum over $\varepsilon > 0$ and G open $\supset E$ we arrive at the result required.

By induction we generalize the inequality for finite families and, finally, for $E_i \in \mathcal{L}$, $i = 1, 2, \ldots$ pairwise disjoint, we have

$$\sum_1^k m^*(E_i) \leq m^*\left(\bigcup_1^k E_i\right) \leq m^*\left(\bigcup_1^\infty E_i\right)$$

The inverse inequality follows from Corollary 3.2.2 (iii). ∎

COROLLARY 3.2.3

Let Z be an arbitrary set such that $m^*(Z) = 0$. Then $Z \in \mathcal{L}$. In other words, all sets of measure zero are Lebesgue measurable.

PROOF Pick an $\varepsilon > 0$. According to the definition of m^*, there exists a G open such that $m(G) < \varepsilon$. It follows from Corollary 3.2.2. (ii) that

$$m^*(G - Z) \leq m^*(G) = m(G) < \varepsilon$$

and therefore Z is Lebesgue measurable. ∎

The set function m^* restricted to Lebesgue measurable sets $\mathcal{L}$ will be called the *Lebesgue measure*. Notice that the same symbol m has been used for both the Lebesgue measure and "measure" of open sets. This practice is justified due to the fact that m^* is an extension of m (for open sets the two notions coincide with each other).

We know already that the σ-algebra of Lebesgue measurable sets contains both Borel sets and sets of measure zero (which are not necessarily Borel). A question is, does $\mathcal{L}$ include some other, different sets? Recalling the definition of G_δ and F_σ sets (see Exercise 1.18.10), we put forward the following answer.

PROPOSITION 3.2.4

The following families of sets coincide with $\mathcal{L}(\mathbb{R}^n)$:

(i) $\{H - Z : H \text{ is } G_\delta\text{-type}, m^*(Z) = 0\}$.

(ii) $\{J \cup Z : J \text{ is } F_\sigma\text{-type}, m^*(Z) = 0\}$.

(iii) $S(\mathcal{B}(\mathbb{R}^n) \cup \{Z : m^*(Z) = 0\})$.

PROOF $\mathcal{L} \subset$ (i). Let $E \in \mathcal{L}$. Thus for every i there exists a G_i open such that

$$m^*\left(G_i - E\right) \leq \frac{1}{i}$$

Define $H = \bigcap_1^\infty G_i$ (is G_δ-type). Obviously,

$$m^*\left(H - E\right)) = \lim_{k\to\infty} m\left(\bigcap_1^k G_i - E\right) = 0$$

and

$$E = H - (H - E)$$

from which the result follows. Use Proposition 3.2.3 (i) to prove the inverse inclusion.

Proofs of two other identities are very similar and we leave them as an exercise (cf. Exercise 3.2.2.). ∎

Before we conclude this section with some fundamental facts concerning Cartesian products of Lebesgue measurable sets we shall prove the following lemma.

LEMMA 3.2.2

Let $Z \subset \mathbb{R}^n$, $m^*(Z) = 0$. Then for every, not necessarily Lebesgue measurable, set $F \subset \mathbb{R}^m$

$$m^*(Z \times F) = 0$$

PROOF

Case 1. F bounded. There exists a set H open such that $m(H) < \infty$ and $F \subset H$. Pick an arbitrary small $\varepsilon > 0$ and a corresponding open set $G \supset Z$ such that $m(G) < \varepsilon$. Obviously,

$$Z \times F \subset G \times H$$

We have

$$m^*(Z \times F) \leq m(G \times H) = m(G)m(H) < \varepsilon m(H)$$

Case 2. F arbitrary. For $B_i = B(\mathbf{0}, i)$, the balls centered at the origin, we have

$$\mathbb{R}^m = \bigcup_1^\infty B_i$$

Consequently,

$$F = F \cap \mathbb{R}^m = \bigcup_1^\infty (F \cap B_i)$$

and

$$F \times Z = \bigcup_1^\infty ((F \cap B_i) \times Z)$$

The assertion follows now from the fact that a countable union of zero measure sets is of measure zero, too. ∎

With Lemma 3.2.2 in hand we can prove the following theorem.

THEOREM 3.2.2

Let $E_1 \in \mathcal{L}(\mathbb{R}^n)$, $E_2 \in \mathcal{L}(\mathbb{R}^m)$. *Then*

(i) $E_1 \times E_2 \in \mathcal{L}(\mathbb{R}^{n+m})$ *and*

(ii) $m(E_1 \times E_2) = m(E_1)\, m(E_2)$.

PROOF

(i) Taking advantage of Proposition 3.2.4 (ii), we can represent each of the sets in the form

$$E_i = J_i \cup Z_i$$

where J_i is F_σ type and $m^*(Z_i) = 0$. One has

$$E_1 \times E_2 = (J_1 \times J_2) \cup (J_1 \times Z_2) \cup (Z_1 \times J_2) \cup (Z_1 \times Z_2)$$

But $J_1 \times J_2$ is F_σ type and according to Lemma 3.2.2, three other sets are of measure zero.

(ii) *Step 1.* Assume each of E_i is G_δ type and bounded, i.e.,

$$E_i = \bigcap_{\nu=1}^{\infty} G_i^\nu \ , \ G_i^\nu \text{ -- open}$$

One can always assume (explain why) that $\{G_i^\nu\}_{\nu=1}^\infty$ is decreasing. We have

$$m(E_1 \times E_2) = m\left(\bigcap_1^\infty G_\nu^1 \times G_\nu^2\right) = \lim_{\nu\to\infty} m\left(G_\nu^1 \times G_\nu^2\right)$$

Step 2. Assume each of E_i is an arbitrary G_δ type set. Representing E_i in the form

$$E_i = \bigcap_{\nu=1}^{\infty} G_i^\nu = \bigcap_{\nu=1}^{\infty} G_i^\nu \cap \bigcup_{k=1}^{\infty} B_k = \bigcup_1^\infty \left(\bigcap_{\nu=1}^{\infty} (G_i^\nu \cap B_k)\right)$$

(B_k, as usual, denotes the family of increasing balls centered at the origin) we can always assume that

$$E_i = \bigcup_{\nu=1}^{\infty} E_i^\nu$$

where $\{E_i^\nu\}_{\nu=1}^\infty$ is an increasing family of G_δ-type bounded sets. Making use of Step 1, we get

$$m(E_1 \times E_2) = \lim_{\nu\to\infty} m(E_1^\nu \times E_2^\nu) = \lim m(E_1^\nu) \cdot \lim m(E_2^\nu)$$

$$= m(E_1)\, m(E_2)$$

Step 3. Assume, finally, that each of E_i is of the form

$$E_i = H_i - Z_i \ , \ H_i - G_\delta \text{ type}, \quad m^*(Z_i) = 0$$

It follows immediately from Lemma 3.2.2 that

$$\begin{aligned} m(H_1 \times H_2) &= m(E_1 \times E_2) + m(E_1 \times Z_2) + m(Z_1 \times E_2) + m(Z_1 \times Z_2) \\ &= m(E_1 \times E_2) \end{aligned}$$

But, simultaneously,

$$m(H_1 \times H_2) = m(H_1)\, m(H_2) = m(E_1)\, m(E_2)$$

which ends the proof. ∎

Exercises

3.2.1. Complete proof of Proposition 3.2.1.

3.2.2. Complete proof of Proposition 3.2.4.

3.3 The Fundamental Characterization of Lebesgue Measure

Though the construction of Lebesgue measure is at least up to a certain extent very natural and fits our intuition, an immediate dilemma arises: perhaps there is another "natural" measure we can construct for Borel sets which does not coincide with Lebesgue measure. The present section brings the answer and explains why we have no other choice, i.e., why the Lebesgue measure is *the only natural measure* we can construct in $\mathbb{R}^n$.

Transitive σ-algebra. Let X be a vector space and $S \subset \mathcal{P}(X)$ a σ-algebra. We say that S is transitive if

$$A \in S \Longrightarrow A + \boldsymbol{a} \in S \text{ for every } \boldsymbol{a} \in X$$

In other words, an arbitrary translation keeps us within the family of measurable sets.

COROLLARY 3.3.1

σ-Algebras of Borel sets $\mathcal{B}(\mathbb{R}^n)$ and Lebesgue measurable sets $\mathcal{L}(\mathbb{R}^n)$ are transitive.

PROOF follows immediately from Propositions 3.1.4 and 3.2.4 (iii) and the fact that the translation

$$\tau_{\boldsymbol{a}} : \mathbb{R}^n \ni \boldsymbol{x} \longrightarrow \boldsymbol{x} + \boldsymbol{a} \in \mathbb{R}^n$$

is a continuous bijection in $\mathbb{R}^n$. ∎

Transitive Measures. Let X be a vector space and $S \subset \mathcal{P}(X)$ a transitive σ-algebra. A measure $\mu : S \to [0, +\infty]$ is called transitive if

$$\mu(A) = \mu(A + \boldsymbol{a}) \quad \text{for every} \quad A \in S, \ \boldsymbol{a} \in X$$

COROLLARY 3.3.2

Let Z and X be two vector spaces, μ a transitive measure on $S \subset \mathcal{P}(X)$ and $f : Z \to X$ an affine isomorphism. Then $\mu \circ f$ is a transitive measure on $f^{-1}(S)$.

PROOF One has to prove that

$$(\mu \circ f) \circ \tau_{\boldsymbol{a}} = \mu \circ f$$

where $\tau_{\boldsymbol{a}} : Z \ni \boldsymbol{x} \to \boldsymbol{x} + \boldsymbol{a} \in Z$ is a translation in Z. Let $f = \boldsymbol{c} + g$, where $\boldsymbol{c}$ is a vector in X and g a linear isomorphism. According to Corollary 3.1.2 (i).

$$(\mu \circ f) \circ \tau_{\boldsymbol{a}} = \mu \circ (f \circ \tau_{\boldsymbol{a}})$$

and we have

$$(f \circ \tau_{\boldsymbol{a}})(\boldsymbol{x}) = f(\boldsymbol{x} + \boldsymbol{a}) = \boldsymbol{c} + g(\boldsymbol{x} + \boldsymbol{a}) = \boldsymbol{c} + g(\boldsymbol{x}) + g(\boldsymbol{a}) = f(\boldsymbol{x}) + g(\boldsymbol{a})$$

$$= (\tau_{g(\boldsymbol{a})} \circ f)(\boldsymbol{x})$$

Consequently,

$$\mu \circ (f \circ \tau_{\boldsymbol{a}}) = (\mu \circ \tau_{g(\boldsymbol{a})}) \circ f = \mu \circ f$$

since μ is transitive in X. ∎

We shall prove now that the Lebesgue measure is transitive. We will need first the following lemma.

LEMMA 3.3.1

Let $S = \mathcal{B}(\mathbb{R}^n)$ or $\mathcal{L}(\mathbb{R}^n)$ and let $\mu : S \to [0, +\infty]$ be a measure such that

$$\mu = m \quad \text{on open cubes} \quad (a_1, b_1) \times \ldots \times (a_n, b_n)$$

Then

$$\mu = m \quad \text{on} \quad S.$$

PROOF

Step 1. $\mu = m$ on cubes $[a_1, b_1) \times \ldots \times [a_n, b_n)$. Indeed,

$$[a_1, b_1) \times \ldots \times [a_n, b_n) = \bigcap_{\nu=1}^{\infty} \left(a_1 - \frac{1}{\nu}, b_1\right) \times \ldots \times \left(a_n - \frac{1}{\nu}, b_n\right)$$

and

$$m([a_1, b_1) \times \ldots \times [a_n, b_n)) = \lim_{\nu \to \infty} m\left(\left(a_1 - \frac{1}{\nu}, b_1\right) \times \ldots \times \left(a_n - \frac{1}{\nu}, b_n\right)\right)$$

$$= \lim_{\nu \to \infty} \mu\left(\left(a_1 - \frac{1}{\nu}, b_1\right) \times \ldots \times \left(a_n - \frac{1}{\nu}, b_n\right)\right) = \mu([a_1, b_1) \times \ldots \times [a_n, b_n))$$

Step 2. $\mu = m$ on open sets. Let G open. According to Corollary 3.2.1.(ii)

$$G = \bigcup_1^{\infty} S_i(G)$$

Consequently,

$$m(G) = \lim_{i \to \infty} m(S_i(G)) = \lim_{i \to \infty} \mu(S_i(G)) = \mu(G)$$

Step 3. Let $E \in S$ and $G \supset E$ be an open set. Then

$$\mu(E) \le \mu(G) = m(G)$$

Taking infimum over all open sets G containing E, we get

$$\mu(E) \leq m(E)$$

In order to prove the inverse inequality we have to consider two cases.

Case 1. E is bounded, i.e.,

$$E \subset \sigma = [a_1, b_1] \times \ldots \times [a_n, b_n]$$

with a proper choice of $a_i, b_i, i = 1, \ldots, n$. Obviously, $\sigma - E \in S$ and

$$\mu(\sigma) - \mu(E) = \mu(\sigma - E) \leq m(\sigma - E) = m(\sigma) - m(E)$$

and therefore $\mu(E) \geq m(E)$ and, consequently,

$$\mu(E) = m(E)$$

Case 2. E arbitrary. Let $\mathbb{R}^n = \bigcup_1^\infty B_i$ (increasing balls centered at the origin). Obviously,

$$E = \bigcup_1^\infty (E \cap B_i) \ , \ E \cap B_i \text{ increasing}$$

and, consequently,

$$\mu(E) = \lim_{i \to \infty} \mu(E \cap B_i) = \lim_{i \to \infty} m(E \cap B_i) = m(E)$$

∎

THEOREM 3.3.1

Let μ be a transitive measure on $S = \mathcal{B}(\mathbb{R}^n)$ or $\mathcal{L}(\mathbb{R}^n)$ such that

$$\mu((0,1)^n) = 1$$

Then $\mu = m$.

LEMMA 3.3.2

Let $\lambda : [0, \infty) \to [0, \infty)$ be an additive function, i.e.,

$$\lambda(t+s) = \lambda(t) + \lambda(s) \quad \text{for every} \quad t, s \geq 0$$

Then

$$\lambda(ts) = t\lambda(s) \quad \text{for every} \quad t, s \geq 0$$

PROOF

Step 1. $t = k$, a positive integer

$$\lambda(ks) = \lambda(s + \ldots + s) = \lambda(s) + \ldots + \lambda(s) = k\lambda(s)$$

Step 2. $t = \frac{k}{m}$, rational, k, m positive. We have

$$\lambda(s) = \lambda\left(m\frac{s}{m}\right) = m\lambda\left(\frac{s}{m}\right)$$

or subdividing by m

$$\lambda\left(\frac{s}{m}\right) = \frac{1}{m}\lambda(s)$$

Consequently,

$$\lambda\left(\frac{k}{m}s\right) = \lambda\left(k\frac{s}{m}\right) = k\lambda\left(\frac{s}{m}\right) = \frac{k}{m}\lambda(s)$$

Step 3. $t \in [0, \infty)$. For $t = 0$ the equality is trivial since

$$\lambda(s) = \lambda(0 + s) = \lambda(0) + \lambda(s)$$

and therefore $\lambda(0) = 0$. Assume $t > 0$ and choose an increasing sequence of rational numbers t_n approaching t from the left. We have

$$t_n\lambda(s) = \lambda(t_n s) \leq \lambda(ts)$$

since t is nondecreasing (it follows from additivity). Passing to the limit on the left-hand side we get

$$t\lambda(s) \leq \lambda(ts)$$

Finally, replacing t by $\frac{1}{t}$ and s by (ts), we get

$$\frac{1}{t}\lambda(ts) \leq \lambda(s)$$

from which the equality follows. ∎

PROOF ***of Theorem 3.3.1.***
Consider a cube $(a_1, b_1) \times \ldots \times (a_n, b_n)$. It follows from transivity of μ that

$$\mu((a_1, b_1) \times \ldots \times (a_n, b_n)) = \mu((0, b_1 - a_1) \times \ldots \times (a_n, b_n))$$

Now, define a function

$$\lambda(t) = \mu((0, t) \times \ldots \times (a_n, b_n))$$

λ is additive since

$$\lambda(t + s) = \mu((0, t + s) \times \ldots \times (a_n, b_n)) = \mu((0, t) \times \ldots \times (a_n, b_n)) +$$

$$\mu((t, t + s) \times \ldots \times (a_n, b_n)) = \lambda(t) + \mu((0, s) \ldots (a_n, b_n)) =$$

$$\lambda(t) + \lambda(s)$$

Lemma 3.3.2, implies that

$$\mu((a_1, b_1) \times \ldots \times (a_n, b_n)) = (b_1 - a_1)\mu((0, 1) \times \ldots \times (a_n, b_n))$$

and, consequently,

$$\mu((a_1, b_1) \times \ldots \times (a_n, b_n)) = (b_1 - a_1) \ldots (b_n - a_n) = m((a_1, b_1) \ldots (a_n, b_n))$$

Lemma 3.3.1 finishes the proof. ∎

Nontrivial Transitive Measures. We shall say that a transitive measure on $S = \mathcal{B}(\mathbb{R}^n)$ or $\mathcal{L}(\mathbb{R}^n)$ is nontrivial if $\mu((0, 1)^n) > 0$.

COROLLARY 3.3.3

The following properties hold:

(i) Let μ be a nontrivial transitive measure on $S = \mathcal{B}(\mathbb{R}^n)$ or $\mathcal{L}(\mathbb{R}^n)$. Then

$$\mu = \alpha m \ , \quad \text{where} \quad \alpha = \mu((0, 1)^n)$$

(ii) Let μ and ν be two nontrivial measures on $\mathcal{B}(\mathbb{R}^n)$ $(\mathcal{L}(\mathbb{R}^n))$ and $\mathcal{B}(\mathbb{R}^m)$ $(\mathcal{L}(\mathbb{R}^m))$, respectively. Then there exists a unique nontrivial transitive measure on $\mathcal{B}(\mathbb{R}^{n+m})$ $(\mathcal{L}(\mathbb{R}^{n+m}))$, called the tensor product of μ and ν, denoted $\mu \otimes \nu$ such that

$$(\mu \otimes \nu)(A \times B) = \mu(A)\nu(B)$$

for $A \in \mathcal{B}(\mathbb{R}^n)$ $(\mathcal{L}(\mathbb{R}^n))$, $B \in \mathcal{B}(\mathbb{R}^m)$ $(\mathcal{L}(\mathbb{R}^m))$.

PROOF

(i) $\frac{1}{\alpha}\mu$ is a measure (Remark 3.1.2.) and satisfies the assumptions of Theorem 3.3.1. So $\frac{1}{\alpha}\mu = m$.

(ii) According to (i), $\mu = \alpha m_{\mathbb{R}^n}, \nu = \beta m_{\mathbb{R}^m}$, where $\alpha = \mu((0,1)^n)$, $\beta = \nu((0,1)^m)$. It implies

$$\mu(A)\nu(B) = \alpha\beta m_{\mathbb{R}^n}(A) m_{\mathbb{R}^m}(B) = \alpha\beta m_{\mathbb{R}^{n+m}}(A \times B)$$

The measure on the right-hand side is the unique transitive measure we are looking for. ∎

Substitution of an Affine Isomorphism. Let $f : \mathbb{R}^n \to \mathbb{R}^n$ be an affine isomorphism and $E \subset \mathbb{R}^n$ an arbitrary Borel set.

According to Proposition 3.1.4 (ii), image $f(E)$ is a Borel set as well and the question arises, what is the relation between measures $m(E)$ and $m(f(E))$? The following theorem brings an answer providing a fundamental geometrical interpretation for determinants.

THEOREM 3.3.2

Let $f : \mathbb{R}^n \to \mathbb{R}^n$ be an affine isomorphism, i.e.,

$$f(\boldsymbol{x}) = g(\boldsymbol{x}) + \boldsymbol{a}$$

where g is a linear isomorphism from $\mathbb{R}^n$ into itself and $\boldsymbol{a} \in \mathbb{R}^n$.

Then $m \circ f$ is a nontrivial transitive measure on $S = \mathcal{B}(\mathbb{R}^n)$ or $\mathcal{L}(\mathbb{R}^n)$ and

$$m \circ f = |\det \boldsymbol{G}|\, m$$

where $\boldsymbol{G}$ is the matrix representation of g in any basis in $\mathbb{R}^n$.

PROOF relies on the fundamental result for linear transformations stating that every linear isomorphism $g : \mathbb{R}^n \to \mathbb{R}^n$ can be represented as a composition of a finite number of so-called *simple isomorphisms* $g^{\lambda}_{H,\boldsymbol{c}} : \mathbb{R}^n \to \mathbb{R}^n$, where H is a $n-1$-dimensional subspace of $\mathbb{R}^n$ (the so called *hyperplane*), $\boldsymbol{c} \in \mathbb{R}^n$ and $\lambda \in \mathbb{R}^n$ (cf. Exercise 3.3.1). Representing an arbitrary vector $\boldsymbol{x} \in \mathbb{R}^n$ in the form

$$\boldsymbol{x} = \boldsymbol{x}_0 + \alpha \boldsymbol{c}, \text{ where } \boldsymbol{x}_0 \in H, \alpha \in \mathbb{R}$$

the action of the simple isomorphism is defined as follows:

$$g^{\lambda}_{H,\boldsymbol{c}}(\boldsymbol{x}) = g^{\lambda}_{H,\boldsymbol{c}}(\boldsymbol{x}_0 + \alpha c) = \boldsymbol{x}_0 + \lambda \alpha \boldsymbol{c}$$

In other words, $g^{\lambda}_{H,\boldsymbol{c}}$ reduces to identity on H and elongation along $\mathbb{R}\boldsymbol{c}$.

Given the result, the proof follows now the following steps.

Step 1. Due to the transitiveness of Lebesgue measure m, one can assume $\boldsymbol{a} = \boldsymbol{0}$.

Step 2. Let $g = g_2 \circ g_1$. Given the result for both g_1 and g_2, we can prove it immediately for composition g. Indeed, by the Cauchy theorem for determinants we have

$$m \circ g = m \circ (g_2 \circ g_1) = (m \circ g_2) \circ g_1 = \det \boldsymbol{G}_1 (m \circ g_2)$$

$$= \det \boldsymbol{G}_2 \det \boldsymbol{G}_1 \, m = \det(\boldsymbol{G}_2 \circ \boldsymbol{G}_1) \, m = \det \boldsymbol{G} \, m$$

where $\boldsymbol{G}_1, \boldsymbol{G}_2, \boldsymbol{G}$ are matrix representations for g_1, g_2 and g, respectively. Thus it is sufficient to prove the theorem for a single simple isomorphism $g^{\lambda}_{H,\boldsymbol{c}}$.

Step 3. Let $g^{\lambda}_{H,\boldsymbol{c}}$ be a simple isomorphism in $\mathbb{R}^n$ and let $\boldsymbol{a}_1, \ldots, \boldsymbol{a}_{n-1}$ denote any basis in hyperplane H. Completing it with $\boldsymbol{a}_n = \boldsymbol{c}$, we obtain a basis in $\mathbb{R}^n$. Denoting by $\boldsymbol{e}_1, \ldots, \boldsymbol{e}_n$ the canonical basis in $\mathbb{R}^n$, we construct a linear isomorphism $\phi : \mathbb{R}^n \to \mathbb{R}^n$ such that $\phi(\boldsymbol{a}_i) = \boldsymbol{e}_i, i = 1, \ldots, n$. Consequently, $g^{\lambda}_{H,\boldsymbol{c}}$ can be represented as the composition

$$g^{\lambda}_{H,\boldsymbol{c}} = \phi^{-1} \circ g^{\lambda} \circ \phi$$

where

$$g^{\lambda}(x_i) = \begin{cases} x_i, & i = 1, \ldots, n-1 \\ \lambda x_i, & i = n \end{cases}$$

Consequently,

$$g^{\lambda}((0,1)^n) = \begin{cases} (0,1) \times \ldots \times (0,1) \times (0,\lambda) & \text{for } \lambda > 0 \\ (0,1) \times \ldots \times (0,1) \times (-\lambda, 0) & \text{for } \lambda < 0 \end{cases}$$

and therefore by Theorem 3.2.2 (ii),

$$m(g^{\lambda}((0,1)^n) = | \lambda | \, m((0,1)^n) = | \det \boldsymbol{G}^{\lambda} | \; m((0,1)^n)$$

where G^λ is the matrix representation of $\boldsymbol{g}^\lambda$.

By Theorem 3.3.1,

$$\mu \circ g^\lambda =| \det \boldsymbol{G}^\lambda | \mu$$

for any transitive measure on $\mathcal{B}(\mathbb{R}^n)$. In particular, by Corollary 3.3.2, we can take $\mu = m \circ \phi^{-1}$ and therefore

$$m \circ \phi^{-1} \circ g^\lambda =| \det \boldsymbol{G}^\lambda | \mu \circ \phi^{-1}$$

and, consequently,

$$m \circ \phi^{-1} \circ g^\lambda \circ \phi =| \det \boldsymbol{G}^\lambda | \mu \circ \phi^{-1} \circ \phi =| \det \boldsymbol{G}^\lambda | \mu$$

Finally, $\det \boldsymbol{G}^\lambda = \det \boldsymbol{G}^\lambda_{H,\boldsymbol{c}}$, where $\boldsymbol{G}^\lambda_{H,\boldsymbol{c}}$ is the matrix representation of simple isomorphism $g^\lambda_{H,\boldsymbol{c}}$ (explain why) which finishes the proof for Borel sets.

Step 4. In order to conclude it for Lebesgue sets one has to prove only that affine isomorphism f prescribes $\mathcal{L}(\mathbb{R}^n)$ into itself. But, according to Proposition 3.2.4, every Lebesgue measurable set can be represented in the form

$$\bigcap_{i=1}^{\infty} G_i - Z$$

where G_i are open and Z is of measure zero. Consequently,

$$f(\bigcap_{i=1}^{\infty} G_i - Z) = \bigcap_{i=1}^{\infty} f(G_i) - f(Z)$$

where $f(G_i)$ are open and it remains to prove only that $f(Z)$ is of measure zero. It follows, however, from proof of Corollary 3.2.3 that

$$Z \subset \bigcap_{i=1}^{\infty} H_i, \quad H_i \text{ open }, \quad m(H_i) < \frac{1}{i}$$

and, consequently,

$$f(Z) \subset \bigcap_{i=1}^{\infty} f(H_i), \quad f(H_i) \text{ open}$$

and therefore, according to the just proven result for Borel sets,

$$m(f(H_i)) =| \det \boldsymbol{G} | m(H_i) <| \det \boldsymbol{G} | \frac{1}{\epsilon}$$

where $\boldsymbol{G}$ is the matrix representation of the linear part of f. Consequently, $m(f(Z)) = 0$, which finishes the proof. ∎

COROLLARY 3.3.4

Let $(0,1)^n$ denote the unit cube in $\mathbb{R}^n$ and $f(\boldsymbol{x}) = g(\boldsymbol{x}) + \boldsymbol{a}$ be an affine isomorphism in $\mathbb{R}^n$ with $\boldsymbol{G}$—the matrix representation of g. Then

$$m(f((0,1)^n)) = |\det \boldsymbol{G}|$$

Exercises

3.3.1. Follow the outlined steps to prove that every linear isomorphism $g : \mathbb{R}^n \to \mathbb{R}^n$ is a composition of simple isomorphisms $g^{\lambda}_{H,\boldsymbol{c}}$.

Step 1. Let H be a hyperplane in $\mathbb{R}^n$, and $\boldsymbol{a}$ and $\boldsymbol{b}$ denote two vectors such that $\boldsymbol{a}, \boldsymbol{b}, \boldsymbol{a} - \boldsymbol{b} \notin H$. Show that there exists a unique simple isomorphism $g^{\lambda}_{H,\boldsymbol{c}}$ such that

$$g^{\lambda}_{H,\boldsymbol{c}}(\boldsymbol{a}) = \boldsymbol{b}$$

Hint: $\boldsymbol{c} = \boldsymbol{b} - \boldsymbol{a}$.

Step 2. Let g be a linear isomorphism in $\mathbb{R}^n$ and consider the subspace $Y = Y(g)$ of $\mathbb{R}^n$ such that $g(\boldsymbol{x}) = \boldsymbol{x}$ on Y. Assume that $Y \neq \mathbb{R}^n$. Let H be any hyperplane containing Y. Show that there exist vectors $\boldsymbol{a}, \boldsymbol{b}$ such that

$$\boldsymbol{a} \in H, \quad g(\boldsymbol{a}) \notin H$$

and

$$\boldsymbol{b} \notin H, \quad \boldsymbol{b} - g(\boldsymbol{a}) \notin H, \quad \boldsymbol{b} - \boldsymbol{a} \notin H$$

Make use of the Step 1 result and consider simple isomorphisms g_1 and h_1 invariant on H and mapping $f(\boldsymbol{a})$ into $\boldsymbol{b}$ and $\boldsymbol{b}$ into $\boldsymbol{a}$, respectively.

Prove that

$$\dim Y(h_1 \circ g_1 \circ g) > \dim Y(g)$$

Step 3. Use induction to argue that after a finite number of steps m

$$\dim Y(h_m \circ g_m \circ \ldots \circ h_1 \circ g_1 \circ g) = n$$

Consequently,

$$h_m \circ g_m \circ \ldots \circ h_1 \circ g_1 \circ g = \mathrm{id}_{\mathbb{R}^n}$$

and

$$g = g_1^{-1} \circ h_1^{-1} \circ \ldots \circ g_m^{-1} \circ h_m^{-1}$$

Finish the proof by arguing that inverse of a simple isomorphism is itself a simple isomorphism, too.

Lebesgue Integration Theory

3.4 Measurable and Borel Functions

Lebesgue Measurable and Borel Functions. We say that a function $\varphi: \mathbb{R}^n \to \overline{\mathbb{R}}$ is (Lebesgue) *measurable* if the following conditions hold:

(i) dom φ is measurable (in $\mathbb{R}^n$).

(ii) the set $\{(\boldsymbol{x}, y) \in \text{ dom } \varphi \times \mathbb{R} : y < \varphi(\boldsymbol{x})\}$ is measurable (in $\mathbb{R}^{n+1}$).

Similarly, we say that function φ is Borel if its domain is a Borel set and the set defined above is Borel. If no confusion occurs, we will use a simplified notation $\{y < \varphi(\boldsymbol{x})\}$ in place of $\{(\boldsymbol{x}, y) \in E \times \mathbb{R} : y < \varphi(\boldsymbol{x})\}$.

Some fundamental properties of measurable and Borel functions are summarized in the following proposition.

PROPOSITION 3.4.1

The following properties hold:

(i) $E \subset \mathbb{R}^n$ *measurable (Borel),* $\varphi: \mathbb{R}^n \to \overline{\mathbb{R}}$ *measurable (Borel)* $\Rightarrow \varphi|_E$ *measurable (Borel).*

(ii) $\varphi_i: E_i \to \overline{\mathbb{R}}$ *measurable (Borel),* E_i *pairwise disjoint* $\Rightarrow \varphi = \bigcup_1^\infty \varphi_i$ *measurable (Borel).*

(iii) φ *measurable (Borel)* $\Rightarrow \lambda\varphi$ *measurable (Borel).*

(iv) $\varphi_i\colon E \to \overline{\mathbb{R}}$ *measurable (Borel)* $\Rightarrow$ *(pointwise)* $\sup \varphi_i$, $\inf \varphi_i$, $\limsup \varphi_i$, $\liminf \varphi_i$ *measurable (Borel). In particular, if* $\lim \varphi_i$ *exists, then* $\lim \varphi_i$ *is measurable (Borel).*

PROOF

(i) follows from

$$\{(\boldsymbol{x}, y) \in E \times \mathbb{R} : y < \varphi|_E(\boldsymbol{x})\} = \{y < \varphi(\boldsymbol{x})\} \cap \{E \times \mathbb{R}\}$$

(ii) follows from

$$\{y < \varphi|_{\cup E_i}(\boldsymbol{x})\} = \bigcup_i \{y < \varphi|_{E_i}(\boldsymbol{x})\}$$

(iii) follows from

$$\{y < \lambda\varphi(\boldsymbol{x})\} = h^{-1}(\{y < \varphi(\boldsymbol{x})\})$$

where $h\colon (\boldsymbol{x}, y) \to (\boldsymbol{x}, y/\lambda)$ is an affine isomorphism.

(iv) It is sufficient to make use of the following identities:

$$\left\{y < \sup_i \varphi_i(\boldsymbol{x})\right\} = \bigcup_i \{y < \varphi_i(\boldsymbol{x})\}$$

$$\left\{y > \inf_i \varphi_i(\boldsymbol{x})\right\} = \bigcup_i \{y > \varphi_i(\boldsymbol{x})\}$$

$$\limsup_{i\to\infty} \varphi_i(\boldsymbol{x}) = \inf_i \left\{\sup_{j\geq i} \varphi_j(\boldsymbol{x})\right\}$$

$$\liminf_{i\to\infty} \varphi_i(\boldsymbol{x}) = \sup_i \left\{\inf_{j\geq i} \varphi_j(\boldsymbol{x})\right\}$$

(cf. Exercise 3.4.1.) ∎

PROPOSITION 3.4.2

Every continuous function $\varphi: E \to \overline{\mathbb{R}}$, E open, is Borel and therefore measurable, too.

PROOF We have

$$\{y < \varphi(\boldsymbol{x})\} = g^{-1}(\{y < z\})$$

where

$$g: E \times \mathbb{R} \ni (x, y) \to (y, \varphi(\boldsymbol{x})) \in \mathbb{R}^2$$

is a continuous function. Since set $\{y < z\}$ is open, the set on the left-hand side must be open, too, and therefore is both Borel and measurable. ∎

We leave as an exercise proof of the following proposition.

PROPOSITION 3.4.3

Let $g: \mathbb{R}^n \to \mathbb{R}^n$ be an affine isomorphism. Then φ is measurable (Borel) if and only if $\varphi \circ g$ is measurable (Borel).

Almost Everywhere Properties. A property $P(x)$, $x \in E$ is said to hold *almost everywhere* on E (often written "a.e." on E) if P fails to hold only on a subset of measure zero. In other words,

$$P(x) \text{ holds a.e. on } A \text{ iff } P(x) \text{ holds for } x \in A - Z \text{ and } m(Z) = 0$$

As an example of an application of the above notion, we have the following simple proposition, proof of which we leave as an exercise.

PROPOSITION 3.4.4

Let $\varphi_i: E \to \overline{\mathbb{R}}, i = 1, 2$ and $\varphi_1 = \varphi_2$ a.e. in E. Then φ_1 is measurable iff φ_2 is measurable.

Exercises

3.4.1. Let $\varphi: \mathbb{R}^n \to \overline{\mathbb{R}}$ be a function such that dom φ is measurable (Borel).

Prove that the following conditions are equivalent to each other:

(i) φ is measurable (Borel)

(ii) $\{(\boldsymbol{x}, y) \in \text{dom } \varphi \times \mathbb{R} : y \le \varphi(\boldsymbol{x})\}$ is measurable (Borel)

(iii) $\{(\boldsymbol{x}, y) \in \text{dom } \varphi \times \mathbb{R} : y > \varphi(\boldsymbol{x})\}$ is measurable (Borel)

(iv) $\{(\boldsymbol{x}, y) \in \text{dom } \varphi \times \mathbb{R} : y \ge \varphi(\boldsymbol{x})\}$ is measurable (Borel)

3.4.2. Prove Proposition 3.4.3. *Hint:* use Theorem 3.3.2.

3.4.3. Prove Proposition 3.4.4.

3.5 Lebesgue Integral of Nonnegative Functions

Definition of Lebesgue Integral. Let $\varphi : \mathbb{R}^n \to \overline{\mathbb{R}}$ be a nonnegative function. Assume that domain of φ, dom φ is measurable (Borel) and define the set

$$S(\varphi) = \{(\boldsymbol{x}, y) \in \text{dom } \varphi \times \mathbb{R} : 0 < y < \varphi(\boldsymbol{x})\}$$

One can easily prove that set $S(\varphi)$ is measurable (Borel) if and only if function φ is measurable (Borel).

The Lebesgue measure (in $\mathbb{R}^{n+1}$) of set $S(\varphi)$ will be called the *Lebesgue integral* of function φ (over its domain) and denoted by

$$\int \varphi dm \quad \text{or} \quad \int \varphi(\boldsymbol{x}) dm(\boldsymbol{x}) \quad \text{or} \quad \int \varphi(\boldsymbol{x}) d\boldsymbol{x}$$

Thus

$$\int \varphi dm = m_{(n+1)}(S(\varphi))$$

where $m_{(n+1)}$ denotes the Lebesgue measure in $\mathbb{R}^{n+1}$.

If E is a measurable set then the integral of function φ over set E is defined as the integral of φ restricted to E, i.e.,

$$\int_E \varphi dm \overset{\text{def}}{=} \int \varphi|_E \, dm = m_{(n+1)}(S(\varphi|_E)) = m_{(n+1)}(S(\varphi) \cap (E \times \mathbb{R}))$$

The concept of the integral is illustrated in Fig. 3.2. The geometrical interpretation is clear.

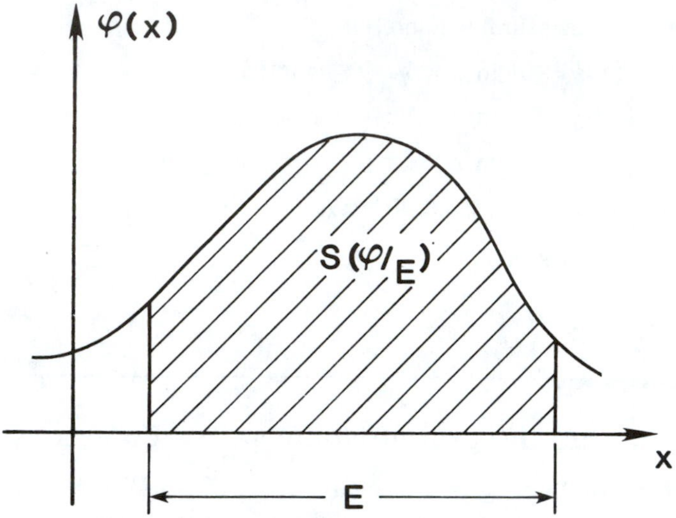

Figure 3.2
Definition of Lebesgue integral for nonnegative functions. Set $S(\varphi\,|_E)$.

A number of properties follow immediately from the definition and the properties of a measure.

PROPOSITION 3.5.1

All considered functions are measurable and nonnegative. The following properties hold:

(i) $m(E) = 0 \Longrightarrow \int_E \varphi\, dm = 0.$

(ii) $\varphi: E \to \overline{\mathbb{R}}, E_i \subset E,\ i = 1, 2, \dots$ *measurable and pairwise disjoint*

$$\int_{\cup E_i} \varphi\, dm = \sum_1^\infty \int_{E_i} \varphi\, dm$$

(iii) $\varphi, \psi: E \to \overline{\mathbb{R}},\ \varphi = \psi$ *a.e. in* E

$$\int_E \varphi\, dm = \int_E \psi\, dm$$

(iv) $c \geq 0$, *E measurable*

$$\int_E c\, dm = cm(E)$$

(v) $\varphi, \psi: E \to \overline{\mathbb{R}}$, $\varphi \leq \psi$ *a.e. in E*

$$\int_E \varphi\, dm \leq \int_E \psi\, dm$$

(vi) $\lambda \geq 0$

$$\int (\lambda\varphi)\, dm = \lambda \int \varphi\, dm$$

PROOF

(i) follows from the inclusion $S(\varphi\, |_E) \subset E \times \mathbb{R}$ and the fact that $m(E \times \mathbb{R}) = 0$ (cf. Lemma 3.2.2).

(ii) is an immediate consequence of measure properties and the decomposition

$$S\left(\varphi|_{\cup E_i}\right) = \bigcup_1^\infty S\left(\varphi|_{E_i}\right)$$

where the sets on the right-hand side are pairwise disjoint.

(iii) follows from (i).

(iv) is a consequence of Theorem 3.2.2 and the formula

$$S(c) = E \times (0, c)$$

(v) follows from the fact that $S(\varphi) \subset S(\psi)$ (except for a set of measure zero).

(vi) For $\lambda = 0$ the formula is trivial. For $\lambda > 0$ we have

$$S(\lambda\varphi) = \{0 < y < \lambda\varphi(\boldsymbol{x})\} = h(S(\varphi))$$

where $h: (\boldsymbol{x}, y) \to (\boldsymbol{x}, \lambda y)$ is an affine isomorphism and $|\det h| = \lambda$. Thus Theorem 3.3.2 implies the result. ∎

The following lemma plays a crucial role in the whole integration theory.

LEMMA 3.5.1

Let $\varphi, \varphi_i: E \to \overline{\mathbb{R}}$, $i = 1, 2, \ldots$ be measurable, nonnegative functions. Then:

(i) $\varphi_i \nearrow \varphi \Longrightarrow \int \varphi_i dm \to \int \varphi dm.$

(ii) $\varphi_i \searrow \varphi$ and $\exists j : \int \varphi_j dm < +\infty \Longrightarrow \int \varphi_i dm \to \int \varphi dm.$

PROOF

(i) We have

$$S(\varphi) = \bigcup_{1}^{\infty} S(\varphi_i)$$

and the family $S(\varphi_i)$ is increasing. An application of Proposition 3.1.6 (v) ends the proof.

(ii) Introduce the set

$$S^1(\varphi) = \{0 < y \le \varphi(\boldsymbol{x})\}$$

We claim that

$$\int \varphi dm = m(S^1(\varphi))$$

Indeed,

$$S(\varphi) \subset S^1(\varphi) \subset S\left(\left(1 + \frac{1}{k}\right)\varphi\right)$$

which implies that

$$\int \varphi dm \le m_{(n+1)}(S^1(\varphi)) \le \left(1 + \frac{1}{k}\right)\int \varphi dm$$

Passing with k to infinity, we get the result.

We have now

$$S^1(\varphi) = \bigcap_{1}^{\infty} S^1(\varphi_i)$$

where $S^1(\varphi_i)$ is decreasing and $m(S^1(\varphi_i)) < +\infty$. Applying Proposition 3.1.6 (vi), we end the proof. ∎

We shall prove now two fundamental results of integration theory. Both of them are simple consequences of Lemma 3.5.1.

THEOREM 3.5.1
(Fatou's Lemma)

Let $f_i: E \to \overline{\mathbb{R}}, i = 1, 2, \ldots$ *be a sequence of measurable, nonnegative functions. Then*

$$\int \liminf f_i dm \leq \liminf \int f_i dm$$

PROOF We have

$$\inf_{\nu \geq i} f_\nu(\boldsymbol{x}) \leq f_i(\boldsymbol{x})$$

and according to Proposition 3.5.1 (v),

$$\int \inf_{\nu \geq i} f_\nu dm \leq \int f_i dm$$

Our main objective now is pass to the limit, or, more precisely, to lim inf, on both sides of this inequality. On the right-hand side we get simply lim inf $\int f_i dm$. On the left-hand side the sequence is increasing, so the lim inf coincides with the usual limit. Moreover,

1. $\lim_{i\to\infty}(\inf_{\nu \geq i} f_\nu(\boldsymbol{x})) = \liminf_{i\to\infty} f_i(\boldsymbol{x})$ (cf. Proposition 1.19.2) and
2. the sequence on the left-hand side is increasing and, therefore, according to Lemma 3.5.1,

$$\int \inf_{\nu \geq i} f_\nu dm \to \int \liminf f_i dm$$

which ends the proof. ∎

THEOREM 3.5.2
(The Lebesgue Dominated Convergence Theorem)

Let $f, f_i: E \to \overline{\mathbb{R}}, i = 1, 2, \ldots$ *be measurable, nonnegative functions. Assume*

(i) $f_i(\boldsymbol{x}) \to f(\boldsymbol{x})$ *for* $\boldsymbol{x} \in E$ *and*

(ii) $f_i(\boldsymbol{x}) \leq \varphi(\boldsymbol{x})$, *where* $\varphi: E \to \overline{\mathbb{R}}$ *is a measurable function such that*

$$\int \varphi dm < +\infty$$

Then

$$\int f_i dm \to \int f\ dm$$

PROOF We have

$$\inf_{\nu \geq i} f_\nu(\boldsymbol{x}) \leq f_i(\boldsymbol{x}) \leq \sup_{\nu \geq i} f_\nu(\boldsymbol{x}) \leq \varphi(\boldsymbol{x})$$

and, consequently,

$$\int \inf_{\nu \geq i} f_\nu dm \leq \int f_i dm \leq \int \sup_{\nu \geq i} f_\nu dm \leq \int \varphi dm$$

Now, looking at the left-hand side,

1. $\liminf_{i \to \infty \nu \geq i} f_\nu = \liminf_{i \to \infty} f_i = \lim_{i \to \infty} f_i = f$
2. sequence $\inf_{\nu \geq i} f_\nu$ is increasing

Thus, according to Lemma 3.5.1. (i) the left-hand side converges to $\int f dm$. Similarly, for the right-hand side,

1. $\limsup_{i \to \infty \nu \geq i} f = \limsup_{i \to \infty} f_i = \lim_{i \to \infty} f_i = f$
2. sequence $\sup_{\nu \geq i} f_\nu$ is decreasing
3. all integrals $\int \sup_{\nu \geq i} f_\nu dm$ are bounded by $\int \varphi dm$

Thus, according to Lemma 3.5.1 (ii), the right-hand side converges to $\int f dm$, too, which proves (the three sequences lemma) that

$$\int f_i dm \to \int f dm$$

∎

REMARK 3.5.1. In both Theorems 3.5.1. and 3.5.2. all the pointwise conditions in E may be replaced by the same conditions, but satisfied a.e. in E only (explain why). ∎

We will conclude this section with a simple result concerning the change of variables in the Lebesgue integral.

PROPOSITION 3.5.2

Let

$g : \mathbb{R}^n \to \mathbb{R}^n$ *be an affine isomorphism and*
$\varphi : \mathbb{R}^n \supset E \to \mathbb{R}$ *a measurable function.*

Then

$$\int_E \varphi dm = \int_{g^{-1}(E)} (\varphi \circ g)|\det g|\, dm$$

PROOF Define the mapping

$$h: \mathbb{R}^{n+1} \ni (\boldsymbol{x}, y) \to (g(\boldsymbol{x}), y) \in \mathbb{R}^{n+1}$$

Obviously, h is an affine isomorphism and $|\det h| = |\det g|$. Also,

$$h(S(\varphi \circ g)) = S(\varphi)$$

Thus, according to Theorem 3.3.2,

$$\int_{g^{-1}(E)} (\varphi \circ g)|\det g| dm = |\det g| m(S(\varphi \circ g))$$

$$= (m \circ h)(S(\varphi \circ g)) = m(h(S(\varphi \circ g)))$$

$$= m(S(\varphi)) = \int_E \varphi dm$$

∎

3.6 Fubini's Theorem for Nonnegative Functions

In this section we derive the third fundamental theorem of integration theory—Fubini's theorem. As in the previous section, we will restrict ourselves to the case of nonnegative functions only, generalizing all the results to the general case in the next paragraph. We will start first with the so-called generic case of Fubini's theorem.

THEOREM 3.6.1
(Fubini's Theorem—The Generic Case)

Let E be a set in $\mathbb{R}^{n+m} = \mathbb{R}^n \times \mathbb{R}^m$. For each $\boldsymbol{x} \in \mathbb{R}^n$ and $\boldsymbol{y} \in \mathbb{R}^m$ we define

$$E^{\boldsymbol{x}} = \{\boldsymbol{y} : (\boldsymbol{x}, \boldsymbol{y}) \in E\}$$

$$E^{\boldsymbol{y}} = \{\boldsymbol{x} : (\boldsymbol{x}, \boldsymbol{y}) \in E\}$$

Sets $E^{\boldsymbol{x}}$ and $E^{\boldsymbol{y}}$ correspond to sections of set E along $\boldsymbol{y}$ and $\boldsymbol{x}$ "axes," respectively (see Fig. 3.3 for the geometrical interpretation). *The following hold:*

(i) *If set E is Borel then*
   1. *$E^{\boldsymbol{x}}$ is Borel for every $\boldsymbol{x}$;*
   2. *$\mathbb{R}^n \ni \boldsymbol{x} \to m_{(m)}(E^{\boldsymbol{x}})$ is a Borel function;*
   3. $m_{(n+m)}(E) = \int m_{(m)}(E^{\boldsymbol{x}}) dm_{(n)}(\boldsymbol{x})$.

(ii) *If set E is measurable then*
   1. *$E^{\boldsymbol{x}}$ is measurable a.e. in $\mathbb{R}^n$;*
   2. *$\mathbb{R}^n \ni \boldsymbol{x} \to m_{(m)}(E^{\boldsymbol{x}})$ (defined a.e.) is measurable;*
   3. $m_{(n+m)}(E) = \int m_{(m)}(E^{\boldsymbol{x}}) dm_{(n)}(\boldsymbol{x})$.

(iii) *If $m_{(n+m)}(E) = 0$ then $m_{(m)}(E^{\boldsymbol{x}}) = 0$ a.e. in $\mathbb{R}^n$.*

Before we start with the proof we will need some auxiliary results. First of all, let us recall the very preliminary definition of cubes introduced in Section 3.2.

$$\sigma = \left[\frac{\nu_1}{2^k}, \frac{\nu_1 + 1}{2^k}\right) \times \ldots \times \left[\frac{\nu_n}{2^k}, \frac{\nu_n + 1}{2^k}\right), \quad k = 1, 2, \ldots$$

Now, let us denote by $\mathcal{T}_k$ a family consisting of all such sets which are finite unions of cubes σ for a given k and consider the family $\mathcal{T} = \bigcup_1^\infty \mathcal{T}_k$. We have an obvious

COROLLARY 3.6.1

If $A, B \in \mathcal{T}$ then $A \cup B,\ A - B \in \mathcal{T}$.

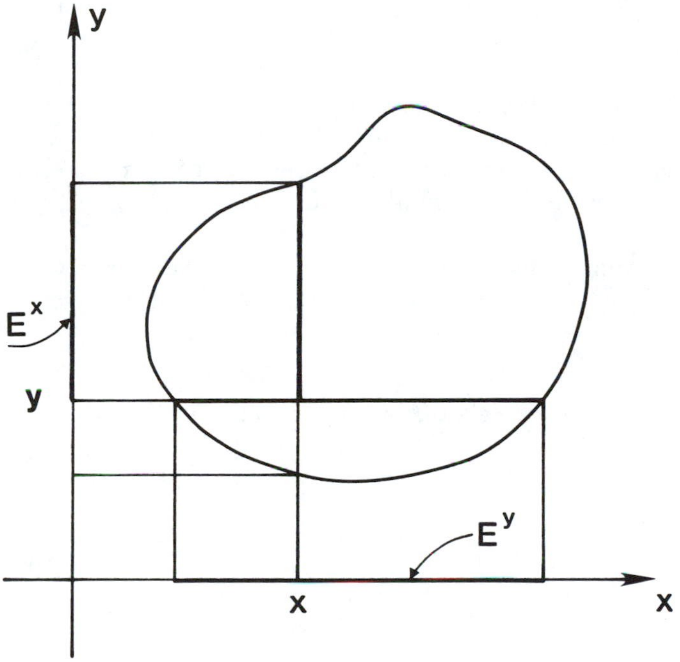

Figure 3.3
Fubini's Theorem—The Generic Case. Sets E^x and E^y

***m*-Class of Sets.** A family $\mathcal{M} \subset \mathcal{P}(\mathbb{R}^n)$ will be called an *m-class of sets* if it is closed with respect to unions of increasing sequences of sets and intersections of decreasing sequences of bounded sets. In other words, the following conditions hold :

$$A_i \in \mathcal{M}\ ,\ A_1 \subset A_2 \subset \ldots \Longrightarrow \bigcup_1^\infty A_i \in \mathcal{M}$$

$$A_1 \in \mathcal{M}\ ,\ A_1 \supset A_2 \supset \ldots,\ A_i \text{ bounded } \Longrightarrow \bigcap_1^\infty A_i \in \mathcal{M}$$

We have an immediate

COROLLARY 3.6.2

(i) A σ-algebra of sets is an m-class.

(ii) For every class $\mathcal{C}$ there exists the smallest m-class $\mathcal{M}(\mathcal{C})$ containing $\mathcal{C}$.

PROOF (i) follows from definition of σ-algebra and Proposition 3.1.1 (iii). To prove (ii) it is sufficient to notice that a common part of m-classes is an m-class and next consider the common part of all m-classes containing $\mathcal{C}$ (the family is nonempty; why?). ∎

The following lemma is crucial in the proof of Theorem 3.6.1.

LEMMA 3.6.1

The family of Borel sets $\mathcal{B}(\mathbb{R}^n)$ is the smallest m-class containing family $\mathcal{T}$, i.e.,

$$\mathcal{B}(\mathbb{R}^n) = \mathcal{M}(\mathcal{T})$$

PROOF Since $\mathcal{B}(\mathbb{R}^n)$ as a σ-algebra is an m-class and it contains $\mathcal{T}$, we have immediately

$$\mathcal{B}(\mathbb{R}^n) \supset \mathcal{M}(\mathcal{T})$$

To prove the inverse inclusion we will prove that

1. open sets are contained in $\mathcal{M}(\mathcal{T})$
2. $\mathcal{M}(\mathcal{T})$ is a σ-algebra

According to Corollary 3.2.1 (ii) every open set $G = \bigcup_{1}^{\infty} S_k(G)$, which proves the first assertion. We claim that in order to prove the second result it suffices to prove that $A, B \in \mathcal{M}(\mathcal{T})$ implies that $A - B \in \mathcal{M}(\mathcal{T})$. Indeed, $\mathbb{R}^n \in \mathcal{M}(\mathcal{T})$ and therefore for $A \in \mathcal{M}(\mathcal{T})$, complement $A' = \mathbb{R}^n - A$ belongs to $\mathcal{M}(\mathcal{T})$. Representing a union of two sets A and B in the form

$$A \cup B = (A' - B)'$$

we see that $A, B \in \mathcal{M}(\mathcal{T})$ implies that $A \cup B \in \mathcal{M}(\mathcal{T})$. And, finally, from the representation

$$\bigcup_{1}^{\infty} A_i = \bigcup_{i=1}^{\infty} \left(\bigcup_{k=1}^{i} A_k \right)$$

and the fact that $\bigcup_{k=1}^{i} A_k$ is increasing, it follows that infinite unions of sets belonging to $\mathcal{M}(\mathcal{T})$ belong to the same class as well. Thus it is sufficient to prove that for $A, B \in \mathcal{M}(\mathcal{T})$ the difference $A - B \in \mathcal{M}(\mathcal{T})$.

Step 1. Pick an $A \in \mathcal{T}$ and consider the class

$$\{B : A - B \in \mathcal{M}(\mathcal{T})\}$$

We claim that the family above is an m-class containing $\mathcal{T}$. Indeed:

1. According to Corollary 3.6.1, it contains $\mathcal{T}$.
2. Let $B_1 \subset B_2 \subset \dots$ be an increasing sequence of sets belonging to the class, i.e., $A - B_i \in \mathcal{M}(\mathcal{T})$. Moreover, the sets $A - B_i$ are bounded (A is bounded) and sequence $A - B_i$ is decreasing.. Thus
$$A - \bigcup_{1}^{\infty} B_i = \bigcap_{1}^{\infty} (A - B_i) \in \mathcal{M}(\mathcal{T})$$
3. Let $B_1 \supset B_2 \supset \dots$ be bounded. Again, from the identity
$$A - \left(\bigcap_{1}^{\infty} B_i \right) = \bigcup_{1}^{\infty} (A - B_i)$$
it follows that $\bigcup_{1}^{\infty} B_i$ belongs to $\mathcal{M}(\mathcal{T})$.

Thus the considered class must contain $\mathcal{M}(\mathcal{T})$, which implies that for $A \in \mathcal{T}$, $B \in \mathcal{M}(\mathcal{T})$ the difference $A - B \in \mathcal{M}(\mathcal{T})$.

Step 2. Pick a $B \in \mathcal{M}(\mathcal{T})$ and consider the class

$$\{A : A - B \in \mathcal{M}(\mathcal{T})\}$$

In the identical manner we prove that this is an m-class, and according to Step 1 it contains $\mathcal{M}(\mathcal{T})$.

Thus we have come to the conclusion that $A, B \in \mathcal{M}(\mathcal{T})$ implies that $A - B \in \mathcal{M}(\mathcal{T})$, which finishes the proof. ∎

PROOF *of Theorem 3.6.1.*

Part (i). For every $\boldsymbol{x} \in \mathbb{R}^n$, define the function

$$i_{\boldsymbol{x}}: \boldsymbol{y} \to (\boldsymbol{x}, \boldsymbol{y})$$

Function $i_{\boldsymbol{x}}$ is obviously continuous and $E^{\boldsymbol{x}} = i_{\boldsymbol{x}}^{-1}(E)$. Thus (cf. Proposition 3.1.4) if E is Borel then $E^{\boldsymbol{x}}$ is Borel as well. Now, pick a set E and define the function

$$\eta_E: \boldsymbol{x} \to m_{(m)}(E^{\boldsymbol{x}})$$

We shall prove that the family

$$\mathcal{F} = \left\{ E \text{ Borel } : \eta_E \text{ Borel and } m_{(n+m)}(E) = \int \eta_E dm_{(n)} \right\}$$

is an m-class containing $\mathcal{J}$.

Step 1. Let $E_1 \subset E_2 \ldots$. Denote $E = \bigcup_1^\infty E_i$. Obviously,

$$E^{\boldsymbol{x}} = \bigcup_1^\infty E_i^{\boldsymbol{x}} \,,\; E_1^{\boldsymbol{x}} \subset E_2^{\boldsymbol{x}} \subset \ldots$$

Thus, according to Proposition 3.1.6 (v),

$$m(E^{\boldsymbol{x}}) = \lim_{i\to\infty} m(E_i^{\boldsymbol{x}})$$

i.e., $\eta_{E_i}(\boldsymbol{x}) \to \eta_E(\boldsymbol{x})$, which implies that η_E is Borel (cf. Proposition 3.4.1 (iv)) and, due to the fact that η_{E_i} is increasing, Lemma 3.5.1 (i) together with Proposition 3.1.6 (v) yield

$$m_{(n+m)}(E) = \lim m_{(n+m)}(E_i) = \lim \int \eta_{E_i} dm_{(n)} = \int \eta_E dm_{(n)}$$

Step 2. is analogous to Step 1. Pick a decreasing sequence of bounded sets $E_1 \supset E_2 \supset \ldots$ and proceed to prove that for $E = \bigcap_1^\infty E_i$, η_E is Borel and $m(E) = \int \eta_E dm$.

Step 3. $\mathcal{F}$ contains class $\mathcal{J}$. Indeed, sets E from $\mathcal{J}$ are obviously Borel. Now let $E \in \mathcal{J}$. One can always represent set E in the form

$$E = \bigcup_1^\infty (E_i \times F_i)$$

where E_i are pairwise disjoint cubes σ in $\mathbb{R}^n$ and F_i are unions of cubes in $\mathbb{R}^m$. Then

$$\eta_E(\boldsymbol{x}) = \begin{cases} m_{(m)}(F_i) & \text{if } \boldsymbol{x} \in E_i \\ 0 & \text{otherwise} \end{cases}$$

(see Fig. 3.4 for geometrical interpretation). Thus η_E as a union of Borel functions (explain why) is Borel.

Finally,

$$\int \eta_E dm_{(n)} = \sum_1^\infty m_{(n)}(E_i) m_{(m)}(F_i) = \sum_1^\infty m_{(n+m)}(E_i \times F_i) = m_{(n+m)}(E)$$

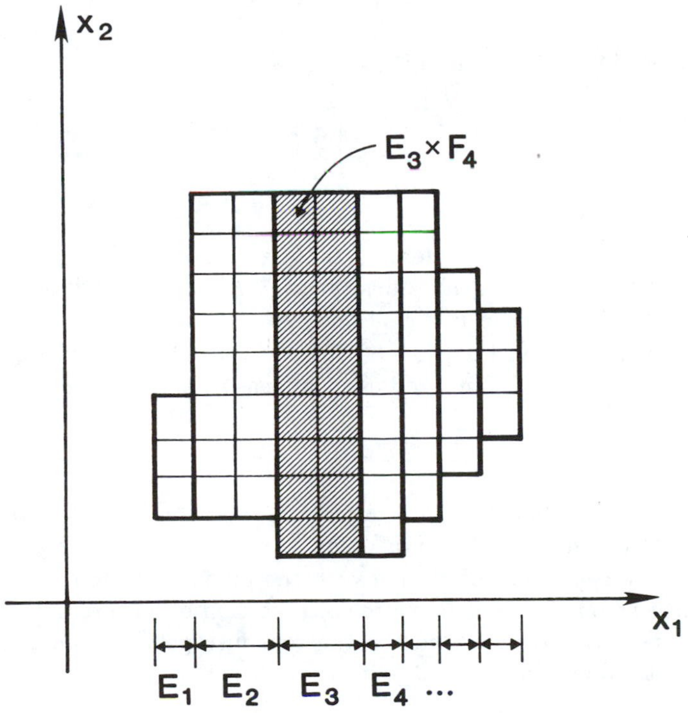

Figure 3.4
Proof of Theorem 3.6.1. Interpretation of sets $E \in \mathcal{J}$.

Part (ii). We shall prove first that if E is Borel of measure zero then $E^{\boldsymbol{x}}$ is of measure zero for almost all $\boldsymbol{x}$. Indeed, $m_{(m)}(E^{\boldsymbol{x}}) = \eta_E(\boldsymbol{x})$ and we already know that

$$\int \eta_E dm_{(n)} = m_{(n+m)}(E) = 0$$

Pick an $\varepsilon > 0$ and consider the set

$$\{\boldsymbol{x} : \eta_E(\boldsymbol{x}) > \varepsilon\}$$

Obviously,

$$\varepsilon m_{(n)}(\{\boldsymbol{x} : \eta_E(\boldsymbol{x}) > \varepsilon\}) \leq \int \eta_E dm_{(n)} = 0$$

which implies that

$$m_{(n)}(\{\eta_E > \varepsilon\}) = 0$$

Making use of the representation

$$\{\eta_E > 0\} = \bigcup_{k=1}^{\infty} \left\{\eta_E > \frac{1}{k}\right\}$$

we get the result.

Now, if E is of measure zero then there exists a Borel set H, also of measure zero, containing E. Consequently, $E^{\boldsymbol{x}} \subset F^{\boldsymbol{x}}$, $F^{\boldsymbol{x}}$ is of measure zero for almost $\boldsymbol{x}$ and so $E^{\boldsymbol{x}}$ is.

Part (iii). We leave the proof of this part as a simple exercise. One has to use Proposition 3.2.4 and represent a measurable set E in the form

$$E = H \cup Z$$

where H is Borel and Z if measure zero. Then the proof follows directly from parts (i) and (iii) of this theorem.

Before we conclude this section with the Fubini's theorem for functions (which turns out to be a simple reinterpretation of Theorem 3.6.1), let us take a break and derive from the theorem just proved some simple, but important, observations. ∎

COROLLARY 3.6.3

The following properties hold:

(i) Let $\varphi: E \rightarrow \overline{\mathbb{R}}$ be a measurable, nonnegative function such that $\int \varphi dm = 0$. Then $\varphi = 0$ a.e.

(ii) Let $\varphi, \psi: E \rightarrow \overline{\mathbb{R}}$ be two measurable, nonnegative functions. Then

$$\int (\varphi + \psi) dm = \int \varphi dm + \int \psi dm$$

PROOF

(i) According to the generic case of Fubini's theorem, Part (iii),

$$\int \varphi dm_{(n)} = m_{(n+1)}(S(\varphi)) = 0$$

implies that

$$m_{(1)}(S(\varphi))^{\boldsymbol{x}} = m((0, \varphi(\boldsymbol{x}))) = \varphi(\boldsymbol{x})$$

is equal zero for almost all $\boldsymbol{x}$.

(ii) Consider the isomorphism

$$g: \mathbb{R}^n \times \mathbb{R} \ni (\boldsymbol{x}, \boldsymbol{y}) \rightarrow (\boldsymbol{x}, -\boldsymbol{y}) \in \mathbb{R}^n \times \mathbb{R}$$

Obviously, $\det g = -1$ and therefore

$$m_{(n+1)}(g(S(\psi))) = m_{(n+1)}(\{-\psi(\boldsymbol{x}) < y < 0\}) = \int \psi dm_{(n)}$$

But $g(S(\psi))$ and $S(\varphi)$ are disjoint and therefore

$$\begin{aligned}
\int \varphi dm + \int \psi dm &= m_{(n+1)}(g(S(\psi)) \cup S(\varphi)) \\
&= \int m_{(1)}(g(S(\psi)) \cup S(\varphi))^{\boldsymbol{x}} dm_{(n)} \\
&= \int m_{(1)}((-\psi(\boldsymbol{x}), 0) \cup (0, \varphi(\boldsymbol{x})) dm_{(n)} \\
&= \int (\varphi + \psi)(\boldsymbol{x}) dm_{(n)}(\boldsymbol{x})
\end{aligned}$$

∎

As the final result of this section we state

THEOREM 3.6.2
(Fubini's Theorem for Nonnegative Functions)

Let $f: \mathbb{R}^n \times \mathbb{R}^m \supset E \ni (\boldsymbol{x}, \boldsymbol{y}) \to f(\boldsymbol{x}, \boldsymbol{y}) \in \overline{\mathbb{R}}$ be a nonnegative function of variables x and y. If f is Borel (measurable) then the following conditions hold:

(i) $\boldsymbol{y} \to f(\boldsymbol{x}, \boldsymbol{y})$ *is Borel for all* $\boldsymbol{x}$ *(is measurable for almost all* $\boldsymbol{x}$*).*

(ii) $\boldsymbol{x} \to \int f(\boldsymbol{x}, \boldsymbol{y}) dm(\boldsymbol{y})$ *is Borel (measurable).*

(iii) $\int f(\boldsymbol{x}, \boldsymbol{y}) dm_{(n+m)} = \int \left(\int f(\boldsymbol{x}, \boldsymbol{y}) dm(\boldsymbol{y}) \right) dm(\boldsymbol{x})$.

PROOF

(i) dom $(\boldsymbol{y} \to f(\boldsymbol{x}, \boldsymbol{y})) = (\text{dom} f)^{\boldsymbol{x}}$ and therefore is Borel (measurable) (Theorem 3.6.1). In the same way,

$$S(\boldsymbol{y} \to f(\boldsymbol{x}, \boldsymbol{y})) = S(f)^{\boldsymbol{x}}$$

and therefore $\boldsymbol{y} \to f(\boldsymbol{x}, \boldsymbol{y})$ is Borel (measurable).

(ii) Apply Theorem 3.6.1 to set $S(f)$ replacing $\boldsymbol{y}$ by $(\boldsymbol{y}, z)$. Thus (ii) follows from Theorem 3.6.1 $(\text{i})_{(2)}$ $(\text{ii})_{(2)}$.

(iii) follows from

$$\begin{aligned} \int f dm_{(n+m)} &= m_{(n+m+1)}(S(f)) = \int m_{(m+1)}(S(f)^{\boldsymbol{x}}) dm_{(n)}(\boldsymbol{x}) \\ &= \int \left(\int f(\boldsymbol{x}, \boldsymbol{y}) dm_{(m)}(\boldsymbol{y}) \right) dm_{(n)}(\boldsymbol{x}) \end{aligned}$$

∎

3.7 Lebesgue Integral of Arbitrary Functions

In this section we generalize the notion of Lebesgue integral to the case of arbitrary functions. As a preliminary step we shall study first the notion of infinite sums.

Infinite Sums. Suppose we are given a sequence $a_i \in \overline{\mathbb{R}}$, $i \in \mathbb{N}$. Note that a_i may take the value of $+\infty$ or $-\infty$. For a given number $a \in \overline{\mathbb{R}}$, we define its positive and negative parts as

$$a^+ = \max\{a, 0\} \ , \ a^- = \max\{-a, 0\}$$

Obviously, only one of the numbers is nonzero and

$$a = a^+ - a^-$$

We will define the *infinite (countable) sum* of a_i as

$$\sum_{\mathbb{N}} a_i \stackrel{\text{def}}{=} \sum_{i=1}^{\infty} a_i^+ - \sum_{i=1}^{\infty} a_i^-$$

provided that at least one of the series on the right-hand side is finite (to avoid the undetermined symbol $+\infty - \infty$).

REMARK 3.7.1. Note the significant difference between infinite sums and infinite series. For instance, the series $\sum_1^{\infty} (-1)^i \frac{1}{i}$ is finite (convergent), but the sum $\sum_{\mathbb{N}} (-1)^i \frac{1}{i}$ is undetermined. ∎

PROPOSITION 3.7.1

Let $a_i \in \overline{\mathbb{R}}$, $i \in \mathbb{N}$. Suppose that $a_i = a_i^1 - a_i^2$, where $a_i^1, a_i^2 \geq 0$ and one of the series

$$\sum_1^{\infty} a_i^1, \ \sum_1^{\infty} a_i^2$$

is finite. Then the sum $\sum_{\mathbb{N}} a_i$ exists and

$$\sum_{\mathbb{N}} a_i = \sum_1^{\infty} a_i^1 - \sum_1^{\infty} a_i^2$$

PROOF

Case 1. Both sequences $\sum_1^{\infty} a_i^1, \sum_1^{\infty} a_1^2$ are finite. This implies that

$$\sum_1^{\infty} a_i^1 - \sum_1^{\infty} a_i^2 = \sum_1^{\infty} (a_i^1 - a_i^2) = \sum_1^{\infty} a_i$$

But $a_i^+ \le a_i^1$ and $a_i^- \le a_i^2$, which implies that both $\sum_1^\infty a_i^+$ and $\sum_1^\infty a_i^-$ are finite, too. Thus, for the same reasons,

$$\sum_{\mathbb{N}} a_i = \sum_1^\infty a_i^+ - \sum_1^\infty a_i^- = \sum_1^\infty a_i$$

from which the equality follows.

Case 2. Suppose $\sum_1^\infty a_i^1 = +\infty$ and $\sum_1^\infty a_i^2 < +\infty$. Then, for the same reasons as before, $\sum_1^\infty a_i^- < \infty$ and therefore

$$\sum_1^\infty (a_i^2 - a_i^-) = \sum_1^\infty a_i^2 - \sum_1^\infty a_i^- < +\infty$$

If $\sum_1^\infty a_i^+$ were finite then according to

$$\sum_1^\infty a_i^1 = \sum_1^\infty a_i^+ + \sum_1^\infty (a_i^1 - a_i^+) = \sum_1^\infty a_i^+ + \sum_1^\infty (a_i^2 - a_i^-)$$

sum $\sum_1^\infty a_i^1$ would have to be also finite, which proves that $\sum_1^\infty a_i^+ = +\infty$. Thus both $\sum_{\mathbb{N}} a_i$ and $\sum_1^\infty a_i^2 - \sum_1^\infty a_i^-$ are equal to $+\infty$, from which the equality follows. The case $\sum_1^\infty a_i^1 < +\infty$ and $\sum_1^\infty a_i^2 = +\infty$ is proved in the same way. ∎

A number of useful properties of infinite sums will be summarized in the following proposition.

PROPOSITION 3.7.2

Let $a_i, b_i \in \overline{\mathbb{R}}$ be arbitrary sequences. The following properties hold:

(i) $\sum_{\mathbb{N}} \alpha a_i = \alpha \sum_{\mathbb{N}} a_i$ *for $\alpha \in \mathbb{R}$. Both sides exist simultaneously.*

(ii) $a_i \le b_i \Rightarrow \sum_{\mathbb{N}} a_i \le \sum_{\mathbb{N}} b_i$ *if both sides exist.*

(iii) $\sum_{\mathbb{N}}(a_i + b_i) = \sum_{\mathbb{N}} a_i + \sum_{\mathbb{N}} b_i$ *if the right-hand side exists (i.e., both sums exist and the symbols* $+\infty - \infty$ *or* $-\infty + \infty$ *are avoided).*

(iv) $\left|\sum_{\mathbb{N}} a_i\right| \leq \sum_{\mathbb{N}} |a_i|$ *if the left-hand side exists.*

PROOF

(i) Case $\alpha = 0$ is trivial. Assume $\alpha > 0$. Then

$$(\alpha a_i)^+ = \alpha a_i^+ \quad (\alpha a_i)^- = \alpha a_i^-$$

and the equality follows from the definition. Case $\alpha < 0$ is analogous.

(ii) $a_i \leq b_i$ implies that

$$a_i^+ \leq b_i^+ \text{ and } a_i^- \geq b_i^-$$

Thus

$$\sum_1^\infty a_i^+ \leq \sum_1^\infty b_i^+ \text{ and } \sum_1^\infty a_i^- \geq \sum_1^\infty b_i^-$$

from which the inequality follows.

(iii) Suppose the right-hand side exists. Then

$$\begin{aligned}\sum_{\mathbb{N}} a_i + \sum_{\mathbb{N}} b_i &= \sum_1^\infty a_i^+ - \sum_1^\infty a_i^- + \sum_1^\infty b_i^+ - \sum_1^\infty b_i^- \\ &= \sum_1^\infty \left(a_i^+ + b_i^+\right) - \sum_1^\infty \left(a_i^- + b_i^-\right)\end{aligned}$$

But $a_i + b_i = (a_i^+ + b_i^+) - (a_i^- + b_i^-)$ and $a_i^+ + b_i^+ \geq 0$, $a_i^- + b_i^- \geq 0$. Thus, according to Proposition 3.7.1, the sum exists and is equal to the right-hand side.

(iv) Case 1. Both sums $\sum_1^\infty a_i^+$ and $\sum_1^\infty a_i^-$ are finite. Then $\sum_1^\infty |a_i| = \sum_1^\infty (a_i^+ + a_i^-)$ is also finite and the result follows from the inequality

$$-|a_i| \leq a_i \leq |a_i|$$

Case 2. If any of the sums is infinite then $\sum_1^\infty |a_i| = +\infty$ and the equality follows. ∎

We leave as an exercise the proof of the following:

COROLLARY 3.7.1

Let $a_i \in \overline{I\!R}$. The sum $\sum_{I\!N} a_i$ is finite if and only if $\sum_{I\!N} |a_i|$ is finite. In such a case

$$\sum_{I\!N} a_i = \sum_1^\infty a_i$$

Definition of Lebesgue Integral for Arbitrary Functions. Let $f: I\!R^n \to I\!R^n$ be a measurable function. Define functions

$$f^+(\boldsymbol{x}) = \max\{f(\boldsymbol{x}), 0\} \quad f^-(\boldsymbol{x}) = \max\{-f(\boldsymbol{x}), 0\}$$

According to Proposition 3.4.1, both functions f^+ and f^- are measurable. We say that function f is *integrable* and define the *integral of* f as

$$\int f dm = \int f^+ dm - \int f^- dm$$

if at least one of the integrals on the right-hand side is finite.

In a manner identical to the proof of Proposition 3.7.1, we prove the following:

COROLLARY 3.7.2

Let $f: E \to \overline{I\!R}$ be measurable and $f = f_1 + f_2$, where $f_1, f_2 \geq 0$ are measurable. Assume that at least one of the integrals $\int f_1 dm, \int f_2 dm$ is finite. Then f is integrable and

$$\int f dm = \int f_1 dm - \int f_2 dm$$

A number of useful properties will be summarized in the following proposition. Please note the similarities between integrals and infinite sums.

PROPOSITION 3.7.3

Let functions f and g be measurable. The following properties hold:

(i) $m(E) = 0 \Longrightarrow \int_E f dm = 0.$

(ii) $f: E \Longrightarrow \overline{I\!R}$, $E_i \subset E$, $i = 1, 2, \ldots$ *measurable and pairwise disjoint* $\Rightarrow$

$$\int_{\cup E_i} f dm = \sum_{I\!N} \int_{E_i} f dm$$

if the left-hand side exists.

(iii) $f, g: E \to \overline{I\!R}$ *integrable,* $f = g$ *a.e. in* $E \Longrightarrow$

$$\int_E f dm = \int_E g dm$$

(iv) $c \in I\!R$, E *measurable* $\Longrightarrow$

$$\int_E c \, dm = c \, m(E)$$

(v) $f, g: E \to \overline{I\!R}$ *integrable,* $f \leq g$ *a.e. in* $E \Longrightarrow$

$$\int_E f dm \leq \int_E g dm$$

(vi) $\lambda \in I\!R$, $f: E \to \overline{I\!R}$ *integrable* $\Longrightarrow$

$$\int_E \lambda f dm = \lambda \int_E f dm$$

(vii) $f, g: E \to \overline{I\!R}$ *integrable* $\Longrightarrow$

$$\int_E (f + g) dm = \int_E f dm + \int_E g dm$$

if the right-hand side exists and function $f + g$ *is determined a.e. in* E.

(viii) $f: E \to \overline{\mathbb{R}}$ *integrable* $\Longrightarrow$

$$\left| \int_E f dm \right| \leq \int_E |f| dm$$

(ix) *Let* $g: \mathbb{R}^n \to \mathbb{R}^n$ *be an affine isomorphism. Then*

$$\int (f \circ g) | \det g | dm = \int f dm$$

and both sides exist simultaneously.

PROOF The whole proof follows from the definition of integral, properties of integral of nonnegative functions and properties of infinite sums.

(i) follows from the definition and Proposition 3.5.1 (i).

(ii) follows from

$$\int_{\cup E_i} f dm = \int_{\cup E_i} f^+ dm - \int_{\cup E_i} f^- dm = \sum_1^\infty \int_{E_i} f^+ dm - \sum_1^\infty \int_{E_i} f^- dm =$$
$$\text{(Proposition 3.7.1)} \;= \sum_{\mathbb{N}} \int_{E_i} f dm$$

(iii) See Proposition 3.5.1 (iii).

(iv) See Proposition 3.5.1 (iv).

(v) See Proposition 3.5.1 (v).

(vi) See Proposition 3.5.1 (vi).

(vii) It follows from Corollary 3.6.3 (ii) that

$$\int f dm + \int g dm = \int f^+ dm - \int f^- dm + \int g^+ dm - \int g^- dm =$$
$$\int (f^+ + g^+) dm - \int (f^- + g^-) dm$$

But $f + g = (f^+ + g^+) - (f^- + g^-)$ and both components are nonnegative, thus, according to Corollary 3.7.2,

$$\int (f^+ + g^+) dm - \int (f^- + g^-) dm = \int (f + g) dm$$

(viii) follows from the inequalities

$$-|f| \le f \le |f|$$

(ix) Apply the definition of integral and Proposition 3.5.2.

∎

Summable Functions. Let $f: E \to \overline{\mathbb{R}}$ be an integrable function. We say that f is *summable* if $\int f dm$ is finite, i.e.,

$$\left| \int f dm \right| < +\infty$$

We leave as an exercise proofs of the following two simple propositions.

PROPOSITION 3.7.4

The following conditions are equivalent:

(i) f *is summable.*

(ii) $\int f^+ dm, \int f^- dm < +\infty.$

(iii) $\int |f| dm < +\infty.$

PROPOSITION 3.7.5

All functions are measurable. The following properties hold:

(i) f *summable,* E *measurable* $\to f|_E$ *summable.*

(ii) $f, \varphi: E \to \overline{\mathbb{R}}$, $|f| \le \varphi$ *a.e. in* E, φ *summable* $\to f$ *summable.*

(iii) $f_1, f_2: E \to \overline{\mathbb{R}}$ *summable* $\Rightarrow \alpha_1 f_1 + \alpha_2 f_2$ *summable for* $\alpha_1, \alpha_2 \in \mathbb{R}$.

We conclude this section with three fundamental theorems of integration theory.

THEOREM 3.7.1
(The Lebesgue Dominated Convergence Theorem)

Let:

$$f_i: E \to \overline{\mathbb{R}}, i = 1, 2, \ldots \text{ integrable}$$

$$f_i(\boldsymbol{x}) \to f(\boldsymbol{x}) \text{ a.e. in } E$$

$$|f_i(\boldsymbol{x})| \le \varphi(\boldsymbol{x}) \text{ a.e. in } E, \text{ where } \varphi: E \to \overline{\mathbb{R}} \text{ is summable}$$

Then

1. *f is summable and*
2. $\displaystyle\int f_i dm \to \int f dm.$

PROOF Let

$$f_i = f_i^+ - f_i^- \ , \ f = f^+ - f^-$$

Since function max is continuous

$$f_i^+ \to f^+ \text{ and } f_i^- \to f^-$$

Obviously, both $f_i^+ \le \varphi$ and $f_i^- \le \varphi$. Thus, according to Theorem 3.5.2,

$$\int f_i^+ dm \to \int f^+ dm \text{ and } \int f_i^- dm \to \int f^- dm$$

and, finally,

$$\int f_i dm = \int f_i^+ dm - \int f_i^- dm \to \int f^+ dm - \int f^- dm$$

Since both $f_i^+ \le \varphi$ and $f_i^- \le \varphi$, both integrals on the right-hand side are finite, which ends the proof. ∎

We leave as an exercise proof of the following:

THEOREM 3.7.2
(Fubini's Theorem)

Let $f: \mathbb{R}^n \times \mathbb{R}^m \to \overline{\mathbb{R}}$ be summable (and Borel). Then the following properties hold:

(i) $\boldsymbol{y} \to f(\boldsymbol{x}, \boldsymbol{y})$ *is summable for almost all* $\boldsymbol{x}$ *(Borel for all* $\boldsymbol{x}$*).*

(ii) $\boldsymbol{x} \to \int f(\boldsymbol{x}, \boldsymbol{y}) dm(\boldsymbol{y})$ *is summable (and Borel).*

(iii) $\int f dm = \int \int (f(\boldsymbol{x}, \boldsymbol{y}) dm(\boldsymbol{y})) dm(\boldsymbol{x})$.

The last, *change of variables theorem*, is a generalization of Proposition 3.7.3 (ix). The proof of it is quite technical and exceeds the scope of this book.

THEOREM 3.7.3
(Change of Variables Theorem)

Let $G, H \subset \mathbb{R}^n$ *be two open sets and* $f : G \to H$ *a* C^1 *bijection from set* G *onto set* H. *Let* $\varphi : H \to \overline{\mathbb{R}}$ *be a function on* H.

Then

(i) φ *is measurable* $\Leftrightarrow \varphi \circ f$ *is measurable and*

(ii)

$$\int_H \varphi(\boldsymbol{x})\, dm(\boldsymbol{x}) = \int_G (\varphi \circ f)(\boldsymbol{x}) \mid \mathrm{jac} f(\boldsymbol{x}) \mid\, dm(\boldsymbol{x})$$

with the two sides existing simultaneously and $\mathrm{jac} f(\boldsymbol{x})$ *denoting the Jacobian of transformation* $f = (f_1, \ldots, f_n)$

$$\mathrm{jac} f(\boldsymbol{x}) = \det(\frac{\partial f_i}{\partial x_j}(\boldsymbol{x}))$$

Exercises

3.7.1. Complete proof of Proposition 3.7.1.

3.7.2. Prove Corollary 3.7.1.

3.7.3. Prove Corollary 3.7.2.

3.7.4. Prove Propositions 3.7.4 and 3.7.5.

3.7.5. Prove Theorem 3.7.2.

3.8 Lebesgue Approximation Sums. Riemann Integrals

We continue our considerations on Lebesgue integration theory with a geometrical characterization of the integral showing particularly the essential difference between Lebesgue and Riemann integrals. We will find also when the two types of integrals coincide with each other.

Lebesgue's Sums. Let $f: \mathbb{R}^n \supset E \to \overline{\mathbb{R}}$ be a measurable function. Pick an $\varepsilon > 0$ and consider a partition of real line $\mathbb{R}$,

$$\ldots < y_{-1} < y_0 < y_1 < \ldots$$

such that $y_{-i} \to -\infty$; $y_i \to +\infty$ and $|y_i - y_{i-1}| \leq \varepsilon$. Define

$$E_i = \{\boldsymbol{x} \in E \, : \, y_{i-1} \leq f(\boldsymbol{x}) < y_i\}$$

Sets E_i are measurable (explain why). The series

$$s = \sum_{-\infty}^{+\infty} y_{i-1} m(E_i), \quad S = \sum_{-\infty}^{+\infty} y_i m(E_i)$$

are called the lower and upper Lebesgue sums, respectively.

We have the following:

THEOREM 3.8.1

Assume additionally that $m(E)$ is finite and consider a sequence $\varepsilon_k \to 0$ with a corresponding family of partitions and Lebesgue sums s_k and S_k. Then:

(i) If f is summable then both s_k and S_k are absolutely convergent and

$$\lim s_k = \lim S_k = \int_E f dm$$

(ii) If one of the Lebesgue's sums s_k or S_k is absolutely convergent, then f is summable and, therefore, according to (i), the other sum converges as well and both limits are equal to the integral.

PROOF

(i) Since E_i are pairwise disjoint one can define the following two functions:

$$\left.\begin{aligned}\varphi(\boldsymbol{x}) &= y_{i-1}\\ \psi(\boldsymbol{x}) &= y_i\end{aligned}\right\} \quad \text{if } \boldsymbol{x} \in E_i$$

Both φ and ψ are measurable (explain why) and

$$\varphi \leq f \leq \psi$$

Moreover, it follows from the definition of the Lebesgue integral that

$$s = \int \varphi dm \quad \text{and} \quad S = \int \psi dm$$

Also,

$$\lim_{k\to\infty} \varphi_k = \lim_{k\to\infty} \psi_k = f$$

Assume now that f is summable. Since both

$$|\varphi|, \ |\psi| \leq f + \varepsilon$$

and $m(E) < \infty$, we have according to the Lebesgue dominated convergence theorem that

$$s_k, S_k \to \int f dm$$

(ii) Obviously,

$$|f| \leq |\varphi| + \varepsilon \quad \text{and} \quad |f| \leq |\psi| + \varepsilon$$

Thus if one of the Lebesgue's sum is absolutely convergent then φ or ψ is summable and consequently f is summable. ∎

The concept of Lebesgue's approximation sums is illustrated in Fig. 3.5. Let us emphasize that contrary to Riemann's approximation sums where the domain of a function is partitioned *a priori* (usually into regular cubes or intervals in a one-dimensional case), in the Lebesgue's construction the partition of E follows from the initial partition of image of function f. The difference between the two concepts has been beautifully illustrated by the anecdote quoted by Ivar Stakgold (see [1], page 36). "A shopkeeper can determine a day's total receipts either by adding the individual transactions (Riemann) or by sorting bills and coins according to their denomination and then adding the respective contributions (Lebesgue). Obviously the second approach is more efficient!"

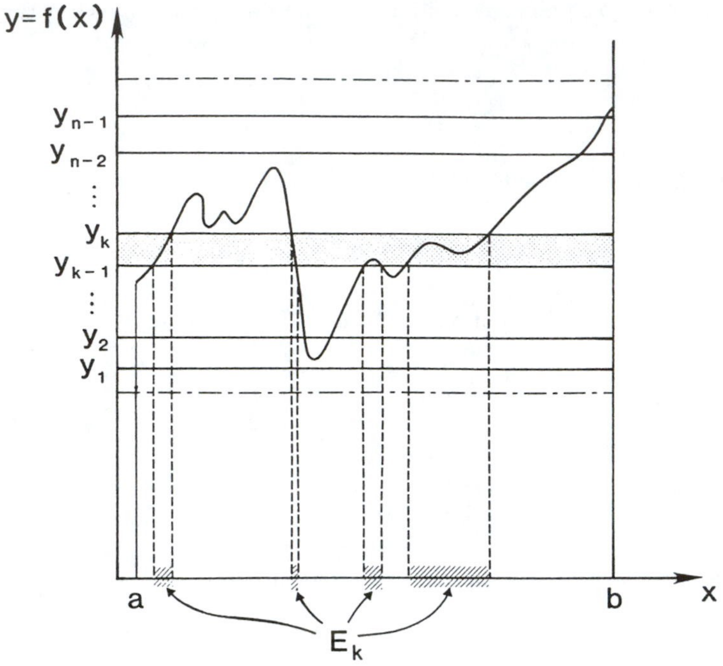

Figure 3.5
Interpretation of Lebesgue sums.

We will devote the rest of this chapter to a study of the Rieman integral and its connection with the Lebesgue concept.

Riemann Integral. Recall that by an (open) cube σ in $\mathbb{R}^n$ we understand the set

$$\sigma = (a_1, b_1) \times \ldots \times (a_n, b_n)$$

Let E be a cube in $\mathbb{R}^n$ and $f: E \to \mathbb{R}$ be a function. By a *partition* $\mathcal{P}$ of E we understand a finite family of cubes σ such that $\sigma \subset E$ and

$$\overline{E} = \bigcup \overline{\sigma}$$

The number

$$r(\mathcal{P}) = \sup_{\sigma \in \mathcal{P}} r(\sigma)$$

where $r^2(\sigma) = \sum_1^n (b_i - a_i)^2$ will be called the *radius of the partition.* Choosing for every cube $\sigma \in \mathcal{P}$ a point $\boldsymbol{\xi}_\sigma \in \sigma$, we define the Riemann approximation sum as

$$R = R(\mathcal{P}, \boldsymbol{\xi}) = \sum_{\sigma \in \mathcal{P}} f(\boldsymbol{\xi}_\sigma) m(\sigma)$$

If for every sequence of partitions $\mathcal{P}_n$ such that $r(\mathcal{P}_n) \to 0$ and an arbitrary choice of points $\boldsymbol{\xi} \in \sigma$, the Riemann sum converges to a common limit J then J is called the *Riemann integral of function* f over cube E. Function f is said to be *Riemann integrable* over E.

We have the following fundamental result:

THEOREM 3.8.2

Let E be a cube in $\mathbb{R}^n$ and $f: E \to \mathbb{R}$ be a bounded function. Then f is Riemann integrable if and only if f is continuous almost everywhere in E. In such a case Riemann and Lebesgue integrals are equal to each other:

$$\text{(Riemann)} \int f(\boldsymbol{x}) d\boldsymbol{x} = \text{(Lebesgue)} \int f(\boldsymbol{x}) dm(\boldsymbol{x})$$

PROOF

Step 1. Assume we are given a partition $\mathcal{P}$ and define on E three functions:

$$\underline{h}_\mathcal{P}(\boldsymbol{x}), \overline{h}_\mathcal{P}(\boldsymbol{x}), r_\mathcal{P}(\boldsymbol{x}) = \begin{cases} \inf_\sigma f, \ \sup_\sigma f, \ \operatorname{osc}_\sigma f & \text{if } \boldsymbol{x} \in \sigma \\ 0 \ \text{ if } \ \boldsymbol{x} \in E - \bigcup_\mathcal{P} \sigma \end{cases}$$

where $\operatorname{osc}_\sigma f = \sup_{\boldsymbol{x}, \boldsymbol{y} \in \sigma} (f(\boldsymbol{x}) - f(\boldsymbol{y}))$ (oscillation of f on σ).

All three functions are summable (explain why). Introducing the lower and upper approximation sums

$$\underline{\sum}(\mathcal{P}) \overset{\text{def}}{=} \sum_{\sigma \in \mathcal{P}} \inf_\sigma f \ m(\sigma) \ = \int_E \underline{h}_\mathcal{P} \, dm$$
$$\overline{\sum}(\mathcal{P}) \overset{\text{def}}{=} \sum_{\sigma \in \mathcal{P}} \sup_\sigma f \ m(\sigma) = \int_E \overline{h}_\mathcal{P} \, dm$$

we get the obvious results

$$\omega(\mathcal{P}) \overset{\text{def}}{=} \sum_{\sigma \in \mathcal{P}} \operatorname{osc}_\sigma f m(\sigma) = \int_E r_\mathcal{P} dm = \overline{\sum}(\mathcal{P}) - \underline{\sum}(\mathcal{P})$$

and

$$\underline{\sum}(\mathcal{P}) \leq \begin{matrix} R(\mathcal{P},\boldsymbol{\xi}) \\ \\ \int_E f\, dm \end{matrix} \leq \overline{\sum}(\mathcal{P})$$

Step 2. We claim that, for all $\mathcal{P}_i$ such that $r(\mathcal{P}_i) \to 0$, and an arbitrary choice of $\boldsymbol{\xi}$, sum $R(\mathcal{P},\boldsymbol{\xi})$ converges to a common limit if and only if $\omega(\mathcal{P}_i) \to 0$. Indeed, choose a sequence of partitions $\mathcal{P}_i$, $r(\mathcal{P}_i) \to 0$, such that $R(\mathcal{P}_i,\boldsymbol{\xi})$ converges to a common limit, for every choice of $\boldsymbol{\xi}$. Thus for an $\varepsilon > 0$ and two choices of $\boldsymbol{\xi}$, say $\boldsymbol{\xi}_1$ and $\boldsymbol{\xi}_2$, one can always find such an index I that for $i \geq I$

$$\sum_\sigma |f(\boldsymbol{\xi}_1) - f(\boldsymbol{\xi}_2)|\, m(\sigma) < \varepsilon$$

which implies that $\omega(\mathcal{P}_i) \to 0$.

Conversely, if $\omega(\mathcal{P}_i) \to 0$ then both $\overline{\sum}(\mathcal{P}_i)$ and $\underline{\sum}(\mathcal{P}_i)$ have the same limit, which proves that $R(\mathcal{P}_i,\boldsymbol{\xi})$ converges. Moreover, if $\int_E f dm$ exists then $R(\mathcal{P}_i,\boldsymbol{\xi})$ converges to the integral.

Step 3. We claim that $\omega(\mathcal{P}_i) \to 0$ for all partitions $\mathcal{P}_i$ such that $r(\mathcal{P}_i) \to 0$ if and only if function f is continuous a.e. in E. So, consider a sequence of partitions $\mathcal{P}_i$ such that $r(\mathcal{P}_i) \to 0$. Fatou's lemma implies that

$$\int_E \liminf r_{\mathcal{P}_i}(\boldsymbol{x}) dm(\boldsymbol{x}) \leq \lim \int_E r_{\mathcal{P}_i}(\boldsymbol{x}) dm(\boldsymbol{x}) = 0$$

Since $r_{\mathcal{P}_i} \geq 0$ it implies that

$$\liminf r_{\mathcal{P}_i}(\boldsymbol{x}) = 0 \quad \text{a.e. in} \quad E$$

But

$$\inf_\sigma f \leq \liminf f \leq \limsup f \leq \sup_\sigma f$$

which means that

$$\liminf \ f = \limsup f \text{ a.e. in } E$$

and therefore f is continuous a.e. in E.

Conversely, if f is continuous a.e. in E, then for every $\boldsymbol{x} \in E$ except for a set of measure zero

$$\forall\, \varepsilon > 0\ \exists \delta > 0\ |\boldsymbol{x} - \boldsymbol{x}'| < \delta \to |f(\boldsymbol{x}) - f(\boldsymbol{x}')| < \varepsilon$$

Thus for $\boldsymbol{x}'$ and $\boldsymbol{x}''$ belonging to a sufficiently small cube σ containing $\boldsymbol{x}$

$$|f(\boldsymbol{x}') - f(\boldsymbol{x}'')| \leq |f(\boldsymbol{x}') - f(\boldsymbol{x})| + |f(\boldsymbol{x}) - f(\boldsymbol{x}'')| < 2\varepsilon$$

which proves that $r_{\mathcal{P}} < 2\varepsilon$ a.e. in E. Since ε is arbitrary small, $\lim r_{\mathcal{P}} = 0$ a.e. and according to the Lebesgue dominated convergence theorem

$$\omega(\mathcal{P}_i) \rightarrow 0$$

∎

We conclude this section with a rather standard example of a function f which is Lebesgue integrable, but not Riemann integrable.

Example 3.8.1

Consider the function of the Dirichlet type $f\colon (0,1) \rightarrow \mathbb{R}$,

$$f(x) = \begin{cases} 3 & \text{if } x \text{ is rational} \\ 2 & \text{if } x \text{ is irrational} \end{cases}$$

□

Then f is continuous nowhere and therefore a Riemann integral does not exist. Simultaneously, since the set of rational numbers is of measure zero, $f = 2$ a.e. in $(0,1)$ and therefore the Lebesgue integral exists and

$$\int_0^1 f\, dm = 2.$$

Exercises

3.8.1. Consider function f from Example 3.8.1. Construct explicitly Lebesgue and Riemann approximation sums and explain why the first sum converges while the other does not.

L^p Spaces

3.9 Hölder and Minkowski Inequalities

We will conclude this chapter with two fundamental integral inequalities and a definition of some very important vector spaces.

THEOREM 3.9.1
(Hölder Inequality)

Let $\Omega \subset \mathbb{R}^n$ be a measurable set. Let $f, g\colon \Omega \to \overline{\mathbb{R}}$ be measurable such that

$$\int_\Omega |f|^p dm \quad \text{and} \quad \int_\Omega |g|^q dm, \; p, q > 1 \; , \; \frac{1}{p} + \frac{1}{q} = 1$$

are finite. Then the integral $\int_\Omega fg$ is finite and

$$\left| \int_\Omega fg dm \right| \leq \left(\int_\Omega |f|^p dm \right)^{\frac{1}{p}} \left(\int_\Omega |g|^q dm \right)^{\frac{1}{q}}$$

PROOF

Step 1. Since

$$\left| \int_\Omega fg dm \right| \leq \int_\Omega |fg| dm = \int_\Omega |f||g| dm$$

it is sufficient to prove the inequality for nonnegative functions only.

Step 2. Assume additionally that $\|f\|_p = \|g\|_q = 1$ where we have introduced the notation

$$\|f\|_p \stackrel{\text{def}}{=} \left(\int_\Omega |f|^p dm \right)^{\frac{1}{p}} \; ; \; \|g\|_q \stackrel{\text{def}}{=} \left(\int_\Omega |g|^q dm \right)^{\frac{1}{q}}$$

So, it is sufficient to prove that

$$\int_\Omega fg dm \leq 1$$

Denote $\alpha = 1/p$, $\beta = 1/q = 1 - \alpha$ and consider function $y = x^\alpha, \alpha < 1$. Certainly, the function is concave. A simple geometrical interpretation (cf. Fig. 3.6) implies that

$$x^\alpha \leq \alpha x + (1 - \alpha) = \alpha x + \beta$$

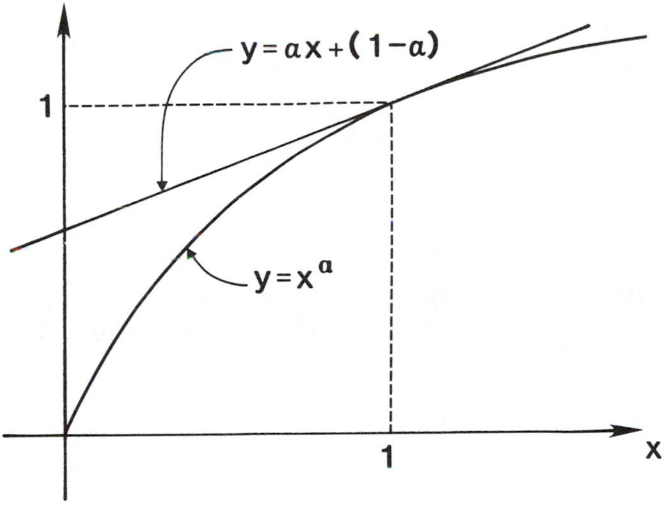

Figure 3.6
Concavity of function $y = x^\alpha, \alpha < 1$.

Replacing x with t/v we get

$$\left(\frac{t}{v}\right)^{\alpha} \le \alpha\left(\frac{t}{v}\right) + \beta$$

or

$$t^\alpha v^\beta \le \alpha t + \beta v$$

Substitituting $f^{\frac{1}{\alpha}}$ for t and $g^{\frac{1}{\beta}}$ for v, one gets

$$fg = \left(f^{\frac{1}{\alpha}}\right)^{\alpha}\left(g^{\frac{1}{\beta}}\right)^{\beta} \le \alpha f^{\frac{1}{\alpha}} + \beta g^{\frac{1}{\beta}}$$

Finally, integrating over Ω, we obtain

$$\int_\Omega fg\,dm \le \alpha\int_\Omega f^p dm + \beta\int_\Omega g^q dm = \alpha + \beta = 1$$

Step 3. If f or g are zero a.e. then both sides are equal zero and the inequality is trivial. So, assume that both $\|f\|_p$ and $\|g\|_q$ are different from zero. Set (normalize)

$$\overline{f} = \frac{f}{\|f\|_p} \quad \text{and} \quad \overline{g} = \frac{g}{\|g\|_q}$$

Obviously, $\|\overline{f}\|_p = \|\overline{g}\|_q = 1$ and it is sufficient to apply the result from Step 2. ∎

As an immediate corollary we get:

THEOREM 3.9.2
(Minkowski's Inequality)

Let $\Omega \subset I\!R^n$ be a measurable set and $f, g\colon \Omega \to \overline{I\!R}$ two measurable functions such that $\|f\|_p$ and $\|g\|_p$ (see proof of the Hölder inequality for notation) are finite, where $p > 1$. Then $\|f + g\|_p$ is also finite and

$$\|f + g\|_p \le \|f\|_p + \|g\|_p$$

PROOF

Step 1. Since

$$|f + g| \le |f| + |g|$$

it is sufficient to prove the inequality for nonnegative functions.

Step 2. One has

$$\int_\Omega (f+g)^p dm = \int_\Omega f(f+g)^{p-1} dm + \int_\Omega g(f+g)^{p-1} dm \le$$

$$\left(\int_\Omega f^p dm\right)^{\frac{1}{p}} \left(\int_\Omega (f+g)^p dm\right)^{\frac{p-1}{p}} + \left(\int_\Omega g^p dm\right)^{\frac{1}{p}} \left(\int_\Omega (f+p)^p dm\right)^{\frac{p-1}{p}}$$

since $1/q = (p-1)/p$. Dividing both sides by the last factor on the right-hand side, we get the result required. ∎

Functions Essentially Bounded. Let $f\colon E \to \overline{I\!R}$ be a function. We say that f is *essentially bounded* on E if there exists a constant $c > 0$ such that

$$|f| \le c \quad \text{a.e. in} \quad E$$

In other words, there exists a set $Z \subset E$ of measure zero such that

$$|f(\boldsymbol{x})| \le c \text{ for all } \boldsymbol{x} \in E - Z$$

Let $f\colon E \to \overline{I\!R}$ be essentially bounded. The number

$$\operatorname*{ess\,sup}_{\boldsymbol{x} \in E} |f(\boldsymbol{x})| = \inf_{m(Z)=0} \sup_{\boldsymbol{x} \in E - Z} |f(\boldsymbol{x})|$$

is called the essential supremum of function f over set E. We introduce the notation

$$\|f\|_\infty \overset{\text{def}}{=} \begin{cases} \operatorname*{ess\,sup}_{\boldsymbol{x} \in E} |f(\boldsymbol{x})| \text{ if } f \text{ is essentially bounded} \\ +\infty \text{ otherwise} \end{cases}$$

The following simple observation follows directly from the definition.

PROPOSITION 3.9.1

Let $f\colon E \to \overline{\mathbb{R}}$ be essentially bounded. Then

$$|f(\boldsymbol{x})| \leq \|f\|_\infty \textit{ a.e. in } E$$

PROOF Indeed, let c_k be a decreasing sequence of positive numbers such that

$$c_k \searrow \|f\|_\infty$$

The sets

$$C_k = \{\boldsymbol{x} \in E : |f(\boldsymbol{x})| > c_k\}$$

are of measure zero and therefore the set

$$C = \{\boldsymbol{x} \in E : |f(\boldsymbol{x})| > \|f\|_\infty\} = \bigcup_{k=1}^{\infty} C_k$$

as a countable union of measure zero sets is of measure zero, too. ∎

L^p **Spaces.** For a given domain Ω in $\mathbb{R}^n$ and a positive number $p \in [1, +\infty]$, we introduce the following set of functions:

$$L^p(\Omega) \overset{\text{def}}{=} \{f : \Omega \to \overline{\mathbb{R}} \text{ measurable } : \|f\|_p < +\infty\}$$

PROPOSITION 3.9.2

Let $\Omega \subset \mathbb{R}^n$ be a measurable set and $p \in [1, +\infty]$. The following properties hold:

(i) $\|\alpha f\|_p = |\alpha|\|f\|_p$ *for every* $f \in L^p(\Omega)$.

(ii) $\|f + g\|_p \leq \|f\|_p + \|g\|_p$ *for every* $f, g \in L^p(\Omega)$.

PROOF

Part (i) follows directly from the definitions.

(ii) One has to prove only the case $p = 1$ and $p = \infty$ (see Minkowski's inequality for the other case). Integrating the inequality

$$|f(\boldsymbol{x}) + g(\boldsymbol{x})| \leq |f(\boldsymbol{x})| + |g(\boldsymbol{x})| \quad \text{for } \boldsymbol{x} \in \Omega$$

we get the result for $p = 1$. Also, according to Proposition 3.9.1,

$$|f(\boldsymbol{x})| \leq \|f\|_\infty \text{ a.e. in } \Omega \text{ and } |g(\boldsymbol{x})| \leq \|g\|_\infty \text{ a.e. in } \Omega$$

and therefore

$$|f(\boldsymbol{x}) + g(\boldsymbol{x})| \leq \|f\|_\infty + \|g\|_\infty \text{ a.e. in } \Omega$$

which ends the proof for $p = \infty$. ∎

COROLLARY 3.9.1

The set $L^p(\Omega)$ for $p \in [1, +\infty]$ is a vector space.

PROOF Indeed, it follows from Proposition 3.9.2 that $L^p(\Omega)$ is closed with respect to both vector addition and multiplication by a scalar, which ends the proof. ∎

The L^p spaces form a fundamental class of vector spaces, which we shall study throughout most of this book. To begin with, let us conclude this section with the following proposition investigating the relation between L^p functions for a domain Ω of finite measure.

PROPOSITION 3.9.3

Let $\Omega \subset \mathbb{R}^n$ *be an open set,* $m(\Omega) < +\infty$. *Then the following properties hold:*

(i) $L^p(\Omega) \subset L^q(\Omega)$ *for* $1 \leq q < p \leq +\infty$.

(ii) *If $f \in L^\infty(\Omega)$ then $f \in L^p(\Omega)$ for $p \geq 1$ and*

$$\lim_{p\to\infty} \|f\|_p = \|f\|_\infty$$

PROOF

(i) *Case 1.* $1 \leq q < p < +\infty$. Apply the Hölder inequality for function $|f|^q$ and a function g identically equal 1 on Ω. We have

$$\left|\int_\Omega |f|^q dm\right| \leq \left(\int_\Omega |f|^p dm\right)^{\frac{q}{p}} \left(\int_\Omega dm\right)^{\frac{p-1}{p}}$$

Raising both sides to the power $1/q$ we get the result.

Case 2. $1 \leq q < p = +\infty$ follows directly from Proposition 3.9.1.

(ii) If $||f||_\infty = 0$ then $f = 0$ a.e. in Ω and therefore $||f||_p = 0$ for every $p > 1$; hence $||f||_p = ||f||_\infty$.

Assume $||f||_\infty > 0$. Pick an $\varepsilon > 0$ and define the set

$$\Omega_\varepsilon = \{\boldsymbol{x} \in \Omega : |f(\boldsymbol{x})| \geq \|f\|_\infty - \varepsilon\}$$

Obviously, $m(\Omega_\varepsilon) > 0$. The following inequality follows:

$$(\|f\|_\infty - \varepsilon)(m(\Omega_\varepsilon))^{\frac{1}{p}} \leq \|f\|_p \leq \|f\|_\infty (m(\Omega))^{\frac{1}{p}}$$

Passing with p to infinity, we get

$$\|f\|_\infty - \varepsilon \leq \liminf \|f\|_p \leq \limsup \|f\|_p \leq \|f\|_\infty$$

from which the result follows. ∎

Exercises

3.9.1. Prove the generalized Hölder inequality:

$$\int uvw dm \leq ||u||_p ||v||_q ||w||_r$$

where $p, q, r \geq 1$, $1/p + 1/q + 1/r = 1$.

3.9.2. Prove that the Hölder inequality turns into an equality if and only if two constants, α and β exist such that

$$\alpha|f|^p + \beta|g|^q = 0 \text{ a.e. in } \Omega$$

3.9.3. (i) Show that integral

$$\int_0^{\frac{1}{2}} \frac{dx}{x \ln^2 x}$$

is finite, but, for any $\varepsilon > 0$, integral

$$\int_0^{\frac{1}{2}} \frac{dx}{[x \ln^2 x]^{1+\varepsilon}}$$

is infinite.

(ii) Use the property above to construct an example of a function $f : (0,1) \to \mathbb{R}$ which belongs to space $L^p(0,1), 1 < p < \infty$, but does not belong to any $L^q(0,1)$ for $q > p$.

3.9.4. Let $f_n, \varphi \in L^p(\Omega)$ such that

(i) $|f_n(\boldsymbol{x})| \le \varphi(\boldsymbol{x})$ a.e. in Ω and

(ii) $f_n(\boldsymbol{x}) \to f(\boldsymbol{x})$ a.e. in Ω.

Prove that

(i) $f \in L^p(\Omega)$ and

(ii) $||f_n - f||_p \to 0$.

4

Topological and Metric Spaces

Elementary Topology

4.1 Topological Structure—Basic Notions

When introducing the concept of topology, one faces the common problem of the choice of a particular path of reasoning, or, equivalently, the particular definition of topology. Mathematics is full of such logical or rather didactic problems. When two statements describing properties of the same object are equivalent to each other, then one can be selected as a definition, whereas the other can be deduced as a consequence. For instance, we may call upon the equivalence of the *axiom of choice* and the *Kuratowski-Zorn lemma* discussed in Chapter 1. The two statements are equivalent to each other and, indeed, it is a matter of a purely arbitrary choice that the axiom of choice bears the name of an *axiom* while the Kuratowski-Zorn lemma serves as a theorem. One of course may argue that it is easier to accept intuitively the axiom rather than the lemma, but such reasoning has very little to do with (formal) logic, of course.

The concept of equivalent paths in developing a theory is not new to engineering students. It is well known, for instance, that, under customary assumptions, Newton's equations of motion, Lagrange equations or Hamilton principle are equivalent to each other. It is again the simplicity of Newton's equations, when compared with the other formulations, which motivates lecturers to introduce them as *axioms* or *laws* and derive the other results in form of theorems.

Sometimes the choice is less obvious, especially when one does not deal with equivalent statements, but rather two different approaches leading to the same (in some sense) object. Our discussion on equivalence of the concepts of equivalence relations and equivalence classes and that of a partition of a set, presented in Chapter 1, provides a good example. In what sense are these two concepts equivalent? Certainly, an equivalence relation and a partition of a set are different things! The equivalence of the two concepts may be summarized making the following points:

- Every equivalence relation R on a set X induces the corresponding partition of X into equivalence classes with respect to relation R

$$\mathcal{P}_R = \{[x]_R,\ x \in X\}$$

- Every partition $\mathcal{P}$ of X

$$X_\iota,\ \iota \in I\ ,\ \cup_{\iota \in I} X_\iota = X$$

  induces the corresponding equivalence relation on X defined as

$$xR_{\mathcal{P}}y \overset{\text{def}}{\Leftrightarrow} \exists \iota \in I\ :\ x, y \in X_\iota$$

- Equivalence relation $R_{\mathcal{P}_R}$ corresponding to partition $\mathcal{P}_R$ corresponds in turn to equivalence relation R, and coincides with the original relation R, i.e.,

$$R_{\mathcal{P}_R} = R$$

- Partition $\mathcal{P}_{R_{\mathcal{P}}}$ corresponding to equivalence relation $R_{\mathcal{P}}$ induced by partition $\mathcal{P}$ coincides with partition $\mathcal{P}$, i.e.,

$$\mathcal{P}_{R_{\mathcal{P}}} = \mathcal{P}$$

The final point is that the two structures in X-partition and equivalence relation always coexist and it is only a matter of convenience or taste as to whether the two objects are constructed in X by setting up an equivalence relation or introducing a partition.

Exactly the same situation is encountered when introducing the concept of topology, except that it is more complicated. The complication comes from the fact that there exists more than just two equivalent objects as in our example with equivalence relations and partitions. Roughly speaking, constructing a topology on a set X consists in introducing in X several objects like

- open sets,
- closed sets,
- the interior operation,

- the closure operation, and
- neigborhoods of points x, for every point $x \in X$,

and others (this list is by no means complete!). *Every* two objects from the list are equivalent to each other in the sense discussed earlier. This means that once any of the objects (with some specific properties to hold, of course) is introduced in set X, the rest of them will be induced in X automatically, as all these objects always coexist simultaneously. Often, we say that the topology in X has been introduced, for instance, through open sets or neigborhoods of points, etc. Some authors go a little bit further and *identify* the notion of topology with one of the particular ways of introducing it in X. Thus, depending on one's taste, the notion of a topology may be identified with a family of open sets, with systems (filters) of neighborhoods of points in X, etc. This identification is sometimes confusing, as it leaves the reader with an impression that there is more than one notion of topology.

In our presentation we shall focus on two equivalent ways of introducing a topology in X, one based on *open sets* and the other one on *neigborhoods* of points. The open sets concept is certainly the most common one in textbooks, whereas introducing a topology by identifying neighborhoods of vectors (points) or just the zero vector is the most natural and convenient approach in the context of the theory of topological vector spaces. By showing the equivalence of two approaches, it is then somewhat easier to appreciate other ways of constructing a topology on a set.

We shall start our considerations with a simple, but useful, algebraic relation between families of sets.

Stronger and Equivalent Families of Sets. Let X be an arbitrary set and $\mathcal{A}, \mathcal{B} \subset \mathcal{P}(X)$ denote two families of subsets of X. We say that $\mathcal{A}$ is *stronger* than $\mathcal{B}$, denoted $\mathcal{A} \succ \mathcal{B}$, if for every set $B \in \mathcal{B}$ there exists a set $A \in \mathcal{A}$ contained in B, i.e.,

$$\mathcal{A} \succ \mathcal{B} \quad \overset{\text{def}}{\Leftrightarrow} \quad \forall B \in \mathcal{B} \quad \exists A \in \mathcal{A} \quad A \subset B$$

If $\mathcal{A} \succ \mathcal{B}$ and $\mathcal{B} \succ \mathcal{A}$ we say that the two families are equivalent and write

$$\mathcal{A} \sim \mathcal{B}$$

Example 4.1.1

Let $f : A \to B$ be a function from set A into set B. We introduce the notation

$$f(\mathcal{A}) = \{f(A) : A \in \mathcal{A}\}$$

i.e., $f(\mathcal{A})$ is the class of all image sets of the function f on sets in the class $\mathcal{A}$. Since $f(\mathcal{A})$ is a class, we can compare its "strongness" with other classes in the spirit of the symbolism $\succ$ defined above.

This leads us to a simple way of expressing symbolically the idea of continuity of f. Suppose that

$$\mathcal{B}_{\boldsymbol{x}} = \text{ the class of all balls centered at } \boldsymbol{x} \in X = \mathbb{R}^n$$

Then $f\colon X \to Y \subset \mathbb{R}^m$ is continuous at $\boldsymbol{x}_0 \in X$ if, $\forall\ B \in \mathcal{B}_{f(\boldsymbol{x}_0)}$ $\exists$ a ball $A \in \mathcal{B}_{\boldsymbol{x}_0}$ such that $A \subset f^{-1}(B)$; i.e., $f(A) \subset B$. Thus, the condition that f is continuous at $\boldsymbol{x}_0$ can be written

$$f\left(\mathcal{B}_{\boldsymbol{x}_0}\right) \succ \mathcal{B}_{f(\boldsymbol{x}_0)}$$

□

Base. A nonempty class of sets $\mathcal{B} \subset \mathcal{P}(X)$ is called a *base* if the following conditions are satisfied:

(i) $\emptyset \notin \mathcal{B}$;

(ii) $\forall A, B \in \mathcal{B} \quad \exists\ C \in \mathcal{B}$ such that $C \subset A \cap B$.

Example 4.1.2

Let $X = \{\alpha, \beta, \gamma, \rho\}$. The family of sets

$$\mathcal{B} = \{\{\alpha\},\ \{\alpha, \beta\},\ \{\alpha, \beta, \rho\}\}$$

is a base, as is easily checked.

□

Example 4.1.3

Every nonempty family of decreasing nonempty sets in $\mathbb{R}^n$ is a base. In particular, the family of balls centered at $\boldsymbol{x}_0 \in \mathbb{R}^n$

$$\mathcal{B} = \{B(\boldsymbol{x}_0, \varepsilon),\ \varepsilon > 0\}$$

is a base.

An example of a *trivial base* is a family consisting of a single nonempty set. □

Filter. A nonempty family of sets $\mathcal{F} \subset \mathcal{P}(X)$ is called a *filter* if the following conditions are satisfied:

(i) $\emptyset \notin \mathcal{F}$;

(ii) $A, B \in \mathcal{F} \Leftrightarrow A \cap B \in \mathcal{F}$;

(iii) $A \in \mathcal{F}, A \subset B \Rightarrow B \in \mathcal{F}$.

Let $A, B \in \mathcal{F}$, but $C = A \cap B \subset A \cap B$. Thus *every filter is a base.*

Let $\mathcal{B} \subset \mathcal{P}(X)$ be a base. We will denote by $\mathcal{F}(\mathcal{B})$ a family of all supersets of sets from the base $\mathcal{B}$, i.e.,

$$C \in \mathcal{F}(\mathcal{B}) \Leftrightarrow \quad \exists B \in \mathcal{B} \quad B \subset C$$

It follows immediately from the definitions that $\mathcal{F} = \mathcal{F}(\mathcal{B})$ is a filter. We say that $\mathcal{B}$ is a *base of filter* $\mathcal{F}$ or, equivalently, that $\mathcal{B}$ *generates* $\mathcal{F}$. Note that, in particular, every filter is a base of itself.

We have the following simple observation:

PROPOSITION 4.1.1

Let $\mathcal{B}$ and $\mathcal{C}$ denote two bases. The following holds:
$\mathcal{B} \succ \mathcal{C} \Leftrightarrow \quad \mathcal{F}(\mathcal{B}) \supset \mathcal{F}(\mathcal{C})$.
In particular, two equivalent bases generate the same filter
$\mathcal{B} \sim \mathcal{C} \Leftrightarrow \quad \mathcal{F}(\mathcal{B}) = \mathcal{F}(\mathcal{C})$.

PROOF The proof follows immediately from definitions. ∎

Topology through Open Sets. Let X be an arbitrary, nonempty set. We say that a *topology is introduced in X through open sets* if a class $\mathcal{X} \subset \mathcal{P}(X)$ of subsets of X, satisfying the following conditions, has been identified.

(i) X and $\emptyset$ belong to $\mathcal{X}$.

(ii) The union of any number of members of $\mathcal{X}$ belongs to $\mathcal{X}$.

(iii) The intersection of any two members (and, therefore, by induction, any finite number of members) of $\mathcal{X}$ belongs to $\mathcal{X}$.

The sets forming $\mathcal{X}$ are called *open sets*, and the family of open sets $\mathcal{X}$ is frequently itself called the *topology* on X. We emphasize that as long as the three conditions are satisfied, *any* family $\mathcal{X}$ can be identified as open sets and, at least at this point, the notion of open sets has nothing to do with the notion of open sets discussed in Chapter 1, except that our abstract open sets satisfy (by definition) the same properties as open sets in $\mathbb{R}^n$ (Proposition 1.18.1). Set X with family $\mathcal{X}$ is called a *topological space.* We shall use the slighty abused notations $(X, \mathcal{X})$ or $\mathcal{X}$ or simply X to refer to the topological space, it generally being understood that X is the *underlying set* for the topology $\mathcal{X}$ characterizing the *topological space* $(X, \mathcal{X})$. Different classes of subsets of $\mathcal{P}(X)$ will define different topologies on X and, hence, define different topological spaces.

Neighborhoods of a Point. Let x be an arbitrary point of a topological space X. The collection of all open sets A containing point x, denoted $\mathcal{B}_x^o$, is called the *base of open neigborhoods* of x

$$\mathcal{B}_x^o = \{A \in \mathcal{X} \,:\, x \in A\}$$

As intersections of two open sets remain open, conditions for a base are immediately satisfied. Filter $\mathcal{F}_x = \mathcal{F}(\mathcal{B}_x^o)$ generated by base $\mathcal{B}_x^o$ is called the *filter* or *system of neighborhoods* of point x. Elements of $\mathcal{F}_x$ are called simply *neighborhoods* of x. Thus, according to the definition, any set B containing an open set A, containing, in turn, the point x, is a neighborhood of x. Consequently, of course, every neighborhood B of x must contain x.

Topology through Neighborhoods. Let X be an arbitrary, nonempty set. We say that a *topology is introduced on* X *through neighborhoods* if, for each $x \in X$, a corresponding family $\mathcal{F}_x$ of subsets of X exists, called the *neighborhoods of* x, which satisfies the following conditions:

(i) $x \in A, \ \forall A \in \mathcal{F}_x$
(consequently elements A of $\mathcal{F}_x$ are nonempty!).

(ii) $A, B \in \mathcal{F}_x \Rightarrow A \cap B \in \mathcal{F}_x$.

(iii) $A \in \mathcal{F}_x,\ A \subset B \Rightarrow B \in \mathcal{F}_x$.

(iv) $A \in \mathcal{F}_x \ \Rightarrow \overset{\circ}{A} \overset{\text{def}}{=} \{y \in A \,:\, A \in \mathcal{F}_y\} \in \mathcal{F}_x$.

The first three conditions guarantee that family $\mathcal{F}_x$ is a filter and, for that reason, $\mathcal{F}_x$ is called the *filter* (or *system*) *of neighborhoods of point* x. Condition (iv) states that the subset of neighborhood A of x, consisting of *all* points y for which A is a neighborhood as well, must itself be a neighborhood of point x. Later on we will reinterpret this condition as the requirement that with every neighborhood A of x, the *interior* is a neighborhood of x as well.

Mapping

$$X \ni x \to \mathcal{F}_x \subset \mathcal{P}(X)$$

prescribing for each x in X the corresponding filter of neighborhoods is frequently itself called the *topology* on X and X is called again a *topological space.*

We emphasize again that the neighborhoods discussed here, once they satisfy the four axioms, are completely arbitrary and may not necessarily coincide with the neighborhoods defined earlier using open sets.

In practice, instead of setting $\mathcal{F}_x$ directly, we may introduce first *bases of neighborhoods* $\mathcal{B}_x$ *of points* x and set the corresponding filters $\mathcal{F}_x$ as filters generated by these bases, $\mathcal{F}_x = \mathcal{F}(\mathcal{B}_x)$. More precisely, families $\mathcal{B}_x$ must satisfy the following conditions:

(i) $x \in A, \forall A \in \mathcal{B}_x$;

(ii) $A, B \in \mathcal{B}_x \Rightarrow \exists C \in \mathcal{B}_x \, : \, C \subset A \cap B$;

(iii) $\forall B \in \mathcal{B}_x \quad \exists C \in \mathcal{B}_x \, : \, \forall y \in C \quad \exists D \in \mathcal{B}_y \, : \, D \subset B$.

Indeed, the first two conditions imply that $\mathcal{B}_x$ is a base and, consequently, $\mathcal{F}_x = \mathcal{F}(\mathcal{B}_x)$ satisfy the first three conditions for filters of neighborhoods of x. To prove the fourth condition, pick an arbitrary $A \in \mathcal{F}(\mathcal{B}_x)$. By definition of a filter generated by a base, there exists $B \in \mathcal{B}_x$ such that $B \subset A$. It follows now from condition (iii) for the base of neighborhoods that there exists $C \in \mathcal{B}_x$ such that

$$y \in C \Rightarrow \exists D \in \mathcal{B}_y \, : \, D \subset B$$

or, equivalently,

$$y \in C \Rightarrow B \in \mathcal{F}_y$$

which implies

$$y \in C \Rightarrow A \in \mathcal{F}_y$$

Thus $C \in \mathcal{B}_x$ is a subset of $\overset{\circ}{A} = \{y \in A \, : \, A \in \mathcal{F}_y\}$ which implies that $\overset{\circ}{A} \in \mathcal{F}_y$.

Conversely, let $\mathcal{F}_x$ satisfy the four conditions for the filter of neighborhoods of x, and $\mathcal{B}_x$ be *any base generating* $\mathcal{F}_x$. Obviously, $\mathcal{B}_x$ satisfies the first two conditions for the base of neighborhoods and it remains to show only the third condition. Toward this goal, let $B \in \mathcal{B}_x$. Consequently, $\mathcal{B} \in \mathcal{F}_x$ and, by condition (iv) for filters and definition of the base, there exists a $C \in \mathcal{B}_x$ such that

$$\begin{aligned} C \subset \overset{\circ}{B} &= \{y \in B \, : \, B \in \mathcal{F}_y\} \\ &= \{y \in B \, : \, \exists D \in \mathcal{B}_y \, : \, D \subset B\} \end{aligned}$$

or, in another words,

$$y \in C \Rightarrow \exists D \in \mathcal{B}_y \, : \, D \subset B$$

which ends the proof.

We mention also that if $\mathcal{B}_x$ is the base of open neighborhoods $\mathcal{B}_x^o$ discussed earlier, then condition (iii) for the base is trivially satisfied as (by definition) open sets are neigborhoods of all their points and, therefore, it is enough to set $C = D = B$ in condition (iii).

Interior Points. Interior of a Set. Open Sets. Let X be a topological space with topology introduced through filters $\mathcal{F}_x, x \in X$. Consider a set $A \subset X$. A point $x \in A$ is called an *interior point* of A if A contains x together with a neighborhood $C \in \mathcal{F}_x$, i.e.,

$$\exists\, C \in \mathcal{F}_x \;:\; C \subset A.$$

Equivalently, if the topology is set through bases $\mathcal{B}_x$, A must contain a set $B \in \mathcal{B}_x$. Note that a set A is a neighborhood of all its interior points. The collection of all interior points of A, denoted int A, is called the *interior of set* A. Finally, if $A = \text{int}A$, i.e., all points of A are interior, then A is called *open*. We note that set $\overset{\circ}{A}$ used in condition (iv) for filters was precisely the interior of set A, $\overset{\circ}{A} = \text{int}A$.

The following proposition summarizes the fundamental properties of the open sets defined through neighborhoods.

PROPOSITION 4.1.2

(i) A union of an arbitrary family of open sets is an open set.

(ii) A common part of a finite family of open sets is an open set.

PROOF

(i) Assume that $A_\iota, \iota \in I$ are open and let $x \in \bigcup_{\iota \in I} A_\iota$. Then $x \in A_\kappa$ for some κ and therefore there exists a neighborhood C of x such that $C \subset A_\kappa$ and, consequently, $C \subset \bigcup_{\iota \in I} A_\iota$ which proves that every point of $\cup A_\iota$ is interior. Thus $\cup A_\iota$ is open.

(ii) Suppose $A_i, i = 1, 2, \ldots, n$ are open. Let $x \in \bigcap_1^n A_i$ and let $B_i \in \mathcal{F}_x$ be a neighborhood of x such that $B_i \subset A_i$. It follows by induction that $\bigcap_1^n B_i \in \mathcal{F}_x$ as well and, consequently, $\bigcap_1^n B_i \subset \bigcap_1^n A_i$ which proves that x is interior to $\bigcap_1^n A_i$. ∎

At this point we have shown that any family $\mathcal{X}$ of open sets in X induces the corresponding filters $\mathcal{F}_x$ of neighborhoods of points x and, conversely, introducing filters $\mathcal{F}_x$ (or bases $\mathcal{B}_x$) of neighborhoods in X implies existence of the corresponding family of open sets.

We emphasize compatibility of the notions introduced in the two alternative ways. Postulated properties of sets open by definition coincide with (proved) properties of open sets defined through neighborhoods and postulated properties of sets, being neighborhoods by definition, are identical with those for neighborhoods defined through open sets.

In order to prove the equivalence of the two ways of introducing a topology in a set, it remains to show that (recall the discussion in the beginning of this section)

- open sets induced by neighborhoods coincide with open sets in the original class of open sets of a topology on a set X and
- neighborhoods induced by open sets coincide with original neighborhoods in the filters or neighborhood systems of points $x \in X$.

So, let us begin with the first statement. Let $\mathcal{X}$ be a family of open sets introduced as sets in the class $\mathcal{X}$ and $\mathcal{B}_x^o$ the corresponding bases of open neighborhoods. A set A is open with respect to a topology generated by $\mathcal{B}_x^o$ if all its points are interior, i.e.,

$$\forall x \in A \quad \exists B_x \in \mathcal{B}_x^o \quad B_x \subset A$$

This implies that

$$A = \bigcup_{x \in A} B_x$$

and, consequently, set A is the union of open (original) sets B_x and belongs to family $\mathcal{X}$, i.e., every open set in the new sense is also open in the old sense.

Conversely, if $A \in \mathcal{X}$ then A is a neighborhood of every one of its points, i.e., all its points are interior points and, therefore, A is also open in the new sense.

In order to prove the second statement, as both original and defined neighborhoods constitute filters, it is sufficient to show that the two families of sets are equivalent to each other. Thus, let $\mathcal{F}_x$ denote the original filter of neighborhoods and let A be a neighborhood of point x in the new sense. By definition, there exists an open set B, containing x (i.e., $B \in \mathcal{B}_x^o$) such that $B \subset A$. Consequently, by definition of open sets, there exists an original neighborhood $F \in \mathcal{F}_x$ such that $F \subset B$ and, in turn, $F \subset A$. Thus, the original filter of neighborhoods is stronger than the new, induced one.

Conversely, let $A \in \mathcal{F}_x$. By condition (iv) for filters, A contains $\text{int}A$, which is an open neighborhood of x in the new sense and, consequently, the family of new open neighborhoods is stronger than the original one.

We conclude our discussion on two equivalent ways of introducing a topology in a set X with a short summary of properties of the interior $\text{int}A$ of a set A.

PROPOSITION 4.1.3
(Properties of the interior operation)

Let X be a topological space. The following properties hold:

(i) $\text{int}(\text{int}A) = \text{int}A$.

(ii) $\text{int}(A \cap B) = \text{int}A \cap \text{int}B$.

(iii) $\text{int}(A \cup B) \supset \text{int}A \cup \text{int}B$.

(iv) $A \subset B \Rightarrow \text{int}A \subset \text{int}B$.

PROOF

(i) By definition, $\text{int}B \subset B$, so the nontrivial inclusion to be shown is

$$\text{int}A \subset \text{int}(\text{int}A)$$

But this is a direct consequence of the fourth property for filters. Indeed, if $x \in \text{int}A$ then there exists a neighborhood of x, contained in A and, consequently, $A \in \mathcal{F}_x$. It follows from the fourth property for filters $\mathcal{F}_x$ that $B = \text{int}A \in \mathcal{F}_x$. Thus, there exists a neighborhood B of x, namely, $B = \text{int}A$, such that $B \subset \text{int}A$ and therefore $x \in \text{int}(\text{int}A)$.

Proof of the remaining three properties is a straightforward consequence of the definition of interior and we leave it to the reader as an exercise. ∎

REMARK 4.1.1. Interior ($\text{int}A$) of set A is equal to the union of all open sets contained in A

$$\text{int}A = \bigcup \{B \subset A \ : \ B \text{ open}\}$$

Indeed, by property (iv) of the preceeding proposition, $B \subset A$ implies $\text{int}B(= B) \subset \text{int}A$ and therefore the inclusion $\supset$ holds. On the other side, $x \in \text{int}A$ implies that there exists an open neighborhood B_x of x such that

$B_x \subset A$ and A can be represented as

$$A = \bigcup_{x \in B} B_x$$

so the inclusion $\subset$ holds as well.

As the representation above provides for a direct characterization of the interior of a set in terms of open sets, it serves frequently as a definition of the interior, especially when the topology is introduced through open sets. $\blacksquare$

Stronger and Weaker Topologies. It is clear that on the same underlying set X more than one topology can be introduced. We have the following result:

PROPOSITION 4.1.4

Let X be an arbitrary nonempty set, and $\mathcal{X}_1$ and $\mathcal{X}_2$ denote two families of open sets with corresponding

- *filters of neighborhoods $\mathcal{F}_x^1, \mathcal{F}_x^2$,*
- *bases of open neighborhoods $\mathcal{B}_x^{o1}, \mathcal{B}_x^{o2}$, and*
- *any other, arbitrary bases of neighborhoods $\mathcal{B}_x^1, \mathcal{B}_x^2$.*

The following conditions are equivalent to each other:

(i) $\mathcal{X}_1 \supset \mathcal{X}_2$.

(ii) $\mathcal{F}_x^1 \supset \mathcal{F}_x^2$.

(iii) $\mathcal{B}_x^{o1} \succ \mathcal{B}_x^{o2}$.

(iv) $\mathcal{B}_x^1 \succ \mathcal{B}_x^2$.

PROOF Equivalence of conditions (ii), (iii) and (iv) has been proved in Proposition 4.1.1. Note that, in particular, $\mathcal{B}_x^{oi} \sim \mathcal{B}_x^i, i = 1, 2$, as the corresponding filters are the same. Let now $A \in \mathcal{B}_x^{o2}$, i.e., A be an open set from $\mathcal{X}_2$ containing x. By (i), $A \in \mathcal{X}_1$, and therefore $A \in \mathcal{B}_x^{o1}$, which proves that $\mathcal{B}_x^{o1} \succ \mathcal{B}_x^{o2}$. Consequently, (i) implies (iii). Conversely, if $A \in \mathcal{X}_2$ then A is an open neighborhood for each of its points $x \in A$, i.e.,

$$A \in \mathcal{B}_x^{o2} \quad \forall x \in A$$

By (iii), for every $x \in A$ there exists an open set B_x from $\mathcal{X}_1$ such that $B_x \subset A$ and, consequently,

$$A = \bigcup_{x \in A} B_x$$

i.e., A can be represented as a union of open sets from $\mathcal{X}_1$ and therefore must belong to $\mathcal{X}_1$. ∎

In the case described in Proposition 4.1.4, we say that the topology corresponding to $\mathcal{X}_1$, or, equivalently, $\mathcal{F}_x^1$, is *stronger* than topology corresponding to families $\mathcal{X}_2$ or $\mathcal{F}_x^2$. Note that, in particular, *equivalent* bases of neighborhoods imply the *same* topology.

Example 4.1.4

Let $X = \mathbb{R}^n$ and $\mathcal{B}_{\boldsymbol{x}}$ denote the family of open balls centered at $\boldsymbol{x}$

$$\mathcal{B}_{\boldsymbol{x}} = \{B(\boldsymbol{x}, \varepsilon), \varepsilon > 0\}$$

Bases $\mathcal{B}_{\boldsymbol{x}}$ define the *fundamental topology* in $\mathbb{R}^n$. Note that this topology can be introduced through many other, but equivalent, bases, for instance:

- open balls with radii $1/n$,
- closed balls centered at $\boldsymbol{x}$,
- open cubes centered at $\boldsymbol{x}$, $C(\boldsymbol{x}, \varepsilon) = \left\{ \boldsymbol{y} : \sum_{i=1}^{n} |y_i - x_i| < \varepsilon \right\}$,

etc. The key point is that all these families constitute different, but equivalent, bases and therefore the corresponding topology is the same. □

Example 4.1.5

Let X be an arbitrary, nonempty set. The topology induced by single set bases

$$\mathcal{B}_x = \{\{x\}\}$$

is called the *discrete topology* on X. Note that every set C containing x is its neighborhood in this topology. In particular, every point is a neighborhood of itself.

A totally opposite situation takes place if we define a topology by single set bases $\mathcal{B}_x = \{X\}$, for every $x \in X$. Then, obviously, $\mathcal{F}(\mathcal{B}_x) = \{X\}$ and the only neighborhood of every x is the whole set X. The corresponding topology is known as the *trivial topology* on X.

Notice that in the discrete topology *every* set is open, i.e., the family of open sets coincides with the whole $\mathcal{P}(X)$, whereas in the trivial topology the *only* two open sets are the empty set $\emptyset$ and the whole space X. Obviously, the trivial topology is the *weakest* topology on X while the discrete topology is the *strongest* one. $\square$

Example 4.1.6

Let $X = \{\alpha,\ \beta,\ \gamma,\ \rho\}$ and consider the classes of subsets X:

$$\mathcal{X}_1 = \{X, \emptyset, \{\alpha\}, \{\alpha,\beta\}, \{\alpha,\beta,\rho\}\}$$
$$\mathcal{X}_2 = \{X, \emptyset, \{\beta\}, \{\beta,\gamma\}, \{\beta,\gamma,\rho\}\}$$
$$\mathcal{X}_3 = \{X, \emptyset, \{\alpha\}, \{\alpha,\beta\}, \{\beta,\gamma,\rho\}\}$$

Now, it is easily verified that $\mathcal{X}_1$ and $\mathcal{X}_2$ are topologies on X: unions and intersections of subsets from each of the classes are in the same class, respectively, as are X and $\emptyset$. However, $\mathcal{X}_3$ is *not* a topology, since $\{\alpha,\beta\} \cap \{\beta,\gamma,\rho\} = \{\beta\} \notin \mathcal{X}_3$.

Neither topology $\mathcal{X}_1$ nor $\mathcal{X}_2$ is weaker or stronger than the other, since one does not contain the other. In such cases, we say that $\mathcal{X}_1$ and $\mathcal{X}_2$ are *incommensurable.* $\square$

Accumulation Points. Closure of a Set. Closed Sets. As in Chapter 1, point x, not necessarily in set A, is called an *accumulation point* of set A if every neighborhood of x contains at least one point of A, distinct from x:

$$N \cap A - \{x\} \neq \emptyset \quad \forall N \in \mathcal{F}_x$$

The union of set A and the set $\hat{A}$ of all its accumulation points, denoted $\overline{A}$, is called the *closure* of set A. Note that sets A and $\hat{A}$ need not be disjoint. Points in A which are not in $\hat{A}$ are called *isolated points* of A (recall Example 1.18.6).

PROPOSITION 4.1.5
(The Duality Principle)

Let X be a topological space. A set $A \in \mathcal{P}(X)$ is closed if and only if its complement $A' = X - A$ is open.

PROOF See Proposition 1.18.2. ∎

PROPOSITION 4.1.6
(Properties of Closed Sets)

(i) *Intersection of an arbitrary family of closed sets is a closed set.*

(ii) *A union of a finite family of closed sets is closed.*

PROOF See Proposition 1.18.3. ∎

REMARK 4.1.2. Note that a set may be simultaneously open and closed! The whole space X and the empty set $\emptyset$ are the simplest examples of such sets in any topology on X. ∎

Sets of G_δ and F_σ Types. Most commonly, the intersection of an infinite sequence of open sets and the union of an infinite sequence of closed sets are not open or closed, respectively. Sets of this type are called sets of G_δ or F_σ *types*, i.e.,

$$A \text{ is of } G_\delta \text{ type} \quad \text{if } A = \bigcap_1^\infty A_i, A_i \text{ open}$$

$$B \text{ is of } F_\sigma \text{ type} \quad \text{if } B = \bigcup_1^\infty B_i, B_i \text{ closed}$$

Recall that we used this notion already in the context of $\mathbb{R}^n$, in the proof of Fubini's theorem in the preceding chapter.

Before listing properties of the closure of a set, we record the relation between the closure and interior operations.

PROPOSITION 4.1.7

Let A be a set in a topological space X. The following relation holds:

$$(\text{int}A)' = \overline{(A')}$$

PROOF

Inclusion $\subset$. Let $x \in (\text{int}A)'$, i.e., $x \notin \text{int}A$. Consequently, for every neighborhood N of x, $N \not\subset A$ or, equivalently, $N \cap A' \neq \emptyset$. Now, either $x \in A'$ or $x \notin A'$. If $x \notin A'$ then

$$N \cap A' = N \cap A' - \{x\} \neq \emptyset, \quad \forall N \in \mathcal{F}_x$$

which means that x is an accumulation point of A'. Thus either x belongs to A' or x is its accumulation point and, therefore, in both cases it belongs to the closure of A'.

Inclusion $\supset$. Let $x \in \overline{A'}$. Then either $x \in A'$ or x is an accumulation point of A' from outside of A'. If $x \in A'$, then $x \notin A$ and, consequently, $x \notin \text{int}A \subset A$, i.e., $x \in (\text{int}A)'$. If x is an accumulation point of A' and $x \in A$, then

$$N \cap A' - \{x\} = N \cap A' \neq \emptyset, \quad \forall N \in \mathcal{F}_x$$

which implies that $x \notin \text{int}A$. ∎

PROPOSITION 4.1.8
(Properties of the Closure Operation)

The following properties hold:

(i) $\overline{\overline{A}} = \overline{A}$.

(ii) $\overline{(A \cap B)} \subset \overline{A} \cap \overline{B}$.

(iii) $\overline{(A \cup B)} = \overline{A} \cup \overline{B}$.

(iv) $A \subset B \Rightarrow \overline{A} \subset \overline{B}$.

PROOF The proof follows immediately from Propositions 4.1.3 and 4.1.7. ∎

Before we proceed with further examples, we emphasize that all the notions introduced are relative to a given topology. A set which is open with respect to one topology does not need to be open with respect to another one; an accumulation point of a set in one topology may not be an accumulation point of the set in a different topology and so on. It follows, however, directly from the definitions that every interior point of a set remains interior in a stronger topology and every accumulation point of a set remains its accumulation point in any weaker topology as well. Consequently, every open or closed set remains open or closed, respectively, with respect to any stronger topology.

Example 4.1.7

As we have indicated in the beginning of this section, every topological notion we have introduced thus far is a generalization of a corresponding definition of elementary topology in $\mathbb{R}^n$, provided we consider in $\mathbb{R}^n$ topology induced by bases of balls centered at a point. Thus, the elementary topology in $\mathbb{R}^n$ supplies us with the most natural examples of open, closed, F_σ and G_δ sets as well. □

Example 4.1.8

Though we will study function spaces throughout most of this book in a more organized fashion, let us consider a simple example of two different topologies in the space of continuous functions $C(0,1)$ on the interval $(0,1)$.

To define the first topology, pick an arbitrary function $f \in C(0,1)$ and consider for a given positive number ε the set

$$B(f,\varepsilon) \stackrel{\text{def}}{=} \{g \in C(0,1) \ : \ |g(x) - f(x)| < \varepsilon \text{ for every } x \in (0,1)\}$$

It is easy to check that sets $B(f,\varepsilon), \varepsilon > 0$, constitute a base $\mathcal{B}_f$. Bases $\mathcal{B}_f, f \in C(0,1)$, generate the so-called *topology of uniform convergence.* To set the second topology, pick again a function f and consider for a given point $x \in (0,1)$ and a positive number ε the set

$$C(f,\varepsilon,x) \stackrel{\text{def}}{=} \{g \in C(0,1) \ : \ |g(x) - f(x)| < \varepsilon\}$$

Next, for finite sequences $x_1, \dots, x_N \in (0,1)$, define the intersections

$$C(f,\varepsilon,x_1,\dots,x_N) = \bigcap_{k=1}^{N} C(f,\varepsilon,x_k)$$

It is again easy to check that sets $C(f,\varepsilon,x_1,\dots,x_N)$, for $\varepsilon > 0$ and $x_1,\dots,x_N$ arbitrary, but finite, sequences, constitute bases $\mathcal{C}_f$ for different f's which in turn generate the so-called *topology of pointwise convergence* in $C(0,1)$. Obviously,

$$B(f,\varepsilon) \subset C(f,\varepsilon,x_1,\dots,x_N)$$

for any finite sequence $x_1,\dots,x_N$ and, therefore, $\mathcal{B}_f \succ \mathcal{C}_f$ which proves that the uniform convergence topology is *stronger* than the topology of pointwise convergence. In particular, any open or closed set in the pointwise convergence topology is open or closed with respect to uniform convergence topology as well.

To see that the converse, in general, is not true, consider the set of monomials

$$A = \{f(x) = x^n, n \in \mathbb{N}\}$$

We will see later in this chapter that A has no accumulation points in the uniform convergence topology from outside of A. Thus A is closed with respect to this topology. We claim, however, that the zero function $f(x) \equiv 0$ is an accumulation point of A with respect to the pointwise convergence topology and therefore A is not closed with respect to that topology. To see this, pick an arbitrary element from the base of neighborhoods of the zero function $C(0, \varepsilon, x_1, \ldots, x_N)$. It is easy to see that for sufficiently large n,

$$|x_k^n - 0| < \varepsilon \qquad k = 1, \ldots, N$$

and, therefore, $f(x) = x^n$ belongs to $C(0, \varepsilon, x_1, \ldots, x_N)$. Consequently,

$$A \cap C(0, \varepsilon, x_1, \ldots, x_N) \neq \emptyset$$

which proves that zero function is an accumulation point for A. □

Exercises

4.1.1. Let $\mathcal{A}, \mathcal{B} \subset \mathcal{P}(X)$ be two arbitrary families of subsets of a nonempty set X. We define the *trace* $\mathcal{A} \bar{\cap} \mathcal{B}$ of familes $\mathcal{A}$ and $\mathcal{B}$ as the family of common parts

$$\mathcal{A} \bar{\cap} \mathcal{B} \stackrel{\text{def}}{=} \{A \cap B \,:\, A \in \mathcal{A},\ B \in \mathcal{B}\}$$

By analogy, by the *trace* of a family $\mathcal{A}$ on a set C, denoted $\mathcal{A} \bar{\cap} C$, we understand the trace of family $\mathcal{A}$ and the single set family $\{C\}$

$$\mathcal{A} \bar{\cap} C \stackrel{\text{def}}{=} \mathcal{A} \bar{\cap} \{C\} = \{A \cap C \,:\, A \in \mathcal{A}\}$$

Prove the following simple properties:

(i) $\mathcal{A} \succ \mathcal{C},\ \mathcal{B} \succ \mathcal{D} \quad \Rightarrow \quad \mathcal{A} \bar{\cap} \mathcal{B} \succ \mathcal{C} \bar{\cap} \mathcal{D}$.

(ii) $\mathcal{A} \sim \mathcal{C},\ \mathcal{B} \sim \mathcal{D} \quad \Rightarrow \quad \mathcal{A} \bar{\cap} \mathcal{B} \sim \mathcal{C} \bar{\cap} \mathcal{D}$.

(iii) $\mathcal{A} \subset \mathcal{P}(C) \quad \Rightarrow \quad \mathcal{A} \bar{\cap} C = \mathcal{A}$.

(iv) $\mathcal{A} \succ \mathcal{B} \quad \Rightarrow \quad \mathcal{A} \bar{\cap} C \succ \mathcal{B} \bar{\cap} C$.

(v) $B \subset C \quad \Rightarrow \quad \mathcal{A} \bar{\cap} B \succ \mathcal{A} \bar{\cap} C$.

(vi) $\mathcal{A} \subset \mathcal{P}(C) \quad \Rightarrow \quad (\mathcal{A} \succ \mathcal{B} \Leftrightarrow \mathcal{A} \succ \mathcal{B} \bar{\cap} C)$.

4.1.2. Let $\mathcal{A} \subset \mathcal{P}(X)$ and $\mathcal{B} \subset \mathcal{P}(Y)$ denote two arbitrary families of subsets of X and Y, respectively, and let $f : X \to Y$ denote an arbitrary function from X into Y. We define the *image of family* $\mathcal{A}$ *by function* f and *inverse image of family* $\mathcal{B}$ *by function* f as follows:

$$f(\mathcal{A}) \stackrel{\text{def}}{=} \{f(A) \,:\, A \in \mathcal{A}\}$$

$$f^{-1}(\mathcal{B}) \stackrel{\text{def}}{=} \{f^{-1}(B) \,:\, B \in \mathcal{B}\}$$

Prove the following simple properties:

(i) $\mathcal{A} \succ \mathcal{B} \quad \Rightarrow \quad f(\mathcal{A}) \succ f(\mathcal{B})$.

(ii) $\mathcal{A} \sim \mathcal{B} \quad \Rightarrow \quad f(\mathcal{A}) \sim f(\mathcal{B})$.

(iii) $\mathcal{C} \succ \mathcal{D} \quad \Rightarrow \quad f^{-1}(\mathcal{C}) \succ f^{-1}(\mathcal{D})$.

(iv) $\mathcal{C} \sim \mathcal{D} \quad \Rightarrow \quad f^{-1}(\mathcal{C}) \sim f^{-1}(\mathcal{D})$.

(v) $f(\mathcal{A}) \succ \mathcal{C} \quad \Leftrightarrow \quad \mathcal{A} \bar{\cap} \text{dom} f \succ f^{-1}(\mathcal{C})$.

4.1.3. Let $X = \{w, x, y, z\}$. Determine whether or not the following classes of subsets of X are topologies:

$$\mathcal{X}_1 = \{X, \emptyset, \{z\}, \{y, z\}, \{x, y, z\}\}$$

$$\mathcal{X}_2 = \{X, \emptyset, \{w\}, \{w, x\}, \{w, y\}\}$$

$$\mathcal{X}_3 = \{X, \emptyset, \{x, y, z\}, \{x, y, w\}, \{x, y\}\}$$

4.1.4. The class $\mathcal{X} = \{X, \emptyset, \{a\}, \{b\}, \{a, b\}, \{a, b, c\}, \{a, b, d\}\}$ is a topology on $X = \{a, b, c, d\}$.

(i) Identify the closed sets relative to $\mathcal{X}$.

(ii) What is the closure of $\{a\}$? of $\{a, b\}$?

(iii) Identify the interior of $\{a, b, c\}$ and a neighborhood system of b.

4.1.5. Let $\mathcal{A} \subset \mathcal{P}(X)$ be a family of subsets of a set X. Prove that the following conditions are equivalent to each other:

(i) $\forall A, B \in \mathcal{A} \quad \exists C \in \mathcal{A} \,:\, C \subset A \cap B$ (condition for a base).

(ii) $\mathcal{A} \succ \mathcal{A} \bar{\cap} \mathcal{A}$.

(iii) $\mathcal{A} \sim \mathcal{A} \bar{\cap} \mathcal{A}$.

(See Exercise 4.1.1 for notation.)

4.1.6. Let X be a topological space. We say that point x is *a cluster point of set* A if

$$N \cap A \neq \emptyset, \quad \text{for every neighborhood } N \text{ of } x$$

Show that point x is a cluster point of set A iff it belongs to its closure: $x \in \overline{A}$.

4.1.7. Let X be a topological space. Show that closure $\overline{A}$ of a set $A \subset X$ is equal to the intersection of all closed sets containing set A

$$\overline{A} = \bigcap \{B \text{ closed} \, : \, A \subset B\}$$

Consequently, closure $\overline{A}$ is the *smallest* closed set containing set A.

4.2 Topological Subspaces and Product Topologies

In this section we shall complete the fundamental topological notions introduced in the previous section. In particular, we demonstrate how a topology on X induces a topology on every subset of X and how two topologies, one on X, another on Y, generate a topology on the Cartesian product $X \times Y$.

Topological Subspaces. Let X be a topological space and $Y \subset X$ be an arbitrary subset of X. Set Y can be supplied with a natural topology in which neighborhoods are simply the intersections of neighborhoods in X with set Y. More precisely, for every $x \in Y$ we introduce the following base of neighborhoods

$$\mathcal{B}_x^Y = \{B \cap Y \, : \, B \in \mathcal{B}_x\}$$

where $\mathcal{B}_x$ is a base of neighborhoods of $x \in X$. It is easily verified that $\mathcal{B}_x^Y$ satisfies the axioms of a base of neighborhoods. With such an introduced topology, set Y is called the *topological subspace* of X.

PROPOSITION 4.2.1

Let Y be a topological subspace of X and $E \subset Y$ a subset of Y. Then

$$^{Y}\overline{E} = \overline{E} \cap Y$$

where $^{Y}\overline{E}$ denotes closure of E in the topological subspace Y.

PROOF

$\subset$. Let $x \in^{Y} \overline{E}$. Then either $x \in E$ or x is an accumulation point of E from $Y - E$. In the first case x obviously belongs to the right-hand side. In the second case we have

$$B^Y \cap E - \{x\} \neq \emptyset \quad \text{for every} \quad B^Y \in \mathcal{B}_x^Y$$

or, equivalently,

$$B \cap Y \cap E - \{x\} \neq \emptyset \quad \text{for every} \quad B \in \mathcal{B}_x$$

This implies that

$$B \cap E - \{x\} \neq \emptyset \quad \text{for every} \quad B \in \mathcal{B}_x$$

i.e., x is an accumulation point of E in X.

$\supset$. If $x \in \overline{E} \cap Y$ then $x \in Y$ and either $x \in E$ or x is an accumulation point of E from outside of E. It remains to consider only the second case. We have

$$B \cap E - \{x\} \neq \emptyset \quad \text{for every} \quad B \in \mathcal{B}_x$$

But $(B \cap Y) \cap E = B \cap (Y \cap E) = B \cap E (E \subset Y)$ and therefore

$$(B \cap Y) \cap E - \{x\} \neq \emptyset \quad \text{for every} \quad B \in \mathcal{B}_x$$

which means that x is an accumulation point of E in Y. ∎

We have the following fundamental characterization of open and closed sets in topological subspaces.

PROPOSITION 4.2.2

Let $Y \subset X$ be a topological subspace of X. The following hold:

(i) *A set $F \subset Y$ is closed in Y if there exists a closed set F_1 in X such that $F = F_1 \cap Y$.*

(ii) *A set $G \subset Y$ is open in Y if there exists an open set G_1 in X such that $G = G_1 \cap Y$.*

In other words, closed and open sets in topological subspaces Y are precisely the intersections of open and closed sets from X with set Y.

PROOF

(i) Assume that F is closed in Y. Then

$$F \overset{Y}{=} \overline{F} = \overline{F} \cap Y$$

and we can choose simply $F_1 = \overline{F}$.

Conversely, let $F = F_1 \cap Y$, where F_1 is closed in X. Then

$$^Y\overline{F} = \overline{(F_1 \cap Y)} \cap Y \subset \overline{F}_1 \cap \overline{Y} \cap Y = \overline{F}_1 \cap Y = F_1 \cap Y = F$$

which proves that F is closed in Y.

(ii) G is open in Y if and only if $Y - G$ is closed in Y. According to part (i), this is equivalent to saying that there is a closed set F_1 in X such that $Y - G = F_1 \cap Y$.

It follows that

$$G = Y - (Y - G) = Y - (F_1 \cap Y) = Y \cap (F_1')$$

which proves the assertion because $G_1 = F_1'$ is open in X. ∎

Example 4.2.1

Let X be the real line $\mathbb{R}$ with the fundamental topology in $\mathbb{R}$ and let $Y = (0, \infty)$. Set $E = (0, 1]$ is *closed* in Y. Indeed, $E = [a, 1] \cap Y$ for any $a \leq 0$ and interval $[a, 1]$ is closed in $\mathbb{R}$. Note that, however, E is *not* closed in the whole $\mathbb{R}$! □

Product Topologies. Let X and Y be two topological spaces. Introducing on the Cartesian product $X \times Y$ the following bases of neighborhoods,

$$\mathcal{B}_{(x,y)} = \{C = A \times B \ : \ A \in \mathcal{B}_x, B \in \mathcal{B}_y\}$$

where $\mathcal{B}_x$ and $\mathcal{B}_y$ denote bases of neighborhoods of x in X and y in Y, respectively, we generate on $X \times Y$ a topology called the *product topology* of topologies on X and Y.

Of course, the Cartesian product $X \times Y$, as any set, can be supplied with a different topology, but the product topology is the most natural one and we shall always assume that $X \times Y$ is supplied with this topology, unless explicity stated otherwise.

We leave as an exercise proof of the following simple result.

PROPOSITION 4.2.3

Let X and Y be two topological spaces. The following hold:

(i) *A is open in X and B is open in Y iff $A \times B$ is open in $X \times Y$.*

(ii) *A is closed in X and B is closed in Y iff $A \times B$ is closed in $X \times Y$.*

The notion of the product topology can be easily generalized to the case of a Cartesian product of more than two spaces.

Example 4.2.2

Consider the space $\mathbb{R}^{n+m} = \mathbb{R}^n \times \mathbb{R}^m$. The following bases of neighborhoods are equivalent to each other:

$$\mathcal{B}_{\boldsymbol{z}} = \{B(\boldsymbol{z}, \varepsilon), \quad \varepsilon > 0\}$$

$$\mathcal{C}_{\boldsymbol{z}} = \{B(\boldsymbol{x}, \varepsilon_1) \times B(\boldsymbol{y}, \varepsilon_2), \quad \varepsilon_1 > 0, \varepsilon_2 > 0\}$$

where $\boldsymbol{z} = (\boldsymbol{x}, \boldsymbol{y}), \boldsymbol{x} \in \mathbb{R}^n, \boldsymbol{y} \in \mathbb{R}^m$.

Thus, the topology in $\mathbb{R}^{n+m}$ coincides with the product topology from $\mathbb{R}^n$ and $\mathbb{R}^m$. $\square$

Dense Sets. Separable Spaces. A set $Y \subset X$ in a topological space X is said to be *dense in* X if its closure coincides with X, i.e.,

$$\overline{Y} = X$$

A space X is called *separable* if there exists a countable set Y dense in X. Equivalently, for every point $x \in X$ and an arbitrary neighborhood B of x there exists a point $y \in Y$ belonging to B ($B \cap Y \neq \emptyset$).

Example 4.2.3

Rationals $\mathbb{Q}$ are dense in the set of real numbers $\mathbb{R}$. $\square$

Exercises

4.2.1. Let $\mathcal{A} \subset \mathcal{P}(X), \mathcal{B} \subset \mathcal{P}(Y)$ be families of subsets of X and Y, respectively. The *Cartesian product of families* $\mathcal{A}$ and $\mathcal{B}$ is defined as the family of Cartesian products of sets from $\mathcal{A}$ and $\mathcal{B}$

$$\mathcal{A} \bar{\times} \mathcal{B} \stackrel{\text{def}}{=} \{A \times B : A \in \mathcal{A}, B \in \mathcal{B}\}$$

Prove the following properties:

(a) $\mathcal{A} \succ \mathcal{C}, \mathcal{B} \succ \mathcal{D} \quad \Rightarrow \quad \mathcal{A} \bar{\times} \mathcal{C} \succ \mathcal{B} \bar{\times} \mathcal{D}$.

(b) $\mathcal{A} \sim \mathcal{C}, \mathcal{B} \sim \mathcal{D} \quad \Rightarrow \quad \mathcal{A} \bar{\times} \mathcal{B} \succ \mathcal{C} \bar{\times} \mathcal{D}$.

(c) $(f \times g)(\mathcal{A} \bar{\times} \mathcal{B}) = f(\mathcal{A}) \bar{\times} f(\mathcal{B})$.

4.2.2. Recall the topology

$$\mathcal{X} = \{X, \emptyset, \{a\}, \{b\}, \{a,b\}, \{a,b,c\}, \{a,b,d\}\}$$

on a set $X = \{a,b,c,d\}$ from Exercise 4.1.4.

(i) Are the sets $\{a\}$ and $\{b\}$ dense in X?

(ii) Are there any other sets dense in X?

(iii) Is the space $(X, \mathcal{X})$ separable? Why?

4.3 Continuity and Compactness

We begin this section with the fundamental notion of continuous functions. Then we study some particular properties of continuous functions and turn to a very important class of so-called compact sets. We conclude this section with some fundamental relations for compact sets and continuous functions proving, in particular, the generalized Weierstrass theorem.

Continuous Function. Let X and Y be two topological spaces and let $f: X \to Y$ be a function defined on whole X. Consider a point $x \in X$. Recalling the introductory remarks in Section 4.1, we say that function f is continuous at x if

$$f(\mathcal{B}_x) \succ \mathcal{B}_{f(x)} \text{ or, equivalently, } f(\mathcal{F}_x) \succ \mathcal{F}_{f(x)}$$

i.e., every neighborhood of $f(x)$ contains a direct image, through function f, of a neighborhood of x (see Fig. 4.1). In the case of a function f defined on a proper subset $\text{dom} f$ of X, we replace in the definition the topological space X with the domain $\text{dom} f$ treated as a topological subspace of X, or, equivalently, ask for

$$f(\mathcal{B}_x^{\text{dom} f}) = f(\mathcal{B}_x \bar{\cap} \text{dom} f) \succ \mathcal{B}_{f(x)}$$

(see Exercise 4.1.1).

We say that f is *(globally) continuous* if it is continuous at every point in its domain.

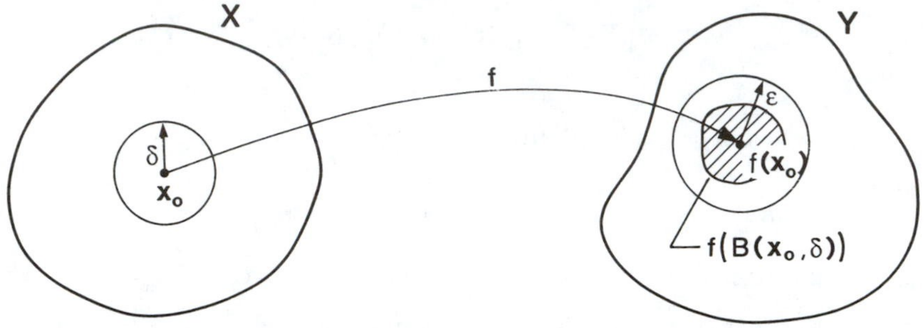

Figure 4.1
Continuity of a function at a point.

Example 4.3.1

The function $f: \mathbb{R} \longrightarrow \mathbb{R}$ given by

$$f(x) = \begin{cases} 0 & x \leq 0 \\ 1 & x > 0 \end{cases}$$

is discontinuous at $x = 0$ and continuous at any other point. To confirm this assertion notice that for any ball $B = B(0, \delta)$ centered at 0, $(B(0,\delta) = (-\delta, \delta))$, $f(B)$ contains the number 1 and, therefore, $f(B)$ cannot be contained in the balls $B(0, \varepsilon) = B(f(0), \varepsilon)$ for $\varepsilon < 1$. $\square$

Example 4.3.2

Let $X = \{1, 2, 3\}$ and $Y = \{a, b, c, d\}$ and consider the following topologies:

$$\mathcal{X} = \{X, \emptyset, \{1\}, \{2\}, \{1,2\}\} \quad \text{and} \quad \mathcal{Y} = \{Y, \emptyset, \{a\}, \{b\}, \{a,b\}, \{b,c,d\}\}$$

Consider the functions F and G from X into Y defined by

F	G
$F(1) = b$	$G(1) = a$
$F(2) = c$	$G(2) = b$
$F(3) = d$	$G(3) = c$

The function F is continuous at 1. Indeed, set $\{1\}$ is a neighborhood of 1 and $f(\{1\}) = \{b\}$ must be contained in all neighborhoods of b. Similarly, F is continuous at 2. The only two neighborhoods of d in Y are Y itself and set $\{b, c, d\}$ and they both contain $f(X)$, with X being the only neighborhood of 3 in X. Thus function F is continuous. Is function G continuous as well? $\square$

The following propositions summarize the fundamental properties of continuous functions.

PROPOSITION 4.3.1

In the following, U, V, X, Y and Z denote topological spaces.

(i) *Let $f: X \to Y$ and $g: Y \to Z$ be continuous. Then the composition $g \circ f: X \to Z$ is continuous as well.*

(ii) *Let $f: X \to Y$ and $g: X \to Z$ be continuous. Then $(f, g): X \ni x \to (f(x), g(x)) \in Y \times Z$ is continuous as well.*

(iii) *Let $f: U \to X$ and $g: V \to Y$ be continuous. Then the Cartesian product of functions f and g*

$$(f \times g): U \times V \ni (u, v) \to (f(u), g(v)) \in X \times Y$$

is continuous.

PROOF The proof follows immediately from definitions and we leave details as an exercise. ∎

PROPOSITION 4.3.2

Let X and Y be two topological spaces and let $f: X \to Y$ denote an arbitrary function defined on the whole $X(\text{dom} f = X)$. Then the following conditions are equivalent to each other:

(i) *f is (globally) continuous.*

(ii) $f^{-1}(G)$ *is open in* X *for every* G *open in* Y.

(iii) $f^{-1}(F)$ *is closed in* X *for every* F *closed in* Y.

PROOF

(i) $\Rightarrow$ (ii). Let G be an open set in Y and let $x \in f^{-1}(G)$. Then $f(x) \in G$ and, consequently, there exists a neighborhood B of $f(x)$ such that $B \subset G$. It follows from continuity at x that there must be a neighborhood C of x such that $f(C) \subset B$, which implies that

$$C \subset f^{-1}(B) \subset f^{-1}(G)$$

(ii) $\Rightarrow$ (i). We use the bases of open neighborhoods $\mathcal{B}_x^o$ and $\mathcal{B}_{f(x)}^o$. Let G be an arbitrary open set containing $f(x)$. Then $f^{-1}(G)$ is open, contains x and, therefore, is a neighborhood of x. Trivially,

$$f(f^{-1}(G)) \subset G$$

which proves that f is continuous at x.

(i) $\Leftrightarrow$ (iii) follows by duality arguments from the identity:

$$f^{-1}(G') = (f^{-1}(G))'$$

∎

Hausdorff Spaces. In what follows, we restrict ourselves to a class of topological spaces called *Hausdorff spaces.* A topological space X is said to be Hausdorff if for every two distinct points x and y there exist neighborhoods B of x and C of y such that $B \cap C = \emptyset$. In other words, every two distinct points can be separated by disjoint neighborhoods. We will see a fundamental consequence of this definition in the next section when we define the limit of a sequence.

Example 4.3.3

Let X be an arbitrary nonempty set. The discrete topology on X is Hausdorff (explain why), the trivial one is not. □

Compact Topological Spaces. Let X be a topological space and $\mathcal{G} \subset \mathcal{P}(X)$ a family of sets. $\mathcal{G}$ is said to be a *covering of space* X if simply

$$X = \bigcup_{G \in \mathcal{G}} G$$

Similarly, if $\mathcal{G}$ contains a subfamily $\mathcal{G}_0$, which is also a covering of X, then $\mathcal{G}_0$ is called a *subcovering*. We say that covering or subcovering is finite if it contains a finite number of sets. Finally, if all sets of $\mathcal{G}$ are open, then we speak of an *open covering*.

We have the following important definition. A Hausdorff space X is said to be *compact* if every open covering of X contains a finite subcovering. In other words, from every family of open sets $G_\iota, \iota \in I$, I being an "index set", such that

$$X = \bigcup_{\iota \in I} G_\iota$$

we can extract a finite number of sets $G_1, \ldots, G_k$ such that

$$X = G_1 \cup \ldots \cup G_k$$

Now, let $\mathcal{B}$ be a base. We say that a point x is a *limit point* of $\mathcal{B}$ if

$$x \in \bigcap_{B \in \mathcal{B}} \overline{B}$$

We have the following important characterization of compact spaces.

PROPOSITION 4.3.3

Let X be a Hausdorff space. The following conditions are equivalent to each other:

(i) *X is compact.*

(ii) *Every base in X possesses a limit point.*

PROOF

(i) $\Rightarrow$ (ii). Let $\mathcal{B}$ be a base in X. It follows from the definition of a base that the family consisting of closures of sets belonging to $\mathcal{B}$ is also a base. Thus, it is sufficient to show that every base of closed sets possesses a limit point. Let $\mathcal{B}$ be such a base and assume, to the contrary, that

$$\bigcap_{B \in \mathcal{B}} B = \emptyset \quad \text{or, equivalently,} \quad \bigcup_{B \in \mathcal{B}} B' = X$$

Since sets B' are open, they form an open covering of X and, according to the definition of compact space, we can find sets $B_1, \ldots, B_k$ such that

$$B'_1 \cup \ldots \cup B'_k = X$$

or, equivalently,

$$B_1 \cap \ldots \cap B_k = \emptyset$$

But this contradicts the assumption that $\mathcal{B}$ is a base, as every finite intersection of elements from a base is nonempty (explain why).

(ii) $\Rightarrow$ (i). Let $\mathcal{G}$ be an open covering of X

$$X = \bigcup_{G \in \mathcal{G}} G$$

or, equivalently,

$$\bigcap_{G \in \mathcal{G}} G' = \emptyset$$

Define the family $\mathcal{B}$ of finite intersections,

$$G_1' \cap \ldots \cap G_k' \qquad G_i \in \mathcal{G}$$

$\mathcal{B}$ is a base provided each of the intersections is nonempty. Assume now, contrary to (i), that there is no finite subcovering in $\mathcal{G}$. This is equivalent to saying that the intersections above are not empty and, consequently, $\mathcal{B}$ is a base of closed sets. According to (ii),

$$\bigcap_{G \in \mathcal{G}} G' \neq \emptyset \quad \text{i.e.} \quad \bigcup_{G \in \mathcal{G}} G \neq X$$

which contradicts that $\mathcal{G}$ is a covering of X. ∎

Compact Sets. A set E in a Hausdorff space X is said to be compact if E, as a topological subspace of X, is compact. Let $\mathcal{G}$ be a family of open sets in X such that

$$E \subset \bigcup_{G \in \mathcal{G}} G$$

We say that $\mathcal{G}$ is an *open covering of set E in space X*. Sets $G \cap E$ are open in E and, therefore, they form an open covering of space (topological subspace) E. If E is compact, then there exists a finite subfamily $G_1, \ldots, G_k$ such that

$$E = (G_1 \cap E) \ \cup \ \ldots \ (G_k \cap E)$$

or, equivalently,

$$E \subset G_1 \ \cup \ \ldots \ \cup G_k$$

Concluding, a set E is compact if every open covering of E in X contains a finite subcovering.

Similarly, reinterpreting Proposition 4.3.3, we establish that a set E is compact if every base $\mathcal{B}$ in E possesses a limit point in E, i.e.,

$$\bigcap_{B\in\mathcal{B}} \overline{B} \cap E \neq \emptyset$$

The following proposition summarizes the fundamental properties of compact sets.

PROPOSITION 4.3.4

(i) *Every compact set is closed.*

(ii) *Every closed subset of a compact set is compact.*

(iii) *Let $f: X \to Y$ be continuous and $E \subset \operatorname{dom} f \subset X$ be compact. Then $f(E)$ is also compact. In other words, a direct image of a continuous function of a compact set is compact.*

(iv) *Cartesian products of compact sets are compact.*

PROOF

(i) Let E be compact and let x be an accumulation point of E. Suppose that x does not belong to E. Consequently,

$$B \cap E = B \cap E - \{x\} \neq \emptyset \quad \text{for every} \quad B \in \mathcal{B}_x$$

But this means that sets $B \cap E$ form a base in E which, by compactness of E, must possess a limit point in E, i.e., the set

$$\bigcap_{B\in\mathcal{B}} \overline{B \cap E} \cap E$$

is nonempty. But

$$\bigcap_{B\in\mathcal{B}} \overline{B \cap E} \cap E \subset \bigcap_{B\in\mathcal{B}} \overline{B} \cap \overline{E} \cap E = \bigcap_{B\in\mathcal{B}} \overline{B} \cap E$$

which implies that $\mathcal{B}_x$ has a limit point in E. But the *only* limit point of $\mathcal{B}_x$ in a Hausdorff space is the point x (explain why) and, therefore, it follows from the compactness of E that x must belong to E, a contradiction.

(ii) Let $F \subset E, F$ closed, E compact. Assume $\mathcal{B}$ is a base in F. Then $\mathcal{B}$ is also a base in E and, therefore, there exists $x \in E$ such that $x \in \bigcap_{B\in\mathcal{B}} \overline{B}$.

But F is closed and therefore $\overline{B} \subset \overline{F} = F$ for every $B \in \mathcal{B}$. Consequently, $\bigcap \overline{B} \subset F$ which proves that x belongs to F, and, therefore, x is a limit point of $\mathcal{B}$ in F.

(iii) Due to the definition of compact sets, it is sufficient to prove the case when $E = X$ and $f : X \to Y$ is a surjection. So, let $\mathcal{B}$ be a base in Y. As in the proof of Proposition 4.3.3, we can assume that $\mathcal{B}$ is a base of closed sets. Consequently, $f^{-1}(B), B \in \mathcal{B}$ form a base of closed sets in X and there exists $x \in X$ such that

$$x \in f^{-1}(B) \qquad \text{for every } B \in \mathcal{B}$$

It follows that

$$f(x) \in f(f^{-1}(B)) = B \qquad \text{for every } B \in \mathcal{B}$$

and, therefore, $f(x)$ is a limit point of $\mathcal{B}$.

(iv) Again, as before, it is sufficient to consider two compact spaces $X$and Y and prove that the product space $X \times Y$ is compact. Denote by i and j the standard projections

$$i : X \times Y \ni (x, y) \to x \in X$$

$$j : X \times Y \ni (x, y) \to y \in Y$$

and pick an arbitrary base $\mathcal{B}$ in $X \times Y$. Consequently, family $\{i(B) : B \in \mathcal{B}\}$ is a base in X and, by compactness of X, there exists a limit point $x \in X$ of this base, i.e.,

$$x \in \overline{i(B)}, \qquad \forall B \in \mathcal{B}$$

or, equivalently (see Exercise 4.3.7),

$$i(B) \cap N \neq \emptyset \qquad \forall B \in \mathcal{B}, \quad \forall N \in \mathcal{B}_x$$

This implies that

$$B \cap i^{-1}(N) \neq \emptyset \qquad \forall B \in \mathcal{B}, \quad \forall N \in \mathcal{B}_x$$

and, consequently, family

$$\{B \cap i^{-1}(N), \quad B \in \mathcal{B}, N \in \mathcal{B}_x\}$$

is a base in $X \times Y$.

Repeating the same argument with this new base in place of the original base $\mathcal{B}$, we obtain

$$B \cap i^{-1}(N) \cap j^{-1}(M) \neq \emptyset \qquad \forall B \in \mathcal{B}, \quad \forall N \in \mathcal{B}_x, \quad \forall N \in \mathcal{B}_y$$

But $i^{-1}(N) \cap j^{-1}(M) = N \times M$ and, consequently,

$$B \cap (N \times M) \neq \emptyset \qquad \forall B \in \mathcal{B}, \quad \forall N \in \mathcal{B}_x, \quad \forall N \in \mathcal{B}_y$$

which implies that (x, y) is a limit point of $\mathcal{B}$. Thus, every base in $X \times Y$ possesses a limit point and, therefore, $X \times Y$ is compact. ∎

We conclude this section with two fundamental theorems concerning compact sets. The first one, the Heine-Borel theorem, characterizes compact sets in $\mathbb{R}$.

THEOREM 4.3.1
(The Heine-Borel Theorem)

A set $E \subset \mathbb{R}$ is compact iff it is closed and bounded.

PROOF Let $E \subset \mathbb{R}$ be compact. According to Proposition 4.3.3 (i), it suffices to prove that E is bounded. Assume, to the contrary, that $\sup E = +\infty$ and consider the family $\mathcal{B}$ of sets of the form

$$[c, \infty) \cap E, \quad c \in \mathbb{R}$$

Obviously, $\mathcal{B}$ is a base of closed sets in E and its intersection is empty, which contradicts that E is compact.

Conversely, to prove that a closed and bounded set E must be compact, it is sufficient to prove that every closed interval [a,b] is compact. Then E, as a closed subset of a compact set, will also have to be compact.

So, let $\mathcal{B}$ be a base of closed sets in a closed interval [a,b]. First of all, any nonempty, closed and bounded set B in $\mathbb{R}$ possesses its maximum and minimum. Indeed, if $b = \sup B$ then, by the boundedness of B, $b < \infty$ and it follows from the definition of supremum that there exists a sequence b_n from B converging to b. Finally, it follows from the closedness of B that $b \in B$ and therefore $\sup B = \max B$. Analogously, $\inf B = \min B$.

Denote now

$$c = \inf_{B \in \mathcal{B}} (\max B)$$

It follows from the definition of infimum that for every $\delta > 0$ there exists a set B_δ from base $\mathcal{B}$ such that

$$\max B_\delta < c + \delta$$

Next, according to the definition of a base, for every $B \in \mathcal{B}$ there exists a $B_1 \in \mathcal{B}$ such that

$$B_1 \subset B \cap B_\delta$$

Consequently,

$$c \leq \max B_1 \leq \max(B \cap B_\delta) \leq \max B_\delta < c + \delta$$

which implies that

$$(c - \delta, c + \delta) \cap B \neq \emptyset \quad \text{for every } B$$

and, since δ was arbitrary,

$$c \in B \quad \text{for every } B$$

which proves that c is a limit point of base $\mathcal{B}$. ∎

COROLLARY 4.3.1

A set $E \subset \mathbb{R}^n$ is compact if and only if it is closed and bounded.

PROOF Let E be compact and denote by i_j the standard projections

$$i_j : \mathbb{R}^n \ni (x_1, \ldots, x_n) \to x_j \in \mathbb{R}$$

Functions i_j are continuous and therefore images $i_j(E)$ in $\mathbb{R}$ for $j = 1,2,\ldots$, n must be compact as well. By the Heine-Borel theorem, $i_j(E)$ are bounded, i.e., there exist $a_j, b_j \in \mathbb{R}$ such that $i_j(E) \subset [a_j, b_j]$. Consequently,

$$E \subset [a_1, b_1] \times \ldots \times [a_n, b_n]$$

and E is bounded as well. Conversely, according to Proposition 4.3.3 (iv), every closed cube $[a_1, b_1] \times \ldots \times [a_n, b_n]$ is compact and, therefore, every closed subset of such a cube is also compact. ∎

We conclude this section with a fundamental theorem characterizing continuous function defined on compact sets. Note that this result generalizes Theorem 1.20.1, established there for functions on sets in $\mathbb{R}^n$, to general topological spaces.

THEOREM 4.3.2
(The Weierstrass Theorem)

Let $E \subset X$ be a compact set and $f\colon E \to \mathbb{R}$ a continuous function. Then f attains on E its maximum and minimum, i.e., there exist $x_{min}, x_{max} \in E$, *such that*

$$f(x_{min}) = \inf_{x \in E} f(x) \quad \textit{and} \quad f(x_{max}) = \sup_{x \in E} f(x)$$

PROOF According to Proposition 4.3.4 (iii), $f(E)$ is compact in $\mathbb{R}$ and therefore the Heine-Borel theorem implies that $f(E)$ is closed and bounded. Thus, both $\sup f(E)$ and $\inf f(E)$ are finite and belong to set $f(E)$ which means that there exist x_{min} and x_{max} such that $f(x_{min}) = \inf f(E)$ and $f(x_{max}) = \sup f(E)$ ∎

Exercises

4.3.1. Let $X = \{1, 2, 3, 4\}$ and consider the topology on X given by $\mathcal{X} = \{X, \emptyset, \{1\}, \{2\}, \}1, 2\{, \{2, 3, 4\}\}$. Show that the function F: $X \to X$ given by

$$F(1) = 2, \quad F(2) = 4, \quad F(3) = 2, \quad F(4) = 3$$

is continuous at 4, but not at 3.

4.3.2. Let A be any subset of a set X and F a function mapping X into Y. Prove that F is continuous iff $F(\overline{A}) \subset \overline{F(A)}$ for every $A \subset X$.

4.3.3. Let F: $X \to Y$. Prove that F is continuous iff $\overline{F^{-1}(B)} \subset F^{-1}(\overline{B})$ for every $B \subset Y$.

4.3.4. Let $\mathcal{X}_1$ and $\mathcal{X}_2$ be two topologies on a set X and let I: $(X, \mathcal{X}_1) \to (X, \mathcal{X}_2)$ be the identity function. Show that I is continuous if and only if $\mathcal{X}_1$ is stronger than $\mathcal{X}_2$; i.e., $\mathcal{X}_2 \subset \mathcal{X}_1$.

4.3.5. Show that the constant function $F(x) = c \; \forall \, x \in X$, where F: $X \to X$, is continuous relative to any topology on X.

4.3.6. Explain why every function is continuous at isolated points of its domain with respect to any topology.

4.3.7. We say that bases $\mathcal{A}$ and $\mathcal{B}$ are *adjacent* if

$$A \cap B \neq \emptyset \qquad \forall A \in \mathcal{A}, \; B \in \mathcal{B}$$

Verify that bases $\mathcal{A}$ and $\mathcal{B}$ are adjacent iff $\mathcal{A} \bar{\cap} \mathcal{B}$ is a base.
Analogously, we say that base $\mathcal{A}$ is *adjacent* to set C if

$$A \cap C \neq \emptyset \qquad \forall a \in \mathcal{A}$$

Verify that base $\mathcal{B}$ is adjacent to set C iff family $\mathcal{A}\tilde{\cap}C$ is a base. Prove the following simple properties:

(a) $\mathcal{A}$ is adjacent to C, $\mathcal{A} \succ \mathcal{B} \quad \Rightarrow \quad \mathcal{B}$ is adjacent to C.

(b) If family $\mathcal{A}$ is a base then (family $f(\mathcal{A})$ is a base iff $\mathcal{A}$ is adjacent to dom$f$).

(c) Family $\mathcal{C}$ being a base implies that ($f^{-1}(\mathcal{C})$ is a base iff $\mathcal{C}$ is adjacent to range $f(X)$).

(d) If families $\mathcal{A}, \mathcal{B}$ are adjacent and families $\mathcal{C}, \mathcal{D}$ are adjacent then families $\mathcal{A} \ \bar{\times} \ \mathcal{C}, \mathcal{B} \ \bar{\times} \ \mathcal{D}$ are adjacent as well.

4.4 Sequences

It turns out that for a class of topological spaces many of the concepts introduced can be characterized in terms of sequences. In the present section, we define notions such as sequential closedness, continuity and compactness and seek conditions under which they are equivalent to the usual concepts of closedness, continuity or compactness.

Convergence of Sequences. A sequence of points x_n in a topological space X *converges* to a point $x \in X$, denoted $x_n \rightarrow x$, if, for every neighborhood B of x, there exists an index $N = N(B)$ such that

$$x_n \in B \qquad \text{for every } n \geq N$$

Sometimes we say that almost all (except for a finite number of) elements of x_n belong to B. Equivalently, we write $x = \lim x_n$ and call x the *limit of sequence* x. Note once again that this notion is a straightforward generalization of the idea of convergence of sequences in $\mathbb{R}^n$ where the word *ball* has been replaced by the word *neighborhood.*

Example 4.4.1

Consider the space of continuous functions $C(0.1)$ of Example 4.1.8, with the topology of pointwise convergence and the sequence of monomials $f_n(x) = x^n$. A careful look at Example 4.1.8 reveals that we have proved there that f_n converges to the zero function. $\square$

It may seem strange, but in an arbitrary topological space a sequence x_n may have more than one limit. To see this, consider an arbitrary nonempty set X with the trivial topology. Since the only neighborhood of any point is just the whole space X, every sequence $x_n \in X$ converges to an arbitrary point $x \in X$. In other words: every sequence is convergent and the set of its limits coincides with the whole X.

The situation changes in a Hausdorff space where a sequence x_n, if convergent, possesses precisely one limit x. To verify this, assume, to the contrary, that x_n converges to another point y distinct from x. Since the space is Hausdorff, there exist neighborhoods, A of x and B of y such that $A \cap B = \emptyset$. Now let N_1 be such an index that $x_n \in A$ for every $n \geq N_1$ and, similarly, let N_2 denote an index for which $x_n \in B$ for every $n \geq N_2$. Thus for $n \geq \max(N_1, N_2)$ all x must belong to $A \cap B = \emptyset$, a contradiction.

Cluster Points. Let x_n be a sequence. We say that x is a *cluster point of sequence* x_n if every neighborhood of x contains an infinite number of elements of sequence x_n, i.e., for every neighborhood B of x and positive integer N, there exists $n \geq N$ such that $a_n \in B$. Trivially, a limit of sequence x_n, if it exists, is its cluster point.

Bases of Countable Type. A base $\mathcal{B}$ is said to be of countable type if it is equivalent to a countable base $\mathcal{C}$.

$$\mathcal{C} = \{C_i, i = 1, 2 \ldots\}$$

Note that sets of the form

$$D_k = C_1 \cap C_2 \cap \ldots \cap C_k \qquad k = 1, 2, \ldots$$

form a new countable base $\mathcal{D} = \{D_k, k = 1, 2, \ldots\}$ such that

$$D_1 \supset D_2 \supset \ldots$$

and $\mathcal{D} \sim \mathcal{C}$. Thus, every base of countable type can be replaced with an equivalent countable base of decreasing sets.

Example 4.4.2

Let $X = \mathbb{R}^n$ and let $\mathcal{B}$ be a base of balls centered at a point $\boldsymbol{x}_0$

$$\mathcal{B} = \{B(\boldsymbol{x}_0, \varepsilon) \quad \varepsilon > 0\}$$

Then $\mathcal{B}$ is of countable type. Indeed, $\mathcal{B}$ is equivalent to its subfamily

$$\mathcal{C} = \{B(\boldsymbol{x}_0, \frac{1}{k}), \qquad k = 1, 2, \ldots\}$$

$\square$

PROPOSITION 4.4.1

Let X be a topological space such that, for every point x, base of neighborhoods $\mathcal{B}_x$ is of countable type. Let x_n be a sequence in X. Then x is a cluster point of x_n iff there exists a subsequence x_{n_k} converging to x.

PROOF Obviously, if x_{n_k} converges to x then x is a cluster point of x_n. Conversely, let x be a cluster point of x_n. Let $B_1 \supset B_2 \supset \ldots$ be a base of neighborhoods of x. Since every B_k contains an infinite number of elements from sequence x_n, for every k one can choose an element $x_{n_k} \in B_k$ different from $x_{n_1}, \ldots, x_{n_{k-1}}$. Since B_k is decreasing, $x_{n_l} \in B_k$ for every $l \geq k$ which implies that $x_{n_k} \to x$. ∎

Sequential Closedness. A set E is said to be *sequentially closed* if every convergent sequence of elements from E possesses a limit in E, i.e.,

$$E \ni x_n \to x \qquad \Rightarrow \qquad x \in E$$

It is easy to notice that closedness implies sequential closedness. Indeed, let $x_n \in E$ converge to x. This means that

$$\forall\, B \in \mathcal{B}_x \;\; \exists N\colon \; x_n \in B \quad \forall\, n \geq N$$

It follows that $B \cap E \neq \emptyset$ for every $B \in \mathcal{B}_x$ and, consequently, $x \in \overline{E} = E$ (cf. Exercise 4.1.6).

PROPOSITION 4.4.2

Let $\mathcal{B}_x$ be of countable type for every $x \in X$. Then set $E \subset X$ is closed iff it is sequentially closed.

PROOF Let x be an accumulation point of E and $B_1 \supset B_2 \supset \ldots$ denote a base of neighborhoods of x. Choosing $x_k \in B_k \cap E - \{x\}$, we get $E \ni x_k \to x$, which implies that $x \in E$. ∎

Sequential Continuity. Let X and Y be Hausdorff spaces and $f : X \to Y$ a function. We say that f is *sequentially continuous* at $x_0 \in \text{dom}\, f$ if for every sequence $x_n \in \text{dom}\, f$ converging to $x_0 \in \text{dom} f$, sequence $f(x_n)$ converges to $f(x_0)$,

$$x_n \to x_0 \quad \Rightarrow \quad f(x_n) \to f(x_0)$$

If f is sequentially continuous at every point in its domain, we say that f is (globally) sequentially continuous.

Continuity implies always sequential continuity. To verify this assertion, pick an arbitrary neighborhood B of $f(x_0)$. If f is continuous at x_0 then there is a neighborhood C of x_0 such that $f(C) \subset B$. Consequently, if x_n is a sequence converging to x one can find an index N such that $x_n \in C$ for every $n \geq N$, which implies that $f(x_n) \in f(C) \subset B$ and therefore $f(x_n) \to f(x_0)$.

The converse is, in general, not true but we have the following simple observation.

PROPOSITION 4.4.3

Let $\mathcal{B}_x$ be of countable type for every $x \in X$. Then function $f: X \to Y$ is continuous iff it is sequentially continuous.

PROOF Let $B_1 \supset B_2 \supset \ldots$ be a base of neighborhoods of $x_0 \in \text{dom } f$. Assume f is sequentially continuous at x_0 and suppose, to the contrary, that f is not continuous at x_0. This means that there exists a neighborhood C of $f(x_0)$ such that

$$f(B_k) \not\subset C \qquad \text{for every } B_k$$

Thus, one can choose a sequence $x_k \in B_k$ such that $f(x_k) \notin C$. It follows that $x_k \to x$ and, simultaneously, $f(x_k) \notin C$ for all k, which implies that $f(x_k)$ does not converge to $f(x_0)$, a contradiction. ∎

Many of the properties of continuous functions hold for sequentially continuous functions as well. For instance, both a composition and Cartesian product of sequentially continuous functions are sequentially continuous.

Sequential Compactness. One of the most important notions in functional analysis is that of sequential compactness. We say that a set E is *sequentially compact* if, from every sequence $x_n \in E$, one can extract a subsequence x_{n_k} converging to an element of E.

The following observation holds.

PROPOSITION 4.4.4

Let $E \subset X$ be a compact set. Then
(i) every sequence in E has a cluster point.
If, additionally, every base of neighborhoods $\mathcal{B}_x$ is of countable type then
(ii) E is sequentially compact.

PROOF

(i) Let x_n be a sequence in E. The family of sets

$$C_k = \{x_k, x_{k+1}, x_{k+2} \ldots\}$$

is a base in E and, therefore, there must be an x such that

$$x \in \overline{C}_k \qquad \text{for every } k$$

But this means that for every B, a neighborhood of x, and for every k there exists an $x_{n_k} (n_k \geq k)$ such that $x_{n_k} \in B$. Thus, an infinite number of elements of sequence x_n belongs to B.

(ii) This follows immediately from Proposition 4.4.1. ∎

In Section 4.8, we prove the famous Bolzano–Weierstrass theorem, which says that, in metric spaces, compactness and sequential compactness are equivalent.

It is interesting to note, however, that many of the results which hold for compact sets can be proved for sets which are sequentially compact in a parallel way, without referring to the notion of compactness. The following proposition is a counterpart of Proposition 4.3.4.

PROPOSITION 4.4.5

The following properties hold:

(i) *every sequentially compact set is sequentially closed;*

(ii) *every sequentially closed subset of a sequentially compact set is sequentially compact;*

(iii) *let* $f: X \to Y$ *be sequentially continuous and let* $E \subset \text{dom } f$ *be a sequentially compact set; then* $f(E)$ *is sequentially compact;*

(iv) *Cartesian product* $A \times B$ *of two sequentially compact sets* A *and* B *is sequentially compact.*

PROOF The proof is left as an exercise. ∎

We also have an equivalent of the Weierstrass theorem.

PROPOSITION 4.4.6

Let $f: X \to Y$ *be sequentially continuous and let* $E \subset \text{dom } f$ *be sequentially compact. Then* f *attains on* E *its supremum and minimum.*

PROOF Let $x_k \in E$ be such that $f(x_k) \to \sup_E f$. Since E is sequentially compact, there exists a subsequence x_{n_k} converging to $x \in E$. It follows that

$$f(x_{n_k}) \to f(x)$$

and, therefore, $f(x) = \sup_E f$.

Similarly, we prove that f attains its minimum on E. ∎

Exercises

4.4.1. Let Φ be a family of subsets of $\mathbb{N}$ of the form

$$\{n, n+1, \ldots\}, \qquad n \in \mathbb{N}$$

(i) Prove that Φ is a base (the so called *fundamental base* in $\mathbb{N}$).

(ii) Characterize sets from filter $\mathcal{F}(\mathbb{N})$.

(iii) Let a_n be a sequence in a topological space X. Prove that the following conditions are equivalent to each other:

i. $a_n \to a_0$ in X,

ii. $a(\Phi) \succ \mathcal{B}_{a_0}$, and

iii. $a(\mathcal{F}(\Phi)) \succ \mathcal{F}_{a_0}$,

where $\mathcal{B}_{a_0}$ and $\mathcal{F}_{a_0}$ are base and filter of neighborhoods of some point $a_0 \in X$, respectively.

4.4.2. Let X be a topological space and a_n a sequence in X. Let Φ be the fundamental base in $\mathbb{N}$ from the previous exercise and $a : \mathbb{N} \ni n \to a_n \in X$ an arbitrary sequence in X. Prove that the following conditions are equivalent to each other:

(i) $a(\Phi) \succ \mathcal{B}_{a_0}$;

(ii) $(a \circ \alpha)(\Phi) \succ \mathcal{B}_{a_0}$, for every injection $\alpha : \mathbb{N} \to \mathbb{N}$

where $\mathcal{B}_{a_0}$ is base of neighborhoods of some point $a_0 \in X$ (sequence a_n converges to a_0 iff its every subsequence converges to a_0).

4.4.3. Let x_n be a sequence in a topological space X and let Φ denote the fundamental base in $\mathbb{N}$ (recall Exercise 4.4.1). Prove that the following conditions are equivalent to each other:

(i) x_0 is a cluster point of x_n;

(ii) bases $x(\Phi)$and $\mathcal{B}_{x_0}$ are adjacent.

4.5 Topological Equivalence. Homeomorphism

Topological Equivalence. In previous chapters, we have frequently encountered the idea of equivalence of various mathematical systems. We have seen that this idea is extremely important to the theory surrounding any particular system. For instance, the notion of an isomorphism provides for "algebraic equivalence" of linear vector spaces or linear algebras: when two such systems are isomorphic, their algebraic properties are essentially the same. Can a parallel concept of equivalence be developed for topological spaces?

It is natural to ask first what common properties we would expect two "topologically equivalent" topological spaces to share. From what we have seen up to this point, the answer is partially clear—the properties of continuity of functions, convergence of sequences, compactness of sets, etc., should be preserved under some correspondence (map) between the two spaces. A simple bijection is not enough—it can only establish a one-to-one and onto correspondence of elements in the underlying sets. For example, suppose X, Y and Z are topological spaces, and $F: X \to Z$ is a continuous function. If X and Y are to be equivalent in some topological sense, we would expect there to exist a bijection map $L: X \to Y$ that preserves the continuity of F in the sense that $F \circ L^{-1}: Y \to Z$ is continuous, too. In other words, the topological equivalence we are looking for is attained if the compositions of the bijections L and L^{-1} with continuous functions are continuous. Such mappings are called *homeomorphisms* (not to be confused with homomorphisms discussed in Chapter 2), and we sum up our observations concerning them by recording the following definition.

Homeomorphic Spaces. Two topological spaces X and Y are said to be *homeomorphic* (or *topologically equivalent*) if and only if there exists a map $L: X \to Y$ such that

(i) L is bijective, and

(ii) L, L^{-1} are continuous.

The map L is called a *homeomorphism* from X to Y.

Any property P of a topological space X is called a *topological property* if every space homeomorphic to X also has property P.

REMARK 4.5.1. Note that, in general, continuity of L does not imply continuity of L^{-1}, even though L is bijective. As an example, consider a finite set X, $1 < \#X < \infty$ with the discrete topology. Let Y be another finite set such that $\#Y = \#X$ and let $L: X \to Y$ denote a bijection. Setting the trivial topology in Y, we easily see that L is continuous, while L^{-1} not (explain why). ∎

We record some of the fundamental properties of homeomorphic spaces in the following proposition.

PROPOSITION 4.5.1

Let $L: X \to Y$ be a homeomorphism. The following properties hold:

(i) *E is open iff $L(E)$ is open.*

(ii) *E is closed iff $L(E)$ is closed.*

(iii) *E is compact iff $L(E)$ is compact.*

PROOF These assertions follow immediately from the corresponding propositions in the previous sections. ∎

Theory of Metric Spaces

4.6 Metric and Normed Spaces. Examples

We now come to the subject of a special type of topological space that shall be of great importance throughout the remainder of this book: metric spaces. A metric on a set amounts to a rather natural generalization of the familiar idea of distance between points.

Metric. Metric Space. Let X be a nonempty set. A function

$$d: X \times X \ni (x, y) \to d(x, y) \in [0, \infty)$$

taking pairs of elements of X into nonnegative real numbers, is called a *metric* on the set X if and only if the following conditions hold:

(i) $d(x, y) = 0$ if and only if $x = y$;

(ii) $d(x, y) = d(y, x)$ for every $x, y \in X$;

(iii) $d(x, z) \leq d(x, y) + d(y, z)$ for every $x, y, z \in X$.

Frequently we refer to $d(x, y)$ as the *distance between points x and y*. Property (i) of the function d characterizes it as *strictly positive* and (ii) as a *symmetric function* of x and y. Property (iii) is known as the *triangle inequality*.

The set X with the metric d, denoted $X = (X, d)$, is called a *metric space*.

Example 4.6.1

Perhaps the most familiar example of a metric space involves the idea of distance between points in the Euclidean plane. Here $X = \mathbb{R}^2$ and the distance between points $x = (x_1, x_2)$ and $y = (y_1, y_2)$ is defined as follows:

$$d(x, y) = ((x_1 - y_1)^2 + (x_2 - y_2)^2)^{\frac{1}{2}}$$

Clearly, $d(x, y)$ is symmetric and strictly positive. To prove the triangle inequality (look at Fig. 4.2 for notation) one has to show that

$$c \leq a + b \text{ or, equivalently, } c^2 \leq a^2 + b^2 + 2ab$$

But it follows from the cosine theorem that

$$c^2 = a^2 + b^2 - 2ab \, \cos\gamma \leq a^2 + b^2 + 2ab$$

which finishes the proof. □

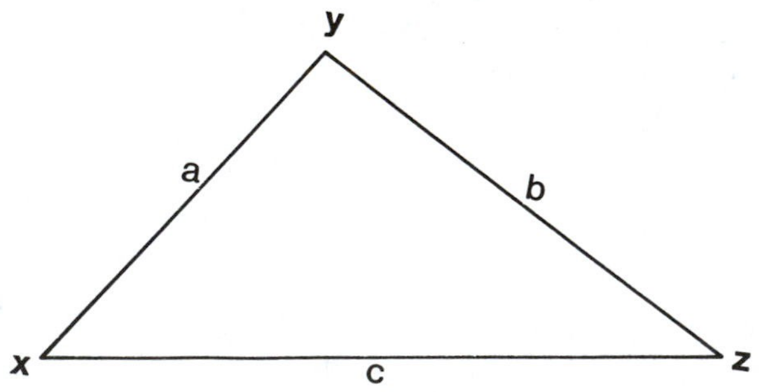

Figure 4.2
Triangle inequality for the Euclidean metric in a plane.

Many of the notions from Euclidean geometry can be easily generalized into the general context of metric spaces. The *(open) ball with center at* x *and radius* $\varepsilon > 0$, denoted $B(x, \varepsilon)$, is defined as follows

$$B(x, \varepsilon) = \{y \in X : d(x, y) < \varepsilon\}$$

If E is a subset of X and x is a point, the *distance between point* x *and set* E, denoted $d(x, E)$, can be defined as the smallest distance from x to members of E. More precisely,

$$d(x, E) = \inf_{y \in E} d(x, y)$$

The number

$$\text{dia } (E) = \sup\{d(x, y) \ : \ x, y, \in E\}$$

is referred to as the diameter of set E. If dia(E) is finite then E is said to be *bounded*, if not, then E is unbounded. Equivalently, E is bounded if there exists a sufficiently large ball which contains E.

Norm. Normed Spaces. Let V be a real or complex vector space. A function

$$||\cdot|| \ : \ V \ni \boldsymbol{v} \to ||\boldsymbol{v}|| \in [0, \infty)$$

prescribing for each vector $\boldsymbol{v}$ a nonnegative real number is called a *norm*, provided the following axioms hold:

(i) $||\boldsymbol{v}|| = 0$ if and only if $\boldsymbol{v} = \boldsymbol{0}$

(ii) $||\lambda \boldsymbol{v}|| = |\lambda| \ ||\boldsymbol{v}||$ for every $\lambda \in \mathbb{R}(\mathbb{C}), \boldsymbol{v} \in V$

(iii) $||\boldsymbol{u} + \boldsymbol{v}|| \leq ||\boldsymbol{u}|| + ||\boldsymbol{v}||$ for every $\boldsymbol{u}, \boldsymbol{v} \in V$

Axiom (i) characterizes the norm as *strictly positive*, property (ii) is frequently referred to as *homogeneity* and (iii) is known as *triangle inequality.*

The norm generalizes the classical notion of the length of a vector. A vector space V with the norm $||\cdot||$, denoted $V = (V, ||\cdot||)$ is called a *normed vector space.* We shall study in depth normed vector spaces in the next chapter. If more than two normed spaces take place simultaneously we will use the notation $||\boldsymbol{u}||_V$ indicating which norm is taken into account.

Let now $V = (V, ||\cdot||)$ be a normed vector space. Define the function

$$d(\boldsymbol{x}, \boldsymbol{y}) = ||\boldsymbol{x} - \boldsymbol{y}||$$

It follows immediately from the axioms of the norm that d is a metric. Thus, every normed space is automatically a metric space with the metric induced by the norm. Let us emphasize that the notion of metric spaces is much more general than that of normed spaces. Metric spaces in general do not involve the algebraic structure of vector spaces.

To reinforce the introduced definitions, we now consider a fairly broad collection of specific examples.

Example 4.6.2

Let $\Omega \subset I\!R^n$ be a measurable set. Consider the space $L^p(\Omega), p \in [1, \infty]$ defined in Section 3.9. According to Proposition 3.9.2, the function

$$
\|f\|_p = \begin{cases} \left(\int_\Omega |f|^p \, dm \right)^{\frac{1}{p}} & 1 \leq p < \infty \\ \\ \operatorname*{ess\,sup}_{\boldsymbol{x} \in \Omega} |f(\boldsymbol{x})| & p = \infty \end{cases}
$$

satisfies the second and third axioms of the norm. It does not, however, satisfy the first axiom. It follows only from Corollary 3.6.3 (i) that f is zero almost everywhere. To avoid this situation we introduce in $L^p(\Omega)$ the subspace M of functions equal zero a.e.,

$$
M = \{f \in L^p(\Omega) : f = 0 \text{ a.e. in } \Omega\}
$$

and consider the quotient space $L^p(\Omega)/M$. Since

$$
\|f\|_p = \|g\|_p
$$

for functions f and g equal a.e. in Ω, function $\| \cdot \|_p$ is well defined on quotient space $L^p(\Omega)/M$. It satisfies again axioms (ii) and (iii) of the norm and also

$$
\| \, [f] \, \|_p = 0 \quad \Rightarrow \quad [f] = M
$$

which proves that $\| \cdot \|_p$ is a norm in quotient space $L^p(\Omega)/M$. If no ambiguity occurs we shall write $L^p(\Omega)$ in place of $L^p(\Omega)/M$ and refer to classes of equivalence $[f] = f + M$ as "functions" from $L^p(\Omega)$. $\square$

Example 4.6.3

Consider the space $I\!R^n$ and define

$$||\boldsymbol{x}||_p = \begin{cases} \left(\sum_1^n |x_i|^p\right)^{\frac{1}{p}} & 1 \le p < \infty \\ \max\{|x_1|, \ldots, |x_n|\} & p = \infty \end{cases}$$

It follows immediately from the definition that $||\cdot||_p$ satisfies the first two axioms of a norm. To prove the triangle inequality we need the following lemma. $\square$

LEMMA 4.6.1
(Hölder and Minkowski Inequalities for Finite Sequences)

Let $\boldsymbol{x}, \boldsymbol{y} \in \mathbb{R}^n(\mathbb{C}^n)$. The following inequalities hold:

(i) $|\sum_1^n x_i y_i| \le ||\boldsymbol{x}||_p ||\boldsymbol{y}||_q$
where $p, q \in [1, \infty]$, $\frac{1}{p} + \frac{1}{q} = 1$

(ii) $||\boldsymbol{x} + \boldsymbol{y}||_p \le ||\boldsymbol{x}||_p + ||\boldsymbol{y}||_p \quad p \in [1, \infty]$

PROOF

Case 1. $p < \infty$. The proof is almost identical to the proof of the Hölder inequality for functions (see Theorem 3.9.1), the only difference being that integrals must be replaced with sums.

Case 2. $p = \infty$.

(i) $|\sum_1^n x_i y_i| \le \sum_1^n |x_i|\,|y_i| \le \sum_1^n ||\boldsymbol{x}||_\infty |y_i| = ||\boldsymbol{x}||_\infty ||\boldsymbol{y}||_1$

(ii) Obviously,

$$|x_i + y_i| \le |x_i| + |y_i| \le ||\boldsymbol{x}||_\infty + ||\boldsymbol{y}||_\infty \quad i = 1, 2, \ldots, n$$

Taking maximum on the left-hand side, we finish the proof. $\blacksquare$

Thus it follows from the Minkowski inequality, which is nothing else than the triangle inequality, that $||\cdot||_p$ is a norm in $\mathbb{R}^n(\mathbb{C}^n)$, which in turn implies that

$$d_p(\boldsymbol{x}, \boldsymbol{y}) = ||\boldsymbol{x} - \boldsymbol{y}||_p \qquad 1 \le p \le \infty$$

is a metric in $\mathbb{R}^n(\mathbb{C}^n)$.

We emphasize that open balls may take on quite different geometrical interpretations for different choices of the metric d_p. Suppose, for example, that $X = \mathbb{R}^2$ is the Euclidean plane, and consider the unit ball centered at the origin

$$B(\mathbf{0}, 1) = \{\boldsymbol{x} \ : \ ||\boldsymbol{x}||_p < 1\}$$

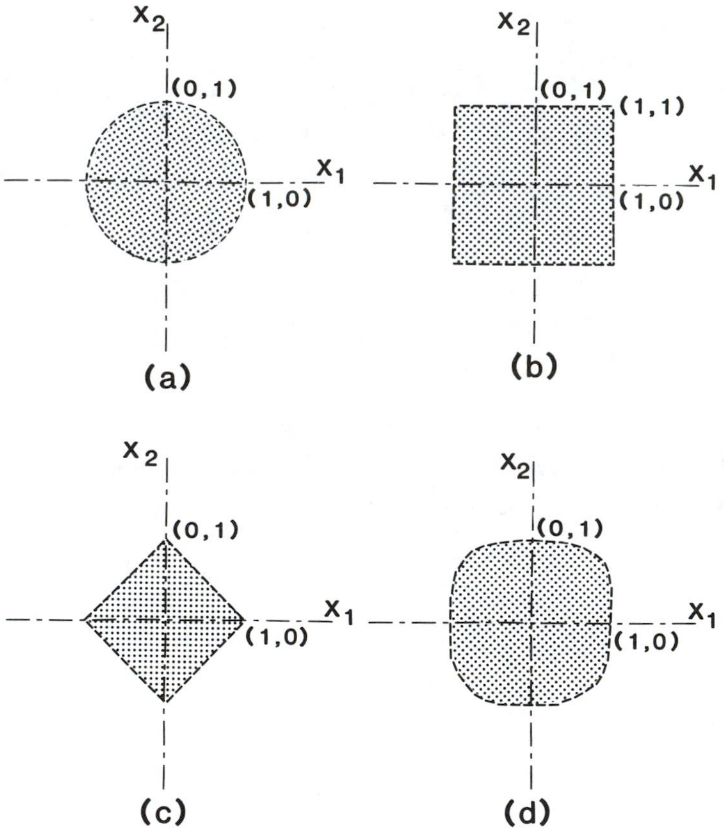

Figure 4.3

Examples of unit balls in $\mathbb{R}^2$ with respect to different metrics.

If we take $p = 2$ (Euclidean norm) then $B(\mathbf{0}, 1)$ is the set of points in the unit circle shown in Fig. 4.3a; if $p = \infty$ then

$$d(\boldsymbol{x}, 0) = ||\boldsymbol{x}||_\infty = \max\{|x_1|, |x_2|\}$$

and the ball coincides with the unit square shown in Fig. 4.3b. The unit ball becomes the diamond-shaped region (a rotated square) shown in Fig. 4.3c if $p = 1$ and, finally, if we select $p > 1, B(\mathbf{0}, 1)$ becomes a figure with curved sides. For example, if $p = 3$ then $B(\mathbf{0}, 1)$ assumes the shape shown in Fig. 4.3d.

Lemma 4.6.1 can be easily generalized for the case of infinite sequences.

LEMMA 4.6.2
(Hölder and Minkowski Inequalities for Infinite Sequences)

Let $\boldsymbol{x} = \{x_i\}_1^\infty, \boldsymbol{y} = \{y_i\}_1^\infty$, be infinite sequences of real or complex numbers. The following inequalities hold:

(i) $|\sum_1^\infty x_i y_i| \le ||\boldsymbol{x}||_p ||\boldsymbol{y}||_p$
where $p, q \in [1, \infty], \frac{1}{p} + \frac{1}{q} = 1$

(ii) $||\boldsymbol{x} + \boldsymbol{y}||_p \le ||\boldsymbol{x}||_p + ||\boldsymbol{y}||_p \quad p \in [1, \infty]$

provided the right-hand sides are finite. In the above,

$$||\boldsymbol{x}||_p = \begin{cases} \left(\sum_1^\infty |x_i|^p\right)^{\frac{1}{p}} & 1 \le p < \infty \\ \max_i |x_i| & p = \infty \end{cases}$$

PROOF The proof follows immediately from Lemma 4.6.1. For instance, according to Lemma 4.6.1, we have

$$\begin{aligned} |\sum_1^n x_i y_i| &\le \left(\sum_1^n |x_i|^p\right)^{\frac{1}{p}} \left(\sum_1^n |y_i|^q\right)^{\frac{1}{q}} \\ &\le \left(\sum_1^\infty |x_i|^p\right)^{\frac{1}{p}} \left(\sum_1^\infty |y_i|^q\right)^{\frac{1}{q}} \end{aligned}$$

Passing with n to infinity on the left-hand side, we get the results required.
∎

Example 4.6.4
(ℓ^p Spaces)

Guided by Lemma 4.6.2, we introduce the sets of infinite sequences

$$\ell_p = \{\boldsymbol{x} = (x_i)_1^\infty \; : \; ||\boldsymbol{x}||_p < \infty\}$$

According to Lemma 4.6.2, ℓ^p is closed with respect to the customary defined operations and, therefore, ℓ^p are vector spaces. It follows also from the definition of $||\cdot||_p$ and Lemma 4.6.2 that $||\cdot||_p$ is a norm in ℓ^p. Thus spaces $\ell^p, p \in [1, \infty]$, are another example of normed and metric spaces. □

Example 4.6.5
(Chebyshev Spaces)

Let $K \subset \mathbb{R}^n$ be a compact set and let $C(K)$ denote, as usual, the space of continuous functions on K. Recalling that every continuous functional attains its maximum on a compact set, we define the following quantity

$$||f||_\infty = \sup_{\boldsymbol{x} \in K} |f(\boldsymbol{x})| = \max_{\boldsymbol{x} \in K} |f(\boldsymbol{x})|$$

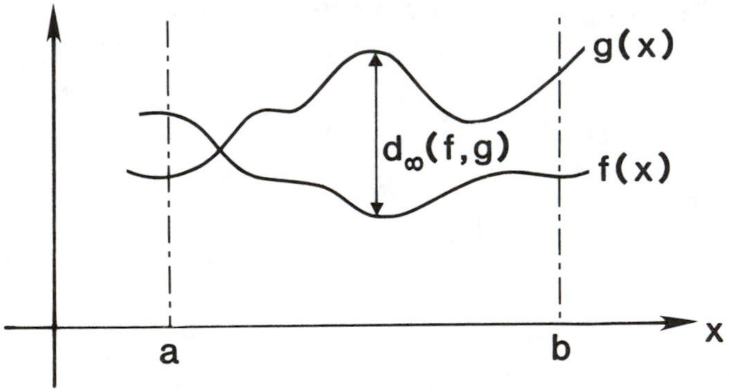

Figure 4.4
Geometrical interpretation of Chebyshev metric.

One can easily verify that $X = C(K)$ is a normed vector space with the norm $||\cdot||_\infty$. The $||\cdot||_\infty$ norm is referred to as the *Chebyshev norm* and X is called the *Chebyshev space*. The resulting metric is known as the *Chebyshev metric*. In fact, X can be considered as a subspace of $L^\infty(K)$ (for continuous functions the essential supremum coincides with the usual supremum).

It is interesting to interpret the Chebyshev metric graphically by considering the two functions $f(x)$ and $g(x)$ in Fig. 4.4. Clearly, $d(f, g)$ is the maximum amount $f(x)$ differs from $g(x)$ in the interval $[a, b]$ as shown. $\square$

We will present now two examples of metric spaces which are not normed spaces.

Example 4.6.6

Let $d(x, y)$ be any metric on a set X. We claim that the function

$$\sigma(x,y) = \frac{d(x,y)}{1 + d(x,y)}$$

is also a metric. Indeed, $\sigma(x,y) = \sigma(y,x), \sigma(x,y) = 0$ iff $x = y$. Finally, let $a \leq b$ be two nonnegative numbers. It follows that

$$\frac{b}{1+b} - \frac{a}{1+a} = \frac{b-a}{(1+b)(1+a)} \geq 0$$

so

$$\frac{a}{1+a} \leq \frac{b}{1+b}$$

Concluding $(a = d(x,z), b = d(x,y) + d(y,z))$,

$$\sigma(x,z) = \frac{d(x,z)}{1+d(x,z)} \leq \frac{d(x,y)+d(y,z)}{1+d(x,y)+d(y,z)}$$

$$= \frac{d(x,y)}{1+d(x,y)+d(y,z)} + \frac{d(y,z)}{1+d(x,y)+d(y,z)}$$

$$\leq \sigma(x,y) + \sigma(y,z)$$

so the triangle inequality holds, and σ is a metric.

It is surprising, but every set E is bounded in the metric $\sigma(x, y)$. Indeed, for arbitrary x and y,

$$\sigma(x,y) = \frac{d(x,y)}{1+d(x,y)} \leq 1$$

$\square$

Example 4.6.7
(The Discrete Metric)

Let X be any set and define the function $d(x, y)$ by

$$d(x,y) = \begin{cases} 1 \; x \neq y \\ \\ 0 \; x = y \end{cases}$$

It is easily verified that d is a metric, generally referred to as the *discrete* or *trivial metric.* □

Product Spaces. Let (X, d) and (Y, ρ) be two metric spaces with metrics d and ρ, respectively. The metric space (Z, σ), where $Z = X \times Y$ and

$$\sigma((x_1, y_1), (x_2, y_2)) = \hat{\sigma}(d(x_1, x_2), \rho(y_1, y_2))$$

where $\hat{\sigma}$ is one of the metrics induced by norms in $\mathbb{R}^2$ discussed in Example 4.6.3, is called the product space of the spaces (X, d) and (Y, ρ). For instance, one can define the metric σ as $(p = 1)$

$$\sigma((x_1, y_1), (x_2, y_2)) = d(x_1, x_2) + \rho(y_1, y_2)$$

Exercises

4.6.1. Let (X, d) be a metric space. Show that

$$\rho(x, y) = \min\{1, d(x, y)\}$$

is also a metric on X.

4.6.2. Show that any two norms $\|\cdot\|_p$ and $\|\cdot\|_q$ in $\mathbb{R}^n$, $1 \leq p, q \leq \infty$, are equivalent, i.e., there exist constants $C_1 > 0$ and $C_2 > 0$ such that

$$\|\boldsymbol{x}\|_p \leq C_1 \|\boldsymbol{x}\|_q \quad \text{and} \quad \|\boldsymbol{x}\|_q \leq C_2 \|\boldsymbol{x}\|_p$$

for any $\boldsymbol{x} \in \mathbb{R}^n$. Try to determine optimal (minimum) constants C_1 and C_2.

4.7 Topological Properties of Metric Spaces

Let $X = (X, d)$ be a metric space. Defining, for every $x \in X$, the family $\mathcal{B}_x$ of neighborhoods of x as the family of open balls centered at x,

$$\mathcal{B}_x = \{B(x, \varepsilon), \varepsilon > 0\}$$

we introduce in X a topology induced by the metric d. Thus every metric space is a topological space with the topology induced by the metric. Two immediate corollaries follow:

1. Bases $\mathcal{B}_x$ are of countable type.
2. The metric topology is Hausdorff.

The first observation follows from the fact that $\mathcal{B}_x$ is equivalent to its subbase of the form

$$\left\{B\left(x, \frac{1}{k}\right), \quad k = 1, 2, \ldots, \right\}$$

To prove the second assertion consider two distinct points $x \neq y$. We claim that balls $B(x, \varepsilon)$ and $B(y, \varepsilon)$ where $\varepsilon = \frac{1}{2}d(x, y)$ are disjoint. Indeed, if z were a point belonging to the balls simultaneously, then

$$d(x, y) \leq d(x, z) + d(z, y) < \varepsilon + \varepsilon = d(x, y)$$

a contradiction.

Thus all the results we have derived in the first five sections of this chapter for Hausdorff topological spaces with bases of neighborhoods of countable type hold also for metric spaces.

Let us briefly review some of them.

Open and Closed Sets in Metric Spaces. A set $G \subset X$ is open if and only if, for every point x of G, there exists a ball $B(x, \varepsilon)$, centered at x, that is contained in G. A point x is an accumulation point of a set F if every ball centered at x contains points from F which are different from x, or, equivalently, there exists a sequence x_n points of F converging to x.

Note that a sequence x_n converges to x if an only if

$$\forall\, \varepsilon > 0\; \exists\, N = N(\varepsilon)\ :\ d(x_n, x) < \varepsilon \quad \forall\, n \geq N$$

Finally, a set is closed if it contains its all accumulation points.

Continuity in Metric Spaces. Let (X, d) and (Y, ρ) be two metric spaces. Recall that a function $f \colon X \to Y$ is continuous at x_0 if

$$f(\mathcal{B}_{x_0}) \succ \mathcal{B}_{f(x_0)}$$

or, equivalently,

$$\forall \varepsilon > 0 \quad \exists \delta > 0 : f(B(x_0, \delta)) \subset B(f(x_0), \varepsilon)$$

The last condition can be put into a more familiar form of the definition of continuity for metric spaces ($\varepsilon - \delta$ continuity):

Function $f: X \to Y$ is continuous at x_0 if and only if for every $\varepsilon > 0$ there is a $\delta = \delta(\varepsilon, x_0)$ such that

$$\rho(f(x), f(x_0)) < \varepsilon \quad \text{whenever} \quad d(x, x_0) < \delta$$

Note that number δ generally depends not only on ε, but also upon the choice of point x_0. If δ happens to be independent of x_0 for all x_0 from a set E, then f is said to be *uniformly continuous on* E. Let us recall also that, since bases of neighborhoods are of countable type, continuity in metric spaces is equivalent to sequential continuity: a function $f: X \to Y$ is continuous at x_0 if and only if

$$f(x_n) \to f(x_0) \quad \text{whenever} \quad x_n \to x_0$$

Suppose now that there exists a constant $C > 0$, such that

$$\rho(f(x), f(y)) \leq Cd(x, y) \quad \text{for every } x, y \in E$$

Functions f satisfying such a condition are called *Lipschitz continuous on* E. Note that every Lipschitz continuous function on E is also uniformly continuous on E. Indeed, choosing $\delta < \frac{\varepsilon}{C}$ (independent of x) we get

$$\rho(f(x), f(x_0)) \leq Cd(x, x_0) < C\frac{\varepsilon}{C} = \varepsilon$$

Example 4.7.1

Let $X = Y = \mathbb{R}$ with the natural metric $d(x, y) = |y - x|$. Every C^1 function $f: \mathbb{R} \to \mathbb{R}$ is Lipschitz continuous on every closed interval [a,b]. Indeed, according to the Lagrange theorem

$$f(y) - f(x) = f'(c)(y - x) \quad \text{for every } a \leq x < y \leq b$$

where $c \in (x, y) \subset [a, b]$. But f' is continuous and the interval $[a, b]$ is compact in $\mathbb{R}$, so, according to the Weierstrass theorem, there exists such an x_0 that

$$C = |f'(x_0)| = \max_{a \le c \le b} |f'(c)|$$

Consequently,

$$|f(y) - f(x)| \le C|y - x|$$

which proves that f is Lipschitz continuous on $[a, b]$. $\square$

Example 4.7.2

Choose again $X = Y = \mathbb{R}$ and consider the function

$$f(x) = \frac{1}{x} \quad x \in (0, \infty)$$

If f were uniformly continuous in $(0, \infty)$ then, for every $\varepsilon > 0$, one could choose a $\delta > 0$ such that $|f(x) - f(x_0)| < \varepsilon$ whenever $|x - x_0| < \delta$. Choose now $x_0 = \frac{1}{n}, x = \frac{1}{2n}$. Then $|f(x) - f(x_0)| = 2n - n$ while $|x - x_0| = \frac{1}{2n}$. In other words, for sufficiently large n, $|x - x_0|$ can be arbitrarily small while $|f(x) - f(x_0)|$ can be arbitrarily large, which proves that $\frac{1}{x}$ is *not* uniformly continuous. $\square$

Example 4.7.3

A nontrivial example of a continuous, nonlinear mapping in function spaces is given by the theorem of Krasnoselski. $\square$

A mapping $g: \Omega \times \mathbb{R} \to \mathbb{R}$, where $\Omega \subset \mathbb{R}^n$ is an open set, is called a *Caratheodory mapping* of $\Omega \times \mathbb{R}$ into $\mathbb{R}$ if:

(i) for every $\xi \in \mathbb{R}$, $\boldsymbol{x} \to g(\boldsymbol{x}, \xi)$ is a measurable function and

(ii) for almost all $\boldsymbol{x} \in \Omega, \xi \to g(\boldsymbol{x}, \xi)$ is a continuous function.

Now, for each measurable function $u: \Omega \to \mathbb{R}$, let $G(u)$ denote the measurable function

$$\Omega \ni \boldsymbol{x} \to g(\boldsymbol{x}, u(\boldsymbol{x}))$$

Recall that the operator G is called the Nemytskii operator (cf. Example 2.5.5). It turns out that if G maps $L^p(\Omega)$ into $L^r(\Omega)$, for some $1 \le p, r < \infty$, then G is continuous in L^p and L^r metrics, respectively. In other words, a well-defined Nemytskii operator (i.e., mapping $L^p(\Omega)$ *into* $L^r(\Omega)$) is automatically continuous.

Let us note also that both norm and metric are continuous in metric topology. Indeed, let $V = (V, \|\cdot\|)$ be a normed space. It follows from the triangle inequality that

$$\|\boldsymbol{x}\| = \|\boldsymbol{y}+\boldsymbol{x}-\boldsymbol{y}\| \le \|\boldsymbol{y}\| + \|\boldsymbol{x}-\boldsymbol{y}\|$$

or, equivalently,

$$\|\boldsymbol{x}\| - \|\boldsymbol{y}\| \le \|\boldsymbol{x}-\boldsymbol{y}\|$$

Similarly,

$$\|\boldsymbol{y}\| - \|\boldsymbol{x}\| \le \|\boldsymbol{y}-\boldsymbol{x}\| = \|\boldsymbol{x}-\boldsymbol{y}\|$$

so

$$|\,\|\boldsymbol{x}\| - \|\boldsymbol{y}\|\,| \le \|\boldsymbol{x}-\boldsymbol{y}\|$$

which proves that norm is Lipschitz continuous with constant equal 1.

Similarly, one can prove that

$$|\,\rho(x,y) - \rho(\hat{x},\hat{y})\,| \le \rho(x,\hat{x}) + \rho(y,\hat{y})$$

which proves that metric is Lipschitz continuous with constant equal 1, provided the product space $X \times X$ is supplied with the metric

$$\rho((x,y),(\hat{x},\hat{y})) = \rho(x,\hat{x}) + \rho(y,\hat{y})$$

Dense Sets. Separable Metric Spaces. As in general topological spaces, a set E is *dense* in a metric space (X,d), if $\overline{E} = X$, or, equivalently,

$$\forall x \in X \quad \exists x_n \in E \text{ such that } x_n \to x$$

The space (X,d) is said to be *separable* if there exists a countable set E dense in X.

Topological Equivalence. Two metric spaces $X = (X,d)$ and $Y = (Y,\rho)$ are *topologically equivalent* if X and Y, with topologies induced by metrics d and ρ, are homeomorphic.

PROPOSITION 4.7.1

Let $(X,d), (Y,\rho)$ be metric spaces and $L: X \to Y$ be a bijection such that there exist constants $\mu_1, \mu_2 > 0$ such that

$$\mu_1 d(x,y) \le \rho(L(x), L(y)) \le \mu_2 d(x,y)$$

Then (X, d) and (Y, ρ) are topologically equivalent.

PROOF It follows from the inequality above that both L and L^{-1} are Lipschitz continuous, and therefore X and Y are homeomorphic. ∎

REMARK 4.7.1. Note that the inequality

$$\mu_1 d(x, y) \leq \rho(L(x), L(y))$$

implies in particular that L is one-to-one. Indeed, if $L(x) = L(y)$ then $d(x, y) = 0$, which implies that $x = y$. ∎

The converse of Proposition 4.7.1 in general is *not true*. To see this consider $\mathbb{R}^n$ with a metric d induced by any of the norms discussed in Example 4.6.3 and let $\sigma = d/(1 + d)$ be the metric introduced in Example 4.6.6 on the same underlying set X. It is easy to check that the two metrics are topologically equivalent. Obviously,

$$\mu_1 \sigma(x, y) \leq d(x, y) \quad \text{for } \mu_1 = 1$$

but the second inequality does not hold since $d(x, y)$ may be arbitrarily large while σ remains bounded ($\sigma \leq 1$).

Thus, the topological equivalence of two metrics defined on the same set X does not imply that they must be bounded by each other in the sense of the discussed inequalities.

The situation is less complicated in the case of metrics induced by norms. It can be shown that *norms in the same vector space are topologically equivalent (generate the same topology) if and only if they are bounded by each other.* Moreover, it turns out that *in a finite-dimensional vector space any two norms are topologically equivalent.* This means, in particular, that the norm-induced topology in $\mathbb{R}^n$ is unique, i.e., can be introduced only in one way! These and other facts concerning the normed spaces will be considered in the next chapter.

Metric Equivalence—Isometries. At this point we have established a notion of topological equivalence of two metric spaces, but this has been in the broad setting of topological spaces. We also saw that every metric space is a topological space, so while the notion of topological equivalence does apply to metric spaces, it provides too general a means of comparison to depict equivalence of purely metric properties, i.e., the concept of distance between points. What is needed to construct a more specialized idea is, as usual, a means of comparing points in two metric spaces (e.g., a bijective map from one space onto the other) and the equivalence of distances between points in each space. These properties are covered by the concept of an isometry:

Two metric spaces (X, d) and (Y, ρ) are said to be *isometric* (or *metrically equivalent*) if and only if there exists a bijection $G: (X, d) \to (Y, \rho)$ such that

$$d(x, y) = \rho(G(x), G(y)) \quad \text{for every } x, y \in X$$

The mapping G with this property is called an *isometry.*

Obviously, if G is an isometry then both G and G^{-1} are Lipschitz continuous with constant 1 and, therefore, G is a homeomorphism. Thus two isometric spaces are homeomorphic.

Example 4.7.4

Recall the classical theorem in elementary geometry: Every isometry $G: \mathbb{R}^n \to \mathbb{R}^n (n = 2, 3, \mathbb{R}^n$ with Euclidean metric) is a composition of a translation and a rotation about a point. Equivalently, in the language of mechanics: A rigid body motion is always a composition of a translation and a rotation about a point. $\square$

Exercises

4.7.1. Prove that $F : (X, d) \to (Y, \rho)$ is continuous if and only if the inverse image of every closed ball $\overline{B}(y, \varepsilon)$ in Y is a closed set in X.

4.7.2. Let $X = C^\infty[a, b]$ be the space of infinitely differentiable functions with Chebyshev metric

$$d(f, g) = \sup_{x \in [a,b]} |f(x) - g(x)|$$

Let $F : X \to X$ be defined such that $F(f) = df/dx$. Is F a continuous mapping on X?

4.8 Completeness and Completion of Metric Spaces

Cauchy Sequences. Recall that every convergent sequence (x_n) in $\mathbb{R}$ satisfies the so-called *Cauchy condition*

$$\forall \varepsilon > 0 \quad \exists N : |x_n - x_m| < \varepsilon \quad \text{whenever } n, m \geq N$$

Roughly speaking, when a sequence converges in $\mathbb{R}$, the entries $x_i, x_{i+1} \cdots$ get closer and closer together as i increases. In other words,

$$\lim_{i,j \to \infty} |x_i - x_j| = 0$$

A sequence which satisfies the Cauchy condition is called a *Cauchy sequence*. Thus, every convergent sequence in $\mathbb{R}$ is a Cauchy sequence. It is well known that the converse is also true. Every Cauchy sequence in $\mathbb{R}$ is convergent.

The notion of a Cauchy sequence can be easily generalized to a general metric space (X, d). A sequence (x_n) is said to be a *Cauchy sequence* iff

$$\forall \varepsilon > 0 \quad \exists N : d(x_n, x_m) < \varepsilon \quad \text{whenever } n, m \geq N$$

As in the case of real numbers, every convergent sequence in a metric space (X, d) is a Cauchy sequence. To see this, suppose that $\lim x_n = x_0$. Since

$$d(x_m, x_n) \leq d(x_m, x_0) + d(x_n, x_0)$$

and $d(x_m, x_0) < \varepsilon/2$ and $d(x_n, x_0) < \varepsilon/2$ for $m, n \geq$ some N,

$$d(x_m, x_n) < \varepsilon \quad \text{for } m, n \geq N$$

Hence, (x_n) is a Cauchy sequence.

In general, however, the converse is not true. In many cases Cauchy sequences do not converge. We shall examine this phenomenon here in some detail.

Example 4.8.1

The sequence $(1/n)$ is convergent in $\mathbb{R}$ and therefore is a Cauchy sequence in $\mathbb{R}$. This implies that $(1/n)$ is also a Cauchy sequence in any subset of $\mathbb{R}$, e.g., the interval $(0, 1)$. Clearly, $\lim 1/n = 0 \notin (0, 1)$ and therefore $(1/n)$ is not convergent in $(0, 1)$. $\square$

Example 4.8.2

Let $X = C([-2,2])$ be the space of continuous functions on the interval $[-2,2]$ with L^1 metric given by

$$d(f,g) = \int_{-2}^{2} |f(x) - g(x)|dx$$

Now consider the sequence of functions $(f_n(x))$, where

$$f_n(x) = \begin{cases} 0 & \text{for } -2 \le x \le 1 - \frac{1}{n} \\ nx + 1 - n & \text{for } 1 - \frac{1}{n} \le x \le 1 \\ 1 & \text{for } 1 \le x \le 2 \end{cases}$$

(f_n) is a Cauchy sequence. Indeed, if $m > n$, then

$$d(f_m, f_n) = \int_{-2}^{2} |f_m(x) - f_n(x)|dx = \frac{1}{2}(\frac{1}{n} - \frac{1}{m})$$

which tends to 0 as $m, n \to \infty$.

However, this sequence does not converge in X. Indeed, suppose that $f_n \to g$ in L^1 metric. It follows from Fatou's lemma that

$$\int_{-2}^{2} \liminf |(f_n - g)| \le \liminf \int_{-2}^{2} |(f_n - g)| = 0$$

which implies that $g = f = \liminf f_n = \lim f_n$ a.e. in $[-2,2]$, where

$$f(x) = \begin{cases} 0 & x > 1 \\ 1 & x \le 1 \end{cases}$$

It is easy to see that no such a continuous function exists. Thus (f_n) is not convergent in X. $\square$

When metric spaces do not have deficiencies of the type illustrated in these examples, we say that they are complete.

Complete Metric Spaces. A metric space (X, d) is said to be *complete* if every Cauchy sequence in (X, d) is convergent.

Example 4.8.3

The ℓ^p normed spaces, $p \in [1, \infty]$, are complete. To prove this, let (x_n) be a Cauchy sequence in ℓ^p, i.e., for every $\varepsilon > 0$ there exists a N such that

$$||x_n - x_m|| = \begin{cases} \left(\sum_{i=1}^{\infty} | x_n^i - x_m^i |^p\right)^{\frac{1}{p}} & \text{for } p \in [1, \infty] \\ \max_i | x_n^i - x_m^i | & \text{for } p = \infty \end{cases}$$

is less that ε, for every $m, n \geq N$. This implies that

$$|x_n^i - x_m^i| < \varepsilon \quad n, m \geq N$$

for every $i = 1, 2, \ldots$ and therefore x_n^i is convergent for every $i = 1, 2, \ldots$. Denote $x^i = \lim_{n\to\infty} x_n^i$. Passing to the limit with $n \to \infty$, we get

$$\left.\begin{matrix} \left(\sum_{i=1}^{\infty} | x^i - x_m^i |^p\right)^{\frac{1}{p}} & \text{for } p \in [1, \infty] \\ \max | x^i - x_m^i | & \text{for } p = \infty \end{matrix}\right\} < \varepsilon \quad \text{for } m \geq N$$

which proves that

1. $x = (x^i)$ belongs to ℓ^p; we have from the triangle inequality

$$\|x\|_p \;\leq\; \|x - x_n\|_p + \|x_n\|_p \;\leq\; \varepsilon + \|x_n\|_p \quad \text{for any } n \geq N$$

2. x_n converges to x.

Thus, spaces ℓ^p are complete. $\square$

Example 4.8.4

Let K be a compact set in $\mathbb{R}^n$. Consider the space $C(K)$ of continuous functions on K with the Chebyshev norm

$$||f||_\infty = \sup_{\boldsymbol{x} \in K} |f(\boldsymbol{x})|$$

The space $C(K)$ is complete. To prove it, consider a Cauchy sequence (f_n) in $C(K)$. Thus, for every $\varepsilon > 0$, there is a N such that for all $\boldsymbol{x} \in K$

$$|f_n(\boldsymbol{x}) - f_m(\boldsymbol{x})| \leq ||f_n - f_m|| < \varepsilon \quad \text{for } m, n \geq N$$

Then, for an arbitrary fixed $\boldsymbol{x} \in K$, $f_n(\boldsymbol{x})$ is a Cauchy sequence in $\mathbb{R}$ and therefore convergent to (say) the number $f(\boldsymbol{x})$. Passing with n to infinity in the equality above, we get

$$|f(\boldsymbol{x}) - f_m(\boldsymbol{x})| \leq \varepsilon \quad \text{for } m \geq N$$

which proves that

$$||f - f_m|| \to 0$$

It remains to prove that f belongs to space $C(K)$.

We show that if f_n converges uniformly to $f(x)$, then $f(x)$ is continuous. Pick an $\varepsilon > 0$. It is clear that

$$|f(\boldsymbol{x}) - f_n(\boldsymbol{x})| < \frac{\varepsilon}{3} \quad \text{for } n \geq \text{ some } N, \text{ for every } \boldsymbol{x} \in K$$

Thus

$$|f(\boldsymbol{x}) - f(\boldsymbol{y})| \leq |f(\boldsymbol{x}) - f_n(\boldsymbol{x})| + |f_n(\boldsymbol{x}) - f_n(\boldsymbol{y})| + |f_n(\boldsymbol{y}) - f(\boldsymbol{y})|$$

$$< 2\tfrac{\varepsilon}{3} + |f_n(\boldsymbol{x}) - f_n(\boldsymbol{y})| \quad \text{for } n \geq N$$

Since f_n is continuous, there exists δ such that

$$|f_n(\boldsymbol{x}) - f_n(\boldsymbol{y})| < \frac{\varepsilon}{3} \quad \text{whenever } ||x - y|| < \delta$$

i.e., f is continuous. $\square$

Example 4.8.5

Let $\Omega \subset \mathbb{R}^n$ be an open set. The normed spaces $L^p(\Omega), p \in [1, \infty]$ are complete. Consider first the case $p < \infty$. Let (f_n) be a Cauchy sequence in $L^p(\Omega)$, i.e.,

$$\forall \varepsilon > 0 \quad \exists N \, : \, ||f_n - f_m||_p < \varepsilon \quad \text{for } n, m \geq N$$

It follows that one can always extract a subsequence f_{n_k} such that

$$||f_{n_k} - f_{n_{k-1}}||_p < (\frac{1}{2})^k \qquad k = 1, 2 \ldots$$

We will show that this subsequence converges to a limit which turns out to be the limit of the entire Cauchy sequence.

Define

$$g_k = f_{n_k} - f_{n_{k-1}}\ ,\ k = 1, 2\ldots$$

and consider the following two series

$$s_k = g_1 + g_2 + \ldots + g_k \quad \text{and} \quad S_k = |g_1| + |g_2| + \ldots + |g_k|$$

Notice that

$$s_m = \sum_{k=1}^{m} g_k = f_{n_m} - f_{n_0}$$

and, therefore, convergence of the k-th partial sum s_k will imply (is, in fact, equivalent to) convergence of the subsequence f_{n_k}.

We have

$$\int_\Omega (S_k)^p = \| |g_1| + |g_2| + \ldots + |g_k| \|_p^p \le (||g_1||_p + ||g_2||_p + \ldots + ||g_k||_p)^p$$

$$\le \left(\tfrac{1}{2} + \tfrac{1}{2^2} + \ldots + \tfrac{1}{2^k}\right)^p \le 1$$

For every $\boldsymbol{x} \in \Omega$, denote

$$S^p(\boldsymbol{x}) = \lim_{k\to\infty} S_k^p(\boldsymbol{x})$$

As a nondecreasing sequence of nonnegative numbers, $S_k^p(\boldsymbol{x})$ converges to a positive number or to ∞. It follows, however, from Fatou's lemma (Theorem 3.5.1) that

$$\int_\Omega S^p = \int_\Omega \lim_{k\to\infty} S_k^p \le \liminf_{k\to\ \infty} \int_\Omega S_k^p \le 1$$

which proves that $S^p(\boldsymbol{x})$ and, therefore, $S(\boldsymbol{x})$ is finite a.e. in Ω. In other words $S_k(\boldsymbol{x})$ converges a.e. in Ω, which in turn implies that $s_k(\boldsymbol{x})$ converges a.e. in Ω to a finite value, too. Denote

$$f_0(\boldsymbol{x}) = f_{n_0}(\boldsymbol{x}) + \lim_{k\to\infty} s_k(\boldsymbol{x}) = \lim_{k\to\infty} f_{n_k}(\boldsymbol{x})$$

We claim that

1. $f_0 \in L^p(\Omega)$
2. $f_n \to f_0$ in the L^p norm

Indeed, the Cauchy condition and Fatou's lemma imply that

$$\int_\Omega |f_m(\boldsymbol{x}) - f_0(\boldsymbol{x})|^p \leq \liminf_{k\to\infty} \int_\Omega | f_m(\boldsymbol{x}) - f_{n_k}(\boldsymbol{x}) |^p < \varepsilon$$

provided $m \geq$ some N. This proves that

$$f_m \to f_0 \quad \text{in } L^p(\Omega)$$

Finally, from the inequality

$$\|f_0\|_p \leq \|f_m\|_p + \|f_0 - f_m\|_p \leq \|f_m\|_p + \varepsilon$$

for m sufficiently large, it follows that $f_0 \in L^p(\Omega)$.

Thus $L^p(\Omega), 1 \leq p < \infty$, is complete.

Proof of the case $p = \infty$ is very similar to the proof of the completeness of $C(K)$ from the previous example and we leave it as an exercise. ∎

The following proposition characterizes two fundamental properties of complete spaces.

PROPOSITION 4.8.1

(i) A subspace (Y, d) of a complete metric space (X, d) is complete iff Y is a closed set.

(ii) Let (X, d) and (Y, ρ) be two metric spaces that are isometric to each other. Then (Y, ρ) is complete iff (X, d) is complete.

PROOF

(i) Assume that Y is closed and let (y_n) be a Cauchy sequence in Y. Since $y_n \in X$ (because $Y \subset X$), it has a limit y_0 in X. However, every convergent sequence in a closed set has a limit in the set; thus $y_0 \in Y$ and (Y, d) is complete.

Now assume that (Y, d) is complete. Let y be an accumulation point of Y. Equivalently, there exists a sequence $y_n \in Y$ converging to y. As a convergent sequence (in X), y_n is a Cauchy sequence (both in X and in Y) and therefore it has a limit y_0 in Y which, according to the uniqueness of limits in a metric space, must coincide with y. Thus $y \in Y$.

(ii) Let $H: X \to Y$ be an isometry and assume that X is complete. Consider a Cauchy sequence $y_n \in Y$, i.e.,

$$\forall \varepsilon > 0 \quad \exists N \, : \, \rho(y_n, y_m) < \varepsilon \quad \text{whenever } n, m \geq N$$

It follows from the definition of isometry that

$$\forall \varepsilon > 0 \quad \exists N : d(H^{-1}(y_n), H^{-1}(y_m)) < \varepsilon \quad \text{whenever } n, m \geq N$$

Hence, $H^{-1}(y_n)$ is a Cauchy sequence in X and therefore possesses a limit x in X, which in turn implies that $y = H(x) = \lim y_n$. Thus Y is complete.

The reverse process, interchanging X and Y, shows that if (Y, ρ) is complete, then (X, d) is complete, and this completes the proof. ∎

Notice that all isometric spaces are homeomorphic, but all homeomorphic spaces are not necessarily isometric (since a homeomorphism need not preserve distances). Hence, it does not follow from Proposition 4.8.1 that if two spaces are homeomorphic, one is complete if and only if the other is. In other words, completeness is not necessarily preserved under homeomorphisms. Therefore, completeness is not a topological property.

Now, in the examples of complete and incomplete metric spaces considered earlier, we note that each incomplete space is "immersed" in a larger metric space that is complete. For example, according to Proposition 4.8.1, any open set in $\mathbb{R}^n$ is not complete. But its closure is. In this case we are able to "complete" the space by merely adding all accumulation points to the set. In more general situations, we will not always be able to find an incomplete space as the subspace of a complete one, but, as will be shown, it will always be possible to identify a "larger" space that is very similar to any given incomplete space X that is itself complete. Such larger complete spaces are called completions of X.

Completion of a Metric Space. Let (X, d) be a metric space. A metric space $(X^\#, d^\#)$ is said to be a *completion* of (X, d) if and only if

(i) there exists a subspace $(Z, d^\#)$ of $(X^\#, d^\#)$ which is dense in $(X^\#, d^\#)$ and isomorphic to (X, d) and

(ii) $(X^\#, d^\#)$ is complete.

Thus, (X, d) may not necessarily be a subspace of a completion, as was the case in the example cited previously, but it is at least isometric to a dense subspace. Hence, in questions of continuity and convergence, (X, d) may be essentially the same as a subspace of its completion.

The question arises of when such completions exist and, if they exist, how many completions there are for a given space. Fortunately, they always exist, and all completions of a given metric space are isometric (i.e., completions are essentially unique). This fundamental fact is the basis of the following theorem.

THEOREM 4.8.1

Every metric space (X, d) has a completion and all of its completions are isometric.

The proof of this theorem is lengthy and involved; so, to make it more digestible, we shall break it into a number of steps. Before launching into these steps, it is informative to point out a few difficulties encountered in trying to develop completions of a given space too hastily. First of all, (X, d) must be isometric to a space $(Z, d^{\#})$ dense in a completion $(X^{\#}, d^{\#})$. This suggests that $(X^{\#}, d^{\#})$ might consist of the Cauchy sequences of (X, d) with $d^{\#}$ defined as a metric on limit points. This is almost the case, but the problem with this idea is that if two Cauchy sequences (x_n) and (y_n) in X have the property that $\lim d(x_n, y_n) = 0$, then we cannot conclude that $(x_n) = (y_n)$ $(x_n \neq y_n$, but $\lim |x_n - y_n| = 0)$. To overcome this problem, we use the notion of equivalence classes.

We will need two following lemmas.

LEMMA 4.8.1

Let (x_n) be a Cauchy sequence in a metric space (X, d). Let (y_n) be any other sequence in X such that $\lim d(x_n, y_n) = 0$. Then

(i) (y_n) is also a Cauchy sequence and

(ii) (y_n) converges to a point $y \in X$ iff (x_n) converges to y.

PROOF

(i) follows from the inequality

$$d(y_n, y_m) \leq d(y_n, x_n) + d(x_n, x_m) + d(x_m, y_m)$$

(ii) Since

$$d(x_n, y) \leq d(x_n, y_n) + d(y_n, y)$$

$\lim y_n = y$ implies $\lim x_n = y$. The converse is obtained by interchanging x_n and y_n. ∎

LEMMA 4.8.2

Let (X, d) and (Y, ρ) be two complete metric spaces. Suppose that there exist two dense subspaces, $\mathcal{X}$ of X and $\mathcal{Y}$ of Y, which are isometric to each other. Then X and Y are isometric, too.

PROOF Let $H : \mathcal{X} \to \mathcal{Y}$ be an isometry from $\mathcal{X}$ onto $\mathcal{Y}$. Pick an arbitrary point $x \in X$. From the density of $\mathcal{X}$ in X it follows that there exists a sequence x_n converging to x. Thus (x_n) is a Cauchy sequence, which in turn implies that also $y_n = H(x_n)$ is a Cauchy sequence. From the completeness of Y it follows that (y_n) has a limit y. Define

$$\widehat{H}(x) = y$$

We claim that $\widehat{H}$ is a well-defined extension of H. Indeed, let $\hat{x}_n$ be another sequence from $\mathcal{X}$ converging to x. From the triangle inequality

$$d(x_n, \hat{x}_n) \leq d(x_n, x) + d(\hat{x}_n, x)$$

it follows that $\lim \rho(H(x_n), H(\hat{x}_n)) = \lim d(x_n, \hat{x}_n) = 0$, which, according to the previous lemma, implies that $\lim H(\hat{x}_n) = y$. Thus $H(x)$ is independent of the choice of (x_n). Choosing $(x_n) = (x, x \ldots)$ for $x \in \mathcal{X}$, we easily see that $\widehat{H}(x) = H(x)$ for $x \in \mathcal{X}$. Thus $\widehat{H}$ is an extension of H.

Next we prove that $\widehat{H}$ is an isometry. Indeed, we have

$$d(x_n, y_n) = \rho(H(x_n), H(y_n))$$

for every $x_n, y_n \in \mathcal{X}$. Taking advantage of continuity of metrics d and ρ, we pass to the limit with $x_n \to x$ and $y_n \to y$, getting the result required. ∎

PROOF ***of Theorem 4.8.1.***

Step 1. Let $C(X)$ denote the set of all Cauchy sequences of points in X. We introduce a relation R on $C(X)$ defined so that two distinct Cauchy sequences (x_n) and (y_n) are related under R if and only if the limit of the distance between corresponding terms in each is zero:

$$(x_n)R(y_n) \quad \text{iff} \quad \lim d(x_n, y_n) = 0$$

The relation R is an equivalence relation. Indeed, by inspection, R is clearly reflexive and symetric owing to the symmetry of metric d. If $(x_n)R(y_n)$ and $(y_n)R(z_n)$, then $(x_n)R(z_n)$, because

$$d(x_n, z_n) \leq d(x_n, y_n) + d(y_n, z_n)$$

and

$$\lim d(x_n, y_n) = 0 \text{ and } \lim d(y_n, z_n) = 0 \text{ implies } \lim d(x_n, z_n) = 0$$

Hence R is transitive and is, therefore, an equivalence relation.

Step 2. Now we recall that an equivalence relation partitions a set $C(X)$ into equivalence classes; e.g., $[(x_n)]$ is the set of all Cauchy sequences related to (x_n) under the equivalence relation R. Let $X^\#$ denote the quotient set $C(X)/R$ and let $d^\#([(x_n)],[(y_n)])$ be defined by

$$d^\#\left([(x_n)],[(y_n)]\right) = \lim_{n\to\infty} d(x_n, y_n)$$

The function $d^\#$ is a well-defined metric on $X^\#$. By well-defined we mean that the limit appearing in the definition of $d^\#$ exists and is unambiguous. Denoting

$$s_n = d(x_n, y_n)$$

we have

$$\begin{aligned}|s_n - s_m| &= |d(x_n, y_n) - d(x_m, y_m)| \\ &= |d(x_n, y_n) - d(y_n, x_m) + d(y_n, x_m) - d(x_m, y_m)| \\ &\leq d(x_n, x_m) + d(y_n, y_m)\end{aligned}$$

from which follows that (s_n) is a Cauchy sequence in $\mathbb{R}$. Since the set of real numbers is complete, (s_n) has a limit s in $\mathbb{R}$.

To show that this limit does not depend upon which representative we pick from the equivalence classes $[(x_n)]$ and $[(y_n)]$ choose

$$(x_n), (\overline{x}_n) \in [(x_n)] \ ; \ (y_n), (\overline{y}_n) \in [(y_n)]$$

Then

$$\begin{aligned}|d(x_n, y_n) - d(\overline{x}_n, \overline{y}_n)| &= |d(x_n, y_n) - d(y_n, \overline{x}_n) + d(y_n, \overline{x}_n) - d(\overline{x}_n, \overline{y}_n)| \\ &\leq |d(x_n, y_n) - d(y_n, \overline{x}_n)| + |d(y_n, \overline{x}_n) - d(\overline{x}_n, \overline{y}_n)| \\ &\leq d(x_n, \overline{x}_n) + d(y_n, \overline{y}_n)\end{aligned}$$

By the definition of $[(x_n)]$ and $[(y_n)]$, $d(x_n, \overline{x}_n)$ and $d(y_n, \overline{y}_n) \to 0$ as $n \to \infty$. Hence

$$\lim d(x_n, y_n) = \lim d(\overline{x}_n, \overline{y}_n)$$

Now, by construction, $d^{\#}([(x_n)],[(y_n)])$ is zero iff (x_n) and (y_n) belong to the same equivalence class and in this case $[(x_n)] = [(y_n)]$. The symmetry of $d^{\#}$ is obvious. From the triangle inequality

$$d(x_n, z_n) \leq d(x_n, y_n) + d(y_n, z_n)$$

for any Cauchy sequences $(x_n), (y_n), (z_n)$, passing to the limit, we obtain

$$d^{\#}([(x_n)],[(z_n)]) \leq d^{\#}([(x_n)],[(y_n)]) + d^{\#}([(y_n)],[(z_n)])$$

Hence, $(X^{\#}, d^{\#})$ is a metric space.

Step 3. Suppose that x_0 is a point in X. It is a simple matter to construct a Cauchy sequence whose limit is x_0. For example, the constant sequence

$$(x_0, x_0, x_0, \ldots)$$

is clearly Cauchy and its limit is x_0. We can construct such sequences for all points x in X, and for each sequence so constructed, we can find an equivalence class of Cauchy sequences in X. Let Z denote the set of all such equivalence classes constructed in this way; i.e.,

$$Z = \{z(x) \,:\, x \in X, z(x) = [(x, x, x, \ldots)]\}$$

We claim that (X, d) is isometric to $(Z, d^{\#})$. Indeed, for $x, y \in X$,

$$d^{\#}(z(x), z(y)) = d^{\#}([(x, x, \ldots)],[(y, y, \ldots)]) = \lim d(x, y) = d(x, y)$$

Hence, $z\colon X \to Z$ is an isometry.

Step 4. $(Z, d^{\#})$ is dense in $(X^{\#}, d^{\#})$.

Let $[(x_n)]$ be an arbitrary point in $X^{\#}$. This means that (x_n) is a Cauchy sequence of points in X. For each component of this sequence there is a corresponding class of sequences equivalent to a constant Cauchy sequence in Z; i.e.,

$$\begin{aligned} z(x_1) &= [(x_1, x_1, \ldots,)] \\ z(x_2) &= [(x_2, x_2, \ldots,)] \\ &\vdots \\ z(x_n) &= [(x_n, x_n, \ldots,)] \end{aligned}$$

Consider the sequence of equivalence classes $(z(x_1), z(x_2), \ldots) = (z(x_n))$ in Z. Since (x_n) is Cauchy, we have

$$\lim_{n\to\infty} d^{\#}([(x_n)], z(x_n)) = \lim_{n\to\infty}(\lim_{m\to\infty} d(x_m, x_n)) = \lim_{n,m\to\infty} d(x_n, x_m) = 0$$

Therefore, the limit of $(z(x_n))$ is $[(x_n)]$. Thus Z is dense in $X^{\#}$.

Step 5. $(X^{\#}, d^{\#})$ is complete.

Let $(\boldsymbol{x}_1, \boldsymbol{x}_2, \ldots)$ be a Cauchy sequence in $X^{\#}$. Since Z is dense in $X^{\#}$, for every n there is an element $z(x_n) \in Z$ such that

$$d^{\#}(\boldsymbol{x}_n, z(x_n)) < 1/n$$

Consequently, $d^{\#}(\boldsymbol{x}_n, z(x_n)) \to 0$. It follows from Lemma 4.8.1 that $(z(x_n))$ is Cauchy, which in turn implies that (x_n) is Cauchy in X. Hence, by construction of $X^{\#}$, $[(x_n)]$ is a limit of $z(x_n)$ (cf. Step 4) and, according to Lemma 4.8.1, of $(\boldsymbol{x}_1, \boldsymbol{x}_2, \ldots)$, too.

Step 6. Uniqueness of the completion.

Let (X^0, d^0) be another completion of (X, d). It follows from the definition of completion that both $X^{\#}$ and X^0 contain dense subspaces which are isometric to space X and therefore are also isometric to each other. Thus, according to Lemma 4.8.2, $X^{\#}$ and X^0 are isometric to each other, too.

This last result completes the proof. ∎

The Baire Categories. It is convenient at this point to mention a special property of complete metric spaces. A subspace A of a topological space X is said to be *nowhere dense* in X if the interior of its closure is empty: $\text{int}\overline{A} = \emptyset$. For example, the integers $\mathbb{Z}$ are nowhere dense in $\mathbb{R}$.

A topological space X is said to be of the*first category* if X is the countable union of nowhere dense subsets of X. Otherwise, X is of *the second category.* These are called the *Baire categories.* For example, the rationals $\mathbb{Q}$ are of the first category, because the singleton sets $\{q\}$ are nowhere dense in $\mathbb{Q}$, and $\mathbb{Q}$ is the countable union of such sets. The real line $\mathbb{R}$ is of the second category. Indeed, that every complete metric space is of the second category is the premise of the following basic theorem.

THEOREM 4.8.2
(The Baire Category Theorem)

Every complete metric space (X, d) is of the second category.

PROOF Suppose, to the contrary, that X is of the first category, i.e.,

$$X = \bigcup_1^\infty M_i, \quad \text{int}\ \overline{M}_i = \emptyset$$

Replacing M_i with their closures $\overline{M}_i$ we can assume from the very beginning that M_i are closed. Hence, complements M_i' are open.

Consider now the set M_1. We claim that

$$\overline{M_1'} = X$$

Indeed, suppose to the contrary that there is an $x \notin \overline{M_1'}$. Equivalently, there exists a neighborhood B of x such that $B \cap M_1' = \emptyset$, which in turn implies that $B \subset M_1$. Thus $x \in \text{int } M_1$, a contradiction.

Since M_1' is open, there exists a closed ball

$$S_1 = (x \, : \, d(x_1, x) \leq r_1) \subset M_1'$$

centered at a point x_1 which, according to the fact that $\overline{M_1'} = X$, can be arbitrarily close to any point of X. Obviously, we may assume that $r_1 < 1/2$.

By the same arguments M_2' contains a closed ball

$$S_2 = \{x \, : \, d(x_2, x) \leq r_2\}$$

contained in S_1 (we can locate center x_2 arbitrarily close to any point in X) such that $r_2 < 1/2^2$.

Proceeding in this way, we obtain a sequence S_n of closed balls with the properties

$$r_n < 1/2^n, S_{n+1} \subset S_n, S_n \cap M_n = \emptyset$$

Obviously, the sequence of centers x_n is a Cauchy sequence and therefore possesses a limit x_0. Since $x_k \in S_n$, for every $k \geq n$ and for every S_n, and S_n are closed, $x_0 \in S_n$, for every n, which in turn implies that $x_0 \in \bigcap_1^\infty S_n$ and, consequently,

$$x_0 \notin \left(\bigcap_1^\infty S_n\right)' = \bigcup_1^\infty S_n' \supset \bigcup_1^\infty M_n = X$$

a contradiction. Hence X is of the second category. ∎

Exercises

4.8.1. Let $\Omega \subset \mathbb{R}^n$ be a measurable set and $(C(\Omega), \| \, \|_p)$ denote the (incomplete) metric space of continuous, real-valued functions on Ω

with metric induced by the L^p norm. Construct arguments supporting the fact that the completion of this space is $L^p(\Omega)$. *Hint:* See Exercises 4.9.3 and 4.9.4 and use Theorem 4.8.1.

4.8.2. Let x_{n_k} be a subsequence of a Cauchy sequence x_n. Show that if x_{n_k} converges to x, so does x_n.

4.9 Compactness in Metric Spaces

Since in a metric space every point possesses a countable base of neighborhoods, according to Proposition 4.4.4, every compact set is sequentially compact. It turns out that in the case of a metric space the converse is also true.

THEOREM 4.9.1
(Bolzano-Weierstrass Theorem)

A set E in a metric space (X, d) is compact if and only if it is sequentially compact.

Before we prove this theorem, we shall introduce some auxiliary concepts.

ε-Nets and Totally Bounded Sets. Let Y be a subset of a metric space (X, d) and let ε be a positive real number. A finite set

$$Y_\varepsilon = \{y_\varepsilon^1, \ldots, y_\varepsilon^n\} \subset X$$

is called an *ε-net for Y* if

$$Y \subset \bigcup_{j=1}^{n} B\left(y_\varepsilon^j, \varepsilon\right)$$

In other words, for every $y \in Y$ there exists a point $y_\varepsilon^j \in Y_\varepsilon$ such that

$$d\left(y, y_\varepsilon^j\right) < \varepsilon$$

A set $Y \subset X$ is said to be *totally bounded in X* if for each $\varepsilon > 0$ there exists in X an ε-net for Y. If Y is totally bounded in itself, i.e., it contains the ε-nets, we say that Y is *totally bounded.* Note that, in particular, every set Y totally bounded in X is bounded. Indeed, denoting by M_ε the maximum distance between points in ε-net Y_ε

$$M_\varepsilon = \max\{d(x,y) \ : \ x, y \in Y_\varepsilon\}$$

we have

$$d(x,y) \le d(x,x^\varepsilon) + d(x^\varepsilon, y^\varepsilon) + d(y^\varepsilon, y) \le M_\varepsilon + 2\varepsilon, \text{ for every } x, y \in Y$$

where x^ε and y^ε are points from ε-net Y_ε such that

$$d(x, x^\varepsilon) < \varepsilon \text{ and } d(y, y^\varepsilon) < \varepsilon$$

Consequently, $\text{dia}Y \le M_\varepsilon + 2\varepsilon$, which proves that Y is bounded.

COROLLARY 4.9.1

Every totally bounded metric space (X, d) is separable.

PROOF Consider a family of $1/n$-nets Y_n and define

$$Y = \bigcup_1^\infty Y_n$$

As a union of a countable family of finite sets, Y is countable. Now, it follows from the definition of ε-net that for a given $x \in X$ and for every $n = 1, 2 \ldots$ there exists a $y_n \in Y_n \subset Y$ such that

$$d(x, y_n) < 1/n$$

Thus, $\lim_{n\to\infty} y_n = x$, which proves that Y is dense in X. ∎

LEMMA 4.9.1

Every sequentially compact set E in a metric space (X, d) is totally bounded.

PROOF Pick $\varepsilon > 0$ and an arbitrary point $a_1 \in E$. If

$$E \subset B(a_1, \varepsilon)$$

then $\{a_1\}$ is the ε-net for E. If not, then there exists an $a_2 \in E - B(a_1, \varepsilon)$. Proceeding in the same manner, we prove that either E contains ε-net of points $\{a_1, \ldots, a_n\}$ or there exists an infinite sequence $(a_i)_1^\infty$ such that

$$d(a_i, a_j) \geq \varepsilon$$

This contradicts the fact that E is sequentially compact. Indeed, let (a_{i_k}) be a subsequence of (a_i) convergent to an element a. Then for sufficiently large i,

$$d(a, a_i) < \frac{\varepsilon}{3}$$

which in turn implies that

$$d(a_i, a_j) \leq d(a_i, a) + d(a, a_j) < 2\frac{\varepsilon}{3}$$

a contradiction. ∎

PROOF ***of the Bolzano-Weierstrass Theorem.***

Since every subset E of a metric space (X, d) is a metric space itself, it is sufficient to prove that every sequentially compact metric space is compact. So let (X, d) be a sequentially compact metric space and let $G_\iota, \iota \in I$ be an open covering of X. We claim that there exists an $\varepsilon > 0$ such that every open ball with radius ε (centered at arbitrary point) is entirely contained in one of the sets G_ι. Indeed, suppose to the contrary that for every n there exists a point $a_n \in X$ such that none of the sets G_ι contains ball $B(a_n, 1/n)$. Since X is sequentially compact, there exists a subsequence a_{n_k} and a point a such that $a_{n_k} \to a$. Obviously, $a \in G_\kappa$, for some $\kappa \in I$ and therefore there exists a ball $B(a, \alpha)$ such that

$$B(a, \alpha) \subset G_\kappa$$

However, for sufficiently large n_k, $d(a_{n_k}, a) \leq \alpha/2$ and, simultaneously, $1/n_k \leq \alpha/2$. It follows that

$$d(x, a) \leq d(x, a_{n_k}) + d(a_{n_k}, a) \leq \frac{\alpha}{2} + \frac{\alpha}{2} = \alpha$$

for every $x \in B(a_{n_k}, 1/n_k)$. Thus

$$B(a_{n_k}, 1/n_k) \subset B(a, \alpha) \subset G_\kappa$$

a contradiction.

Now, according to Lemma 4.9.1, X is totally bounded. Let Y_ε be an ε-net corresponding to the value of ε for which every ball $B(y_\varepsilon, \varepsilon)$, $y_\varepsilon \in Y_\varepsilon$, is contained entirely in the corresponding set G_{y_ε} from the covering $G_\iota, \iota \in I$. We have

$$X = \bigcup_{y \in Y\varepsilon} B(y, \varepsilon) \subset \bigcup_{y \in Y_\varepsilon} G_{y_\varepsilon}$$

which proves that $G_{y_\varepsilon}, y_\varepsilon \in Y_\varepsilon$ form a finite subcovering of $G_\iota, \iota \in I$. Thus X is compact. ∎

Recall that every (sequentially) compact set is (sequentially) closed. According to Lemma 4.9.1, every compact (or, equivalently, sequentially compact) set in a metric space is totally bounded and therefore bounded. Thus compact sets in metric spaces are both closed and bounded. The converse, true in $\mathbb{R}$ (the Heine-Borel Theorem), in general is false. The following is an example of a set in a metric space which is both closed and bounded, but not compact.

Example 4.9.1

Consider the closed unit ball in the space ℓ^2, centered at zero vector:

$$\overline{B} = \overline{B(\mathbf{0}, 1)} = \{(x_i)_{j=1}^\infty \ : \ \left(\sum_1^\infty x_i^2\right)^{\frac{1}{2}} \leq 1\}$$

Obviously, $\overline{B}$ is bounded (dia $\overline{B} = 2$). Since $\overline{B}$ is an inverse image of the closed interval $[0, 1]$ in $\mathbb{R}$ through the norm in ℓ^2 which is continuous, $\overline{B}$ is also closed. However, $\overline{B}$ is not compact. To see it, consider the sequence

$$\boldsymbol{e}_i = (0, \ldots, 1, 0, \ldots) \quad i = 1, 2, \ldots$$

Obviously, $\boldsymbol{e}_i \in \overline{B}$ and $d(\boldsymbol{e}_i, \boldsymbol{e}_j) = \sqrt{2}$, for every $i \neq j$. Thus sequence e_i cannot be contained in any finite union of balls with radii smaller than $\sqrt{2}$, which proves that $\overline{B}$ is not totally bounded and therefore not compact, too. □

The situation changes if we restrict ourselves to complete metric spaces, replacing the condition of boundedness with that of total boundedness.

THEOREM 4.9.2

Let (X, d) be a complete metric space. A set $E \subset X$ is compact (equivalently, sequentially compact) if and only if it is closed and totally bounded.

PROOF Every compact set is closed and, according to Lemma 4.9.1, compact sets in metric spaces are totally bounded.

Conversely, assume that E is closed and totally bounded. We shall prove that E is sequentially compact. So let (x_n) be an arbitrary sequence of points in E and consider a collection of ε-nets corresponding to choices of ε of $\varepsilon_1 = 1, \varepsilon_2 = 1/2, \ldots, \varepsilon_n = 1/n, \ldots$. Since E is totally bounded, for each of these choices we can construct a finite family of balls of radius ε that cover E. For example, if $\varepsilon = 1$, we construct a collection of a finite number of balls of radius 1. One of these balls, say B_1, contains an infinite subsequence of (x_n), say $(x_n^{(1)})$. Similarly, about each point in the 1/2-net, we form balls of radius 1/2, one of which (say B_2) contains an infinite subsequence of $(x_n^{(1)})$, say $(x_n^{(2)})$. Continuing in this manner, we develop the following set of infinite subsequences:

$$
\begin{array}{ll}
\varepsilon_1 = 1 & \left(x_1^{(1)}, x_2^{(1)}, \ldots, x_n^{(1)}, \ldots\right) \subset B_1 \\
\varepsilon_2 = 1/2 & \left(x_1^{(2)}, x_2^{(2)}, \ldots, x_n^{(2)}, \ldots\right) \subset B_2 \\
\cdot & \\
\cdot & \\
\cdot & \\
\varepsilon_n = 1/n & \left(x_1^{(n)}, x_2^{(n)}, \ldots, x_n^{(n)}, \ldots\right) \subset B_n
\end{array}
$$

Selecting the diagonal sequence $(x_n^{(n)})$, we get a subsequence of the original sequence (x_n), which satisfies the Cauchy condition. This follows from the fact that all $x_m^{(m)}$ for $m \geq n$ are contained in ball B_n of radius $1/n$. Since X is complete, $x_m^{(m)}$ converges to a point x which, according to the assumption that E is closed, belongs to E. Thus E is sequentially compact. ∎

REMARK 4.9.1. Note that the assumption of completeness of metric space (X, d) in Theorem 4.9.2 is not very much restrictive since every compact set E in a metric space (X, d) (not necessarily complete) is itself complete. To see it, pick a Cauchy sequence (x_n) in E. As every sequence in $E, (x_n)$ contains a convergent subsequence (x_{n_k}) to an element of E, say x_0. It follows from the triangle inequality that

$$d(x_0, x_n) \leq d(x_0, x_{n_k}) + d(x_{n_k}, x_n)$$

and therefore the whole sequence must converge to x_0. Thus E is complete. ∎

REMARK 4.9.2. The method of selecting the subsequence of functions converging on a countable set of points, used in the proof of Theorem 4.9.2, is known as "the diagonal choice method." We will use it frequently in this book. ∎

Precompact Sets. A set E in a topological space X is said to be precompact if its closure E is compact. Thus, according to the Bolzano-Weierstrass theorem, a set E in $\mathbb{R}$ is precompact iff it is bounded. Looking back at Theorem 4.9.2, we can characterize precompact sets in complete metric spaces as those which are totally bounded.

Due to the importance of compact sets, one of the most fundamental questions in functional analysis concerns finding criteria for compactness in particular function spaces. We shall conclude this section with two of them: the famous Ascoli-Arzela thorem formulating a criterium for a compactness in the Chebyshev space $C(K)$ and the Frechet-Kolmogorov theorem for spaces $L^p(\mathbb{R})$.

Equicontinuous Classes of Functions. Let $C(X,Y)$ denote a class of continuous functions (defined on the entire X) mapping a metric space (X,d) into a metric space(Y,ρ). Thus $f \in C(X,Y)$ iff

$$\forall x_0 \in X \quad \forall \varepsilon > 0 \quad \exists \delta > 0 : d(x_0, x) < \delta \Rightarrow \rho(f(x_0), f(x)) < \varepsilon$$

Obviously, δ generally depends upon x_0, ε and function f. Symbolically, we could write $\delta = \delta(x_0, \varepsilon, f)$. Recall that if δ is independent of $x_0, \delta = \delta(\varepsilon, f)$ then f is said to be uniformly continuous.

A subclass $\mathcal{F} \subset C(X,Y)$ of functions is said to be *equicontinuous* (on X) if δ happens to be independent of f, for every f from the class $\mathcal{F}$. In other words,

$$\forall\ x_0 \in X \quad \forall \varepsilon > 0 \quad \exists \delta = \delta(x_0, \varepsilon) : d(x_0, x) < \delta \Rightarrow \rho(f, x_0), f(x)) < \varepsilon$$

for every function $f \in \mathcal{F}$.

Uniformly Bounded Real-Valued Function Classes. Recall that a function (functional) $f : X \to \mathbb{R}$, defined on an arbitrary set X, is bounded if there exists a constant $M > 0$ such that

$$|f(x)| \le M \text{ for every } x \in X$$

A class $\mathcal{F}$ of such functions is said to be *uniformly bounded* if constant M happens to be independent of function $f \in \mathcal{F}$, i.e.,

$$|f(x)| \le M \text{ for every } x \in X, f \in \mathcal{F}.$$

Consider now the Chebyshev space $C(K)$ of continuous functions defined on a compact set K in $\mathbb{R}^n$ with the Chebyshev norm

$$\|f\| = \sup_{x \in K} |f(x)|$$

and the resulting Chebyshev metric

$$d(f,g) = \sup_{x \in K} |f(x) - g(x)|$$

We will need the following lemmas.

LEMMA 4.9.2
(Dini's Theorem)

Let E be a compact topological space. Suppose we are given a monotone sequence of continuous functions $f_n : E \to \mathbb{R}$ converging pointwise to a continuous function $f : E \to \mathbb{R}$. Then f_n converges uniformly to f.

PROOF

Case 1. Sequence f_n is increasing.

Suppose, to the contrary, that f_n does not converge uniformly to f. Thus there exists an $\varepsilon > 0$ such that the sets

$$E_n = \{x \, : \, f(x) - f_n(x) \geq \varepsilon\}$$

are not empty, for every $n = 1, 2, \ldots$. Thus, as the decreasing family of nonempty sets, E_n forms a base of closed sets. According to the compactness of E, there exists an element $x_0 \in \bigcap_1^\infty E_n$, which in turn implies that

$$f(x_0) - f_n(x_0) \geq \varepsilon \text{ for every } n = 1, 2 \ldots$$

which contradicts the fact that $f_n(x_0)$ converges to $f(x_0)$.

Case 2. The proof for decreasing sequences of functions f_n is identical.

∎

LEMMA 4.9.3

Every continuous function $f\colon X \to Y$ from a compact metric space (X, d) to a metric space (Y, ρ) is *uniformly* continuous.

In particular, every continuous functional defined on a compact metric space must be necessarily *uniformly* continuous.

PROOF Suppose, to the contrary, that there is an $\varepsilon > 0$ such that for every n there exist points $x_n, y_n \in X$ such that

$$d(x_n, y_n) < 1/n \qquad \text{and} \qquad \rho(f(x_n), f(y_n)) \geq \varepsilon$$

Since X is compact, we can choose a subsequence x_{n_k} convergent to a point x in X. Selecting another convergent subsequence from y_{n_k}, we can assume that both x_n and y_n converge to points x and y. But $d(x_n, y_n) \to 0$ and therefore (cf. Lemma 4.8.1) $x = y$. On the other side, passing to the limit with $x_n \to x$ and $y_n \to y$ in

$$\rho(f(x_n), f(y_n)) \geq \varepsilon$$

we obtain from continuity of function f and metric ρ that

$$\rho(f(x), f(y)) \geq \varepsilon$$

a contradiction. ∎

We have the following:

THEOREM 4.9.3
(Ascoli-Arzela Theorem)

A subclass $\mathcal{F}$ of $C(K)$ is precompact if and only if

(i) $\mathcal{F}$ is equicontinuous and

(ii) $\mathcal{F}$ is uniformly bounded.

PROOF We first prove necessity. Assume $\mathcal{F}$ is precompact. Then $\mathcal{F}$ is totally bounded in $C(K)$, which means that, for every $\varepsilon > 0$, we can construct an ε-net $Y_\varepsilon = (f_1, \ldots, f_n)$ in $C(K)$. Denoting

$$M = \max\{\|f_i\|_\infty, i = 1, \ldots, n\}$$

we have for every $f \in \mathcal{F}$

$$|f(\boldsymbol{x})| \leq |f(\boldsymbol{x}) - f_k(\boldsymbol{x})| + |f_k(x)| \leq \varepsilon + M$$

where f_k is a function from ε-net Y_ε such that

$$\|f - f_k\|_\infty < \varepsilon$$

Thus $\mathcal{F}$ is uniformly bounded.

To prove equicontinuity, pick an $\varepsilon > 0$ and consider a corresponding $\frac{\varepsilon}{3}$-net $Y_{\frac{\varepsilon}{3}} = (f_1, \ldots, f_n)$. Since each f_i is uniformly continuous (cf. Lemma 4.9.3), there exists a δ_i such that

$$|f_i(\boldsymbol{x}) - f_i(\boldsymbol{y})| < \frac{\varepsilon}{3} \text{ whenever } \|\boldsymbol{x} - \boldsymbol{y}\| < \delta_i$$

where $\| \cdot \|$ denotes any of the norms in $\mathbb{R}^n$. Setting

$$\delta = \min\,(\delta_i, i = 1, 2, \ldots, n)$$

we have

$$|f_i(\boldsymbol{x}) - f_i(\boldsymbol{y})| < \frac{\varepsilon}{3} \text{ whenever } \| \boldsymbol{x} - \boldsymbol{y} \| < \delta, \quad \text{for every } f_i \in Y_{\frac{\varepsilon}{3}}$$

Now, for an arbitrary function f from $\mathcal{F}$, it follows from the definition of ε-nets that

$$|f(\boldsymbol{x}) - f(\boldsymbol{y})| \leq |f(\boldsymbol{x}) - f_i(\boldsymbol{x})| + |f_i(\boldsymbol{x}) - f_i(\boldsymbol{y})| + |f_i(\boldsymbol{y}) - f(\boldsymbol{y})|$$

$$< \tfrac{\varepsilon}{3} + \tfrac{\varepsilon}{3} + \tfrac{\varepsilon}{3} = \varepsilon$$

whenever $\|\boldsymbol{x} - \boldsymbol{y}\| < \delta$.

To prove sufficiency we assume that $\mathcal{F}$ is uniformly bounded and equicontinuous and show that every sequence $(f_n) \subset \mathcal{F}$ contains a convergent subsequence. Since every compact set in a metric space is separable (see Corollary 4.9.1), there exists a countable set $\mathcal{K} = \{\boldsymbol{x}_1, \boldsymbol{x}_2, \ldots\}$ such that $\overline{\mathcal{K}} = K$. Each of the sequences $\{f_k(\boldsymbol{x}_i)\}_{k=1}^{\infty}$ for $\boldsymbol{x}_1, \boldsymbol{x}_2, \ldots \in \mathcal{K}$ is bounded, so by the diagonal choice method (see Remark 4.9.2), we can extract such a subsequence f_{k_j} that $f_{k_j}(\boldsymbol{x})$ is convergent for every $\boldsymbol{x} \in \mathcal{K}$.

Pick an $\varepsilon > 0$. According to equicontinuity of $\mathcal{F}$, there exists a $\delta > 0$ such that

$$|f_{k_j}(\boldsymbol{x}) - f_{k_j}(\boldsymbol{y})| < \frac{\varepsilon}{3} \text{ whenever } \|\boldsymbol{x} - \boldsymbol{y}\| < \delta$$

Let $\boldsymbol{x}$ be an arbitrary point of K. It follows from density of $\mathcal{K}$ in K that there is a $\boldsymbol{y} \in \mathcal{K}$ such that $\|\boldsymbol{x} - \boldsymbol{y}\| < \delta$. Consequently,

$$|f_{k_i}(\boldsymbol{x}) - f_{k_j}(\boldsymbol{x})| \leq |f_{k_i}(\boldsymbol{x}) - f_{k_i}(\boldsymbol{y})| + |f_{k_i}(\boldsymbol{y}) - f_{k_j}(\boldsymbol{y})|$$

$$+ |f_{k_j}(\boldsymbol{y}) - f_{k_j}(\boldsymbol{x})|$$

$$< \frac{\varepsilon}{3} + \frac{\varepsilon}{3} + \frac{\varepsilon}{3} = \varepsilon$$

for k_i and k_j sufficiently large ($f_{k_i}(\boldsymbol{y})$ is convergent and therefore Cauchy). Concluding, for every $\boldsymbol{x} \in K$, $f_{k_i}(\boldsymbol{x})$ is Cauchy in $\mathbb{R}$ and therefore convergent. Denote

$$f_0(\boldsymbol{x}) = \lim_{k_i \to \infty} f_{k_i}(\boldsymbol{x})$$

It remains to prove that

1. f_0 is continuous and
2. f_{k_i} converges uniformly to f (in norm $\| \cdot \|_\infty$).

It follows from equicontinuity of $\mathcal{F}$ that

$$\forall \varepsilon > 0 \quad \exists \delta > 0 : |f_{k_i}(\boldsymbol{x}) - f_{k_i}(\boldsymbol{y})| < \varepsilon \quad \text{whenever } \|\boldsymbol{x} - \boldsymbol{y}\| < \delta$$

for every f_{k_i}. Passing to the limit with $k_i \to \infty$, we get that f_0 is uniformly continuous in K.

To prove the last assertion consider the functions

$$\varphi_{k_i}(\boldsymbol{x}) = \inf_{k_j \geq k_i} f_{kj}(\boldsymbol{x}) \qquad \psi_{k_i}(\boldsymbol{x}) = \sup_{k_j \geq k_i} f_{kj}(\boldsymbol{x})$$

It follows from equicontinuity of f_{k_i} that both φ_{k_i} and ψ_{k_i} are continuous on K. Now, φ_{k_i} is increasing, ψ_{k_i} is decreasing and

$$\lim \varphi_{k_i}(\boldsymbol{x}) = \liminf f_{k_i} = \lim f_{k_i}(\boldsymbol{x}) = f_0(\boldsymbol{x})$$

together with

$$\lim \psi_{k_i}(\boldsymbol{x}) = \limsup f_{k_i}(\boldsymbol{x}) = \lim f_{k_i}(\boldsymbol{x}) = f_0(\boldsymbol{x})$$

and therefore Lemma 4.9.2 implies that both φ_{k_i} and ψ_{k_i} converge uniformly to f.

Finally, from the inequality

$$\varphi_{k_i}(\boldsymbol{x}) \leq f_{k_i}(\boldsymbol{x}) \leq \psi_{k_i}(\boldsymbol{x})$$

it follows that f_{k_i} converges uniformly to f_0, too. ∎

THEOREM 4.9.4
(Frechet-Kolmogorov Theorem)

A family $\mathcal{F} \subset L^p(\mathbb{R}), 1 \leq p < \infty$, is precompact in $L^p(\mathbb{R})$ iff the following conditions hold:

(i) $\mathcal{F}$ *is uniformly bounded, i.e., there exists an* $M > 0$ *such that*

$$\|f\|_p \leq M, \text{ for every } f \in \mathcal{F}$$

(ii) $\displaystyle\lim_{t\to 0} \int_{\mathbb{R}} |f(t+s) - f(s)|^p ds = 0$ *uniformly in* $\mathcal{F}$*;*

(iii) $\displaystyle\lim_{n\to\infty} \int_{|s|>n} |f(s)|^p ds = 0$ *uniformly in* $\mathcal{F}$*.*

PROOF First of all, we claim that for every $f \in L^p(\mathbb{R})$, limits defined in (ii) and (iii) are zero. (Thus the issue is not convergence to zero, but the assertion of uniform convergence.) Indeed, one can prove (see Exercises 4.9.3–4.9.5) that the space of continuous functions with compact support $C_0(\mathbb{R})$ is dense in $L^p(\mathbb{R}), 1 \leq p < \infty$. Thus for an arbitrary $\varepsilon > 0$ there exists a $g \in C_0(\mathbb{R})$ such that

$$\|f - g\|_p \leq \frac{\varepsilon}{3}$$

Now, since g is continuous and bounded (explain why), by the Lebesgue dominated convergence theorem, we have

$$\left(\int_{\mathbb{R}} |g(t+s) - g(s)|^p \, ds \right)^{\frac{1}{p}} \to 0 \quad \text{for } t \to 0$$

and, consequently, $\|T_t g - g\|_p < \frac{\varepsilon}{3}$ for $t \leq$ some t_0, where T_t is the translation operator

$$T_t g(s) = g(t+s)$$

Finally,

$$\|f - T_t f\|_p \leq \|f - g\|_p + \|g - T_t g\|_p + \|T_t g - T_t f\|_p < \varepsilon$$

whenever $t \leq t_0$, since the norm $\|\cdot\|_p$ is invariant under the translation T_t. Thus

$$\lim_{g\to 0} \int_{\mathbb{R}} |f(t+s) - f(s)|^p ds = 0 \quad \text{for every } f \in L^p(\mathbb{R})$$

The second assertion follows immediately from the Lebesgue dominated convergence theorem and pointwise convergence of *truncations*

$$f_n(x) = \begin{cases} f(x), & |x| \leq n \\ 0, & |x| > n \end{cases}$$

to function f.

Assume now that $\mathcal{F}$ is precompact. Thus $\mathcal{F}$ is totally bounded in $L_p(\mathbb{R})$ and therefore bounded. To prove the second assertion consider an $\frac{\varepsilon}{3}$-net $(f_1, \ldots, f_n) \subset L^p(\mathbb{R})$ for $\mathcal{F}$. According to our preliminary considerations,

$$\|T_t f_i - f_i\|_p < \frac{\varepsilon}{3} \text{ whenever } t \leq t_0 = t_0(f_i)$$

with $t_0 = t_0(f_i)$ depending on f_i. By choosing, however,

$$t_0 = \min(t_0(f_i), i = 1, 2, \ldots, n)$$

we get

$$\|T_t f_i - f_i\|_p < \frac{\varepsilon}{3} \text{ whenever } t \leq t_0$$

Consequently,

$$\|T_t f - f\|_p \leq \|T_t f - T_t f_i\|_p + \|T_t f_i - f_i\|_p + \|f_i - f\|_p < \varepsilon$$

for $t \leq t_0$. This proves (ii). Using exactly the same technique, we prove that (iii) holds.

To prove the converse we shall show that conditions (i)–(iii) imply total boundedness of $\mathcal{F}$ in $C(K)$ (cf. Theorem 4.9.2).

First of all, condition (ii) is equivalent to saying that

$$\|T_t f - f\|_p \to 0 \text{ uniformly in } \mathcal{F}$$

Define the mean-value operator as

$$(M_a f)(s) = (2a)^{-1} \int_{-a}^{a} T_t f(s) dt$$

It follows from the Hölder inequality and Fubini's theorem that

$$
\begin{aligned}
\|M_a f - f\|_p &= \left\{ \int_{-\infty}^{\infty} \left| \int_{-a}^{a} (2a)^{-1} f(t+s)dt - f(s) \right|^p ds \right\}^{\frac{1}{p}} \\
&\leq \left\{ \int_{-\infty}^{\infty} \left[\int_{-a}^{a} (2a)^{-1} |f(t+s) - f(s)| \, dt \right]^p ds \right\}^{\frac{1}{p}} \\
&\leq (2a)^{-1} \left\{ \int_{-\infty}^{\infty} (2a)^{\frac{p}{q}} \int_{-a}^{a} |f(t+s) - f(s)|^p \, dtds \right\}^{\frac{1}{p}} \\
&= (2a)^{-1+\frac{1}{q}} \left\{ \int_{-a}^{a} \int_{-\infty}^{\infty} |f(t+s) - f(s)|^p \, dsdt \right\}^{\frac{1}{p}} \\
&= (2a)^{-1+\frac{1}{q}+\frac{1}{p}} \sup_{|t|\leq a} \left\{ \int_{-\infty}^{\infty} |f(t+s) - f(s)|^p \, ds \right\}^{\frac{1}{p}} \\
&= \sup_{|t|\leq a} \|T_t f - f\|_p
\end{aligned}
$$

where $\frac{1}{p} + \frac{1}{q} = 1$.

Thus, according to (ii),

$$
M_a f \to f \text{ uniformly in } \mathcal{F}
$$

We shall show now that, for a fixed $a > 0$, functions $M_a f, f \in \mathcal{F}$ are uniformly bounded and equicontinuous. It follows from the Hölder inequality again that

$$
\begin{aligned}
|(M_a f)(s_1) - (M_a f)(s_2)| &\leq (2a)^{-1} \int_{-a}^{a} |f(s_1+t) - f(s_2+t)| \, dt \\
&\leq (2a)^{\frac{1}{p}} \left(\int_{-a}^{a} |f(s_1+t) - f(s_2+t)|^p \, dt \right)^{\frac{1}{p}} \\
&= (2a)^{-\frac{1}{p}} \|T_{s_1-s_2} f - f\|_p
\end{aligned}
$$

But $\|T_t f - f\|_p \to 0$ while $t \to 0$ uniformly in $f \in \mathcal{F}$ and therefore $M_a f, f \in \mathcal{F}$ are equicontinuous. Similarly,

$$|(M_a f)(s)| \le (2a)^{-1} \int_{-a}^{a} |f(t+s)|\,dt \le (2a)^{-1} \int_{s-a}^{s+a} |f(t)|\,dt$$

$$\le (2a)^{-\frac{1}{p}} \left(\int_{s-a}^{s+a} |f(t)|^p\,dt \right)^{\frac{1}{p}} \le (2a)^{-\frac{1}{p}} \|f\|_p$$

and therefore

$$\sup_{s\in\mathbb{R}} |M_a f(s)| < \infty \text{ uniformly in } \mathcal{F}$$

Thus, by the Ascoli-Arzela theorem, for a fixed n (note that $\mathbb{R}$ is not compact and therefore we restrict ourselves to finite intervals $[-n,n]$) and a given ε there exists an ε-net of function $g_j \in C[-n,n])$ such that for any $f \in L^p(\mathbb{R})$

$$\sup_{x\in[-n,n]} |M_a f(s) - g_j(s)| < \varepsilon \text{ for some } g_j$$

Denoting by $\hat{g}_j$ the zero extension of g_j outside $[-n,n]$, we obtain

$$\int_{\mathbb{R}} |f(s) - \hat{g}_j(s)|^p\,ds = \int_{|s|>n} |f(s)|^p\,ds + \int_{|s|\le n} |f(s) - g_j(s)|^p\,ds$$

$$\le \int_{|s|>n} |f(s)|^p\,ds + \int_{|s|\le n} (|f(s) - M_a f(s)| + |M_a f(s) - g_j(s)|)^p\,ds$$

$$\le \int_{|s|>n} |f(s)|^p\,ds + 2^p \int_{|s|\le n} |f(s) - M_a f(s)|^p\,ds$$

$$+ \int_{|s|\le n} |M_a f(s) - g_j(s)|^p\,ds$$

Now, pick an arbitrary $\varepsilon > 0$. By condition (iii) we can select an n such that the first term on the right-hand side is bounded by $\frac{\varepsilon}{3}$ uniformly in $\mathcal{F}$. Next, from the inequality

$$2^p \int_{|s|\le n} |f(s) - M_a f(s)|^p\,ds \le 2^p \|f - M_a f\|_p^p$$

it follows that we can select sufficiently small a such that the second term is bounded by $\frac{\varepsilon}{3}$. Finally, considering the corresponding $\hat{\varepsilon}$-net of function $g_j \in C([-n,n])$, we have

$$\int_{|s|\le n} |M_a f(s) - g_j(s)|^p \, ds \le (\sup |M_a f - g_j|)^p 2n \le \frac{\varepsilon}{3}$$

for $\hat{\varepsilon} = (\varepsilon/3/2n)^{\frac{1}{p}}$.

Consequently, for every $f \in L^p(\mathbb{R})$

$$\int_{\mathbb{R}} |f(s) - \hat{g}_j(s)|^p \, ds < \frac{\varepsilon}{3} + \frac{\varepsilon}{3} + \frac{\varepsilon}{3} = \varepsilon$$

for some extensions $\hat{g}_j$. Thus $\mathcal{F}$ is totally bounded in $L^p(\mathbb{R})$ and therefore precompact in $L^p(\mathbb{R})$. ∎

REMARK 4.9.3. The mean-value operator $M_a f$ can be equivalently defined as

$$M_a f = f * \varphi$$

where the star $*$ denotes the so-called convolution operation

$$(f * \varphi)(s) = \int_{\mathbb{R}} f(s-t)\varphi(t) \, dt$$

provided φ is defined as follows

$$\varphi(t) = \begin{cases} \frac{1}{2a} & \text{for } |t| \le a \\ 0 & \text{otherwise} \end{cases}$$

∎

Functions φ of this type are called *mollifiers* and the convolution above is known as the *mollification* or *regularization* of function f. It turns out that by taking more regular mollifiers, we obtain more regular mollifications. In particular, for a C^∞ mollifier the corresponding mollifications are also C^∞ functions.

Exercises

4.9.1. Let $E \subset \mathbb{R}^n$ be a Lebesgue measurable set. Function

$$\chi_E(\boldsymbol{x}) = \begin{cases} 1 & \boldsymbol{x} \in E \\ 0 & \boldsymbol{x} \notin E \end{cases}$$

is called the *characteristic function of set* E. Prove that there exists a sequence of continuous functions $\phi_n : \mathbb{R}^n \to [0,1]$, converging to χ_E in the L^p norm, for any $1 \le p \le \infty$.
Hint: Pick $\varepsilon = 1/n$ and consider a closed set F and an open set G such that $F \subset E \subset G$ and $m(G - F) < \varepsilon$ (recall Proposition 3.2.3 (ii)). Then set

$$\phi_n(\boldsymbol{x}) = \phi_\varepsilon(\boldsymbol{x}) \stackrel{\text{def}}{=} \frac{d(\boldsymbol{x}, G')}{d(\boldsymbol{x}, G') + d(\boldsymbol{x}, F)}$$

where $d(\boldsymbol{x}, A)$ denotes the distance from point $\boldsymbol{x}$ to set A

$$d(\boldsymbol{x}, A) \stackrel{\text{def}}{=} \inf_{\boldsymbol{y} \in A} d(\boldsymbol{x}, \boldsymbol{y})$$

4.9.2. Let $f : \Omega \to \overline{\mathbb{R}}$ be a measurable function. Function $\phi : \Omega \to \mathbb{R}$ is called a *simple function* if $\Omega = \cup_{i=1}^{\infty} E_i$, where E_i are measurable and pairwise disjoint, $i = 1, 2, \ldots$, and the restriction of ϕ to E_i is constant. In other words,

$$\phi = \cup_{i=1}^{\infty} a_i \chi_{E_i}$$

where $a_i \in \mathbb{R}, i = 1, 2 \ldots$. Prove that, for every $\varepsilon > 0$, there exists a simple function $\phi_\varepsilon : \Omega \to \mathbb{R}$ such that

$$\|f - \phi_\varepsilon\|_{L^\infty(\Omega)} \le \varepsilon$$

Hint: Use the Lebesgue approximation sums.

4.9.3. Let $\Omega \subset \mathbb{R}^n$ be a measurable set of finite measure, $m(\Omega) < \infty$, and $f \in L^p(\Omega), 1 \le p \le \infty$. Prove that there exists a sequence of *continuous* functions $\phi_n : \Omega \to \mathbb{R}$ converging to function f in the L^p norm.
Hint: Use results of Exercises 4.9.1 and 4.9.2.

4.9.4. Extend the result of Exercise 4.9.3 to sets Ω of infinite measure.
Hint: Distinguish between the cases $p < \infty$ and $p = \infty$. For $p < \infty$ and function $f : \Omega \to \mathbb{R}$ consider its *truncations*

$$f_n(\boldsymbol{x}) = \begin{cases} f(\boldsymbol{x}), & \boldsymbol{x} \in \Omega \cap B_n \\ 0, & \text{otherwise} \end{cases}$$

where $B_n = B(\mathbf{0}, n)$ are balls of radii n, centered at $\mathbf{0}, n = 1, 2, \ldots$. Prove that sequence f_n converges to function f in the L^p norm and use the result of Exercise 4.9.3 for functions f_n.

4.9.5. Argue that, in the results of Exercises 4.9.3 and 4.9.4, one can assume that functions f_n have compact support.

4.9.6. Let $\mathcal{F}$ be a uniformly bounded class of functions in Chebyshev space $C[a, b]$, i.e.,

$$\exists M > 0 : |f(x)| \leq M \quad \forall x \in [a, b], \quad \forall f \in \mathcal{F}$$

Let $\mathcal{G}$ be the corresponding class of primitive functions

$$F(x) = \int_a^x f(s)\, ds, \quad f \in \mathcal{F}$$

Show that $\mathcal{G}$ is precompact.

4.10 Contraction Mappings and Fixed Points

The ideas of a contraction mapping and of the fixed point of a function are fundamental to many questions in applied mathematics. We shall outline briefly in this section the essential ideas.

Fixed Points of Mappings. Let $F: X \to X$. A point $x \in X$ is called a fixed point of F if

$$x = F(x)$$

Contraction Mapping. Let (X, d) be a metric space and F a mapping of X into itself. The function F is said to be a contraction or a contraction mapping if there is a real number $k, 0 \leq k < 1$, such that

$$d(F(x), F(y)) \leq k d(x, y), \text{ for every } x, y \in X$$

Obviously, every contraction mapping F is uniformly continuous. Indeed, F is Lipschitz continuous with a Lipschitz constant k. The constant k is called the contraction constant for F.

We now arrive at an important theorem known as the principle of contraction mappings or Banach contraction mapping theorem.

THEOREM 4.10.1
(Banach)

Let (X, d) be a complete metric space and $F: X \to X$ be a contraction mapping. Then F has a unique fixed point.

PROOF First, we show that if F has a fixed point, it is unique. Suppose there are two: $x = F(x)$ and $y = F(y); x \neq y$. Since F is a contraction mapping,

$$d(x,y) = d(F(x), F(y)) \le k d(x,y) < d(x,y)$$

which is impossible. Hence, F has at most one fixed point.

To prove the existence we shall use the method of successive approximations. Pick an arbitrary starting point $x_0 \in X$ and define

$$x_1 = F(x_0), x_2 = F(x_1), \ldots, x_n = F(x_{n-1})$$

Since F is contractive, we have

$$\begin{aligned} d(x_2, x_1) &\le k d(x_1, x_0) \\ d(x_3, x_2) &\le k d(x_2, x_1) \le k^2 d(x_1, x_0) \\ &\vdots \\ d(x_{n+1}, x_n) &\le k^n d(x_1, x_0) \end{aligned}$$

and, consequently,

$$\begin{aligned} d(x_{n+p}, x_n) &\le d(x_{n+p}, x_{n+p-1}) + \ldots + d(x_{n+1}, x_n) \\ &\le (k^{p-1} + \ldots + k + 1) k^n d(x_1, x_0) \\ &\le \frac{k^n}{1-k} d(x_1, x_0) \end{aligned}$$

which in turn implies that (x_n) is Cauchy. Since X is complete, there exists a limit $x = \lim_{n\to\infty} x_n$. But F is continuous and therefore passing to the limit in

$$x_{n+1} = F(x_n)$$

we get that

$$x = F(x)$$

This completes the proof of the theorem. ∎

We derive from the proof of Banach contraction mapping theorem an estimate of the error by choosing an arbitrary starting point x_0 and passing to the limit with x_{n+p} in the estimate

$$d(x_{n+p}, x_n) \le \frac{k^n}{1-k} d(x_1, x_0)$$

Defining the error as

$$e_n = \text{ error } = d(x_n, x)$$

we have

$$e_n \leq \frac{k^n}{1-k} d(x_0, F(x_0))$$

Example 4.10.1

Let $F\colon \mathbb{R} \to \mathbb{R}, F(x) = x + 1$
Since

$$|F(x) - F(y)| = |x - y| = 1\ |x - y|$$

then $k = 1$ and F is not a contraction. We observe that $F(x)$ has no fixed point; this fact does not follow from Theorem 4.10.1, however, which gives only sufficient conditions for the existence of fixed points. In other words, there are many examples of operators with fixed points that are not contraction mappings. For example,

$$F(x) = 2x + 1$$

is not a contraction mapping, but it has a unique fixed point, $x = -1$. $\square$

Example 4.10.2

Now suppose $X = (0, \frac{1}{4}], F\colon X \to X, F(x) = x^2$. Thus

$$\begin{aligned} |F(x) - F(y)| = |x^2 - y^2| &\leq (|x| + |y|)\ |x - y| \\ &\leq \tfrac{1}{2}\ |x - y| \end{aligned}$$

Hence, $k = \frac{1}{2}$ and F is a contraction. But F has no fixed points (in X!). This is not a contradiction of Theorem 4.10.1 because X is not complete. $\square$

Example 4.10.3

Let $F\colon [a, b] \to [a, b], F$ differentiable at every $x \in (a, b)$ and $|F'(x)| \leq k < 1$. Then, by the mean-value theorem, if $x, y \in [a, b]$, there is a point ξ between x and y, such that

$$F(x) - F(y) = F'(\xi)(x - y)$$

Then

$$|F(x) - F(y)| \;=\; |F'(\xi)|\,|x-y| \leq\; k|x-y|$$

Hence F is a contraction mapping. ▯

Example 4.10.4

Let $F\colon [a,b] \to \mathbb{R}$. Assume that there exist constants μ and γ such that $\mu < \frac{1}{\gamma}$ and $0 < \mu \leq F'(x) \leq \frac{1}{\gamma}$ and assume that $F(a) < 0 < F(b)$ (i.e., we have only one zero between a and b). How do we find the zero of $F(x)$? That is, how can we solve

$$F(x) = 0 \text{ in } [a,b]$$

To solve this problem, we transform it into a different problem: Consider a new function

$$\widehat{F}(x) = x - \gamma F(x)$$

Clearly, the fixed points of $\widehat{F}(x)$ are the zeros of $F(x)$. Observe that

$$\widehat{F}(a) = a - \gamma F(a) > a$$

$$\widehat{F}(b) = b - F(b) < b$$

Also,

$$\widehat{F}'(x) = 1 - \gamma F'(x) \geq 0, \qquad \widehat{F}'(x) \leq 1 - \mu\gamma < 1$$

Hence $\widehat{F}$ transforms $[a,b]$ into itself and $|\widehat{F}'(x)| \leq 1 - \mu\gamma < 1$, for every $x \in [a,b]$. In view of our results in the previous example, $\widehat{F}(x)$ is a contraction mapping. ▯

Example 4.10.5
(Kepler's Equations)

In orbital mechanics, we encounter the equation

$$\xi = \eta - e\sin\eta$$

where e is the eccentricity of an orbit of some satellite and η is the central angle from perigee (if P = period, t = time for perigee then $\xi = 2\pi t/P$). We wish to solve for η for a given ξ, $\xi < 2\pi$. Toward this end, define

$$F(\eta) = \eta - e\sin\eta - \xi$$

We must now solve for the zeros of the function $F(\eta)$. Suppose that $0 \leq \eta \leq 2\pi$; note that $F(0) = -\xi < 0, F(2\pi) = 2\pi - \xi > 0$. Moreover, $1 - e \leq F'(\eta) \leq 1 + e$, since $F'(\eta) = 1 - e\cos\eta$. Thus, using the results of the previous example, set $\mu = 1 - e, \gamma = 1/(1+e)$, and

$$\widehat{F}(\eta) = \eta - \frac{1}{1+e}F(\eta) = \eta - \frac{1}{1+e}(\eta - e\sin\eta - \xi)$$

or

$$\widehat{F}(\eta) = \frac{e\eta + (\xi + e\sin\eta)}{1+e}$$

Hence

$$k = 1 - \frac{1-e}{1+e} = \frac{2e}{1+e}$$

We can solve this problem by successive approximations when $e < 1$. □

Example 4.10.6
(Fredholm Integral Equation)

Consider the integral equation

$$f(x) = \varphi(x) + \lambda \int_a^b K(x,y)f(y)dy$$

wherein

$$K(x,y) \text{ is continuous on } [a,b] \times [a,b]$$

$$\varphi(x) \text{ is continuous on } [a,b]$$

Then, according to the Wierstrass theorem, there is a constant M such that $|K(x,y)| < M$ for every $x, y \in [a,b]$.

Consider now the Chebyshev space $C([a,b])$ and the mapping θ from $C([a,b])$ into itself, $\theta(g) = h$, defined by

$$h(x) = \varphi(x) + \lambda \int_a^b K(x,y)g(y)dy$$

A solution of the integral equation is a fixed point of θ.

We have

$$
\begin{aligned}
d(\theta(f), \theta(g)) &= \sup_{x \in [a,b]} |\theta(f(x)) - \theta(g(x))| \\
&= \sup_{x \in [a,b]} \left| \lambda \int_a^b K(x,y) f(y) dy - \lambda \int_a^b K(x,y) g(y) dy \right| \\
&= \sup_{x \in [a,b]} \left| \lambda \int_a^b K(x,y)(f(y) - g(y)) dy \right| \\
&\leq |\lambda| \, M \left| \int_a^b (f(y) - g(y)) dy \right| \\
&\leq |\lambda| \, M(b-a) \sup_{y \in [a,b]} |f(y) - g(y)| \\
&\leq |\lambda| \, M(b-a) d(f,g)
\end{aligned}
$$

Thus the method of successive approximations will produce a (*the*) solution to the Fredholm integral equation if there exists a $k < 1$ such that $|\lambda| \leq k/M(b-a)$. □

Example 4.10.7
(A Dynamical System—Local Existence and Uniqueness of Trajectories)

An important and classical example of an application of the contraction mapping principle concerns the study of the local existence and uniqueness of trajectories $q(t) \in C^1(0,t)$ that are solutions of nonlinear ordinary differential equations of the form

$$
\frac{dq(t)}{dt} = F(t, q(t)), \quad 0 < t \leq T, \qquad q(0) = q_0
$$

Here $F(t,q)$ is a function continuous in first argument and uniformly (with respect to t) Lipschitz continuous with respect to q: there exists an $M > 0$ such that

$$
|F(t,q_1) - F(t,q_2)| \leq M|q_1 - q_2| \quad \text{for every } t \in [0,T]
$$

As a continuous function on compact set $[0,T] \times [0,Q]$, $|F(t,q)|$ attains its maximum and, therefore, is bounded. Assume that

$$
|F(t,q)| < k \text{ for } t \in [0,T], q \in [0,Q]
$$

Now, we select t_0 (the time interval) so that $t_0 M < 1, t_0 \le T$ and consider the set C in the space $C([0,T])$ of continuous functions on $[0, t_0]$:

$$C = \{q : [0,T] \to I\!\!R \,:\, |q(t) - q_0| \le kt_0 \quad \text{for } 0 \le t \le t_0\}$$

As a closed subset of the complete space $C([0,T]), C$ is complete.

We transform the given problem into the form of a fixed-point problem by setting

$$q(t) = q_0 + \int_0^t F(s, q(s))ds$$

Then, if we set

$$q_0(t) = q_0, \quad q_{n+1}(t) = q_0 + \int_0^t F(s, q_n(s))ds$$

we may obtain a sequence of approximations to the original problem if the integral operator indicated is a contraction mapping. This method for solving nonlinear differential equations is known as *Picard's method.*

We shall show now that we, in fact, have a contraction mapping. Let $q(t) \in C$. Then denoting

$$\psi(t) = q_0 + \int_0^t F(s, q(s))ds$$

we have

$$|\psi(t) - q_0| = \left| \int_0^t F(t, q(t))dt \right| \le kt_0$$

Thus the considered mapping maps C into itself.

Moreover,

$$\begin{aligned} |\psi_1(t) - \psi_2(t)| &\le \int_o^t |F(s, q_1(s)) - F(s, q_2(s))| \, ds \\ &\le Mt_0 d_\infty(q_1, q_2) \end{aligned}$$

Since $Mt_0 < 1$, the mapping is a contraction mapping. Hence the nonlinear equation has one and only one solution on the interval $[0, t_0]$. $\square$

Exercises

4.10.1. Reformulate Example 4.10.6 concerning the Fredholm Integral Equation using the L^p spaces. What would be the natural regularity assumption on kernel function $K(x, y)$. Does it have to be bounded?

4.10.2. Using the Pickard method described in Example 4.10.7, construct a sequence of approximations of the nonlinear equation

$$\begin{cases} \frac{dq}{dt} & = q^2 + 2, \quad 0 < t < T \\ q(0) = 0 \end{cases}$$

Determine T for which a unique solution is guaranteed and evaluate explicitly the first three terms in the sequence.

4.10.3. Show that $f(x) = \frac{1}{2}(x + \frac{3}{2})$ is a contraction mapping with the fixed point $x = 3/2$. If $x_0 = 2$ is the starting point of a series of successive approximations, show that the error after m iterations is $\leq (0.5)^{m+1}$.

4.10.4. Use the idea of contraction mappings and fixed points to compute an approximate value of $\sqrt[3]{5}$.

5

Banach Spaces

Topological Vector Spaces

5.1 Topological Vector Spaces—An Introduction

The most important mathematical systems encountered in applications of mathematics are neither purely topological (i.e., without algebraic structure, such as metric spaces) nor purely algebraic (without topological structure, such as vector spaces); rather, they involve some sort of natural combinations of both. In this chapter, we study such systems, beginning with the concept of a topological vector space and quickly passing on to normed vector spaces.

Topological Vector Space. V is called a *topological vector space* (t.v.s.) iff

(i) V is a vector space (real or complex),

(ii) the underlying set of vectors, also denoted V, is endowed with a topology so that the resulting topological space is a Hausdorff topological space, also denoted V, and

(iii) vector addition

$$V \times V \ni (\boldsymbol{u}, \boldsymbol{v}) \to \boldsymbol{u} + \boldsymbol{v} \in V$$

and multiplication by a scalar

$$\mathbb{R}(\text{or } \mathbb{C}) \times V \ni (\alpha, \boldsymbol{u}) \to \alpha \boldsymbol{u} \in V$$

are continuous operations.

Example 5.1.1

Every normed vector space is a t.v.s. As normed vector spaces are metric spaces, it is sufficient to prove that both operations of vector addition and scalar multiplication are *sequentially continuous.*

Let $\boldsymbol{u}_n \to \boldsymbol{u}$ and $\boldsymbol{v}_n \to \boldsymbol{v}$. It follows from the triangle inequality that

$$\|(\boldsymbol{u}_n + \boldsymbol{v}_n) - (\boldsymbol{u} + \boldsymbol{v})\| \leq \|\boldsymbol{u}_n - \boldsymbol{u}\| + \|\boldsymbol{v}_n - \boldsymbol{v}\|$$

and, consequently, $\boldsymbol{u}_n + \boldsymbol{v}_n \to \boldsymbol{u} + \boldsymbol{v}$, which proves that vector addition is continuous.

Similarly, if $\alpha_n \to \alpha$ and $\boldsymbol{u}_n \to \boldsymbol{u}$ then

$$\|\alpha_n \boldsymbol{u}_n - \alpha \boldsymbol{u}\| = \|\alpha_n \boldsymbol{u}_n - \alpha \boldsymbol{u}_n + \alpha \boldsymbol{u}_n - \alpha \boldsymbol{u}\|$$

$$\leq |\alpha_n - \alpha| \|\boldsymbol{u}_n\| + |\alpha| \|\boldsymbol{u}_n - \boldsymbol{u}\|$$

Since $\|\boldsymbol{u}_n\|$ is bounded (explain why), the right-hand side converges to zero, which proves that $\alpha_n \boldsymbol{u}_n \to \alpha \boldsymbol{u}$. $\square$

In a topological vector space *translations*

$$T_{\boldsymbol{u}} : V \ni \boldsymbol{v} \to T_{\boldsymbol{u}}(\boldsymbol{v}) \stackrel{\text{def}}{=} \boldsymbol{u} + \boldsymbol{v} \in V$$

are homeomorphisms. Indeed:

1. $T_{\boldsymbol{u}}$ is a bijection. Its inverse is equal to $T_{-\boldsymbol{u}}$

$$(T_{\boldsymbol{u}})^{-1} = T_{-\boldsymbol{u}}$$

2. Both $T_{\boldsymbol{u}}$ and $T_{-\boldsymbol{u}}$ are continuous, since the vector addition is continuous.

Similarly, the transformations

$$T_\alpha : V \ni \boldsymbol{v} \to T_\alpha(\boldsymbol{v}) \stackrel{\text{def}}{=} \alpha \boldsymbol{v} \in V \ , \ \alpha \neq 0$$

are homeomorphisms, too.

This leads to an observation that if $\mathcal{B}_{\boldsymbol{0}}$ denotes a base of neighborhoods of zero vector and $\mathcal{B}_{\boldsymbol{u}}$ is a base of an arbitrary vector $\boldsymbol{u}$ then

$$\mathcal{B}_{\boldsymbol{u}} \sim \boldsymbol{u} + \mathcal{B}_{\boldsymbol{0}}$$

where

$$\boldsymbol{u} + \mathcal{B}_{\boldsymbol{O}} \stackrel{\text{def}}{=} \{\boldsymbol{u} + \mathcal{B} : B \in \mathcal{B}_o\}$$

Similarly,

$$\mathcal{B}_{\boldsymbol{O}} \sim \alpha \mathcal{B}_{\boldsymbol{O}}$$

where

$$\alpha \mathcal{B}_{\boldsymbol{O}} \stackrel{\text{def}}{=} \{\alpha B : B \in \mathcal{B}_{\boldsymbol{O}}\} \ , \ \alpha \neq 0$$

The practical conclusion from these observations is that when constructing a topological vector space, one can start by introducing a base of neighborhoods for the zero vector (which must be invariant under multiplication by scalars according to the second of the equivalence relations). One next defines neighborhoods for arbitrary vectors by "shifting" the base for the zero vector and, finally, verifying that the topological vector space axioms hold.

This is precisely the way in which we construct the important case of *locally convex topological vector spaces* discussed in the next section.

Exercises

5.1.1. Let V be a t.v.s. and let $\mathcal{B}_{\boldsymbol{O}}$ denote a base of neighborhoods for the zero vector. Show that $\mathcal{B}_{\boldsymbol{O}}$ is equivalent to $\alpha\mathcal{B}_o$ for $\alpha \neq 0$.

5.2 Locally Convex Topological Vector Spaces

Seminorm. Let V be a vector space. Recall that a function $p: V \to [0, \infty)$ is called a *seminorm* iff

(i) $p(\alpha \boldsymbol{u}) = |\alpha| p(\boldsymbol{u})$ (homogeneity) and

(ii) $p(\boldsymbol{u} + \boldsymbol{v}) \leq p(\boldsymbol{u}) + p(\boldsymbol{v})$ (triangle inequality)

for every scalar α and vectors $\boldsymbol{u}, \boldsymbol{v}$. Obviously, every norm is a seminorm, but not conversely.

Example 5.2.1

Let $V = \mathbb{R}^2$. Define

$$p(\boldsymbol{x}) = p\left((x_1, x_2)\right) = |x_1|$$

Then p is a seminorm, but not a norm since $p(\boldsymbol{x}) = 0$ implies that only the first component of $\boldsymbol{x}$ is zero. □

The assumption that seminorms p are nonnegative is not necessary as it follows from the following proposition.

PROPOSITION 5.2.1

Let V be a vector space and p any real-valued function defined on V such that p satisfies the two conditions for a seminorm. Then

(i) $p(\mathbf{0}) = 0$ *and*

(ii) $|p(\boldsymbol{u}) - p(\boldsymbol{v})| \leq p(\boldsymbol{u} - \boldsymbol{v})$.

In particular, taking $\boldsymbol{v} = \mathbf{0}$ in the second inequality one gets $p(\boldsymbol{u}) \geq |p(\boldsymbol{u})|$, which proves that p must take on only nonnegative values.

PROOF (i) follows from the first property of seminorms by substituting $\alpha = 0$. Inequality (ii) is equivalent to

$$-p(\boldsymbol{u} - \boldsymbol{v}) \leq p(\boldsymbol{u}) - p(\boldsymbol{v}) \leq p(\boldsymbol{u} - \boldsymbol{v})$$

or, equivalently,

$$p(\boldsymbol{v}) \leq p(\boldsymbol{u} - \boldsymbol{v}) + p(\boldsymbol{u}) \text{ and } p(\boldsymbol{u}) \leq p(\boldsymbol{v}) + p(\boldsymbol{u} - \boldsymbol{v})$$

Both inequalities follow directly from the triangle inequality and homogeneity of seminorms. ∎

Recall that by a ball centered at zero with radius c and corresponding to a particular norm $\|\cdot\|$, one means a collection of all vectors bounded in the norm by c. The following proposition investigates properties of more general sets of this type, using seminorms rather than norms.

PROPOSITION 5.2.2

Let V be a vector space and p a seminorm defined on V. Define

$$M_c \stackrel{\text{def}}{=} \{\boldsymbol{v} \in V : p(\boldsymbol{v}) \leq c\} \quad c > 0$$

The following properties hold:

(i) $\mathbf{0} \in M_c$

(ii) M_c *is convex, i.e.,*

$$\boldsymbol{u}, \boldsymbol{v} \in M_c \Longrightarrow \alpha\boldsymbol{u} + (1-\alpha)\boldsymbol{v} \in M_c \text{ for every } \; 0 \leq \alpha \leq 1$$

(iii) M_c *is "balanced"*

$$\boldsymbol{u} \in M_c, \; |\alpha| \leq 1 \Longrightarrow \alpha\boldsymbol{u} \in M_c$$

(iv) M_c *is "absorbing"*

$$\forall\, \boldsymbol{u} \in V \exists \alpha > 0 : \alpha^{-1}\boldsymbol{u} \in M_c$$

(v) $p(\boldsymbol{u}) = \inf\{\alpha c : \alpha > 0, \; \alpha^{-1}\boldsymbol{u} \in M_c\}$

PROOF (i) follows from Proposition 5.2.1 (i). Next,

$$p(\alpha\boldsymbol{u} + (1-\alpha)\boldsymbol{v}) \leq \alpha p(\boldsymbol{u}) + (1-\alpha)p(\boldsymbol{v}) \leq \alpha c + (1-\alpha)c = c$$

which proves convexity of M_c. Property (iii) is a direct consequence of homogeneity of seminorms. To prove (iv), it is sufficient to take $\alpha = p(\boldsymbol{u})/c$ as

$$p(\alpha^{-1}\boldsymbol{u}) = \alpha^{-1}p(\boldsymbol{u}) = c$$

Finally, $\alpha^{-1}\boldsymbol{u} \in M_c$ implies that

$$p(\alpha^{-1}\boldsymbol{u}) = \alpha^{-1}p(\boldsymbol{u}) \leq c \Longrightarrow p(\boldsymbol{u}) \leq \alpha c$$

and the infimum on the right-hand side of (v) is attained for $\alpha = p(\boldsymbol{u})/c$. ∎

Locally Convex Topological Vector Space (Bourbaki). Let V be a vector space and p_i, $i \in I$, a family (not necessarily countable) of seminorms satisfying the following *axiom of separation:*

$$\forall\, \boldsymbol{u} \neq \mathbf{0} \; \exists \kappa \in I : p_\kappa(\boldsymbol{u}) \neq 0$$

We begin by constructing a base of neighborhoods for the zero vector. Consider the family $\mathcal{B} = \mathcal{B}_{\boldsymbol{0}}$ of all sets B of the form

$$B = B(I_o, \varepsilon) \stackrel{\text{def}}{=} \{\boldsymbol{u} \in V \, : \, p_i(\boldsymbol{u}) \le \varepsilon, \; i \in I_o\}$$

where I_o denotes any *finite* subset of I.

The following properties of sets B are easily observed:

(i)

$$B(I_o, \varepsilon) = \bigcap_{i \in I_o} M_\varepsilon^i$$

where

$$M_\varepsilon^i = \{\boldsymbol{v} \in V \, : \, p_i(\boldsymbol{v}) \le \varepsilon\}$$

(ii) $B(I_o, \varepsilon)$ are convex, balanced and absorbing.

Since sets B are nonempty (why?) and

$$B(I_1, \varepsilon_1) \cap B(I_2, \varepsilon_2) \supset B\left(I_1 \cup I_2, \min(\varepsilon_1, \varepsilon_2)\right)$$

it follows that $\mathcal{B}$ is a base. Since each of the sets contains the zero vector, the family can be considered as a base of neighborhoods for the zero vector.

Following the observations from the previous section, we proceed by defining the base of neighborhoods for an arbitrary vector $\boldsymbol{u} \neq \boldsymbol{O}$ in the form

$$\mathcal{B}_{\boldsymbol{u}} \stackrel{\text{def}}{=} \boldsymbol{u} + \mathcal{B} = \{\boldsymbol{u} + B \, : \, B \in \mathcal{B}\}$$

Vector space V with topology induced by bases $\mathcal{B}_{\boldsymbol{u}}$ is identified as a *locally convex topological vector space*. To justify the name it remains to show that the topology is Hausdorff and that the operations in V are continuous.

The first property follows from the axiom of separation. Let $\boldsymbol{u} \neq \boldsymbol{v}$ be two arbitrary vectors. There exists a seminorm p_κ such that

$$p_\kappa(\boldsymbol{v} - \boldsymbol{u}) > 0$$

Take $2\varepsilon < p_\kappa(\boldsymbol{v} - \boldsymbol{u})$ and consider neighborhoods of $\boldsymbol{u}$ and $\boldsymbol{v}$ in the form

$$\boldsymbol{u} + M_\varepsilon^\kappa, \; \boldsymbol{v} + M_\varepsilon^\kappa, \; M_\varepsilon^\kappa = \{\boldsymbol{w} \, : \, p_\kappa(\boldsymbol{w}) \le \varepsilon\}$$

If there were a common element $\boldsymbol{w}$ of both sets then

$$p_\kappa(\boldsymbol{v} - \boldsymbol{u}) \le p_\kappa(\boldsymbol{v} - \boldsymbol{w}) + p_\kappa(\boldsymbol{w} - \boldsymbol{u}) < 2\varepsilon$$

a contradiction.

In order to show that vector addition is continuous we pick two arbitrary vectors $\boldsymbol{u}$ and $\boldsymbol{v}$ and consider a neighborhood of $\boldsymbol{u}+\boldsymbol{v}$ in the form

$$\boldsymbol{u}+\boldsymbol{v}+B(I_o,\varepsilon)$$

We claim that for each $\boldsymbol{u}_1 \in \boldsymbol{u}+B(I_o,\frac{\varepsilon}{2})$ and $\boldsymbol{v}_1 \in \boldsymbol{v}+B(I_o,\frac{\varepsilon}{2})$, $\boldsymbol{u}_1+\boldsymbol{v}_1$ is an element of the neighborhood, which shows that vector addition is continuous. This follows easily from the triangle inequality

$$p_i\left(\boldsymbol{u}_1+\boldsymbol{v}_1-(\boldsymbol{u}+\boldsymbol{v})\right) \le p_i(\boldsymbol{u}_1-\boldsymbol{u})+p_i(\boldsymbol{v}_1-\boldsymbol{v}) \le \varepsilon$$

for every $i \in I_o$.

Similarly, taking neighborhood of $\alpha\boldsymbol{u}$ in the form

$$\alpha\boldsymbol{u}+B(I_o,\varepsilon)$$

for each $\alpha_1 \in (\alpha-\beta,\alpha+\beta)$ and $\boldsymbol{u}_1 \in \boldsymbol{u}+B(I_0,\delta)$ where we select δ and β such that

$$\beta p_i(\boldsymbol{u}) \le \frac{\varepsilon}{2}\ \forall\, i \in I_o \quad \text{and} \quad \max(|\alpha-\beta|,|\alpha+\beta|)\delta \le \frac{\varepsilon}{2}$$

we have

$$\begin{aligned} p_i\left(\alpha_1\boldsymbol{u}_1-\alpha\boldsymbol{u}\right) &\le p_i\left(\alpha_1\left(\boldsymbol{u}_1-\boldsymbol{u}\right)+\left(\alpha_1-\alpha\right)\boldsymbol{u}\right) \\ &\le |\alpha_1| p_i\left(\boldsymbol{u}_1-\boldsymbol{u}\right)+|\alpha_1-\alpha| p(\boldsymbol{u}) \\ &\le \frac{\varepsilon}{2}+\frac{\varepsilon}{2}=\varepsilon \end{aligned}$$

for every $i \in I_o$, which proves that multiplication by a scalar is continuous.

REMARK 5.2.1. In a nontrivial case the family of seminorms inducing the locally convex topology must be infinite. If I were finite then we could introduce a single function

$$p(\boldsymbol{u}) = \max_{i\in I} p_i(\boldsymbol{u})$$

which, by the axiom of separation, would have been a norm with a corresponding topology identical to the locally convex topology. Thus, only in the case of infinite families of seminorms do locally convex topological vector spaces provide us with a nontrivial generalization of normed vector spaces. ∎

Example 5.2.2

Recall the definition of the topology of pointwise convergence discussed in Example 4.1.8. Identifying with each point $x \in (0,1)$ a corresponding seminorm

$$p_x(f) \stackrel{\text{def}}{=} |f(x)|$$

we easily see that the family of such seminorms $p_x, x \in (0,1)$ satisfies the axiom of separation. The corresponding topology is *exactly* the previously discussed topology of pointwise convergence in $C(0,1)$. □

According to Proposition 5.2.2, for every seminorm p and a constant $c > 0$, we can construct the corresponding set $M_c = M_c(p)$ consisting of all vectors bounded in p by c and proved to be convex, balanced, and absorbing. In property (v) from the same proposition we also have established a direct representation of the seminorm p in terms of the set M_c. It turns out that once we have a convex, balanced, and absorbing set, the set defines a seminorm.

Minkowski's Functional. Let M be a convex, balanced, and absorbing set in a vector space V. We define the *Minkowski functional of* M as

$$p_M(\boldsymbol{u}) \stackrel{\text{def}}{=} \inf\{\alpha > 0 : \alpha^{-1}\boldsymbol{u} \in M\}$$

PROPOSITION 5.2.3

The Minkowski functional p_M *is a seminorm. Moreover,*

$$\{\boldsymbol{u} : p_M(\boldsymbol{u}) < 1\} \subset M \subset \{\boldsymbol{u} : p_M(\boldsymbol{u}) \leq 1\}$$

PROOF

Step 1. M absorbing implies that set

$$\{\alpha > 0 : \alpha^{-1}\boldsymbol{u} \in M\}$$

is nonempty and therefore p_M is well-defined (takes on real values).

Step 2. p_M is homogeneous.

$$p_M(\lambda\boldsymbol{u}) = \inf\left\{\alpha > 0 : \alpha^{-1}\lambda\boldsymbol{u} \in M\right\}$$

$$= \inf \{\alpha > 0 : \alpha^{-1}|\lambda|\boldsymbol{u} \in M\} \quad (M \text{ is balanced})$$

$$= \inf \{\beta|\lambda| : \beta > 0, \ \beta^{-1}\boldsymbol{u} \in M\}$$

$$= |\lambda| \inf \{\beta > 0 : \beta^{-1}\boldsymbol{u} \in M\}$$

$$= |\lambda| p_M(\boldsymbol{u})$$

Step 3. p_M satisfies the triangle inequality.

Let $\alpha, \beta > 0$ denote arbitrary positive numbers such that $\alpha^{-1}\boldsymbol{u} \in M$, $\beta^{-1}\boldsymbol{v} \in M$. By convexity of M,

$$\frac{\alpha}{\alpha+\beta}\alpha^{-1}\boldsymbol{u} + \frac{\beta}{\alpha+\beta}\beta^{-1}\boldsymbol{v} = (\alpha+\beta)^{-1}(\boldsymbol{u}+\boldsymbol{v})$$

is also an element of M and, consequently,

$$p_M(\boldsymbol{u}+\boldsymbol{v}) = \inf \{\gamma > 0 : \gamma^{-1}(\boldsymbol{u}+\boldsymbol{v}) \in M\}$$

$$\leq \alpha + \beta \ , \ \alpha^{-1}\boldsymbol{u} \in M, \ \beta^{-1}\boldsymbol{v} \in M$$

It remains to take the infimum with respect to α and β on the right-hand side of the inequality.

Finally, the relation between set M and sets of vectors bounded in p_M by one follows directly from the definition of p_M. ∎

Thus, by means of the Minkowski functional, one can establish a one-to-one correspondence between seminorms and convex, balanced, and absorbing sets. We summarize now these observations in the following proposition.

PROPOSITION 5.2.4

Let V be a topological vector space. The following conditions are equivalent to each other:

(i) V is a locally convex space topologized through a family of seminorms satisfying the axiom of separation.

(ii) There exists a base of neighborhoods for the zero vector consisting of convex, balanced, and absorbing sets.

PROOF (i) $\Longrightarrow$ (ii) follows from the construction of the locally convex topology and Proposition 5.2.2. Conversely, with every set M from the base we can associate the corresponding Minkowski functional, which, by Proposition 5.2.3, is a seminorm. Since each of the sets is absorbing, the family of seminorms trivially satisfies the axiom of separation. Finally, by the relation from Proposition 5.2.3 between sets M and sets of vectors bounded in p_M, the topology induced by seminorms p_M is identical to the original topology (the bases are equivalent). ∎

We shall use the just-established equivalence to discuss a very important example of a locally convex topological vector space, the space of test functions, in the next section.

Exercises

5.2.1. Show that each of the seminorms inducing a locally convex topology is *continuous* with respect to this topology.

5.2.2. Show that replacing the weak equality in the definition of set M_c with a strict one does not change the properties of M_c.

5.2.3. Show that by replacing the weak equality in the definition of sets $B(I_o, \varepsilon)$ with a strict one, one obtains bases of neighborhoods *equivalent* to the original ones and therefore the *same* topology.

5.2.4. Show that seminorms are convex functionals.

5.2.5. Prove the following characterization of continuous linear functionals.

Proposition

Let V be a locally convex t.v.s. A linear functional f on V is continuous iff there exists a continuous seminorm $p(\boldsymbol{u})$ on V (not necessarily one of the family inducing the topology) such that

$$|f(\boldsymbol{v})| \leq p(\boldsymbol{v})$$

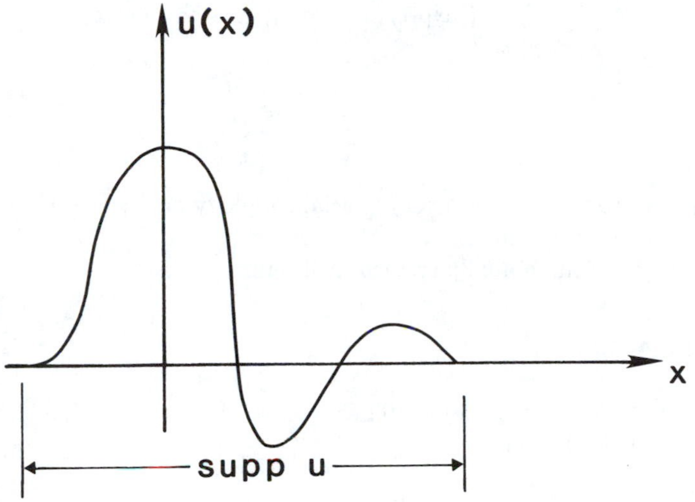

Figure 5.1
Support of a function.

5.3 Space of Test Functions

Functions with Compact Support. Let $\Omega \subset \mathbb{R}^n$ be an open set and f any real- (or complex-) valued function defined on Ω. The closure of the set of all points $\boldsymbol{x} \in \Omega$ for which f takes nonzero values is called the *support* of f:

$$\text{supp } f \overset{\text{def}}{=} \overline{\{x \in \Omega : f(x) \neq 0\}}$$

Note that, due to the closure operation, the support of a function f may include the points at which f vanishes (see Fig. 5.1)

The collection of all infinitely differentiable functions defined on Ω, whose supports are compact (i.e., bounded) and contained in Ω will be denoted as

$$C_0^\infty(\Omega) \overset{\text{def}}{=} \{f \in C^\infty(\Omega) : \text{supp } f \subset \Omega, \quad \text{supp } f \text{ compact}\}$$

Obviously, $C_0^\infty(\Omega)$ is a vector subspace of $C^\infty(\Omega)$.

Example 5.3.1

A standard example of a function in $C_0^\infty(\mathbb{R})$ is

$$\phi(x) = \begin{cases} \exp[1/(x^2 - a^2)], \ |x| < a \ (a \in \mathbb{R}) \\ 0, \qquad\qquad\qquad\qquad |x| \geq a \end{cases}$$

□

We shall construct now a very special topology on $C_0^\infty(\Omega)$, turning it into a topological vector space.

We begin with an auxiliary technical lemma.

LEMMA 5.3.1

Let $\Omega \subset \mathbb{R}^n$ be an open set. There exists always a sequence of compact sets $K_i \subset \Omega$ such that

(i) $K_i \subset \text{int } K_{i+1}$ and

(ii) $\displaystyle\bigcup_1^\infty K_i = \Omega.$

PROOF Consider the set of all closed balls with rational coordinates of their centers and rational radii, contained in Ω. The set is countable (why?) and therefore can be put into a sequential form

$$\overline{B}_1, \overline{B}_2, \overline{B}_3, \ldots$$

Also, by the definition of open sets,

$$\bigcup_1^\infty \overline{B}_i = \Omega \quad \text{and} \quad \bigcup_1^\infty B_i = \Omega$$

where $B_i = \text{int } \overline{B}_i$ are the corresponding open balls. Next, set

$$\begin{aligned} K_1 &= \overline{B}_1 \\ K_2 &= \overline{B}_1 \cup \overline{B}_2 \\ &\vdots \\ K_n &= \overline{B}_1 \cup \ldots \cup \overline{B}_n \end{aligned}$$

Each of sets K_i is compact; they form an increasing sequence ($K_1 \subset K_2 \subset \ldots$) and

$$\text{int } K_i \supset \bigcup_{j=1}^{i} \text{int } \overline{B}_j = \bigcup_{j=1}^{i} B_j$$

Consequently,

$$\bigcup_{i=1}^{\infty} \text{int } K_i \supset \bigcup_{i=1}^{\infty} \bigcup_{j=1}^{i} B_j = \bigcup_{j=1}^{\infty} B_j = \Omega$$

As *int* K_i are also increasing ($A \subset B \Longrightarrow$ *int* $A \subset$ *int* B), we have

$$\bigcup_{j=i+1}^{\infty} \text{int } K_i = \Omega \ , \quad \text{for every} \quad i$$

which proves that for each compact set K_i, sets int K_j, $j \geq i+1$ form an open covering of K_i. Thus one can always find a finite number of them covering K_i. Taking the largest one (the sequence int K_i is increasing) we see that for each K_i we can always select an index $j > i$ such that $K_i \subset$ int K_j, which, by the principle of mathematical induction, finishes the proof. ∎

The Space of Test Functions. Let $\Omega \subset \mathbb{R}^n$ be an open set. For each compact subset $K \subset$ int Ω we introduce the space of C^∞ functions with supports in K

$$C_0^\infty(K) \stackrel{\text{def}}{=} \{u \in C_0^\infty(\Omega) : \text{supp } u \subset K\}$$

Introducing a sequence of seminorms

$$p_n(u) = \sup\left\{|D^{\boldsymbol{\alpha}}u(\boldsymbol{x})| : \boldsymbol{x} \in K, \ |\boldsymbol{\alpha}| = n\right\}$$

we equip $C_0^\infty(K)$ with the corresponding locally convex topology. With this topology the space $C_0^\infty(K)$ is frequently called *the space of test functions with supports in* K and denoted $\mathcal{D}(K)$.

Let $\mathcal{B}_K$ denote a corresponding base of convex, balanced, and absorbing neighborhoods for the zero function $\mathbf{0}$.

Consider now the family $\mathcal{B}$ of all nonempty *convex, balanced* sets W from $C_0^\infty(\Omega)$ such that

$$\forall \ \text{compact} \ \ K \subset \Omega \ \exists \ V \in \mathcal{B}_K : V \subset W \cap C_0^\infty(K)$$

PROPOSITION 5.3.1

$\mathcal{B}$ is a base of a locally convex topology on $C_0^\infty(\Omega)$.

PROOF First of all, sets W from $\mathcal{B}$ are *absorbing*. Indeed, if $u \in C_0^\infty(\Omega)$, then by definition, taking $K = \text{supp}\, u$, we can find an absorbing set $V \in \mathcal{B}_K$ such that $V \subset W \cap C_0^\infty(K) \subset W$, which proves that W is absorbing.

Next, if $W_1, W_2 \in \mathcal{B}$, then simply $W = W_1 \cap W_2$ is also an element of $\mathcal{B}$. Indeed, W is convex and balanced and if $V_i \in \mathcal{B}_K$, $i = 1, 2$ denote sets such that

$$V_i \subset W_i \cap C_0^\infty(K) \ , \ i = 1, 2$$

then, since $\mathcal{B}_K$ is a base, there exists $V \in \mathcal{B}_K$ such that

$$V \subset V_1 \cap V_2 \subset W_1 \cap W_2 \cap C_0^\infty(K) = W \cap C_0^\infty(K)$$

which proves that $W \in \mathcal{B}$.

Finally, $\mathcal{B}$ is nonempty as it contains at least the entire space $C_0^\infty(\Omega)$. ∎

The space $C_0^\infty(\Omega)$ equipped with the just-defined topology is called *the space of test functions on* Ω and denoted $\mathcal{D}(\Omega)$.

REMARK 5.3.1. The condition defining base $\mathcal{B}$ can be written in a more concise form using notation from Chapter 4.

$$\mathcal{B}_K \succ \mathcal{B} \cap C_0^\infty(K) \ \text{ for every compact } \ K \subset \Omega$$

This is equivalent to saying that the inclusion

$$i_K : C_0^\infty(K) \hookrightarrow C_0^\infty(\Omega)$$

is continuous at zero. The topology of the space of test functions on Ω can be identified as the *strongest* topology in $C_0^\infty(\Omega)$ in which all such inclusions are continuous and is frequently called the *inductive topological limit* of topologies in $\mathcal{D}(K)$. ∎

The space of test functions is a basis for developing the theory of distributions, introduced by L. Schwartz in 1948. We conclude this section by presenting one of its most crucial properties characterizing convergence of sequences in $\mathcal{D}(\Omega)$.

PROPOSITION 5.3.2

Let φ_n be a sequence of functions from $\mathcal{D}(\Omega)$. The following conditions are equivalent to each other:

(i) $\varphi_n \to 0$ *in* $\mathcal{D}(\Omega)$.

(ii) *There exists a compact set* $K \subset \Omega$ *such that* supp $\varphi_n \subset K$ *and* $D^{\alpha}\varphi_n \to 0$ *uniformly in* K.

PROOF (ii) $\Longrightarrow$ (i) follows immediately from Remark 5.3.1 and the definition of topology in $\mathcal{D}(K)$. To prove the converse we need only demonstrate the existence of K. Let K_n, $n = 1, 2\ldots$ be the sequence of compact sets discussed in Lemma 5.3.1 and satisfying

$$K_i \subset \text{int } K_{i+1}\ , \quad \bigcup_i K_i \left(= \bigcup_i \text{int } K_i \right) = \Omega$$

We proceed by contradicting the existence of such a set K. Accordingly, we can find an index n_1 and a corresponding point $\boldsymbol{x}_1 \in K_{n_1}$ such that $\varphi_{n_1}(\boldsymbol{x}) \neq 0$. Similarly,

$$\exists n_2 > n_1, \boldsymbol{x}_2 \in K_{n_2} - K_{n_1} \ : \ \varphi_{n_2}(\boldsymbol{x}_2) \neq 0$$

and by induction

$$\exists n_i > n_{i-1},\ \boldsymbol{x}_i \in K_{n_i} - K_{n_{i-1}} \ : \ \varphi_{n_i}(\boldsymbol{x}_i) \neq 0$$

Finally, consider the set

$$W = \left\{ \varphi \in C_0^{\infty}(\Omega) : \varphi(\boldsymbol{x}_i) < i^{-1}\varphi_{n_i}(\boldsymbol{x}_i),\ i = 1, 2, \ldots \right\}$$

We claim that W is an element of $\mathcal{B}$, the base of neighborhoods for the zero vector in $\mathcal{D}(\Omega)$. Indeed, W is convex and balanced. Moreover, if K is a compact subset of Ω, there exists a set K_i from the sequence such that $K \subset K_i$ and, consequently, only a *finite* number of points $\boldsymbol{x}_i$ is in K. This implies that for every compact K we can always find $\delta > 0$ such that

$$\sup_{\boldsymbol{x} \in K} |\varphi(\boldsymbol{x})| < \delta \quad \text{implies} \quad \varphi \in W$$

Thus W is a well-defined neighborhood of the zero vector. But this contradicts the convergence of φ_n to zero vector as for $\varepsilon = 1$ and every i

$$\varphi_{n_i}(\boldsymbol{x}_i) \geq i^{-1}\varphi_{n_i}(\boldsymbol{x}_i) \Longrightarrow \varphi_{n_i} \notin W$$

∎

Hahn-Banach Extension Theorem

5.4 The Hahn-Banach Theorem

In this section we establish a fundamental result concerning the extension of linear functionals on infinite dimensional vector spaces, the famous Hahn-Banach theorem. The result will be obtained in a general setting of arbitrary vector spaces and later on specialized in a more specific context.

Sublinear Functionals. Let V be a real vector space. A functional $p : V \to \mathbb{R}$ is said to be *sublinear* iff

(i) $p(\alpha\boldsymbol{u}) = \alpha p(\boldsymbol{u}) \quad \forall \alpha > 0$ and

(ii) $p(\boldsymbol{u}+\boldsymbol{v}) \leq p(\boldsymbol{u}) + p(\boldsymbol{v}) \quad$ (p is subadditive)

for arbitrary vectors $\boldsymbol{u}$ and $\boldsymbol{v}$. Obviously, every linear functional is sublinear and every seminorm is sublinear as well.

THEOREM 5.4.1
(The Hahn-Banach Theorem)

Let

X be a real vector space

$p : X \to \mathbb{R}$ a sublinear functional on X

$M \subset X$ a subspace of X

$f : M \to \mathbb{R}$ a linear functional on M ($f \in M^*$)

dominated by p on M; i.e.,

$$f(\boldsymbol{x}) \leq p(\boldsymbol{x}) \qquad \forall\, \boldsymbol{x} \in M$$

Then, there exists such a linear functional $F : X \to \mathbb{R}$ defined on the whole X such that *(i)* $F|_M \equiv f$ *and*

(ii) $F(\boldsymbol{x}) \leq p(\boldsymbol{x}) \quad \forall\, \boldsymbol{x} \in X.$

In other words, F is an extension of f dominated by p on the whole X.

PROOF Let us pick an element $\boldsymbol{u}_o$ of X, not in M, and consider the subspace

$$M_1 = M + \mathbb{R}\boldsymbol{u}_o = \{\boldsymbol{x} = \boldsymbol{m} + \alpha\boldsymbol{u}_o : \boldsymbol{m} \in M,\ \alpha \in \mathbb{R}\}$$

The possible extension of f to M_1 must have the form

$$F(\boldsymbol{x}) = F(\boldsymbol{m} + \alpha\boldsymbol{u}_o) = F(\boldsymbol{m}) + \alpha F(\boldsymbol{u}_o) = f(\boldsymbol{m}) + \alpha c$$

where we have used linearity of F and $c \overset{\text{def}}{=} F(\boldsymbol{u}_o)$. We now determine if it is possible to choose c such that $F(\boldsymbol{x}) \leq p(\boldsymbol{x})$ on M_1. Then we have extended f to the space M_1 of dimension "one larger" than M.

We will have

$$F(\boldsymbol{m} + \alpha\boldsymbol{u}_o) = f(\boldsymbol{m}) + \alpha c \leq p(\boldsymbol{m} + \alpha\boldsymbol{u}_o)$$

and this is equivalent to

$$f\left(-\alpha^{-1}\boldsymbol{m}\right) - p\left(-\alpha^{-1}\boldsymbol{m} - \boldsymbol{u}_o\right) \leq c \quad \text{for } \alpha < 0$$

$$c \leq p\left(\alpha^{-1}\boldsymbol{m} + \boldsymbol{u}_o\right) - f\left(\alpha^{-1}\boldsymbol{m}\right) \qquad \text{for } \alpha > 0$$

Thus, it is sufficient to check whether a constant c exists such that

$$f(\boldsymbol{m}') - p(\boldsymbol{m}' - \boldsymbol{u}_o) \leq c \leq p(\boldsymbol{m}'' + \boldsymbol{u}_o) - f(\boldsymbol{m}'')$$

for every $\boldsymbol{m}', \boldsymbol{m}'' \in M$, or in other words, is it true that

$$\sup_{\boldsymbol{m}' \in M} \{f(\boldsymbol{m}') - p(\boldsymbol{m}' - \boldsymbol{u}_o)\} \leq \inf_{\boldsymbol{m}'' \in M} \{p(\boldsymbol{m}'' + \boldsymbol{u}_o) - f(\boldsymbol{m}'')\}$$

The answer is yes, as

$$f(\boldsymbol{m}') + f(\boldsymbol{m}'') = f(\boldsymbol{m}' + \boldsymbol{m}'') \leq p(\boldsymbol{m}' + \boldsymbol{m}'')$$

$$= p(\boldsymbol{m}' - \boldsymbol{u}_o + \boldsymbol{m}'' + \boldsymbol{u}_o)$$

$$\leq p(\boldsymbol{m}' - \boldsymbol{u}_o) + p(\boldsymbol{m}'' + \boldsymbol{u}_o)$$

Moving the terms with $\boldsymbol{m}'$ to the left side and those with $\boldsymbol{m}''$ to the right side and taking the supremum and infimum, respectively, we get the result required.

Thus, we have shown that the extension F to M_1 exists. We could now repeat this process for a larger space $M_2 = M_1 \oplus \mathbb{R}\boldsymbol{u}_1$, etc. Continuing in this way, we would produce an increasing family of spaces along with corresponding extensions of f dominated by p. The question is whether this process could be used to eventually cover the whole space.

We proceed by appealing to the Kuratowski-Zorn lemma: If in a partially ordered set every linearly ordered subset (a chain) has an upper bound, then there exists a maximum element in the set.

Step 1. Define a family

$$\begin{aligned} \mathcal{F} = \{(Y, f_Y) : \ & Y \text{ is a subspace of } X, M \subset Y \\ & f_Y : Y \to \mathbb{R} \text{ is a linear extension of } f \\ & f_Y(\boldsymbol{y}) \leq p(\boldsymbol{y}) \quad \forall \boldsymbol{y} \in Y\} \end{aligned}$$

Thus $\mathcal{F}$ is a family of all possible extensions of f along with their domains of definition—subspaces of X, containing M. According to the first part of this proof, $\mathcal{F}$ is nonempty.

Step 2. Introduce a relation on $\mathcal{F}$

$$(Y, f_Y) \leq (Z, f_Z) \overset{\text{def}}{\Longleftrightarrow} Y \subset Z \text{ and } f_Z|_Y = f_Y$$

It is a simple exercise (see Exercise 5.4.1) that relation "$\leq$" is a partial ordering of $\mathcal{F}$.

Step 3. Let $\mathcal{G}$ be a linearly ordered subset of $\mathcal{F}$

$$\mathcal{G} = \{(Y_\iota, f_{Y_\iota}) : \iota \in I\}$$

where I is a set of indices. Recall that $\mathcal{G}$ being linearly ordered means that any two elements of $\mathcal{G}$ are comparable with each other, i.e.,

$$(Y, f_Y), (Z, f_Z) \in \mathcal{G} \Longrightarrow (Y, f_Y) \leq (Z, f_Z) \text{ or } (Z, f_Z) \leq (Y, f_Y)$$

The question is, does $\mathcal{G}$ have an upper bound in $\mathcal{F}$? Define

$$Y = \bigcup_{\iota \in I} Y_\iota$$

$$f_Y : Y \rightarrow \mathbb{R}$$

$$f_Y(\boldsymbol{x}) \stackrel{\text{def}}{=} f_{Y_\iota}(\boldsymbol{x}), \text{ where } \boldsymbol{x} \in Y_\iota, \text{ for some } \iota \in I$$

It is left as an exercise (see Exercise 5.4.2) to prove that (Y, f_Y) is a well-defined upper bound for $\mathcal{G}$.

Step 4. By the Kuratowski-Zorn lemma, family $\mathcal{F}$ has a maximal element, say (Z, f_Z). We claim that it must be $Z = X$. Indeed, if there were an element left, $\boldsymbol{u}_o \in X - Z$, then by the procedure discussed in the first part of this proof, we could have extended f_Z further to $Z \oplus \mathbb{R}\boldsymbol{u}_o$, which contradicts the maximality of (Z, f_Z). This finishes the proof. ∎

Exercises

5.4.1. Prove that relation $\leq$ introduced in the proof of the Hahn-Banach theorem is a partial ordering of the family $\mathcal{F}$.

5.4.2. Prove that element (Y, f_Y) of $\mathcal{F}$ defined in the proof of the Hahn-Banach theorem

(i) is well defined, i.e.,

i. Y is a linear supspace of X and

ii. value $f_Y(\boldsymbol{x}) \stackrel{\text{def}}{=} f_{Y_\iota}(\boldsymbol{x})$ is well defined, i.e., independent of the choice of index ι, and

(ii) is an upper bound for the chain $\mathcal{G}$.

5.5 Extensions and Corollaries

In this section we generalize the Hahn-Banach theorem to the case of complex vector spaces and provide a number of important corollaries.

We start with a simple corollary.

COROLLARY 5.5.1

Let

X be a real vector space

$p : X \to [0, \infty]$ a seminorm

$M \subset X$ a subspace of X

$f : M \to \mathbb{R}$ a linear functional on $M (f \in M^*)$ such that

$$|f(\boldsymbol{x})| \leq p(\boldsymbol{x}) \qquad \boldsymbol{x} \in M$$

Then, there exists a linear extension $F : X \to \mathbb{R}$ of f such that

$$|F(\boldsymbol{x})| \leq p(\boldsymbol{x}) \qquad \boldsymbol{x} \in X$$

PROOF Obviously, p satisfies assumptions of the Hahn-Banach theorem and

$$f(\boldsymbol{x}) \leq |f(\boldsymbol{x})| \leq p(\boldsymbol{x}) \quad \boldsymbol{x} \in M$$

i.e., f is dominated by p on M. Let $F : X \to \mathbb{R}$ be an extension of f to the whole X, dominated by p, i.e.,

$$F(\boldsymbol{x}) \leq p(\boldsymbol{x}) \qquad \boldsymbol{x} \in X$$

Replacing $\boldsymbol{x}$ with $-\boldsymbol{x}$, we get

$$-F(\boldsymbol{x}) = F(-\boldsymbol{x}) \leq p(-\boldsymbol{x}) = p(\boldsymbol{x}) \quad \boldsymbol{x} \in X$$

which implies that

$$-p(\boldsymbol{x}) \leq F(\boldsymbol{x}) \quad \boldsymbol{x} \in X$$

and, consequently,

$$|F(\boldsymbol{x})| \leq p(\boldsymbol{x}) \qquad \forall \ \boldsymbol{x} \in X$$

∎

We proceed now with the generalization for complex spaces.

THEOREM 5.5.1
(Bohnenblust-Sobczyk)

Let

X *be a complex vector space*

$p : X \to [0, \infty)$ *a seminorm*

$M \subset X$ *a subspace of* X

$f : M \to \mathbb{C}$ *a linear functional on* $M (f \in M^*)$ *such that*

$$|f(\boldsymbol{x})| \leq p(\boldsymbol{x}) \qquad \forall\, \boldsymbol{x} \in M$$

Then, there exists a linear functional $F : X \to \mathbb{C}$ *defined on the whole* X *such that*

(i) $F|_M \equiv f$ *and*

(ii) $|F(\boldsymbol{x})| \leq p(\boldsymbol{x}) \qquad \forall\, \boldsymbol{x} \in X.$

PROOF Obviously, X is also a real space when the multiplication of vectors by scalars is restricted to real numbers. Functional f on M has the form

$$f(\boldsymbol{x}) = g(\boldsymbol{x}) + ih(\boldsymbol{x}) \quad \boldsymbol{x} \in M$$

where both g and h are linear, real-valued functionals defined on M. Note that g and h are not independent of each other, as f is complex–linear, which in particular implies that

$$f(i\boldsymbol{x}) = if(\boldsymbol{x})$$

or

$$g(i\boldsymbol{x}) + ih(i\boldsymbol{x}) = i(g(\boldsymbol{x}) + ih(\boldsymbol{x})) = -h(\boldsymbol{x}) + ig(\boldsymbol{x})$$

and therefore

$$h(\boldsymbol{x}) = -g(i\boldsymbol{x}) \qquad \forall\, \boldsymbol{x} \in M$$

Also,

$$|g(\boldsymbol{x})| \leq |f(\boldsymbol{x})| \left(= \sqrt{g(\boldsymbol{x})^2 + h(\boldsymbol{x})^2}\right) \leq p(\boldsymbol{x})$$

By the preceding corollary, there exists a real-valued linear extension G of g to the whole X such that

$$|G(\boldsymbol{x})| \leq p(\boldsymbol{x})$$

Define:

$$F(\boldsymbol{x}) = G(\boldsymbol{x}) - iG(i\boldsymbol{x}) \qquad \boldsymbol{x} \in X$$

Clearly, F is an extension of f.

To prove that F is complex–linear, it is sufficient to show (why?) that

$$F(i\boldsymbol{x}) = iF(\boldsymbol{x}) \qquad \forall\, \boldsymbol{x} \in X$$

But

$$F(i\boldsymbol{x}) = G(i\boldsymbol{x}) - iG(ii\boldsymbol{x}) = G(i\boldsymbol{x}) - iG(-\boldsymbol{x})$$

$$= iG(\boldsymbol{x}) + G(i\boldsymbol{x}) = i(G(\boldsymbol{x}) - iG(i\boldsymbol{x}))$$

$$= iF(\boldsymbol{x})$$

It remains to prove that F is bounded by seminorm p on X. Representing $F(\boldsymbol{x})$ in the form

$$F(\boldsymbol{x}) = r(\boldsymbol{x})e^{-i\theta(\boldsymbol{x})}, \quad \text{where} \quad r(\boldsymbol{x}) = |F(\boldsymbol{x})|$$

we have

$$|F(\boldsymbol{x})| = e^{i\theta(\boldsymbol{x})}F(\boldsymbol{x}) = F\left(e^{i\theta(\boldsymbol{x})}\boldsymbol{x}\right) = G\left(e^{i\theta(\boldsymbol{x})}\boldsymbol{x}\right)$$

since $|F(\boldsymbol{x})|$ is real.

Finally,

$$\begin{aligned} G\left(e^{i\theta(\boldsymbol{x})}\boldsymbol{x}\right) &\leq \left|G\left(e^{i\theta(\boldsymbol{x})}\boldsymbol{x}\right)\right| \leq p\left(e^{i\theta(\boldsymbol{x})}\boldsymbol{x}\right) \\ &\leq \left|e^{i\theta(\boldsymbol{x})}\right| p(\boldsymbol{x}) = p(\boldsymbol{x}) \end{aligned}$$

which finishes the proof. ∎

COROLLARY 5.5.2

Let $(U, \|\cdot\|_U)$ be a normed space and let $\boldsymbol{u}_o \in U$ denote an arbitrary nonzero vector.

There exists a linear functional f on U such that

(i) $f(\boldsymbol{u}_o) = \|\boldsymbol{u}_o\| \neq 0$ and

(ii) $|f(\boldsymbol{u})| \leq \|\boldsymbol{u}\| \quad \forall\, \boldsymbol{u} \in \boldsymbol{U}$.

PROOF Take $p(\boldsymbol{u}) = \|\boldsymbol{u}\|$ and consider the one-dimensional subspace M of U spanned by $\boldsymbol{u}_o$

$$M = \mathbb{R}\boldsymbol{u}_o \text{ (or } \mathbb{C}\boldsymbol{u}_o) = \{\alpha \boldsymbol{u}_o \,:\, \alpha \in \mathbb{R} \text{ (or } \mathbb{C})\}$$

Define a linear functional f on M by setting $f(\boldsymbol{u}_o) = \|\boldsymbol{u}_o\|$

$$f(\alpha \boldsymbol{u}_o) = \alpha f(\boldsymbol{u}_o) = \alpha \|\boldsymbol{u}_o\|$$

Obviously,

$$|f(\boldsymbol{u})| \leq p(\boldsymbol{u})$$

(in fact, the equality holds). Use the Hahn-Banach theorem to extend f to all of U and thus conclude the assertion. ∎

REMARK 5.5.1. Using the language of the next two sections one can translate the corollary above into the assertion that for every nonzero vector $\boldsymbol{u}_0$ from a normed space U there exists a linear and continuous functional f such that $\|f\| = 1$ and $f(\boldsymbol{u}_o) = \|\boldsymbol{u}_o\| \neq 0$. $\|f\|$ denotes here a norm on f as a member of the "topological dual space" of U, a concept we shall address more carefully in Section 5.7 and thereafter. This in particular will imply that the topological duals U' are nonempty. We shall return to this remark after reading the next two sections. ∎

Bounded (Continuous) Linear Operators on Normed Spaces

5.6 Fundamental Properties of Linear Bounded Operators

We developed most of the important algebraic properties of linear transformations in Chapter 2. We now expand our study to linear transformations on normed spaces. Since the domains of such linear mappings now have topological structure, we can also apply many of the properties of functions on metric spaces. For example, we are now able to talk about continuous linear transformations from one normed linear space into another. It is not uncommon to use the term *operator* to refer to a mapping or function on sets that have both algebraic and topological structure. Since all of our subsequent work involves cases in which this is so, we henceforth use the term *operator* synomously with *function*, *mapping*, and *transformation*.

To begin our study, let $(U, \|\cdot\|_U)$ and $(V, \|\cdot\|_V)$ denote two normed linear spaces over the same field $\mathbb{F}$, and let A be an operator from U into V. We recall that an operator A from U into V is *linear* if and only if it is homogeneous (i.e., $A(\alpha \boldsymbol{u}) = \alpha A\boldsymbol{u} \ \forall\ \boldsymbol{u} \in U$ and $\alpha \in \mathbb{F}$) and additive (i.e., $A(\boldsymbol{u}_1 + \boldsymbol{u}_2) = A(\boldsymbol{u}_1) + A(\boldsymbol{u}_2) \ \forall\ \boldsymbol{u}_1, \boldsymbol{u}_2 \in U$). Equivalently, $A : U \to V$ is linear if and only if $A(\alpha \boldsymbol{u}_1 + \beta \boldsymbol{u}_2) = \alpha A(\boldsymbol{u}_1) + \beta A(\boldsymbol{u}_2) \ \forall\ \boldsymbol{u}_1, \boldsymbol{u}_2 \in U$ and $\forall\ \alpha, \beta \in \mathbb{F}$. When A does *not* obey this rule, it is called a *nonlinear operator*. In the sequel we shall always take the field $\mathbb{F}$ to be real or complex numbers: $\mathbb{F} = \mathbb{R}$ or $\mathbb{F} = \mathbb{C}$.

Recall that the *null space*, $\mathcal{N}(A)$, of a linear operator $A : U \to V$ is defined by $\mathcal{N}(A) = \{\boldsymbol{u} : Au = 0, \boldsymbol{u} \in U\}$ and is a subspace of U, and the range $\mathcal{R}(A)$ of a linear operator $A : U \to V$ is defined to be $\mathcal{R}(A) = \{v : A\boldsymbol{u} = \boldsymbol{v} \in V,$ for $\boldsymbol{u} \in U\}$ and $\mathcal{R}(A) \subset V$. We note here that the operator A is one-to-one if and only if the null space $\mathcal{N}(A)$ is trivial, $\mathcal{N}(A) = \{0\}$.

Thus far we have introduced only algebraic properties of linear operators. To talk about boundedness and continuity of linear operators, we use the topological structure of the normed spaces U and V.

We begin with the fundamental characterization of linear continuous operators on normed spaces.

PROPOSITION 5.6.1

Let $(U, \|\cdot\|_U)$ and $(V, \|\cdot\|_V)$ be two normed vector spaces over the same field and $T : U \to V$ a linear transformation defined on U with the values in $V (T \in L(U, V))$. The following conditions are equivalent to each other:

(i) *T is continuous (with respect to norm topologies).*

(ii) *T is continuous at $\mathbf{0}$.*

(iii) *T is bounded, i.e., T maps bounded sets in U into bounded sets in V.*

(iv) $\exists C > 0 : \|T\boldsymbol{u}\|_V \leq C\|\boldsymbol{u}\|_U \ \forall\ \boldsymbol{u} \in U$.

PROOF

(i) $\Longrightarrow$ (ii) trivial.

(ii) $\Longrightarrow$ (iii) Let A be a bounded set in U, i.e., there exists a ball $B(\mathbf{0}, r)$ such that $A \subset B(\mathbf{0}, r)$. T being continuous at $\mathbf{0}$ means that

$$\forall\, \varepsilon > 0\ \exists\, \delta > 0 : \ \|\boldsymbol{u}\|_U < \delta \Longrightarrow \|T\boldsymbol{u}\|_V < \varepsilon$$

Selecting $\varepsilon = 1$, we get

$$\exists\, \delta > 0 : \ \|\boldsymbol{u}\|_U < \delta \Longrightarrow \|T\boldsymbol{u}\|_V < 1$$

Let $\boldsymbol{u} \in A$ and therefore $||\boldsymbol{u}||_U \leq r$. Consequently,

$$\left\| \frac{\delta}{r}\boldsymbol{u} \right\|_U = \frac{\delta}{r}\, \|\boldsymbol{u}\|_U \leq \delta$$

which implies that

$$\left\| T\left(\frac{\delta}{r}\boldsymbol{u}\right) \right\|_V \leq 1 \Longrightarrow \|T(\boldsymbol{u})\|_V \leq \frac{r}{\delta}$$

which is equivalent to saying that $T(A) \subset B(\boldsymbol{O}, \frac{r}{\delta})$ and therefore is bounded.

(iii) $\Longrightarrow$ (iv) From the boundedness of $T(B(\mathbf{0}, 1))$ follows that

$$\exists C > 0 : \ \|\boldsymbol{u}\|_U \leq 1 \Longrightarrow \|T\boldsymbol{u}\|_V \leq C$$

Consequently, for every $\boldsymbol{u} \neq \mathbf{0}$

$$\left\| T\left(\frac{\boldsymbol{u}}{\|\boldsymbol{u}\|_U}\right) \right\|_V \leq C$$

or, equivalently,

$$\|T(\boldsymbol{u})\|_V \leq C\, \|\boldsymbol{u}\|_U$$

(iv) $\Longrightarrow$ (i) It is sufficient to show sequential continuity.
Let $\boldsymbol{u}_n \to \boldsymbol{u}$. Then

$$\|T\boldsymbol{u}_n - T\boldsymbol{u}\|_V = \|T\left(\boldsymbol{u}_n - \boldsymbol{u}\right)\|_V \leq C\, \|\boldsymbol{u}_n - \boldsymbol{u}\|_U \to 0$$

∎

Operator Norm. According to the definition just given, we can always associate with any bounded linear operator $A: U \to V$ a collection of positive numbers C such that

$$\|A\boldsymbol{u}\|_V \leq C\, \|\boldsymbol{u}\|_U \quad \forall\, \boldsymbol{u} \in U$$

If we consider the infimum of this set, then we effectively establish a correspondence N between the operator A and the nonnegative real numbers

$$N(A) = \inf \{C : \|A\boldsymbol{u}\|_V \le C \|\boldsymbol{u}\|_U \quad \forall\, \boldsymbol{u} \in U\}$$

Remarkably, the function N determined in this way satisfies all the requirements for a norm, and we denote $N(A)$ by $\|A\|$ and refer to it as the *norm of the operator* A:

$$\|A\| = \inf \{C : \|A\boldsymbol{u}\|_V \le C \|\boldsymbol{u}\|_U \quad \forall\, \boldsymbol{u} \in U\}$$

Notice that passing with C to the infimum in

$$\|A\boldsymbol{u}\|_V \le C \|\boldsymbol{u}\|_U$$

we immediately get the inequality

$$\|A\boldsymbol{u}\|_V \le \|A\| \ \|\boldsymbol{u}\|_U$$

We will demonstrate later that the notation $\|A\|$ is justified. First, we develop now some alternative forms for defining $\|A\|$.

PROPOSITION 5.6.2

Let A be a bounded linear operator from $(U, \|\cdot\|_U)$ into $(V, \|\cdot\|_V)$. Then

(i) $\displaystyle \|A\| = \sup_{\boldsymbol{u}\in U} \frac{\|A\boldsymbol{u}\|_V}{\|\boldsymbol{u}\|_U} \quad \boldsymbol{u} \neq \boldsymbol{0}$

(ii) $\displaystyle \|A\| = \sup_{\boldsymbol{u}\in U} \{\|A\boldsymbol{u}\|_V,\ \|\boldsymbol{u}\|_U \le 1\}$

(iii) $\displaystyle \|A\| = \sup_{\boldsymbol{u}\in U} \{\|A\boldsymbol{u}\|_V,\ \|\boldsymbol{u}\|_U = 1\}$

PROOF

(i) $\|A\boldsymbol{u}\|_V \le C \|\boldsymbol{u}\|_U$ implies that

$$\frac{\|A\boldsymbol{u}\|_V}{\|\boldsymbol{u}\|_U} \le C$$

Thus, taking the supremum over all $\boldsymbol{u} \neq \boldsymbol{0}$ on the left-hand side and the infimum over all C's on the right side, we get

$$\sup_{\boldsymbol{u}\neq\boldsymbol{0}} \frac{\|A\boldsymbol{u}\|_V}{\|\boldsymbol{u}\|_U} \le \|A\|$$

On the other side, for an arbitrary $\boldsymbol{w} \neq \boldsymbol{0}$,

$$\frac{\|A\boldsymbol{w}\|_V}{\|\boldsymbol{w}\|_U} \leq \sup_{\boldsymbol{u}\neq\boldsymbol{0}} \frac{\|A\boldsymbol{u}\|_V}{\|\boldsymbol{u}\|_U} \stackrel{\text{def}}{=} C_0$$

and, consequently,

$$\|A\boldsymbol{w}\|_V \leq C_0 \|\boldsymbol{w}\|_U$$

so that

$$\begin{aligned}\|A\| &= \inf\{C : \|A\boldsymbol{u}\|_V \leq C\|\boldsymbol{u}\|_U, \quad \forall \boldsymbol{u} \in U\} \\ &\leq C_0 = \sup_{\boldsymbol{u}\neq\boldsymbol{0}} \frac{\|A\boldsymbol{u}\|_V}{\|\boldsymbol{u}\|_U}\end{aligned}$$

This proves (i).

(iii) follows directly from (i) as

$$\frac{\|A\boldsymbol{u}\|_V}{\|\boldsymbol{u}\|_U} = \left\|A\left(\frac{\boldsymbol{u}}{\|\boldsymbol{u}\|_U}\right)\right\|_V$$

and $\left\|\|\boldsymbol{u}\|_U^{-1}\boldsymbol{u}\right\|_U = 1$.

(ii) As

$$\|A\boldsymbol{u}\|_V \leq \|A\| \; \|\boldsymbol{u}\|_U \quad \forall\, \boldsymbol{u}$$

we immediately have

$$\sup_{\|\boldsymbol{u}\|_U \leq 1} \|A\boldsymbol{u}\|_V \leq \|A\|$$

The inverse inequality follows directly from (iii) (supremum is taken over a larger set). ∎

It is not difficult now to show that the function $\|A\|$ satisfies the norm axioms.

1. In view of the definition, if $\|A\| = 0$, then $\|A\boldsymbol{u}\|_V = 0 \;\forall\, \boldsymbol{u}$. But this is not possible unless $A \equiv 0$.

2. $\|\lambda A\| = \sup_{\boldsymbol{u}\neq\boldsymbol{0}} \dfrac{\|\lambda A\boldsymbol{u}\|_V}{\|\boldsymbol{u}\|_U} = |\lambda| \sup_{\boldsymbol{u}\neq\boldsymbol{0}} \dfrac{\|A\boldsymbol{u}\|_V}{\|\boldsymbol{u}\|_U} = |\lambda| \; \|A\|.$

3. $$\|A+B\| = \sup_{\boldsymbol{u}\neq \boldsymbol{0}} \frac{\|A\boldsymbol{u}+B\boldsymbol{u}\|_V}{\|\boldsymbol{u}\|_U}$$

$$\leq \sup_{\boldsymbol{u}\neq \boldsymbol{0}} \frac{\|A\boldsymbol{u}\|_V + \|B\boldsymbol{u}\|_V}{\|\boldsymbol{u}\|_U} \leq \|A\| + \|B\|.$$

4. An additional property of the norm of a bounded (continuous) linear operator can be identified that is often useful: If AB denotes the composition of two bounded operators, then

$$\|AB\boldsymbol{u}\| \leq \|A\|\ \|B\boldsymbol{u}\| \leq \|A\|\ \|B\|\ \|\boldsymbol{u}\|$$

Consequently,

$$\|AB\| \leq \|A\|\ \|B\|$$

Space $\mathcal{L}(U,V)$. We recall that the class $L(U,V)$ of all linear transformations from a linear vector space U into a linear vector space V is, itself, a linear space. The results we have just obtained lead us to an important observation: Whenever U and V are equipped with a norm, it is possible to identify a subspace

$$\mathcal{L}(U,V) \subset L(U,V)$$

consisting of all bounded linear operators from U into V, which is also a normed space equipped with the operator norm $\|A\|$ defined above. The norm $\|A\|$ of a bounded operator can be viewed as a measure of the stretch, distortion, or amplification of the elements in its domain.

Example 5.6.1

Consider the operator A from a space U into itself defined as

$$A\boldsymbol{u} = \lambda \boldsymbol{u}\ ,\ \lambda \in \mathbb{R}(\mathbb{C})$$

The norm of A in this case is

$$\|A\| = \sup_{\boldsymbol{u}\neq \boldsymbol{0}} \frac{\|A\boldsymbol{u}\|_U}{\|\boldsymbol{u}\|_U} = \sup_{\boldsymbol{u}\neq \boldsymbol{0}} \frac{|\lambda|\|\boldsymbol{u}\|}{\|\boldsymbol{u}\|} = |\lambda|$$

□

Example 5.6.2

Let $\boldsymbol{A}$ be a matrix operator from $\mathbb{R}^n$ into itself and $\|\cdot\|$ denote the Euclidean norm in $\mathbb{R}^n$, i.e.,

$$\|\boldsymbol{x}\| = \|(x_1, \ldots, x_n)\| = \left(\sum_{i=1}^{n} x_i^2 \right)^{\frac{1}{2}}$$

Then the problem of finding the norm of $\boldsymbol{A}$ reduces to finding the maximum eigenvalue of the composition $\boldsymbol{A}^T \boldsymbol{A}$.

Indeed, finding the norm of $\boldsymbol{A}$ is equivalent to solving a constrained maximization problem in the form

$$\|\boldsymbol{A}\boldsymbol{x}\|^2 = \sum_i \left(\sum_j A_{ij} x_j \right)^2 \to \max$$

subjected to the constraint

$$\sum_i x_i^2 = 1$$

Using the method of Lagrange multipliers, we arrive at the necessary condition in the form

$$\sum_i \left(\sum_j A_{ij} x_j \right) A_{ik} - \lambda x_k = 0 \quad k = 1, 2, \ldots, n$$

with λ being the Lagrange multiplier, or, equivalently,

$$\boldsymbol{A}^T \boldsymbol{A} \boldsymbol{x} = \lambda \boldsymbol{x}$$

Thus $\|\boldsymbol{A}\boldsymbol{x}\|^2$ attains its maximum at one of the eigenvectors λ of $\boldsymbol{A}^T \boldsymbol{A}$ and, consequently,

$$\frac{\|\boldsymbol{A}^T \boldsymbol{A} \boldsymbol{x}\|}{\|\boldsymbol{x}\|} = \frac{\|\lambda \boldsymbol{x}\|}{\|\boldsymbol{x}\|} = |\lambda|$$

which implies that the norm $\|\boldsymbol{A}\|$ is equal to the square root of the maximum (in modulus) eigenvalue of $\boldsymbol{A}^T \boldsymbol{A}$. Square roots μ_i of the nonnegative eigenvalues of $\boldsymbol{A}^T \boldsymbol{A}$ ($\boldsymbol{A}^T \boldsymbol{A}$ is positive semidefinite)

$$\mu_i^2 = \lambda_i \left(\boldsymbol{A}^T \boldsymbol{A} \right)$$

are frequently called the *singular values of* $\boldsymbol{A}$. For symmetric matrices the singular values of $\boldsymbol{A}$ coincide with absolute values of eigenvalues of $\boldsymbol{A}$, since for an eigenvalue λ of $\boldsymbol{A}$ and a corresponding eigenvector $\boldsymbol{x}$ one has

$$A^T A x = A^T \lambda x = \lambda A^T x = \lambda A x = \lambda^2 x$$

and therefore λ^2 is an eigenvalue of $A^T A = A^2$.

Consider, for instance, the matrix

$$A = \begin{bmatrix} 2 & 1 \\ 1 & 2 \end{bmatrix}$$

from $\mathbb{R}^2$ into $\mathbb{R}^2$.

Now, it is easily verified that the eigenvalues of A are

$$\lambda_1 = 3, \quad \lambda_2 = 1$$

Clearly, in this particular case

$$\|A\| = \max\{|\lambda_1|, \ |\lambda_2|\} = 3$$

We emphasize that $\max \lambda \neq \|A\|$ in general. For instance, for matrix

$$A = \begin{bmatrix} 1 & 1 \\ 0 & 1 \end{bmatrix}$$

the singular values are

$$\mu_{1,2}^2 = \frac{3 \overset{+}{-} \sqrt{5}}{2}$$

and, therefore,

$$\|A\| = \max\{|\frac{3-\sqrt{5}}{2}|^{\frac{1}{2}}, |\frac{3+\sqrt{5}}{2}|^{\frac{1}{2}}\} \approx 1.618$$

whereas the matrix A has only a single eigenvalue $\lambda = 1$.

It is clear, however, that if $A : \mathbb{R}^n \to \mathbb{R}^n$, then

$$\|A\| = \sup_{u \neq 0} \frac{\|Au\|}{\|u\|} \geq \max_{1 \leq i \leq n} |\lambda_i(A)|$$

where $\lambda_i(A)$ are the eigenvalues of A (why?). $\square$

Example 5.6.3

We wish to emphasize that the character of the norm assigned to a bounded operator $A : U \to V$ depends entirely on the choice of norms used in U and V.

Suppose, for example, that $U = V = \mathbb{R}^n$ and A is identified with a given $n \times n$ matrix $[A_{ij}]$. Among possible choices of norms for $\mathbb{R}^n$ are the following:

$$\|\boldsymbol{u}\|_\infty = \max_{1 \le j \le n} |u_j| \quad \|\boldsymbol{u}\|_2 = \left(\sum_{j=1}^{n} |u_j|^2 \right)^{\frac{1}{2}}$$

$$\|\boldsymbol{u}\|_1 = \sum_{j=1}^{n} |u_j| \quad \|\boldsymbol{u}\|_p = \left(\sum_{j=1}^{n} |u_j|^p \right)^{\frac{1}{p}}$$

$$1 \le p < \infty$$

Depending upon the choice of norm in U and V, the operation A has a different, corresponding norm. For example,
if $A : (\mathbb{R}^n, \|\cdot\|_\infty) \to (\mathbb{R}^n, \|\cdot\|_1)$, then

$$\|\boldsymbol{A}\|_{\infty,1} = \sum_{i,j=1}^{n} |A_{ij}|$$

if $A : (\mathbb{R}^n, \|\cdot\|_1) \to (\mathbb{R}^n, \|\cdot\|_\infty)$, then

$$\|\boldsymbol{A}\|_{1,\infty} = \max_{1 \le i \le n} \max_{1 \le j \le n} |A_{ij}|$$

if $A : (\mathbb{R}^n, \|\cdot\|_\infty) \to (\mathbb{R}^n, \|\cdot\|_\infty)$, then

$$\|\boldsymbol{A}\|_\infty = \max_{1 \le i \le n} \sum_{j=1}^{n} |A_{ij}|$$

if $A : (\mathbb{R}^n, \|\cdot\|_1) \to (\mathbb{R}^n, \|\cdot\|_1)$, then

$$\|\boldsymbol{A}\|_1 = \max_{1 \le j \le n} \sum_{i=1}^{n} |A_{ij}|$$

and so forth. If the Euclidean norm is used, i.e.,

$$A : (\mathbb{R}^n, \|\cdot\|_2) \longrightarrow (\mathbb{R}^n, \|\cdot\|_2)$$

then

$$\|\boldsymbol{A}\|_2 = \sqrt{\rho\left(\boldsymbol{A}^T\boldsymbol{A}\right)}$$

where $\boldsymbol{A}^T$ is the transpose of $\boldsymbol{A}$ and ρ is the *spectral radius* of $\boldsymbol{A}^T\boldsymbol{A}$, and the spectral radius of any square matrix $\boldsymbol{B}$ is defined by

$$\rho(\boldsymbol{B}) = \max_s |\lambda_s(\boldsymbol{B})|$$

where $\lambda_s(\boldsymbol{B})$ is the s-th eigenvalue of $\boldsymbol{B}$. When $\boldsymbol{A}$ is symmetric,

$$\|\boldsymbol{A}\|_2 = \max_s |\lambda_s(\boldsymbol{A})|$$

(compare this with the previous example). □

Example 5.6.4
(An Unbounded Operator)

Consider the differential operator

$$Du = \frac{d}{dx}(u(x))$$

defined on the set of differentiable functions with the norm $\|u\| = \sup |u(x)|, x \in [0,1]$. We shall show that D is *not* bounded in $C[0,1]$. Toward this end, let

$$u_n(x) = \sin(nx)$$

Clearly, $\|u_n\| = \sup\limits_{x\in[0,1]} |u_n(x)| = 1$ for all n, and $Du_n = n\cos(nx)$, $\|Du_n\| = n$. Since $\|u_n\| = 1$ and Du_n increases infinitely for $n \to \infty$, there is no constant M such that $\|Du\| < M\|u\|$ for all $u \in C[0,1]$. Thus, D is not bounded.

We also note that D is not defined everywhere in $C[0,1]$. However, if D is considered as an operator from $C^1[0,1]$ into $C[0,1]$ with $\|u\| = \max(\sup\limits_{x\in[0,1]} |u(x)|, \sup\limits_{x\in[0,1]} |Du(x)|)$, then it can be shown to be bounded. In general, a linear differential operator of order m with continuous coefficients can be considered as a bounded operator from $C^m[0,1]$ into $C[0,1]$ if we select an appropriate norm; e.g., $\|u\| = \max\limits_{0\le k\le m} \sup\limits_{0\le x\le 1} |D^k u(x)|$. □

Exercises

5.6.1. Verify the assertions given in Example 5.6.3.

5.6.2. Let $A : (\mathbb{R}^2, \|\cdot\|_a) \to (\mathbb{R}^2, \|\cdot\|_b)$ and $B : (\mathbb{R}^2, \|\cdot\|_a) \to (\mathbb{R}^2, \|\cdot\|_b)$, where $a, b = 1, 2, \infty$, be linear operators represented by the matrices

$$A = \begin{bmatrix} 2 & 1 \\ 3 & -2 \end{bmatrix}, \quad B = \begin{bmatrix} 4 & 2 \\ 2 & 1 \end{bmatrix}$$

Determine $\|A\|$ and $\|B\|$ for all choices of a and b.

5.6.3. Construct an example of a matrix A in $\mathbb{R}^2 \times \mathbb{R}^2$ such that

$$\|A\| \neq \max_{i=1,2} |\lambda_i|$$

where the operator norm is calculated with respect to the Euclidean norm in $\mathbb{R}^2$.

5.6.4. Let A be an invertible matrix in $\mathbb{R}^n \times \mathbb{R}^n$, and $\mu_i > 0$ denote its singular values (see Example 5.6.2). Show that with $\|A\|$ calculated with respect to the Euclidean norm in $\mathbb{R}^n$,

$$\|A^{-1}\| = \frac{1}{\min\limits_{1 \le i \le n} \mu_i}$$

5.7 The Space of Continuous Linear Operators

In this section, we will investigate closer the space $\mathcal{L}(U, V)$ of all continuous operators from a normed space U into a normed space V. We have already learned that $\mathcal{L}(U, V)$ is a subspace of the space $L(U, V)$ consisting of all linear (and not necessarily continuous) operators from U to V and that it can be equipped with the norm

$$\|A\| = \|A\|_{\mathcal{L}(U,V)} = \sup_{u \neq 0} \frac{\|Au\|_V}{\|u\|_U}$$

In the case of a finite-dimensional space U, the space $\mathcal{L}(U, V)$ simply coincides with $L(U, V)$ as every linear operator on U is automatically continuous. In order to show this, consider an arbitrary basis

$$e_i, \ i = 1, 2, \ldots, n$$

for U and a corresponding norm,

$$\|u\| = \sum_{i=1}^{n} |u_i|, \quad \text{where} \quad u = \sum_{1}^{n} u_i e_i$$

As any two norms are equivalent in a finite-dimensional space, it is sufficient to show that any linear operator on U is continuous with respect to this particular norm. This follows easily from

$$\|A\boldsymbol{u}\|_V = \left\| A\left(\sum_1^n u_i \boldsymbol{e}_i\right)\right\| \leq \sum_1^n |u_i| \, \|A\boldsymbol{e}_i\|_V$$

$$\leq \left(\max_i \|A\boldsymbol{e}_i\|_V\right) \sum_1^n |u_i|$$

REMARK 5.7.1. The notion of the space $\mathcal{L}(U, V)$ can easily be generalized to arbitrary topological vector spaces U and V. Indeed, if A and B are two *continuous* linear operators from U to V, then $\alpha A + \beta B$, because of continuity of the operations of addition and scalar multiplication in V, is also continuous, and therefore the set of all continuous linear operators is closed with respect to vector space operations. Obviously, in general, $\mathcal{L}(X, Y)$ cannot be equipped with a norm topology. ∎

Convergence of Sequences in $\mathcal{L}(U, V)$. A sequence $\{A_n\}$ of operators in $\mathcal{L}(U, V)$ is said to *converge uniformly* to $A \in \mathcal{L}(U, V)$ if simply $A_n \to A$ in the norm topology, i.e.,

$$\lim_{n\to\infty} \|A_n - A\| = 0$$

The sequence $\{A_n\}$ from $\mathcal{L}(U, V)$ is said to *converge strongly* to $A \in \mathcal{L}(U, V)$, denoted $A_n \xrightarrow{s} A$, if

$$\lim_{n\to\infty} \|A_n \boldsymbol{u} - A\boldsymbol{u}\| = 0 \quad \text{for every} \quad \boldsymbol{u} \in U$$

It follows immediately from the inequality

$$\|A_n \boldsymbol{u} - A\boldsymbol{u}\| \leq \|A_n - A\| \; \|\boldsymbol{u}\|$$

that uniform convergence implies strong convergence. The converse is in general not true.

We will prove now an important assertion concerning the completeness of the space $\mathcal{L}(U, V)$.

PROPOSITION 5.7.1

Let U, V be two normed spaces and V be complete, i.e., V is a Banach space. Then $\mathcal{L}(U, V)$ is complete, and is therefore also a Banach space.

PROOF Let $A_n \in \mathcal{L}(U, V)$ be a Cauchy sequence, i.e.,

$$\lim_{n,m\to\infty} \|A_n - A_m\| = 0$$

Since for every $\boldsymbol{u} \in U$,

$$\begin{aligned}\|A_n\boldsymbol{u} - A_m\boldsymbol{u}\|_V &= \|(A_n - A_m)\,\boldsymbol{u}\|_V \\ &\le \|A_n - A_m\| \;\; \|\boldsymbol{u}\|_U\end{aligned}$$

it follows that $A_n\boldsymbol{u}$ is a Cauchy sequence in V and by completeness of V has a limit. Define

$$A\boldsymbol{u} \stackrel{\text{def}}{=} \lim_{n\to\infty} A_n\boldsymbol{u}$$

Then:

Step 1. A is linear. Indeed, it is sufficient to pass to the limit with $n \to \infty$ on both sides of the identity

$$A_n(\alpha\boldsymbol{u} + \beta\boldsymbol{v}) = \alpha A_n(\boldsymbol{u}) + \beta A_n(\boldsymbol{v})$$

Step 2. From the fact that A_n is Cauchy it follows again that

$$\forall\, \varepsilon > 0\; \exists\, N : \forall\, n, m \ge N \quad \|A_n - A_m\| < \varepsilon$$

Combining this with the inequality

$$\|A_n\boldsymbol{u} - A_m\boldsymbol{u}\|_V \le \|A_n - A_m\| \;\; \|\boldsymbol{u}\|_U$$

we get

$$\|A_n\boldsymbol{u} - A_m\boldsymbol{u}\|_V \le \varepsilon\, \|\boldsymbol{u}\|_U \quad \forall\, n, m \ge N$$

with ε independent of $\boldsymbol{u}$.

Passing now with $m \to \infty$ and making use of the continuity of the norm in V, we get

$$\|A_n\boldsymbol{u} - A\boldsymbol{u}\|_V \le \varepsilon\, \|\boldsymbol{u}\|_U \quad \forall\, n \ge N$$

This inequality implies that:

1. A is continuous. Indeed, it follows from the inequality that

$$\sup_{\|\boldsymbol{u}\|\le 1} \|A_n\boldsymbol{u} - A\boldsymbol{u}\|_V$$

is finite and, consequently,

$$\sup_{\|\boldsymbol{u}\|\le 1} \|A\boldsymbol{u}\|_V \le \sup_{\|\boldsymbol{u}\|\le 1} \|A_n\boldsymbol{u} - A\boldsymbol{u}\|_V + \|A_n\|$$

with both terms on the right-hand side being bounded (every Cauchy sequence is bounded). Thus, being a bounded linear operator, A is continuous by Proposition 5.6.1.

2.

$$\sup_{\|\boldsymbol{u}\|\le 1} \|(A_n - A)\,\boldsymbol{u}\|_V = \|A_n - A\| \le \varepsilon \quad \forall\, n \ge N$$

which proves that $A_n \to A$; i.e., this arbitrary Cauchy sequence A_n converges to $A \in \mathcal{L}(U, V)$.

∎

Topological Duals. Let V be a normed space. The space of all continuous and linear functionals $\mathcal{L}(V,\mathbb{R})$ (or $\mathcal{L}(V,\mathbb{C})$) is called the *topological dual* of V, or concisely, the dual space of V if no confusion with the algebraic dual is likely to occur, and is denoted by V'. Obviously:

(i) Topological dual V' is a subspace of algebraic dual V^*.

(ii) For a finite-dimensional space V, both duals are the same.

(iii) The topological dual space of a normed space V is always a Banach space, even if V is not complete (compare the previous proposition).

Example 5.7.1
(Neumann Series)

Let $A : U \to U$ be a continuous linear operator from a Banach space into itself. We wish to make use of the topological ideas discussed so far to compute an inverse of the operator

$$\lambda I - A$$

if it exists, where λ is a scalar and I is the identity operator. In particular, we recall that geometric series

$$\frac{1}{\lambda - a} = \frac{1}{\lambda(1 - a/\lambda)} = \frac{1}{\lambda}\left(1 + \frac{a}{\lambda} + \frac{a^2}{\lambda^2} + \cdots\right)$$

converges if $a/\lambda < 1$, and we wish to derive a similar expansion for operators

$$(\lambda I - A)^{-1} = \frac{1}{\lambda}I + \frac{1}{\lambda^2}A + \frac{1}{\lambda^3}A^2 + \cdots$$

(if possible). $\square$

Toward this end, consider the series

$$\frac{1}{\lambda}\sum_{k=0}^{\infty}\frac{1}{\lambda^k}A^k$$

by which, as in classical analysis, we understand both the sequence of partial sums

$$S_N = \frac{1}{\lambda}\sum_{k=0}^{N}\frac{1}{\lambda^k}A^k$$

and the limit $S = \lim_{N\to\infty} S_N$, if it exists.

This is called a *Neumann series* for the operator A. Since A is continuous, so are the compositions A^k and

$$\|A^k\| \le \|A\|^k$$

From the estimate

$$\begin{aligned}\|S_N - S_M\| &= \left\|\frac{1}{\lambda}\sum_{k=M+1}^{N}\frac{1}{\lambda^k}A^k\right\| \le \frac{1}{|\lambda|}\sum_{k=M+1}^{N}\frac{1}{|\lambda|^k}\|A\|^k \\ &\le \frac{1}{|\lambda|}\left(\frac{\|A\|}{|\lambda|}\right)^{M+1}\sum_{k=0}^{\infty}\left(\frac{\|A\|}{|\lambda|}\right)^k, \quad \text{for} \quad N \ge M\end{aligned}$$

it follows that the sequence S_N is Cauchy if

$$\|A\| < |\lambda|$$

Since U is complete, so is $\mathcal{L}(U,U)$ and therefore S_N has a limit, say $S \in \mathcal{L}(U,U)$.

We proceed now to show that $S = (\lambda I - A)^{-1}$. We have

$$
\begin{aligned}
(\lambda I - A)S_N &= (\lambda I - A)\frac{1}{\lambda}\sum_{k=0}^{N}\frac{1}{\lambda^k}A^k \\
&= \sum_{k=0}^{N}\frac{1}{\lambda^k}A^k - \sum_{k=1}^{N+1}\frac{1}{\lambda^k}A^k \\
&= I - \left(\frac{A}{\lambda}\right)^{N+1}
\end{aligned}
$$

and, consequently,

$$
\|(\lambda I - A)\, S_N - I\| \leq \left(\frac{\|A\|}{|\lambda|}\right)^{N+1}
$$

Passing to the limit with $N \to \infty$, we get

$$
\|(\lambda I - A)S - I\| = 0 \Longrightarrow (\lambda I - A)S = I
$$

which proves that S is a right inverse of $\lambda I - A$. A similar argument reveals that $S(\lambda I - A) = I$. Hence

$$
S = (\lambda I - A)^{-1}
$$

Observe that $\|A\| < |\lambda|$ is only a sufficient condition for $(\lambda I - A)^{-1}$ to exist and to be continuous. Cases exist in which $(\lambda I - A)^{-1} \in \mathcal{L}(U, U)$, but $\|A\| \geq |\lambda|$.

Exercises

5.7.1. Show that the integral operator defined by

$$
Au(y) = \int_0^1 K(x, y)u(x)dx
$$

where $K(x, y)$ is a function continuous on the square $\overline{\Omega} = \{(x, y) \in \mathbb{R}^2 : 0 \leq x, y \leq 1\}$, is continuous on $C[0, 1]$ (with $\|u\| = \sup_{x\in[0,1]} |u(x)|$).

5.7.2. Let U and V be two arbitrary topological vector spaces. Show that a linear operator $A : U \to V$ is continuous iff it is continuous at $\mathbf{0}$.

5.7.3. Discuss why, for linear mappings, continuity and uniform continuity are equivalent concepts.

5.7.4. Show that the null space $\mathcal{N}(A)$ of any continuous linear operator $A \in \mathcal{L}(U,V)$ is a closed linear subspace of U.

5.8 Uniform Boundedness and Banach-Steinhaus Theorems

In some situations, we are interested in determining whether the norms of a given collection of bounded linear operators $\{A_\alpha\} \in \mathcal{L}(U,V)$ have a finite least upper bound or, equivalently, if there is some uniform bound for the set $\{\|A_\alpha\|\}$. Though the norm of each A_α is finite, there is no guarantee that they might not form an increasing sequence. The following theorem is called the *principle of uniform boundedness* and it provides a criterion for determining when such an increasing sequence is not formed.

THEOREM 5.8.1
(The Uniform Boundedness Theorem)

Let U be a Banach space and V a normed space, and let

$$T_\iota \in \mathcal{L}(U,V) \ , \ \iota \in I$$

be a family of linear, continuous operators, pointwise uniformly bounded, i.e.,

$$\forall \ \boldsymbol{u} \in U \ \exists \ C(\boldsymbol{u}) > 0 : \ \|T_\iota \boldsymbol{u}\|_V \leq C(\boldsymbol{u}) \quad \forall \ \iota \in I$$

Then T_ι are uniformly bounded, i.e.,

$$\exists \ c > 0 \quad \|T_\iota\|_{\mathcal{L}(U,V)} \leq c \quad \forall \ \iota \in I$$

PROOF The proof is based on the Baire category theorem (Chapter 4, Theorem 4.8.2) for complete metric spaces.

Define

$$M_k \stackrel{\text{def}}{=} \{\boldsymbol{u} \in U : \|T_\iota \boldsymbol{u}\|_V \leq k \quad \forall \ \iota \in I\}$$

Note that the M_k are closed (why?). Certainly,

$$U = \bigcup_{1}^{\infty} M_k$$

Since U, as a Banach space, is of the second Baire category, one of the sets M_k must have a nonempty interior, i.e., there exists k and a ball $B(\boldsymbol{u}_0, \varepsilon)$ such that

$$B(\boldsymbol{u}_o, \varepsilon) \subset M_k$$

Consequently, for every $\|\boldsymbol{u}\|_U = 1$,

$$\begin{aligned}
\left\|T_\iota\left(\frac{\varepsilon}{2}\boldsymbol{u}\right)\right\|_V &= \left\|T_\iota\left(\frac{\varepsilon}{2}\boldsymbol{u} + \boldsymbol{u}_o - \boldsymbol{u}_o\right)\right\|_V \\
&\le \left\|T_\iota\left(\frac{\varepsilon}{2}\boldsymbol{u} + \boldsymbol{u}_o\right)\right\|_V + \|T_\iota(\boldsymbol{u}_o)\|_V \\
&\le k + C(\boldsymbol{u}_o)
\end{aligned}$$

for every $\iota \in I$, which implies that

$$\|T_\iota\| \le \frac{2}{\varepsilon}\left(k + C(\boldsymbol{u}_o)\right) \quad \forall\, \iota \in I$$

∎

One of the most important consequences of the uniform boundedness theorem is the following Banach-Steinhaus theorem examining properties of pointwise limits of sequences of continuous linear operators defined on Banach spaces.

THEOREM 5.8.2
(The Banach-Steinhaus Theorem)

Let U be a Banach space and V a normed space, and let

$$T_n \in \mathcal{L}(U, V)$$

be a pointwise convergent sequence of continuous, linear operators from U to V, i.e.,

$$\forall\, \boldsymbol{u} \in U \ \exists \lim_{n\to\infty} T_n\boldsymbol{u} = \text{(by definition) } T\boldsymbol{u}$$

Then:

(i) $T \in \mathcal{L}(U, V)$.

(ii) $\|T\| \leq \liminf_{n\to\infty} \|T_n\|$.

PROOF

Step 1. T is linear (essentially follows the proof of Proposition 5.7.1).

Step 2. From the continuity of the norm it follows that

$$\lim_{n\to\infty} \|T_n(\boldsymbol{u})\|_V = \|T\boldsymbol{u}\|_V \quad , \ \forall\, \boldsymbol{u} \in U$$

Consequently, T_n are pointwise uniformly bounded and, by the uniform boundedness theorem, the sequence of the norms $\|T_n\|$ is bounded.

Step 3. Passing to the lim inf on both sides of the inequality (according to Step 2 the limit is finite)

$$\|T_n\boldsymbol{u}\|_V \leq \|T_n\| \; \|\boldsymbol{u}\|_U$$

we get

$$\|T\boldsymbol{u}\|_V = \lim_{n\to\infty} \|T_n\boldsymbol{u}\|_V \leq \left(\liminf_{n\to\infty} \|T_n\|\right) \|\boldsymbol{u}\|_U$$

which proves that

1. T is bounded and
2. $\|T\| \leq \liminf_{n\to\infty} \|T_n\|$.

∎

5.9 The Open Mapping Theorem

Open Functions. Let X and Y be two topological spaces. A function $f : X \to Y$ is said to be *open* iff it maps open sets in X into open sets in Y, i.e.,

$$A \text{ open in } X \implies f(A) \text{ open in } Y$$

Notice that if f is bijective and f^{-1} is continuous, then f is open.

The fundamental result of S. Banach reads as follows:

THEOREM 5.9.1
(The Open Mapping Theorem)

Let X and Y be two Banach spaces and T a nontrivial continuous linear operator from X onto Y such that

$$T \in \mathcal{L}(X, Y)$$

and T is surjective. Then T is an open mapping from X to Y, i.e.,

$$A \text{ open in } X \Longrightarrow T(A) \text{ open in } Y$$

LEMMA 5.9.1

Let X, Y be two normed vector spaces, T a continuous, linear operator from X into Y such that the range of T, $\mathcal{R}(T)$ is of the second Baire category in Y.

Then, for every A, a neighborhood of $\mathbf{0}$ in X, there exists D, a neighborhood of $\mathbf{0}$ in Y, such that

$$D \subset \overline{T(A)}$$

In other words, for every neighborhood A of $\mathbf{0}$ in X, the closure $\overline{T(A)}$ is a neighborhood of $\mathbf{0}$ in Y.

PROOF

Step 1. One can always find a ball $B = B(\mathbf{0}, \varepsilon)$ with radius ε small enough such that

$$B + B \subset A$$

Step 2. Since, for every $\boldsymbol{x} \in X$, $\lim_{n\to\infty} \frac{1}{n}\boldsymbol{x} = \mathbf{0}$, there must exist a large enough n such that $\boldsymbol{x} \in nB$. Consequently,

$$X = \bigcup_{n=1}^{\infty} (nB)$$

which implies that

$$\mathcal{R}(T) = \bigcup_{n=1}^{\infty} T(nB)$$

Step 3. As $\mathcal{R}(T)$ is of the second category, there exists an index n_o such that $\overline{T(n_o B)}$ has a nonempty interior. But since multiplication by a nonzero scalar is a homeomorphism,

$$\overline{T(n_o B)} = \overline{n_o T(B)} = n_o \overline{T(B)}$$

and therefore

$$\text{int}\overline{T(B)} \neq \emptyset$$

which means that there exists a ball $B(\boldsymbol{y}_o, \delta)$ such that

$$B(\boldsymbol{y}_o, \delta) \subset \overline{T(B)}$$

One can always assume that $\boldsymbol{y}_o \in T(B)$, i.e., that $\boldsymbol{y}_o = T\boldsymbol{x}_o$, for some $\boldsymbol{x}_o \in B$ (why?).

Step 4. Consider the ball $D = B(\mathbf{0}, \delta)$. We have

$$\begin{aligned} D = -\boldsymbol{y}_o + B(\boldsymbol{y}_o, \delta) &\subset -\boldsymbol{y}_o + \overline{T(B)} \\ &= T(-\boldsymbol{x}_o) + \overline{T(B)} \\ &= \overline{T(-\boldsymbol{x}_o + B)} \\ &\subset \overline{T(A)} \end{aligned}$$

since $-\boldsymbol{x}_o + B \subset B + B \subset A$. ∎

PROOF ***of the Open Mapping Theorem***

Step 1. Denote by A_ε and B_ε balls centered at $\mathbf{0}$ in X and Y, respectively.

$$A_\varepsilon = B(\mathbf{0}, \varepsilon) \subset X \ , \ B_\varepsilon = B(\mathbf{0}, \varepsilon) \subset Y$$

Pick also an arbitrary $\varepsilon > 0$ and denote $\varepsilon_i = \frac{\varepsilon}{2^i}$. By the lemma,

$$\forall i \quad \exists \eta_i : B_{\eta_i} \subset \overline{T(A_{\varepsilon_i})}$$

One can always assume that $\lim_{i\to\infty} \eta_i = 0$ (why?).

Step 2. Let $\boldsymbol{y} \in B_{\eta_0}$. We claim that there exists an element $\boldsymbol{x} \in A_{2\varepsilon_0}$ such that $T\boldsymbol{x} = \boldsymbol{y}$.

Indeed, from the above inclusion, we know that

$$\exists \ \boldsymbol{x}_0 \in A_{\varepsilon_0} : \ \|\boldsymbol{y} - T\boldsymbol{x}_o\|_Y < \eta_1$$

It follows that $\boldsymbol{y} - T\boldsymbol{x}_o \in B_{\eta_1}$ and, by the same reasoning,

$$\exists\, \boldsymbol{x}_1 \in A_{\varepsilon_1} : \|\boldsymbol{y} - T\boldsymbol{x}_o - T\boldsymbol{x}_1\|_Y < \eta_2$$

By induction, there exists a sequence $\boldsymbol{x}_i \in A_{\varepsilon_i}$ such that

$$\left\| \boldsymbol{y} - T\left(\sum_{i=0}^{n} \boldsymbol{x}_i\right) \right\|_Y < \eta_{n+1}$$

Since

$$\left\| \sum_{k=m+1}^{n} \boldsymbol{x}_k \right\|_X \le \sum_{k=m+1}^{n} \|\boldsymbol{x}_k\|_X \le \sum_{k=m+1}^{n} \varepsilon_k$$

$$\le \left(\sum_{m+1}^{n} 2^{-k} \right) \varepsilon_o$$

the sequence of finite sums

$$\sum_{k=0}^{n} \boldsymbol{x}_k$$

is Cauchy and by the completeness of X has a limit $\boldsymbol{x} \in X$.

Moreover, by the continuity of the norm,

$$\|\boldsymbol{x}\|_X = \lim_{n\to\infty} \left\| \sum_{k=0}^{n} \boldsymbol{x}_k \right\|_X \le \lim_{n\to\infty} \sum_{k=0}^{n} \|\boldsymbol{x}_k\|_X$$

$$\le \left(\sum_{0}^{\infty} 2^{-k} \right) \varepsilon_0 = 2\varepsilon_0$$

Finally, passing to the limit with $n \to \infty$, we get

$$\|\boldsymbol{y} - T\boldsymbol{x}\|_Y = 0 \Longrightarrow \boldsymbol{y} = T\boldsymbol{x}$$

As $\boldsymbol{y}$ was an arbitrary element of B_{η_o}, we have shown that

$$B_{\eta_o} \subset T\left(A_{2\varepsilon}\right)$$

Step 3. Let G be a nonempty open set in X and let $\boldsymbol{x} \in G$. By the openness of G, there exists $\varepsilon > 0$ such that

$$x + A_{2\varepsilon} \subset G$$

Consequently,

$$\begin{aligned} Tx + B_{\eta_o} &\subset Tx + T(A_{2\varepsilon}) = T(x + A_{2\varepsilon}) \\ &\subset T(G) \end{aligned}$$

and therefore $T(G)$ is open. ∎

COROLLARY 5.9.1

Let X and Y be two Banach spaces and $T \in \mathcal{L}(X, Y)$ have a closed range $\mathcal{R}(T)$ in Y. Then T is an open mapping from X onto its range $\mathcal{R}(T)$.

PROOF $\mathcal{R}(T)$ as a closed subspace of a complete space is complete and the assertion follows immediately from the open mapping theorem. Note that T being open from X into its range $\mathcal{R}(T)$ means only that if $G \subset X$ is open, then $T(G)$ is open in $\mathcal{R}(T)$, i.e.,

$$T(G) = H \cap \mathcal{R}(T)$$

for some open H in Y. In general, this does not mean that $T(G)$ is open in Y. ∎

COROLLARY 5.9.2
(The Banach Theorem)

Let X, Y be two Banach spaces and $T \in \mathcal{L}(X, Y)$ a bijective, *continuous, linear operator from X onto Y. Then the inverse T^{-1} is continuous.*

In other words, every bijective, continuous, and linear map from a Banach space onto a Banach space is automatically a homeomorphism.

PROOF T^{-1} exists and T open implies T^{-1} is continuous. ∎

The last observation is crucial for many developments in applied functional analysis.

Exercises

5.9.1. Construct an example of a continuous function from $\mathbb{R}$ into $\mathbb{R}$ which is *not* open.

Closed Operators

5.10 Closed Operators. Closed Graph Theorem

We begin with some simple observations concerning Cartesian products of normed spaces. First of all, recall that if X and Y are vector spaces, then the Cartesian product $X \times Y$ is also a vector space with operations defined by

$$(\boldsymbol{x}_1, \boldsymbol{y}_1) + (\boldsymbol{x}_2, \boldsymbol{y}_2) \overset{\text{def}}{=} (\boldsymbol{x}_1 + \boldsymbol{x}_2, \boldsymbol{y}_1 + \boldsymbol{y}_2)$$

$$\alpha\ (\boldsymbol{x}, \boldsymbol{y}) \overset{\text{def}}{=} (\alpha \boldsymbol{x}, \alpha \boldsymbol{y})$$

where the vector additions and multiplications by a scalar on the right-hand side are those in the X and Y spaces, respectively.

If, additionally, X and Y are normed spaces with norms $\|\cdot\|_X$ and $\|\cdot\|_Y$, respectively, then $X \times Y$ may be equipped with a (not unique) norm of the form

$$\|(\boldsymbol{x}, \boldsymbol{y})\| = (\|\boldsymbol{x}\|_X^p + \|\boldsymbol{y}\|_Y^p)^{\frac{1}{p}} \quad 1 \le p \le +\infty$$

Finally, if X and Y are complete, then $X \times Y$ is also complete. Indeed, if $(\boldsymbol{x}_n, \boldsymbol{y}_n)$ is a Cauchy sequence in $X \times Y$, then $\boldsymbol{x}_n$ is a Cauchy sequence in X and $\boldsymbol{y}_n$ is a Cauchy sequence in Y. Consequently, both $\boldsymbol{x}_n$ and $\boldsymbol{y}_n$ have limits, say $\boldsymbol{x}$ and $\boldsymbol{y}$, and, therefore, by the definition of the norm in $\boldsymbol{X} \times Y$, $(\boldsymbol{x}_n, \boldsymbol{y}_n) \to (\boldsymbol{x}, \boldsymbol{y})$. Thus, if X and Y are Banach spaces, then $X \times Y$ is a Banach space, too.

Operators. Up to this point, all of the linear transformations from a vector space X into a vector space Y have been defined on the *whole* space X, i.e., their domain of definition coincided with the entire space X. In a more general situation, it may be useful to consider linear operators defined on a *proper subspace* of X only (see Example 5.6.4). In fact, some authors reserve the name *operator* to such functions, distinguishing them from *transformations*, which are defined on the whole space.

Thus, in general, a linear operator T from a vector space X into a vector space Y may be defined only on a proper subspace of X, denoted $D(T)$ and called the *domain of definition of* T, or concisely, the *domain of* T:

$$X \supset D(T) \ni \boldsymbol{x} \longrightarrow T\boldsymbol{x} \in Y$$

Note that in the case of *linear* operators, the domain $D(T)$ must be a *vector subspace* of X (otherwise it would make no sense to speak of linearity of T).

Still, the choice of the domain is somehow arbitrary. Different domains with the same rule defining T result formally in different operators in much the same fashion as functions are defined by specifying their *domain, codomain*, and the rule (see Chapter 1).

With every operator T (not necessarily linear) we can associate its *graph*, denoted $G(T)$ and defined as

$$\text{graph } T = G(T) \stackrel{\text{def}}{=} \{(\boldsymbol{x}, T\boldsymbol{x}) : \boldsymbol{x} \in D(T)\} \subset X \times Y$$

(recall the discussion in Chapter 1, Section 1.11).

PROPOSITION 5.10.1

Let X, Y be two vector spaces and $T : X \supset D(T) \to Y$ an operator. Then T is linear iff its graph $G(T)$ is a linear subspace of $X \times Y$.

PROOF

Assume T is linear and let $(\boldsymbol{x}_1, \boldsymbol{y}_1)$, $(\boldsymbol{x}_2, \boldsymbol{y}_2) \in G(T)$, i.e., $\boldsymbol{x}_1, \boldsymbol{x}_2 \in D(T)$ and $\boldsymbol{y}_1 = T\boldsymbol{x}_1$, $\boldsymbol{y}_2 = T\boldsymbol{y}_2$.

Consequently, for every α_1, α_2,

$$\alpha_1 \boldsymbol{x}_1 + \alpha_2 \boldsymbol{x}_2 \in D(T)$$

and

$$\alpha_1 \boldsymbol{y}_1 + \alpha_2 \boldsymbol{y}_2 = \alpha_1 T\boldsymbol{x}_1 + \alpha_2 T\boldsymbol{x}_2 = T\left(\alpha_1 \boldsymbol{x}_1 + \alpha_2 \boldsymbol{x}_2\right)$$

which proves that

$$\alpha_1 \left(\boldsymbol{x}_2, \boldsymbol{y}_2\right) + \alpha_2 \left(\boldsymbol{x}_2, \boldsymbol{y}_2\right) \in G(T)$$

and therefore $G(T)$ is a linear subspace of $X \times Y$.

Conversely, assume that $G(T)$ is a vector subspace of $X \times Y$. Let $\boldsymbol{x}_1, \boldsymbol{x}_2 \in D(T)$ and $\boldsymbol{x} = \alpha_1 \boldsymbol{x}_1 + \alpha_2 \boldsymbol{x}_2$. By linearity of $G(T)$,

$$\alpha_1 (\boldsymbol{x}_1, T\boldsymbol{x}_1) + \alpha_2 (\boldsymbol{x}_2, T\boldsymbol{x}_2) = (\boldsymbol{x}, \alpha_1 T\boldsymbol{x}_1 + \alpha_2 T\boldsymbol{x}_2) \in G(T)$$

which means that $\boldsymbol{x} \in D(T)$ (thus $D(T)$ is a linear subspace of X) and

$$T\boldsymbol{x} = \alpha_1 T\boldsymbol{x}_1 + \alpha_2 T\boldsymbol{x}_2$$

∎

Closed Operators. Let X and Y be two normed spaces. A linear operator $T : D(T) \to Y$ is said to be *closed* iff its graph $G(T)$ is a closed subspace of $X \times Y$.

It follows from the definition that every linear and continuous operator defined on a closed subspace of X (in particular the whole X) is automatically closed.

Indeed, if $(\boldsymbol{x}_n, T\boldsymbol{x}_n) \to (\boldsymbol{x}, \boldsymbol{y})$, then $\boldsymbol{x}_n \to \boldsymbol{x}$ and $\boldsymbol{x} \in D(T)$. By continuity of T, $T\boldsymbol{x} = \lim_{n\to\infty} T\boldsymbol{x}_n = \boldsymbol{y}$.

With every *injective* operator

$$T : X \supset D(T) \longrightarrow \mathcal{R}(T) \subset Y$$

we can associate an inverse T^{-1} defined on the range $\mathcal{R}(T)$

$$T^{-1} : Y \supset D\left(T^{-1}\right) = \mathcal{R}(T) \longrightarrow \mathcal{R}\left(T^{-1}\right) = D(T) \subset X$$

As the operation

$$X \times Y \ni (\boldsymbol{x}, \boldsymbol{y}) \longrightarrow (\boldsymbol{y}, \boldsymbol{x}) \in Y \times X$$

is obviously a homeomorphism, it follows immediately from the definition of closed operators that if T is closed and injective then T^{-1} is closed, too.

The following is a simple characterization of closed operators.

PROPOSITION 5.10.2

Let X and Y be two normed spaces and T a linear operator from $D(T) \subset X$ to Y. The following conditions are equivalent to each other:

(i) *T is closed.*

(ii) *For an arbitrary sequence $\boldsymbol{x}_n \in D(T)$, if $\boldsymbol{x}_n \to \boldsymbol{x}$ and $T\boldsymbol{x}_n \to \boldsymbol{y}$ then $\boldsymbol{x} \in D(T)$ and $T\boldsymbol{x} = \boldsymbol{y}$.*

PROOF

(i) $\Longrightarrow$ (ii) Since

$$G(T) \ni (\boldsymbol{x}_n, T\boldsymbol{x}_n) \longrightarrow (\boldsymbol{x}, \boldsymbol{y})$$

and $G(T)$ is closed, $(\boldsymbol{x}, \boldsymbol{y}) \in G(T)$, which means that $\boldsymbol{x} \in D(T)$ and $\boldsymbol{y} = T\boldsymbol{x}$.

(ii) $\Longrightarrow$ (i) Let

$$G(T) \ni (\boldsymbol{x}_n, \boldsymbol{y}_n) \longrightarrow (\boldsymbol{x}, \boldsymbol{y})$$

Then $\boldsymbol{x}_n \to \boldsymbol{x}$ and $T\boldsymbol{x}_n = \boldsymbol{y}_n \to \boldsymbol{y}$, which implies that $\boldsymbol{x} \in D(T)$ and $T\boldsymbol{x} = \boldsymbol{y}$, or, equivalently, $(\boldsymbol{x}, \boldsymbol{y}) \in G(T)$. ∎

Closable Operators. Let X, Y be normed spaces and

$$T : X \supset D(T) \longrightarrow Y$$

a linear operator. Operator T is said to be *closable* (or *preclosed*) iff the closure of graph of T, $\overline{G(T)}$, in $X \times Y$ can be identified as a graph of a (possibly another) linear operator $\overline{T}$. The operator $\overline{T}$ is called the *closure* of T.

PROPOSITION 5.10.3

Let X and Y be normed spaces and T a linear operator from $D(T) \subset X$ to Y. The following conditions are equivalent to each other:

(i) T is closable.

(ii) For an arbitrary sequence $\boldsymbol{x}_n \in D(T)$,

$$\boldsymbol{x}_n \to \boldsymbol{0} \text{ and } T\boldsymbol{x}_n \to \boldsymbol{y}$$

implies $\boldsymbol{y} = \boldsymbol{0}$.

PROOF

(i) $\Longrightarrow$ (ii) $\boldsymbol{x}_n \to \boldsymbol{0}$ and $T\boldsymbol{x}_n \to \boldsymbol{y}$ means that $(\boldsymbol{x}_n, T\boldsymbol{x}_n) \to (\boldsymbol{0}, \boldsymbol{y})$, which implies that

$$(\boldsymbol{0}, \boldsymbol{y}) \in \overline{G(T)} = G(\overline{T})$$

and, consequently, $\boldsymbol{y} = \overline{T}(\boldsymbol{0}) = \boldsymbol{0}$ ($\overline{T}$ is linear).

(ii) $\Longrightarrow$ (i) *Step 1.* Consider the closure $\overline{G(T)}$ and define

$$D(S) = \text{projection of } \overline{G(T)} \text{ on } X$$

$$= \left\{ \boldsymbol{x} \in X : \exists\, \boldsymbol{y} \in Y,\ (\boldsymbol{x}, \boldsymbol{y}) \in \overline{G(T)} \right\}$$

Then $D(S)$ is a linear subspace of X (why?). We claim that for every $\boldsymbol{x} \in D(S)$ there exists only one (a unique) $\boldsymbol{y}$ such that $(\boldsymbol{x}, \boldsymbol{y}) \in \overline{G(T)}$.

Indeed, if there were two, say $\boldsymbol{y}_1$ and $\boldsymbol{y}_2$, then there would have to exist two corresponding sequences $\boldsymbol{x}_n^1$ and $\boldsymbol{x}_n^2$ in $D(T)$ such that

$$\left(\boldsymbol{x}_n^1, T\boldsymbol{x}_n^1\right) \longrightarrow (\boldsymbol{x}, \boldsymbol{y}_1) \text{ and } \left(\boldsymbol{x}_n^2, T\boldsymbol{x}_n^2\right) \longrightarrow (\boldsymbol{x}, \boldsymbol{y}_2)$$

Consequently,

$$\boldsymbol{x}_n = \boldsymbol{x}_n^1 - \boldsymbol{x}_n^2 \longrightarrow \boldsymbol{0} \text{ and } T\boldsymbol{x}_n \longrightarrow \boldsymbol{y}_1 - \boldsymbol{y}_2$$

and by condition (ii), $\boldsymbol{y}_1 - \boldsymbol{y}_2 = \boldsymbol{0}$ or $\boldsymbol{y}_1 = \boldsymbol{y}_2$.

Step 2. Define

$$D(S) \ni \boldsymbol{x} \longrightarrow S\boldsymbol{x} = \boldsymbol{y} \in Y,\ (\boldsymbol{x}, \boldsymbol{y}) \in \overline{G(T)}$$

For arbitrary $\boldsymbol{x}^1, \boldsymbol{x}^2 \in D(S)$ there exist corresponding sequences $\boldsymbol{x}_n^1$, $\boldsymbol{x}_n^2$ such that

$$\left(\boldsymbol{x}_n^1, T\boldsymbol{x}_n^1\right) \rightarrow \left(\boldsymbol{x}^1, S\boldsymbol{x}^1\right) \text{ and } \left(\boldsymbol{x}_n^2, T\boldsymbol{x}_n^2\right) \rightarrow \left(\boldsymbol{x}^2, S\boldsymbol{x}^2\right)$$

Consequently,

$$\left(\alpha_1\boldsymbol{x}_n^1 + \alpha_2\boldsymbol{x}_n^2,\ T\left(\alpha_1\boldsymbol{x}_n^1 + \alpha_2\boldsymbol{x}_n^2\right)\right) \rightarrow \left(\alpha_1\boldsymbol{x}^1 + \alpha_2\boldsymbol{x}^2,\ \alpha_1 S\boldsymbol{x}^1 + \alpha_2 S\boldsymbol{x}^2\right)$$

which proves that

$$S\left(\alpha_1\boldsymbol{x}^1 + \alpha_2\boldsymbol{x}^2\right) = \alpha_1 S\boldsymbol{x}^1 + \alpha_2 S\boldsymbol{x}^2$$

and therefore S is linear. ∎

COROLLARY 5.10.1

Every linear and *continuous* operator T from $D(T)$ in a normed space X to a normed space Y is closable.

PROOF Let $\boldsymbol{x}_n \rightarrow \boldsymbol{0}$ and $T\boldsymbol{x}_n \rightarrow \boldsymbol{y}$. By continuity of T, $T\boldsymbol{x}_n \rightarrow \boldsymbol{0}$ and since T is single-valued, it must be $\boldsymbol{y} = \boldsymbol{0}$. Then the assertion follows from Proposition 5.10.3. ∎

We conclude this section with the fundamental result concerning closed operators due to Banach.

THEOREM 5.10.1
(Closed Graph Theorem)

Let X and Y be Banach spaces and T a linear and closed operator from X to Y with the domain of definition coinciding with the whole X, i.e., $D(T) = X$. Then T is continuous.

PROOF As $X \times Y$ is a Banach space and T is closed, it follows that the graph of T, $G(T)$ is a Banach space, too.

Considering the projection

$$i_X : G(T) \longrightarrow X \ , \ i_X\left((\boldsymbol{x}, T\boldsymbol{x})\right) \stackrel{\text{def}}{=} \boldsymbol{x}$$

we see that

1. i_X is a bijection and
2. i_X is continuous.

Thus, by Corollary 5.9.2 to the open mapping theorem, i_X has a *continuous* inverse i_X^{-1}.

Introducing now the second projection

$$i_Y : G(T) \longrightarrow Y \ , \ i_Y((\boldsymbol{x}, T\boldsymbol{x})) \stackrel{\text{def}}{=} T\boldsymbol{x}$$

we can represent T in the form

$$T = i_Y \circ i_X^{-1}$$

which proves that, as a composition of continuous operators, T must be continuous. ∎

The important message in this theorem is that nontrivial closed operators, i.e., those which are not continuous, are *never* defined on the entire space X.

Exercises

5.10.1. Let A be a closed linear operator from $U \supset D(A)$ into V where U and V are Banach spaces. Show that the vector space $(D(A), \|\cdot\|_A)$ where $\|\boldsymbol{u}\|_A = \|\boldsymbol{u}\|_U + \|A\boldsymbol{u}\|_V$ (the so-called *operator norm* on $D(A)$) is Banach.

5.11 Example of a Closed Operator

Distributional Derivatives. Let $\Omega \subset \mathbb{R}^n$ be an open set, $\boldsymbol{\alpha} = (\alpha_1, \ldots, \alpha_n)$ a multiindex and $u \in L^p(\Omega)$ an arbitrary L^p function. A function $u^{\boldsymbol{\alpha}}$ defined on Ω is called the *distributional derivative* of $\boldsymbol{u}$, denoted $D^{\boldsymbol{\alpha}}u$, iff

$$\int_\Omega u D^{\boldsymbol{\alpha}} \varphi dx = (-1)^{|\boldsymbol{\alpha}|} \int_\Omega u^{\boldsymbol{\alpha}} \varphi dx \quad \forall\, \varphi \in C_0^\infty(\Omega)$$

where $C_0^\infty(\Omega)$ is the space of test functions discussed in Section 5.3. (It is understood that function $u^{\boldsymbol{\alpha}}$ must satisfy sufficient conditions for the right-hand side to exist.)

Notice that the notion of the distributional derivative is a generalization of the classical derivative. Indeed, in the case of a $C^{|\boldsymbol{\alpha}|}$ function u, the formula above follows from the (multiple) integration by parts and the fact that test functions, along with their derivatives, vanish on the boundary $\partial\Omega$.

Example 5.11.1

Let $\Omega = (0, 1) \subset \mathbb{R}$ and $x_0 \in (0, 1)$. Consider a function

$$u(x) = \begin{cases} u_1(x) & 0 < x \le x_0 \\ u_2(x) & x_0 \le x < 1 \end{cases}$$

where each of the branches is C^1 in the corresponding subinterval, including the endpoints, and $u_1(x_0) = u_2(x_0)$ (see Fig. 5.2). Thus u is globally continuous but may not be C^1 (the derivative at x_0 may not exist). For an

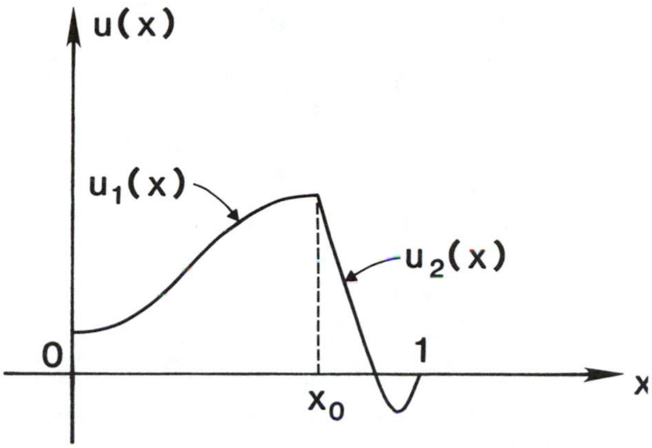

Figure 5.2
An example of a function differentiable in the distributional but not classical sense.

arbitrary (test) function $\varphi \in C_0^\infty(0,1)$ we have

$$\int_0^1 u\varphi' dx = \int_0^{x_0} u_1\varphi' dx + \int_{x_0}^1 u_2\varphi' dx$$

$$= -\int_0^{x_0} u_1'\varphi dx + u_1\varphi|_{x_0}^1 - \int_{x_0}^1 u_2'\varphi dx + u_2\varphi|_{x_0}^1$$

$$= -\int_0^{x_0} u_1'\varphi dx - \int_{x_0}^1 u_2'\varphi dx - [u_2(x_0) - u_1(x_0)]\,\varphi(x_0)$$

since $\varphi(0) = \varphi(1) = 0$.

But $u_2(x_0) = u_1(x_0)$ and therefore

$$\int_0^1 u\varphi' dx = -\left(\int_0^{x_0} u_1'\varphi dx + \int_{x_0}^1 u_2'\varphi dx\right)$$

Introducing a function

$$u'(x) = \begin{cases} u_1'(x) & 0 < x < x_0 \\ c & x = x_0 \\ u_2'(x) & x_0 < x < 1 \end{cases}$$

where c is an arbitrary constant, we see that

$$\int_0^1 u\varphi' dx = -\int_0^1 u'\varphi dx$$

which proves that u' is a distributional derivative of u. As constant c is completely arbitrary in the definition of u', we remind that the Lebesgue integral is *insensitive* to the change of the integrand on a set of measure zero. In fact, in order to be uniquely defined, the distributional derivatives have to be understood as *equivalence classes of functions* equal almost everywhere. □

Sobolev Spaces. Let $\Omega \subset \mathbb{R}^n$ be an open set, m an integer and $1 \leq p \leq \infty$. Consider the set of all L^p-functions on Ω, whose distributional derivatives of order up to m all exist and are themselves L^p functions:

$$W^{m,p}(\Omega) \stackrel{\text{def}}{=} \{u \in L^p(\Omega) : D^{\boldsymbol{\alpha}} u \in L^p(\Omega) \quad \forall\, |\boldsymbol{\alpha}| \leq m\}$$

It can be easily checked (Exercise 5.11.2) that $W^{m,p}(\Omega)$ is a normed space with the norm

$$\|u\|_{m,p} = \|u\|_{W^{m,p}(\Omega)} = \begin{cases} \left(\sum_{|\boldsymbol{\alpha}| \leq m} \|D^{\boldsymbol{\alpha}} u\|^p_{L^p(\Omega)} \right)^{\frac{1}{p}} & \text{for } 1 \leq p < \infty \\ \max_{|\boldsymbol{\alpha}| \leq m} \|D^{\boldsymbol{\alpha}} u\|_{L^\infty(\Omega)} & \text{for } p = \infty \end{cases}$$

PROPOSITION 5.11.1

The Sobolev spaces $W^{m,p}(\Omega)$ are complete and therefore Banach spaces.

PROOF Let u_n be a Cauchy sequence in $W^{m,p}(\Omega)$. The definition of the norm in the space implies that both functions u_n and their derivatives $D^{\boldsymbol{\alpha}} u_n$, for every multiindex $\boldsymbol{\alpha}$, form Cauchy sequences in $L^p(\Omega)$ and therefore, by completeness of $L^p(\Omega)$, converge to some limits u and $u^{\boldsymbol{\alpha}}$, respectively. It remains only to show that the limits $u^{\boldsymbol{\alpha}}$ are distributional derivatives of the limit u. This is done by passing with $n \to \infty$ in the identity

$$\int_\Omega u_n D^{\boldsymbol{\alpha}} \varphi \, dx = (-1)^{|\boldsymbol{\alpha}|} \int_\Omega D^{\boldsymbol{\alpha}} u_n \varphi \, dx$$

It follows from the Hölder inequality that both sides are linear and continuous functionals of u_n and $D^{\boldsymbol{\alpha}} u_n$, respectively, and therefore if $u_n \to u$ and $D^{\boldsymbol{\alpha}} u_n \to u^{\boldsymbol{\alpha}}$ for $n \to \infty$, then in the limit,

$$\int_\Omega u D^{\boldsymbol{\alpha}} \varphi \, dx = (-1)^{|\boldsymbol{\alpha}|} \int_\Omega u^{\boldsymbol{\alpha}} \varphi \, dx$$

which proves that $u^{\boldsymbol{\alpha}} \in L^p(\Omega)$ is a distributional derivative of the limit $u \in L^p(\Omega)$. This holds for every $|\boldsymbol{\alpha}| \leq m$ and therefore $u \in W^{m,p}(\Omega)$. ∎

For $p = 2$, the Sobolev spaces have structure of a Hilbert space (we will study that particular case in the next chapter). A different notation is then used:

$$H^m(\Omega) \stackrel{\text{def}}{=} W^{m,2}(\Omega)$$

At this point we are prepared to give a nontrivial example of a closed operator in Banach spaces.

Example 5.11.2

For an open set $\Omega \subset \mathbb{R}^n$ and $1 \leq p \leq \infty$ consider the Banach space $L^p(\Omega)$ and define the operator T as

$$Tu = \sum_{|\boldsymbol{\alpha}| \leq m} a_{\boldsymbol{\alpha}} D^{\boldsymbol{\alpha}} u$$

with domain $D(T)$ defined as

$$D(T) = \{ u \in L^p(\Omega) \; : \; Tu \in L^p(\Omega) \}$$

and derivatives understood in the distributional sense discussed earlier. Here $a_{\boldsymbol{\alpha}}$ are arbitrary constants, and $m > 0$. By the construction of the domain $D(T)$, operator T is well defined, i.e., it takes on its values in space $L^P(\Omega)$. We will demonstrate now that operator T is closed.

Toward this goal, pick a sequence $u_n \in D(T)$ and assume that

$$u_n \to u \text{ and } Tu_n = \sum_{|\boldsymbol{\alpha}| \leq m} a_{\boldsymbol{\alpha}} D^{\boldsymbol{\alpha}} u_n \to w \text{ in } L^p(\Omega)$$

for some $u, w \in L^p(\Omega)$. Passing with $n \to \infty$ in the identity

$$\int_\Omega u_n \sum_{|\boldsymbol{\alpha}| \le m} a_{\boldsymbol{\alpha}} (-1)^{|\boldsymbol{\alpha}|} D^{\boldsymbol{\alpha}} \varphi \, dx = \int_\Omega \sum_{|\boldsymbol{\alpha}| \le m} a_{\boldsymbol{\alpha}} D^{\boldsymbol{\alpha}} u_n \varphi \, dx$$

for arbitrary test function $\varphi \in C_0^\infty(\Omega)$, we learn that

1. $Tu \in L^p(\Omega)$
2. $Tu = w$

and therefore, by Proposition 5.10.2, the operator T is closed.

At the same time, in general, the operator T is not continuous, or equivalently bounded, as can be seen in the particular case of $\Omega = (0,1) \subset \mathbb{R}$, $D(T) = W^{1,2}(0,1)$, and $Tu = u'$. Taking $u_n(x) = \sqrt{2n+1} x^n$, we easily check that

$$\|u_n\|_{L^2} = 1 \quad \text{but} \quad \|Tu_n\|_{L^2} = \|u_n'\|_{L^2} = n\sqrt{\frac{2n+1}{2n-1}} \longrightarrow \infty$$

as $n \to \infty$. $\square$

Exercises

5.11.1. Consider $X = L^2(0,1)$ and define a linear operator $Tu = u'$, $D(T) = C^\infty([0,1])$. Show that T is closable. Can you suggest what would be the closure of T?

5.11.2. Show that the Sobolev space $W^{m,p}(\Omega)$ is a normed space.

Topological Duals. Weak Compactness

5.12 Examples of Dual Spaces. Representation Theorem for Topological Duals of L^p Spaces

In Section 7 of this chapter, for every normed space U, we introduced its topological dual U', defined as the space of all linear and *continuous* functionals from U to $\mathbb{R}$ (or $\mathbb{C}$).

Since there are many linear functionals on U that are not continuous, the topological dual is a smaller space than the algebraic dual $U^* = L(U,\mathbb{R})$ described in Chapter 2, and, algebraically, $\mathcal{L}(U,\mathbb{R}) \subset L(U,\mathbb{R})$. But we have little need for $L(U,\mathbb{R})$ in discussions of normed spaces. Unless some specific distinction is needed, we shall henceforth refer to U' as *the* dual of U.

Let $f \in U' = \mathcal{L}(U,\mathbb{R})$. As in Chapter 2, it is customary to represent the functional f as a *duality pairing*; i.e., we usually write

$$f(\boldsymbol{u}) = \langle f, \boldsymbol{u} \rangle, \quad f \in U', \quad \boldsymbol{u} \in U$$

Then the symbol $\langle \cdot, \cdot \rangle$ can be regarded as a bilinear map from $U' \times U$ into $\mathbb{R}$ or $\mathbb{C}$.

Now, since $f(\boldsymbol{u})$ is a real or complex number, $\|f(\boldsymbol{u})\| = |\langle f, \boldsymbol{u} \rangle|$. Hence, in view of what was said about the norms on spaces $\mathcal{L}(U,V)$ of linear operators, the norm of an element of U' is given by

$$\|f\|_{U'} = \sup_{\boldsymbol{u} \in U} \left\{ \frac{|\langle f, \boldsymbol{u} \rangle|}{\|\boldsymbol{u}\|_U} \ , \ \boldsymbol{u} \neq \boldsymbol{0} \right\}$$

Hence we always have

$$|\langle f, \boldsymbol{u} \rangle| \leq \|f\|_{U'} \|\boldsymbol{u}\|_U \quad f \in U', \ \boldsymbol{u} \in U$$

which in particular implies that the duality pairing is continuous (why?).

Before we proceed with some general results concerning dual spaces, we present in this section a few nontrivial examples of dual spaces in the form of so-called *representation theorems.* The task of a representation theorem is to identify elements from a dual space (i.e., linear and continuous functionals defined on a normed space) with elements from some other space, for instance, some other functions, through a *representation formula* relating functionals with those functions. The representation theorems not only provide meaningful characterizations of dual spaces, but are also of great practical value in applications.

The main result we present in this chapter is the representation theorem for the duals of the spaces $L^p(\Omega)$, $1 \leq p > \infty$.

LEMMA 5.12.1

Let $1 < q \leq 2$. There exists a positive number $c > 0$ such that

$$|1 + u|^q \geq 1 + qu + c\theta(u) \quad \forall \ u \in \mathbb{R}$$

where

$$\theta(u) = \begin{cases} |u|^2 & |u| < 1 \\ |u|^q & |u| \geq 1 \end{cases}$$

PROOF Define

$$\psi(u) \stackrel{\text{def}}{=} |1+u|^q - 1 - qu \ , \ u \in \mathbb{R}$$

$$\chi(u) \stackrel{\text{def}}{=} \frac{\psi(u)}{\theta(u)} \ , \ u \neq 0$$

As (check it)

$$\lim_{u \to 0} \chi(u) = \frac{q(q-1)}{2} \ , \ \lim_{|u| \to \infty} \chi(u) = 1$$

it follows from the continuity of χ that there exist positive constants $c_1, \delta, \Delta > 0$ such that

$$\chi(u) \geq c_1 \text{ for } |u| \leq \delta \text{ or } |u| \geq \Delta$$

At the same time,

$$\psi'(u) = q\,|1+u|^{q-1}\,\text{sgn}(1+u) - q = 0 \Longrightarrow u = 0$$

and

$$\psi''(u) = q(q-1)\,|1+u|^{q-2} \Longrightarrow \chi''(0) > 0$$

which implies that $\psi(u)$ has a unique minimum attained at 0 (equal 0) and therefore must be bounded away from 0 for $\delta < |u| < \Delta$ (why?). As the function $\theta(u)$ is bounded from above in the same range of u (why?), function $\chi(u)$ is bounded away from zero as well, i.e.,

$$\exists\, c_2 > 0 : \chi(u) \geq c_2 \text{ for } \delta < |u| < \Delta$$

It is sufficient now to take $c = \min(c_1, c_2)$ ∎

THEOREM 5.12.1
(Representation Theorem for $(L^p(\Omega))'$)

Let $\Omega \subset \mathbb{R}^N$ be an open set and $1 \leq p < \infty$. For every linear and continuous functional f defined on the space $L^p(\Omega)$, there exists a unique function $\varphi \in L^q(\Omega)$, $\frac{1}{p} + \frac{1}{q} = 1$, such that

$$f(u) = \langle f, u \rangle = \int_\Omega \varphi u \, dx \qquad \forall \, u \in L^p(\Omega)$$

Moreover,

$$\|f\|_{(L^p)'} = \|\varphi\|_{L^q}$$

REMARK 5.12.1. Consider the linear mapping

$$F : L^q(\Omega) \longrightarrow (L^p(\Omega))' \, , \; \varphi \longrightarrow F(\varphi) = f$$

where $f(u) = \int_\Omega \varphi u \, dx$.

Denoting by $\|\cdot\|_p$ the L^p norm, we have (from the Hölder inequality)

$$|f(u)| = \left| \int_\Omega \varphi u \, dx \right| \leq \|\varphi\|_q \, \|u\|_p$$

which proves that

(i) F is well defined and

(ii) $\|F(\varphi)\| \leq \|\varphi\|_q$.

At the same time taking $u = |\varphi|^{q-1} \text{sgn } \varphi$, we check that

(i) $u \in L^p(\Omega)$ and

(ii) $f(u) = \int_\Omega |\varphi|^q dx = \|\varphi\|_q \|u\|_p$.

which proves that $\|F(\varphi)\| = \|\varphi\|_q$.

Thus F defines a linear, norm-preserving (and therefore injective) map from $L^q(\Omega)$ into the dual of $L^p(\Omega)$. What the representation theorem says is that *all* functionals from the dual space are of this form and consequently F is a (norm-preserving) isomorphism between $L^q(\Omega)$ and the dual of $L^p(\Omega)$. ∎

PROOF ***of Theorem 5.12.1.***

We shall restrict ourselves to the case of bounded Ω.

Case 1 (a unit cube). $\Omega = Q \overset{\text{def}}{=} (0,1)^N = (0,1) \times \ldots \times (0,1)$ (N times). $2 \leq p < \infty$.

Step 1. For an arbitrary positive integer n, divide Q into cubes (with disjoint interiors)

$$Q_k^{(n)}\,, \qquad k = 1, 2, \ldots, (2^n)^N$$

Let $\chi_k^{(n)}$ denote the characteristic function of cube $Q_k^{(n)}$, i.e.,

$$\chi_k^{(n)} = \begin{cases} 1 \text{ on } Q_k^{(n)} \\ 0 \text{ otherwise} \end{cases}$$

Obviously, $\chi_k^{(n)}$ belong to $L^p(Q)$.

Set

$$\varphi_n = \sum_{k=1}^{2^{nN}} \frac{1}{\text{meas}\left(Q_k^{(n)}\right)} f\left(\chi_k^{(n)}\right) \chi_k^{(n)}$$

$$= \sum_k 2^{nN} f\left(\chi_k^{(n)}\right) \chi_k^{(n)}$$

Consequently,

$$\int_Q \varphi_n \chi_k^{(n)} dx = \int_{Q_k^{(n)}} \varphi_n dx = f\left(\chi_k^{(n)}\right)$$

and therefore, by linearity of integrals and functional f,

$$\int_Q \varphi_n u \; dx = f(u)$$

for any u, a linear combination of characteristic functions $\chi_k^{(n)}$.

Selecting

$$u = \frac{|\varphi_n|^q}{\varphi_n} = |\varphi_n|^{q-1} \operatorname{sgn} \varphi_n$$

$$\|u\|_p^p = \int_Q |\varphi_n|^{p(q-1)} \, dx = \int_Q |\varphi_n|^q \, dx$$

we have

$$\int_Q |\varphi_n|^q \, dx = \int_Q \varphi_n u \, dx \le \|f\| \; \|u\|_p$$

$$= \|f\| \left(\int_Q |\varphi_n|^q \, dx \right)^{\frac{1}{p}}$$

which implies

$$\|\varphi_n\|_q \le \|f\|$$

Step 2. $\{\varphi_n\}$ is a Cauchy sequence in $L^q(Q)$. Assume $n \ge m$. Applying Lemma 5.12.1 with

$$u = \frac{\varphi_n - \varphi_m}{\varphi_m}$$

we get

$$\left|\frac{\varphi_n}{\varphi_m}\right|^q \ge 1 + q\frac{\varphi_n - \varphi_m}{\varphi_m} + c\theta\left(\frac{\varphi_n - \varphi_m}{\varphi_m}\right)$$

which upon multiplying by $|\varphi_m|^q$ and integrating over Q yields

$$\int_Q |\varphi_n|^q \, dx \ge \int_Q |\varphi_m|^q \, dx + q\int_Q \frac{|\varphi_m|^q}{\varphi_m}(\varphi_n - \varphi_m)dx$$

$$+ c\int_Q |\varphi_m|^q \, \theta\left(\frac{\varphi_n - \varphi_m}{\varphi_m}\right) dx$$

But, selecting $u = \frac{|\varphi_m|^q}{\varphi_m}$, we have

$$\int_Q \frac{|\varphi_m|^q}{\varphi_m}(\varphi_n - \varphi_m)dx = \int_Q (\varphi_n - \varphi_m)u \, dx = f(u) - f(u) = 0$$

and therefore

$$\int_Q |\varphi_n|^q \, dx \ge \int_Q |\varphi_m|^q \, dx + c\int_Q |\varphi_m|^q \, \theta\left(\frac{\varphi_n - \varphi_m}{\varphi_m}\right) dx$$

In particular, sequence $\int_Q |\varphi_n|^q dx$ is increasing. Since, according to Step 1, it is also bounded, it converges to a finite value.

Consequently,

$$\lim_{n,m\to\infty} \int_Q |\varphi_m|^q \,\theta\left(\frac{\varphi_n - \varphi_m}{\varphi_m}\right) dx = 0$$

Denote now by $e'_{m,n}$ the collection of all points where

$$|\varphi_n - \varphi_m| \geq |\varphi_m| \Longrightarrow \theta\left(\frac{\varphi_n - \varphi_m}{\varphi_m}\right) = \frac{|\varphi_n - \varphi_m|^q}{|\varphi_m|^q}$$

and by $e''_{m,n}$ the set of all x for which

$$|\varphi_n - \varphi_m| \leq |\varphi_m| \Longrightarrow \theta\left(\frac{\varphi_n - \varphi_m}{\varphi_m}\right) = \frac{|\varphi_n - \varphi_m|^2}{|\varphi_m|^2}$$

This leads to the decomposition

$$\int_Q |\varphi_m|^q \,\theta\left(\frac{\varphi_n - \varphi_m}{\varphi_m}\right) dx = \int_{e'_{m,n}} |\varphi_n - \varphi_m|^q \, dx$$

$$+ \int_{e''_{m,n}} |\varphi_n|^{q-2} \, |\varphi_n - \varphi_m|^2 \, dx$$

Since $|\varphi_n - \varphi_m| < |\varphi_n|$ on $e''_{m,n}$, we also have

$$\int_{e''_{m,n}} |\varphi_n - \varphi_m|^q \, dx \leq \int_{e''_{m,n}} |\varphi_m|^{q-1} \, |\varphi_n - \varphi_m| \, dx$$

$$= \int_{e''_{m,n}} \left(|\varphi_m|^{\frac{q-1}{2}} \, |\varphi_n - \varphi_m|\right) |\varphi_m|^{\frac{q}{2}} \, dx$$

(by the Cauchy-Schwarz inequality)

$$\leq \left(\int |\varphi_m|^{q-2} \, |\varphi_n - \varphi_m| \, dx\right)^{\frac{1}{2}} \left(\int |\varphi_m|^q \, dx\right)^{\frac{1}{2}}$$

Concluding,

$$\int_Q |\varphi_n - \varphi_m|^q \, dx \longrightarrow 0 \qquad \text{for } n, m \longrightarrow \infty$$

which means that φ_n is a Cauchy sequence and, by the completeness of $L^q(Q)$, converges to a function $\varphi \in L^q(Q)$.

REMARK 5.12.2. In all inequalities above we have implicitly assumed that $\varphi_m \neq 0$. Technically speaking, one should eliminate from all the corresponding integrals such points and notice that the *final inequalities* are trivially satisifed at points where $\varphi_m = 0$. ∎

Step 3. For any function u, a linear combination of the characteristic functions $\chi_k^{(m)}$,

$$\int_Q \varphi_n u \, dx = f(u) \qquad \forall\, n \geq m$$

Passing to the limit with $n \to \infty$,

$$\int_Q \varphi u \, dx = f(u)$$

Finally, by the density of the characteristic functions in $L^p(Q)$, the equality holds for any $u \in L^p(Q)$.

Case 2. Ω bounded. $2 \leq p < \infty$. By a simple scaling argument, the Case 1 result holds for any cube $Q \subset \mathbb{R}^N$ (not necessarily unit).

Choose now a sufficiently large Q such that $\Omega \subset Q$. Extending functions from $L^p(\Omega)$ by zero to the whole Q, we can identify the $L^p(\Omega)$ space with a subspace of $L^p(Q)$.

$$L^p(\Omega) \subset L^p(Q)$$

By the Hahn-Banach theorem, any linear and continuous functional f defined on $L^p(\Omega)$ can be extended to a linear and continuous functional F defined on the whole $L^p(Q)$. According to the Case 1 result, there exists a function $\Phi \in L^q(Q)$ such that

$$\int_Q \Phi u dx = F(u) \qquad \forall\, u \in L^p(Q)$$

Define $\varphi = \Phi|_\Omega$ (restriction to Ω). Obviously,

$$\int_\Omega \varphi u dx = F(u) = f(u) \qquad \forall\, u \in L^p(\Omega)$$

Case 3. Ω is bounded. $1 \le p < 2$. According to Proposition 3.9.3, $L^2(\Omega)$ is continuously embedded in $L^p(\Omega)$ and therefore any linear and continuous functional f defined on $L^p(\Omega)$ is automatically continuous on $L^2(\Omega)$. By the Case 2 result, specialized for $p = 2$, there exists a function $\varphi \in L^2(\Omega)$, such that

$$\int_\Omega \varphi u dx = f(u) \qquad \forall\, u \in L^2(\Omega)$$

We will show that

1. $\varphi \in L^q(\Omega)$, $\dfrac{1}{p} + \dfrac{1}{q} = 1$ and
2. $\displaystyle\int_\Omega \varphi u dx = f(u) \qquad \forall\, u \in L^p(\Omega)$.

Step 1. Assume $p > 1$.
Define

$$\varphi_n(x) = \begin{cases} \varphi(x) \text{ if } |\varphi(x)| \le n \\ n \quad\;\; \text{if } |\varphi(x)| > n \end{cases}$$

and set

$$u_n = |\varphi_n|^{q-1} \operatorname{sgn}\varphi$$

Obviously, functions u_n are bounded and, therefore, they are elements of $L^p(\Omega)$. We have

$$f(u_n) = \int_\Omega \varphi u_n dx = \int_\Omega \varphi \,|\varphi_n|^{q-1} \operatorname{sgn}\varphi \, dx \ge \int_\Omega |\varphi_n|^q \, dx$$

At the same time,

$$f(u_n) \le \|f\| \; \|u_n\|_p = \|f\| \left[\int_\Omega |\varphi_n|^q dx\right]^{\frac{1}{p}}$$

So

$$\left(\int_\Omega |\varphi_n|^q \, dx\right)^{\frac{1}{q}} \le \|f\|$$

By the Lebesgue dominated convergence theorem ($\varphi_n \to \varphi$),

$$\|\varphi\|_q \le \|f\|$$

By the density argument (L^2 functions are dense in L^p, $1 \leq p < 2$; see Exercise 5.12.1),

$$\int_\Omega \varphi u dx = f(u) \qquad \forall\, u \in L^p(\Omega)$$

Step 2. Case $p = 1$.
Define

$$e_n = \{x \in \Omega : |\varphi(x)| \geq n\}$$

and set

$$u_n = \begin{cases} \text{sgn}\varphi \text{ on } e_n \\ 0 \qquad \text{otherwise} \end{cases}$$

Obviously, $u_n \in L^2(\Omega)$ and

$$f(u_n) = \int_\Omega \varphi u_n dx = \int_{e_n} |\varphi| dx \geq n \text{ meas}(e_n)$$

At the same time,

$$f(u_n) \leq \|f\| \ \|u_n\|_1 = \|f\| \text{ meas}(e_n)$$

and therefore

$$(n - \|f\|) \text{ meas}(e_n) \leq 0$$

which proves that meas $(e_n) = 0$ for $n > \|f\|$ and therefore φ is essentially bounded.

By the density argument, again the representation formula must hold for all $u \in L^1(\Omega)$. ∎

THEOREM 5.12.2
(Representation Theorem for $(\ell^p)'$)

Let $1 \leq p < \infty$. For every linear and continuous functional f defined on the space ℓ^p, there exists a unique sequence $\varphi \in \ell^q$, $\frac{1}{p} + \frac{1}{q} = 1$ such that

$$f(u) = \langle f, u \rangle = \sum_{i=1}^{\infty} \varphi_i u_i \qquad \forall\, u \in \ell^p$$

Moreover,

$$\|f\|_{(\ell^p)'} = \|\varphi\|_{\ell^q}$$

and the map

$$\ell^q \ni \varphi \to \left\{ \ell^p \ni u \to \sum_{i=1}^{\infty} \varphi_i u_i \right\} \in (\ell^p)'$$

is a norm-preserving isomorphism from ℓ^q onto $(\ell^p)'$.

The proof follows the same lines as for the L^p spaces and is left as an exercise.

REMARK 5.12.3. Note that both representation theorems *do not* include the case $p = \infty$. A separate theorem identifies the dual of $L^\infty(\Omega)$ with so-called *functions of bounded variation.* ∎

Distributions. The notion of the topological dual understood as the space of linear and continuous functionals can be generalized to any topological vector space. In particular, the dual of the space of test functions $\mathcal{D}(\Omega)$ (see Section 5.3) introduced by L. Schwartz, denoted $\mathcal{D}'(\Omega)$, is known as the famous *space of distributions.* The elements $q \in \mathcal{D}'(\Omega)$ are called *distributions.*

To test continuity of linear functionals defined on $\mathcal{D}(\Omega)$ we must recall (see Remark 5.3.1) that the topology in $\mathcal{D}(\Omega)$ is identified as the *strongest* topology in $C_0^\infty(\Omega)$ such that all inclusions

$$i_K : \mathcal{D}(K) \hookrightarrow \mathcal{D}(\Omega)$$

are continuous, for every compact set $K \subset \Omega$. Consequently, a linear functional $q : \mathcal{D}(\Omega) \to \mathbb{R}$ is continuous if the composition

$$q \circ i_K : \mathcal{D}(K) \longrightarrow \mathbb{R}$$

is continuous, for every compact $K \subset \Omega$. This leads to the following criterion for continuity: a *linear functional $q : \mathcal{D}(\Omega) \to \mathbb{R}$ is continuous iff to every compact set $K \subset \Omega$ there corresponds a constant $C_K > 0$ and $k > 0$ such that*

$$|q(\varphi)| \leq C_K \sup_{\substack{|\boldsymbol{\alpha}| \leq k \\ \boldsymbol{x} \in K}} \left| D^{\boldsymbol{\alpha}} \varphi(\boldsymbol{x}) \right| \qquad \forall\, \varphi \in \mathcal{D}(K)$$

It is easy to check, using the criterion above, that for any locally integrable function f on Ω, i.e., such that $\int_K f dx < \infty$, for compact $K \subset \Omega$, the linear functional

$$C_0^\infty(\Omega) \ni \varphi \longrightarrow \int_\Omega f\varphi dx \quad \in \mathbb{R}$$

is continuous on $\mathcal{D}(\Omega)$ and therefore defines a distribution. Distributions of this type are called *regular* and are identified with the underlying, locally integrable function f.

Distributions which are not regular are called *irregular*. The most famous example of an irregular distribution is the Dirac delta functional

$$\langle \delta_{\boldsymbol{x}_0}, \varphi \rangle \stackrel{\text{def}}{=} \varphi(\boldsymbol{x}_0)$$

The theory of distributions exceeds significantly the scope of this book. From a number of remarkable properties of distributions, we mention only the definition of the distributional derivative, generalizing the notions discussed in Section 5.11.

For any multiindex $\boldsymbol{\alpha}$, a distribution $q^{\boldsymbol{\alpha}}$ is called the $D^{\boldsymbol{\alpha}}$ derivative of a distribution q, denoted $D^{\boldsymbol{\alpha}}q$, iff

$$\langle q^{\boldsymbol{\alpha}}, \varphi \rangle = (-1)^{|\boldsymbol{\alpha}|} \langle q, D^{\boldsymbol{\alpha}}\varphi \rangle \quad \forall\, \varphi \in \mathcal{D}(\Omega)$$

Suprisingly enough, it can be proved that every distribution $q \in \mathcal{D}'(\Omega)$ possesses derivatives of arbitrary order.

Exercises

5.12.1. Let $\Omega \subset \mathbb{R}^N$ be a bounded set and let $1 \le p < r \le \infty$. Prove that $L^r(\Omega)$ is dense in $L^p(\Omega$.

Hint: For an arbitrary $u \in L^p(\Omega)$ define

$$u_n(x) = \begin{cases} u(x) & u(x) \le n \\ n & \text{otherwise} \end{cases}$$

Show that

1. $u_n \in L^r(\Omega)$ and
2. $\|u_n - u\|_p \to 0$.

5.12.2. Prove Theorem 5.12.2.

5.12.3. Why have we not formulated a representation theorem for linear and continuous functionals on finite-dimensional spaces?

5.12.4. Let $\Omega \subset I\!R^n$ be an open set and $f : \Omega \to I\!R$ a function defined on Ω. Prove that the following conditions are equivalent to each other:

(i) For every $\boldsymbol{x} \in \Omega$ there exists a neighborhood $N(\boldsymbol{x})$ of $\boldsymbol{x}$ (e.g., a ball $B(\boldsymbol{x}, \varepsilon)$ with some $\varepsilon = \varepsilon(\boldsymbol{x}) > 0$) such that

$$\int_{N(\boldsymbol{x})} f dx < +\infty$$

(ii) For every compact $K \subset \Omega$

$$\int_K f dx < +\infty$$

Functions of this type are called *locally integrable* and form a vector space, denoted $L^1_{loc}(\Omega)$.

5.12.5. Prove that the regular distributions and the Dirac delta functional defined in the text are continuous on $\mathcal{D}(\Omega)$.

5.12.6. Consider function $u : (0,1) \to I\!R$ of the form

$$u(x) = \begin{cases} u_1(x) \; 0 < x \leq x_0 \\ u_2(x) \; x_0 < x \leq 1 \; x_0 \in (0,1) \end{cases}$$

where u_1 and u_2 are C^1 functions (see Example 5.11.1), but the global function u is not necessarily continuous at x_0. Follow the lines of Example 5.11.1 to prove that the distributional derivative of the regular distribution q_u corresponding to u is given by the formula

$$q_u' = q_{u'} + [u(x_0)]\delta_{x_0}$$

where u' is the union of the two branches derivatives u_1' and u_2' (see Example 5.11.1), δ_{x_0} is the Dirac delta functional at x_0 and $[u(x_0)]$ denotes the jump of u at x_0,

$$[u(x_0)] = u_2(x_0) - u_1(x_0)$$

5.13 Bidual. Reflexive Spaces

The Bidual Space. Let U be a normed space and U' its topological dual. Then U' equipped with the dual norm is itself a normed space (always complete, even if U is not) and it also makes sense to speak of the space of all continuous linear functionals on U'. The dual of the dual of a normed space U is again a Banach space, denoted U'' and called the *bidual* of U

$$U'' \stackrel{\text{def}}{=} (U')'$$

It turns out that any normed space U is isomorphic, in a natural way, to a subspace of its bidual U''. To see this, let $u \in U$ and $f \in U'$. Then $\langle f, \boldsymbol{u} \rangle = f(u)$ is a linear functional on U (by the choice of f). However, for each fixed u, $\langle f, u \rangle$ is also a linear functional on U' (by definition of vector space operations in U'). More precisely, for each $u \in U$, we define a corresponding linear functional F_u on U', called the *evaluation at* u and defined as

$$U' \ni f \longrightarrow F_u(f) \stackrel{\text{def}}{=} \langle f, u \rangle \in \mathbb{R}(\mathbb{C})$$

From the inequality

$$|F_u(f)| = \langle f, u \rangle \le \|u\|_U \ \|f\|_{U'}$$

it follows that

$$\|F_u\|_{U''} \le \|u\|_U$$

Moreover, by Corollary 5.5.2 to the Hahn-Banach theorem, for any vector $u \in U$, there exists a corresponding functional $f \in U'$, $\|f\|_{U'} = 1$ such that $f(u) = \|u\|$. Let f_1 denote such a functional. Then

$$\|F_u\|_{U''} \ge \frac{|F_u(f_1)|}{\|f_1\|_{U'}} = \frac{|\langle f_1, u \rangle|}{1} = \|u\|_U$$

Therefore,

$$\|F_u\|_{U''} = \|u\|_U$$

Summing that up, the linear map F prescribing for each $u \in U$ the corresponding element F_u of bidual U''

$$U \in u \longrightarrow F_u \in U''$$

is linear (why?) and norm-preserving. Thus F establishes an isometric isomorphism between any normed space U and its range in bidual U''. Note that, in general, F is *not* surjective.

Reflexive Spaces. A normed vector space U is called reflexive if the map F discussed above is surjective, i.e., space U is isomorphic and isometric with its bidual U''.

Before we proceed with a number of properties of reflexive spaces, we need to record the following simple, but important, lemma:

LEMMA 5.13.1
(Mazur Separation Theorem)

Let U be a normed space and $M \subset U$ a *closed* subspace of U. For every nonzero vector $\boldsymbol{u}_0 \notin M$ there exists a continuous linear functional f on U, $f \in U'$, such that

(i) $f|_M \equiv 0$,

(ii) $f(\boldsymbol{u}_0) = \|\boldsymbol{u}_0\| \neq 0$, and

(iii) $\|f\| = 1$.

PROOF Consider the subspace $M_1 = M \oplus \mathbb{R}\boldsymbol{u}_0$ (or $M \oplus \mathbb{C}\boldsymbol{u}_0$) and a corresponding linear functional f defined on M_1 as

$$f(\boldsymbol{u}) = \begin{cases} 0 & \text{on } M \\ \alpha\|\boldsymbol{u}_0\| & \text{for } \boldsymbol{u} = \alpha\boldsymbol{u}_0 \ , \quad \alpha \in \mathbb{R}(\mathbb{C}) \end{cases}$$

We claim that f is continuous. Indeed, suppose to the contrary that there exists an $\varepsilon > 0$ and a sequence

$$\boldsymbol{u}_n = \boldsymbol{m}_n + \alpha_n \boldsymbol{u}_0 \longrightarrow \boldsymbol{0} \quad \boldsymbol{m}_n \in M, \ \alpha_n \in \mathbb{R}(\mathbb{C})$$

such that

$$f(\boldsymbol{u}_n) = \alpha_n \|\boldsymbol{u}_0\| > \varepsilon$$

Consequently, $\alpha_n^{-1} < \varepsilon^{-1}\|\boldsymbol{u}_0\|$ and

$$\alpha_n^{-1}\boldsymbol{u}_n = \alpha_n^{-1}\boldsymbol{m}_n + \boldsymbol{u}_0 \longrightarrow \boldsymbol{0}$$

which proves that $\boldsymbol{u}_0 \in \overline{M} = M$, a contradiction. Finally, taking $p(\boldsymbol{u}) = \|\boldsymbol{u}\|$, we apply the Hahn-Banach theorem and extend f to the whole U. From the inequality

$$|f(\boldsymbol{u})| \leq \|\boldsymbol{u}\|$$

and the fact that $f(\boldsymbol{u}_0) = \|\boldsymbol{u}_0\|$, it follows that $\|f\| = 1$. ∎

REMARK 5.13.1. Defining a hyperplane Π as

$$\Pi = \{\boldsymbol{u} \in U : f(\boldsymbol{u}) = c\}$$

where $f \in U'$ and c is a constant, we can interpret the discussed result as a separation of the subspace M from the point $\boldsymbol{u}_0$ by any hyperplane corresponding to the constructed functional f and any constant $0 < c < \|\boldsymbol{u}_0\|$. Indeed,

$$f(\boldsymbol{u}_0) = \|\boldsymbol{u}_0\| > c > 0 = f(\boldsymbol{m}) \text{ for } \boldsymbol{m} \in M$$

which means that $\boldsymbol{u}_0$ and M stay on "opposite" sides of the hyperplane. This explains why the result is interpreted as a *separation theorem.* ∎

PROPOSITION 5.13.1

(i) *Any reflexive normed space must be complete and, hence, is a Banach space.*

(ii) *A closed subspace of a reflexive Banach space is reflexive.*

(iii) *Cartesian product of two reflexive spaces is reflexive.*

(iv) *Dual of a reflexive space is reflexive.*

PROOF

(i) By definition, U is isomorphic with the *complete* space U'' and therefore must be complete as well.

(ii) Let $V \subset U$ be a closed subspace of a reflexive space U and let g denote an arbitrary linear and continuous functional on the dual V'. Consider now the map

$$i : U' \ni f \longrightarrow i(f) = f|_V \in V'$$

prescribing for each linear and continuous functional f on U its restriction to V. Composition $g \circ i$ defines a linear and continuous functional on U' and therefore, by reflexivity of U, there exists an element $u \in U$ such that

$$(g \circ i)(f) = \langle f, u \rangle \qquad \forall\, f \in U'$$

By the Hahn-Banach theorem u must be an element of V. Indeed, if $u \notin V$ then any continuous functional f vanishing on V and taking a nonzero value at u could be extended to the whole U and, consequently, $\langle f, u \rangle = f(u) \neq 0$, but

$$(g \circ i)(f) = g(f|_V) = g(\mathbf{0}) = 0$$

a contradiction.

Question: Where have we used the assumption that V is closed?

(iii) Let U_i, $i = 1, 2$, be two reflexive spaces and U_i', $i = 1, 2$, their duals. The following map establishes an isomorphism between the dual of the Cartesian product $U_1 \times U_2$ and the Cartesian product of the duals U_1', U_2'.

$$i : (U_1 \times U_2)' \ni f \longrightarrow i(f) = (f_1, f_2) \in U_1' \times U_2'$$

where

$$f_1(u_1) \stackrel{\text{def}}{=} f((u_1, 0))$$

and

$$f_2(u_2) \stackrel{\text{def}}{=} f((0, u_2))$$

Consequently, if $F_i : U_i \to U_i''$ denote the isomorphisms between U_1, U_2 and their biduals, then

$$F_1 \times F_2 : (u_1, u_2) \to \{U_1' \times U_2' \ni (f_1, f_2) \to f_1(u_1) + f_2(u_2) \in \mathbb{R}(\mathbb{C})\}$$

establishes the isomorphism between $U_1 \times U_2$ and the dual to $U_1' \times U_2'$ or, equivalently, $(U_1 \times U_2)'$.

(iv) Assume $F : U \to U''$ is an isomorphism and consider the map

$$G : U' \ni f \longrightarrow G(f) \stackrel{\text{def}}{=} f \circ F^{-1} \in (U'')' \sim (U')''$$

It is a straightforward exercise to prove that G is an isomorphism, too. ∎

REMARK 5.13.2. One can prove that the reflexivity of the dual space U' is not only a necessary, but also a sufficient, condition for the reflexivity of a normed space U. The proof considerably exceeds the scope of this book. ∎

Example 5.13.1

The $L^p(\Omega)$ spaces are reflexive for $1 < p < \infty$. This follows immediately from the representation theorem for duals of L^p spaces. For $1 < p < \infty$, the dual of $L^p(\Omega)$ is identified with $L^q(\Omega)$ and, in turn, the dual of $L^q(\Omega)$ can again be identified with $L^p(\Omega)$. Note that the result holds neither for $L^1(\Omega)$ nor for $L^\infty(\Omega)$ spaces which are *not* reflexive.

The same conclusions apply to ℓ^p spaces. ▯

Example 5.13.2

Every finite-dimensional space is reflexive. Explain why. ▯

Example 5.13.3

Sobolev spaces $W^{m,p}(\Omega)$ (see Section 5.11) are reflexive for $1 < p < \infty$.

For proof, it is sufficient to notice that the space $W^{m,p}(\Omega)$ is isomorphic to a closed subspace of the Cartesian product of reflexive $L^p(\Omega)$ spaces:

$$\left\{\{u^{\boldsymbol{\alpha}} \in L^p(\Omega), |\boldsymbol{\alpha}| \le m\} : \int_\Omega u D^{\boldsymbol{\alpha}} \varphi dx = (-1)^{|\boldsymbol{\alpha}|} \int_\Omega u^{\boldsymbol{\alpha}} \varphi dx, \ \forall\, \varphi \in C_0^\infty(\Omega)\right\}$$

Indeed, the subspace above can be identified as an image of the operator from $W^{m,p}(\Omega)$ into the Cartesian product $L^p(\Omega) \times \ldots \times L^p(\Omega)$ (n times, where $n = \#\{|\boldsymbol{\alpha}| \le m\}$), prescribing for each function u all its distributional derivatives. ▯

Exercises

5.13.1. Explain why every finite-dimensional space is reflexive.

5.13.2. Let $W^{m,p}(\Omega)$ be a Sobolev space for Ω, a smooth domain in $\mathbb{R}^n$. The closure in $W^{m,p}(\Omega)$ of the test functions $C_0^\infty(\Omega)$ (with respect to the $W^{m,p}$ norm), denoted $\dot{W}^{m,p}(\Omega)$,

$$\dot{W}^{m,p}(\Omega = \overline{C_0^\infty(\Omega)}$$

may be identified as a collection of all functions from $W^{m,p}(\Omega)$ which vanish on the boundary together with their derivatives up to $m-1$ order (this is a very nontrivial result based on Lions' trace theorem; see [8,12]). The duals of the spaces $\dot{W}^{m,p}(\Omega)$ are the so-called *negative* Sobolev spaces

$$W^{-m,p}(\Omega) \stackrel{\text{def}}{=} \left(\dot{W}^{m,p}(\Omega)\right)' \quad m > 0$$

Explain why both $\dot{W}^{m,p}(\Omega)$ and $W^{-m,p}(\Omega)$ for $1 < p < \infty$ are reflexive.

5.14 Weak Topologies. Weak Sequential Compactness

The topological properties of normed linear spaces are complicated by the fact that topologies can be induced on such spaces in more than one way. This leads to alternative notions of continuity, compactness, and convergence for a normed space U.

Weak Topology. Let U be a normed space and let U' denote its dual. For each continuous, linear functional $f \in U'$ we introduce a corresponding seminorm p_f on U defined as

$$p_f(\boldsymbol{u}) \stackrel{\text{def}}{=} |f(\boldsymbol{u})| = |\langle f, \boldsymbol{u}\rangle| \ , \ f \in U'$$

By Corollary 5.5.2, for each $\boldsymbol{u} \neq \boldsymbol{0}$ there exists a functional $f \in U'$ taking a nonzero value at $\boldsymbol{u}$, which implies that the family of seminorms

$$p_f : U \to [0, \infty) \ , \ f \in U'$$

satisfies the axiom of separation (see Section 5.2). Consequently, the p_f seminorms can be used to construct a locally convex topology on U. We refer to it as the *weak topology* in contrast to the topology induced by norm and called the *strong topology.*

Indeed, it follows immediately from the definition of locally convex spaces that the weak topology is *weaker* than the one induced by the norm. To see this, consider an arbitrary element from the base of neighborhoods for the zero vector in U:

$$B(I_0, \varepsilon) = \{\boldsymbol{u} \in U : |f(\boldsymbol{u})| \leq \varepsilon \ , \ f \in I_0\}$$

where $I_0 \subset U'$ is *finite.*

By continuity of functionals f, there exist corresponding constants $C_f > 0$ such that

$$|f(\boldsymbol{u})| \leq C_f \|\boldsymbol{u}\|$$

Take $\delta = \min\{ \frac{\varepsilon}{C_f} : f \in I_0\}$ and consider the ball $B(\boldsymbol{0}, \delta)$. It follows that

$$|f(\boldsymbol{u})| \leq C_f \|\boldsymbol{u}\| \leq C_f \delta \leq \varepsilon \ , \quad \text{for every} \quad f \in I_0$$

which proves that $B(\boldsymbol{0}, \delta) \subset B(I_0, \varepsilon)$.

Consequently, the base of neighborhoods for the zero vector in the strong topology (the balls) *is stronger* than the base of neighborhoods in the weak topology (sets $B(I_0, \varepsilon)$). As bases of neighborhoods for nonzero vectors u are obtained by shifting the base for zero to $\boldsymbol{u}$, the same property holds for any vector $\boldsymbol{u}$. This proves that the norm topology is *stronger* than the weak one.

Weak* Topology. A third fundamental topology in a normed space U can be generated when U is identified as the dual of some other normed space V, $U = V'$. For every $\boldsymbol{v} \in V$, seminorms

$$U \ni f \longrightarrow |\langle f, \boldsymbol{v}\rangle| = |f(\boldsymbol{v})| \in [0, \infty)$$

trivially satisfy the axiom of separation (why?) and therefore can be used to induce another locally convex topology, called the *weak* (weak "star") topology* on U. As, in general, elements from V are identified with a *proper* subspace of the bidual $V'' = U'$, the neighborhoods in the weak* topology form a proper subset of the neighborhoods in the weak topology. Consequently, the weak* topology is *weaker* than the weak topology. Notice, however, that for reflexive spaces, $V'' \sim V$ and the two topologies are the same.

In this section we study some basic topological properties of weak topologies including the fundamental notion of the weak sequential compactness.

We begin with a simple characterization for the convergence of sequences in the weak topologies.

PROPOSITION 5.14.1

Let U be a normed space and consider the sequences

$$\{u_n\} \in U, \ \{f_n\} \in U'$$

then

(i) *u_n converges to u in the strong (norm) topology, denoted $u_n \to u$, iff*

$$\|u_n - u\|_U \to 0$$

(ii) *u_n converges to u in the weak topology, denoted $u_n \rightharpoonup u$, iff*

$$\langle f, u_n - u\rangle \longrightarrow 0 \qquad \forall\, f \in U'$$

(iii) *f_n converges to f in the strong (dual) topology, denoted $f_n \to f$, iff*

$$\|f_n - f\|_{U'} \to 0$$

(iv) *f_n converges to f in the weak (dual) topology, denoted $f_n \rightharpoonup f$, iff*

$$\langle g, f_n - f\rangle \to 0 \quad \forall\, g \in U''$$

(v) f_n *converges to* f *in the weak* topology, denoted* $f_n \stackrel{*}{\rightharpoonup} f$*, iff*

$$\langle f_n - f, u\rangle \to 0 \qquad \forall\, u \in U$$

PROOF Proof is a straightforward consequence of the definitions and is left as an exercise. ∎

Example 5.14.1

Many weakly convergent sequences do not converge strongly. For any integer $n > 0$, consider the partition of unit interval $(0,1)$ into 2^n equal subintervals and a corresponding sequence of functions

$$\varphi_n(x) = \begin{cases} 1 \text{ for } (k-1)2^{-n} \le x \le k2^{-n}\ ,\ k \text{ even} \\ \\ -1 \text{ otherwise} \end{cases}$$

(see Fig. 5.3). Obviously, $\varphi_n \in L^2(0,1)$. We will prove later (see Example 5.14.3) that sequence φ_n converges weakly in $L^2(0,1)$ to zero function $\mathbf{0}$. At the same time,

$$\|\varphi_n\|^2_{L^2} = \int_0^1 \varphi_n^2 dx = 1$$

and therefore φ_n *does not* converge strongly to $\mathbf{0}$. □

Example 5.14.2

Recall that $L^1(a,b)$ is not reflexive, and $L^\infty(a,b)$ can be identified with the dual of $L^1(a,b)$. Let $\{\phi_n(x)\}$ be a sequence of functions in $L^\infty(a,b)$. Then $\{\phi_n\}$ converges *weak** to a function $\phi_0(x) \in L^\infty(a,b)$ if

$$\lim_{n\to\infty} \int_a^b f\phi_n dx = \int_a^b f\phi_0 dx, \qquad \forall\, f \in L^1(a,b)$$

The reason that this represents weak* convergence is clear: the ϕ_n represent continuous linear functionals on $L^1(a,b)$ and $L^1(a,b) \subset (L^\infty(a,b))' = L^1(a,b)''$. Hence we construct a functional on $L^1(a,b)'$, using $f \in L^1(a,b)$, and apply the definition. □

The notion of the topological dual, understood as the space of all linear and continuous functionals, can be generalized to any topological vector space. Keeping this in mind, we could speculate what the topological dual

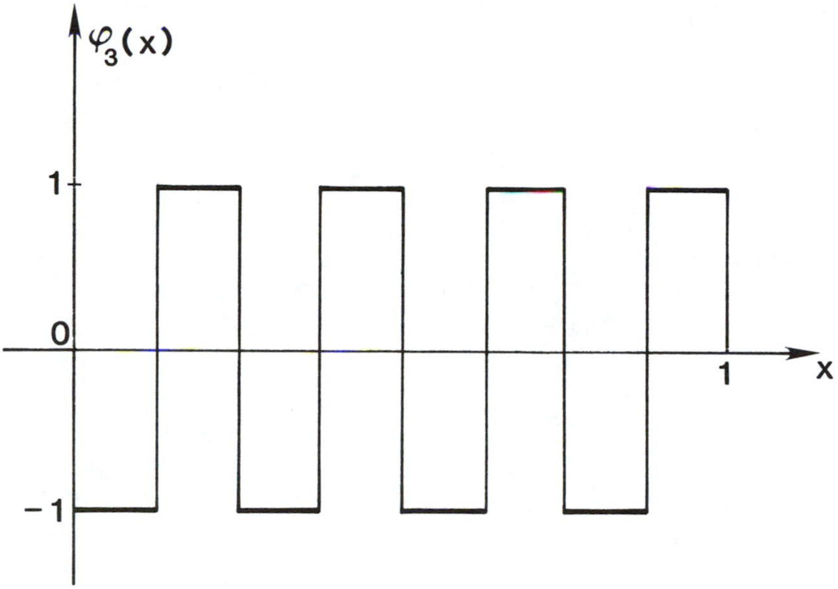

Figure 5.3
Example 5.14.1. Illustration of function $\varphi_n (n = 3)$.

corresponding to the weak topology would look like. As the weak topology is weaker than the strong topology, functionals continuous in the strong topology are automatically continuous in the weak topology. Surprisingly enough, the converse is also true: *any weakly continuous and linear functional is also strongly continuous.* This follows from the definition of the weak topology. Indeed, suppose that $g : U \to \mathbb{R}(\mathbb{C})$ is linear and weakly continuous. This implies that for $\varepsilon = 1$ there exists a neighborhood

$$B = B(I_0, \delta) = \{\boldsymbol{u} \in U : |f(\boldsymbol{u})| \leq \delta \ , \ f \in I_0\}$$

where $I_0 \subset U'$ is finite, such that

$$\boldsymbol{u} \in B \quad \text{implies} \quad |g(\boldsymbol{u})| \leq 1$$

Taking $C = \min\{\|f\|_{U'} : f \in I_0\}$ we have for every $\boldsymbol{u} \in U$

$$\left| f\left(\frac{\delta}{C}\frac{u}{\|u\|}\right)\right| \le \frac{\delta}{C} f\left(\frac{u}{\|u\|}\right) \le \delta\ ,\ \forall\ f \in I_0$$

which means that $\frac{\delta}{C}\frac{u}{\|u\|} \in B(I_0, \delta)$, and therefore

$$g\left(\frac{\delta}{C}\frac{u}{\|u\|}\right) \le 1 \Longrightarrow g(u) \le \frac{C}{\delta}\|u\|$$

which proves that g is bounded.

This simple, but fundamental, result assures that it does not make sense to speak about topological duals with respect to any topology weaker than the norm topology and stronger than the weak one. They all simply *coincide* with the regular dual with respect to the norm topology.

Let U now be a normed space. Consider a sequence of vectors $\boldsymbol{u}_n \in U$ such that

$$f(\boldsymbol{u}_n) \text{ is convergent } \forall\ f \in U'$$

Sometimes we say that $\boldsymbol{u}_n$ is *weakly convergent* (do not confuse this terminology with the weak convergence to a point $\boldsymbol{u}$; there is no such $\boldsymbol{u}$ identified here). The question is, *what can we say about u_n?*

To begin with, let us consider a corresponding sequence of functionals g_n defined on the dual U':

$$g_n : U' \ni f \longrightarrow \langle f, \boldsymbol{u}_n \rangle \in \mathbb{R}(\mathbb{C})$$

The fact that $f(\boldsymbol{u}_n)$ is convergent and therefore bounded implies that functionals g_n are *pointwise uniformly bounded.* By the uniform boundedness theorem they must be *uniformly bounded*, i.e.,

$$\exists\ C > 0 :\ \|g_n\|_{U''} \le C$$

But $\|g_n\|_{U''} = \|\boldsymbol{u}_n\|$ and therefore the first conclusion is that sequence $\boldsymbol{u}_n$ must be *bounded.*

Following the same reasoning, it follows from the Banach-Steinhaus theorem that the functional g,

$$g : U' \ni f \longrightarrow \lim_{n\to\infty} g_n(f) = \lim_{n\to\infty} f(\boldsymbol{u}_n)$$

is a continuous linear functional on U' and

$$\|g\|_{U''} \le \liminf_{n\to\infty} \|g_n\|_{U''} = \liminf_{n\to\infty} \|\boldsymbol{u}_n\|_U$$

Thus, if we additionally assume that U is *reflexive*, then there exists $\boldsymbol{u} \in U$ such that

$$\boldsymbol{u}_n \rightharpoonup \boldsymbol{u} \quad \text{and} \quad \|\boldsymbol{u}\| \leq \liminf_{n\to\infty} \|\boldsymbol{u}_n\|$$

We summarize these observations in the following proposition.

PROPOSITION 5.14.2

Let $\boldsymbol{u}_n$ be a weakly convergent sequence in a normed space U. Then $\boldsymbol{u}_n$ is bounded.

If, additionally, U is reflexive, then $\boldsymbol{u}_n$ converges weakly to an element $\boldsymbol{u} \in U$ and

$$\|\boldsymbol{u}\| \leq \liminf_{n\to\infty} \|\boldsymbol{u}_n\|$$

PROPOSITION 5.14.3
(Sufficient and Necessary Condition for Weak Convergence)

Let $\boldsymbol{u}_n$ be a sequence in a normed space U. The following conditions are equivalent to each other:

(i) $f(\boldsymbol{u}_n)$ *is convergent* $\quad \forall\, f \in U'$.

(ii) *a)* $\boldsymbol{u}_n$ *is bounded and*
b) $f(\boldsymbol{u}_n)$ *is convergent* $\forall\, f \in D$, *where D is a dense subset of U',* $\overline{D} = U'$.

PROOF It remains to prove (ii) $\Longrightarrow$ (i). Let $f \in U'$. By density of D in U', for every $\varepsilon > 0$, there exists a corresponding $f_\varepsilon \in D$ such that $\|f - f_\varepsilon\|_{U'} \leq \varepsilon$. It follows that

$$|f(\boldsymbol{u}_n)| \leq |f(\boldsymbol{u}_n) - f_\varepsilon(\boldsymbol{u}_n) + f_\varepsilon(\boldsymbol{u}_n)|$$

$$\leq \|f - f_\varepsilon\| \;\|\boldsymbol{u}_n\| + |f_\varepsilon(\boldsymbol{u}_n)|$$

which, in view of the boundedness of $\boldsymbol{u}_n$, implies that $f(\boldsymbol{u}_n)$ is bounded in $I\!R$. By compactness argument in $I\!R$, there exists a subsequence $\boldsymbol{u}_{n_k}$ and $a \in I\!R$ such that

$$f(\boldsymbol{u}_{n_k}) \longrightarrow a$$

But again,

$$|f(\boldsymbol{u}_n) - f(\boldsymbol{u}_{n_k})| \le \|f - f_\varepsilon\| \; \|\boldsymbol{u}_n\|$$

$$+ |f_\varepsilon(\boldsymbol{u}_n) - f_\varepsilon(\boldsymbol{u}_{n_k})|$$

$$+ \|f_\varepsilon - f\| \; \|\boldsymbol{u}_{n_k}\|$$

implies that the whole sequence must converge to a. ∎

Example 5.14.3

We are now ready to prove that the sequence of functions φ_n from Example 5.14.1 converges weakly to zero. Toward this goal, recall that piecewise constant functions on the uniform partitions $\mathcal{Q}^n$,

$$\left((k-1)2^{-n},\ k2^n\right)\ ,\ k = 1, 2, \ldots, 2^n\ ,\ n > 0$$

form a dense subset in $L^2(0,1)$. On the other hand, for any piecewise constant function u on the partition $\mathcal{Q}^n$, by definition of φ_m,

$$\int_0^1 \varphi_m u \, dx = 0 \qquad \forall\ m > n$$

and, consequently, $\varphi_m \rightharpoonup 0$ on the dense subset of $L^2(0,1)$. As φ_n is bounded, it follows from Proposition 5.14.2 that $\varphi_n \rightharpoonup \mathbf{0}$. □

We now proceed with the main result of this section.

THEOREM 5.14.1
(Weak Sequential Compactness)

Let U be a reflexive Banach space and $\{\boldsymbol{u}_n\}$ any sequence of U that is bounded in the norm of U; i.e., there exists a constant $M > 0$ such that

$$\|\boldsymbol{u}_n\|_U \le M \ \forall\ n$$

Then there exists a subsequence $\{\boldsymbol{u}_{n_k}\}$ of $\{\boldsymbol{u}_n\}$ that converges weakly to an element $\mathbf{u}$ of U such that $\|\boldsymbol{u}\| \le M$.

In other words, in a reflexive Banach space, closed balls are weakly sequentially compact.

In the proof of the theorem, we will restrict ourselves only to the case of *separable* spaces. Recall that a normed space U is *separable* iff there exists a countable subset of U which is dense in U.

LEMMA 5.14.1

If the dual U' of a normed space U is separable, then so is U.

PROOF Assume D is a countable, dense subset of U' and let D_{ε_0} denote a (countable) subset of D such that

$$D_{\varepsilon_0} = \{f \in D : 1 - \varepsilon_0 \le \|f\| \le 1\}$$

By definition of the dual norm,

$$\|f\| = \sup_{\|\boldsymbol{u}\|=1} |f(\boldsymbol{u})|$$

for ε_0 small enough for each $f \in D_{\varepsilon_0}$, there exists a corresponding $\boldsymbol{u}_f \in U$, $\|\boldsymbol{u}_f\| = 1$, such that

$$\langle f, \boldsymbol{u}_f \rangle = f(\boldsymbol{u}_f) > \frac{1}{2}$$

We claim that the set M of all linear combinations of $\boldsymbol{u}_f$ with rational coefficients (and so is still countable) must be dense in U.

Suppose to the contrary that $\overline{M} \neq U$. Pick $\boldsymbol{u}_0 \in U - \overline{M}$. By the Masur separation theorem (Lemma 5.13.1), there exists $f_0 \in U'$, $\|f_0\| = 1$, vanishing on M and nonzero at $\boldsymbol{u}_0$. Let f_n be a corresponding sequence from D_{ε_0}, converging to f_0. We have

$$\frac{1}{2} < \langle f_n, \boldsymbol{u}_{f_n} \rangle \le \langle f_n - f_0, \boldsymbol{u}_{f_n} \rangle + \langle f_0, \boldsymbol{u}_{f_n} \rangle$$

a contradiction, since the right-hand side converges to zero. ∎

COROLLARY 5.14.1

If a normed space U is reflexive and separable, then so is the dual U'.

PROOF ***of Theorem 5.14.1***

As we have mentioned above, we assume additionally that U is separable. By the preceding corollary, U' is separable, too. Let $\{f_j\}$ be a countable and dense subset of U'. As $\boldsymbol{u}_n$ is bounded it follows that

$$|f_j(\boldsymbol{u}_n)| \leq \|f_j\|_{U'} \|\boldsymbol{u}_n\|_U \leq \|f_j\|_{U'} M \quad j = 1, 2, \ldots$$

and therefore $f_j(\boldsymbol{u}_n)$ is bounded for every j. By the Bolzano-Weierstrass theorem and the diagonal choice method, one can extract a subsequence $\boldsymbol{u}_{n_k}$ such that

$$f_j(\boldsymbol{u}_{n_k}) \quad \text{is convergent for every} \quad j$$

By Proposition 5.14.3, $\boldsymbol{u}_{n_k}$ is weakly convergent and by Proposition 5.14.2, there exists an element $\boldsymbol{u}_0 \in U$ such that $\boldsymbol{u}_{n_k} \rightharpoonup \boldsymbol{u}_0$. Also, by the same proposition,

$$\|\boldsymbol{u}_0\| \leq \liminf_{k\to\infty} \|\boldsymbol{u}_{n_k}\| \leq M$$

∎

Example 5.14.4

Let U be a reflexive Banach space and U' its dual. Then, for every $f \in U'$

$$\|f\|_{U'} = \max_{\|\boldsymbol{u}\|\leq 1} \langle f, \boldsymbol{u} \rangle$$

i.e., there exists an element $\|\boldsymbol{u}\| \leq 1$ such that $\|f\|_{U'} = f(\boldsymbol{u})$.

Indeed, the unit ball $\overline{B} = \overline{B}(\boldsymbol{0}, 1)$ is weakly sequentially compact and f is weakly continuous and, therefore, by the Weierstrass theorem, f attains its maximum on $\overline{B}$. □

Exercises

5.14.1. Prove Proposition 5.14.1.

5.14.2. Let U and V be two normed spaces. Prove that a linear transformation $T \in L(U, V)$ is strongly continuous iff it is weakly continuous, i.e., continuous with respect to weak topologies in U and V.
Hint: Prove first the following:
Lemma

$$T : X \text{ (with strong topology)} \to Y \text{ (weak topology)}$$

is continuous, iff the composition

$$f \circ T : X \to \mathbb{R}(\mathbb{C}) \text{ is continuous } \forall f \in Y'$$

and then follow the discussion in the section about strongly and weakly continuous linear functionals.

5.14.3. The space c_0 containing an infinite sequence of real numbers converging to zero is a normed linear space, equipped with the norm

$$\|\{\xi_l\}\|_{c_0} = \sup_{l \geq 1} |\xi_l| \ , \quad \left(\{\xi_l\} \in c_0 \Longleftrightarrow \lim_{l \to \infty} |\xi_l| = 0 \right)$$

(a) Show that

$$c_0' = \ell_1, \quad c_0'' = \ell_\infty$$

(b) Let

$$e_j = \{0, 0, \ldots, 1, 0, 0, \ldots\}$$

(1 is in the j-th position) be an element of ℓ_1 and, hence, identifiable as an element of c_0'. If $\mathbf{0} = \{0, 0, 0, \ldots, \}$, show that $\boldsymbol{e}_j$ converges to $\mathbf{0}$ weak*, but $\boldsymbol{e}_j$ does not converge weakly to $\mathbf{0}$.

5.14.4. Let U and V be normed spaces, and let either U or V be reflexive. Prove that every operator $A \in \mathcal{L}(U, V)$ has the property that A maps bounded sequences in U into sequences having weakly convergent subsequences in V.

5.14.5. In numerical analysis, one is often faced with the problem of approximating an integral of a given continuous function $f \in C[0, 1]$ by using some sort of numerical quadrature formula. For instance, we might partition $[0, 1]$ into n subintervals $x_1^n < x_2^n < \cdots < x_j^n < \cdots < x_n^n$ and set

$$\int_0^1 f(x)dx \approx \sum_{k=1}^n a_{nk} f(x_k^n) \equiv Q_n(f)$$

where the coefficients a_{nk} are constants satisfying

$$\sum_{k=1}^n |a_{nk}| < M \ , \quad \forall \, n \geq 1$$

Suppose that the quadrature rule $Q_n(f)$ integrates polynomials $p(x)$ of degree $n - 1$ exactly; i.e.,

$$Q_n(p) = \int_0^1 p(x)dx$$

Then

(a) show that

$$\lim_{n\to\infty}\left\{Q_n(f) - \int_0^1 f(x)dx\right] = 0$$

(b) characterize the type of convergence this limit defines in terms of convergence on the space $C[0,1]$; $\|f\| = \sup\{|f(x)|,\ 0 \le x \le 1\}$.

5.15 Compact (Completely Continuous) Operators

We establish here several interesting properties of an important class of operators on normed spaces—the compact operators. We shall show that compact operators behave almost like operators on finite-dimensional spaces and they take sequences that only converge weakly and produce strongly convergent sequences.

Compact and Completely Continuous Operators. Recall that a set K in a topological space is said to be *precompact (relatively compact)* iff its closure $\overline{K}$ is compact.

Consider now two normed spaces U and V and let $T : U \to V$ be any (not necessarily linear) operator from U to V. T is said to be *compact* iff it maps bounded sets in U into precompact sets in V, i.e.,

$$A \text{ bounded in } U \Longrightarrow \overline{T(A)} \text{ compact in } V$$

If, in addition, T is continuous, then T is said to be *completely continuous.* If V is a Banach space (complete), then, according to Theorem 4.9.2, T is compact if and only if it maps bounded sets in U into totally bounded sets in V. This implies that every compact operator is *bounded* and therefore, in particular, every compact *linear* operator is *automatically* completely continuous. Note also that, since in a finite-dimensional space boundedness is equivalent to the total boundedness, every bounded operator with a finite-dimensional range is automatically compact. In particular, every continuous linear operator with a finite-dimensional range is compact. This also implies that every *linear* T operator defined on a finite-dimensional space U is compact. Indeed, T is automatically continuous and the range of T is of finite dimension.

Example 5.15.1

Let $U = V = C[0,1]$ and consider the integral operator

$$(Tu)(\xi) = \int_0^1 K(x, \xi)u(x)dx = v(\xi)$$

where $K(x, \xi)$ is continuous on the square $0 \le x, \xi \le 1$. We shall show that T is compact. Suppose S_N is a bounded set of functions of $C[0,1]$ with $\|u\| \le N$. Obviously, for $u \in S_N$, $Tu(\xi)$ is uniformly bounded and $|Tu(\xi)| \le MN$, where $M = \max_{x,\xi} |K(x, \xi)|$. Since the kernel $K(x, \xi)$ is uniformly continuous, for each $\epsilon > 0$ there exists a δ such that

$$|K(x, \xi_1) - K(x, \xi_2)| < \frac{\epsilon}{N}$$

for $|\xi_1 - \xi_2| < \delta$ and $\forall\, x \in [0, 1]$. Then

$$|v(\xi_1) - v(\xi_2)| \le \int_0^1 |K(x, \xi_1) - K(x, \xi_2)|\ |u(x)|dx < \epsilon$$

for all $u \in S_N$. Hence the functions $Tu(\xi)$ are equicontinuous. By Arzela's theorem, the set $T(S_N)$ with the metric of $C[0,1]$ is precompact. This proves that the operator T is compact. $\square$

Example 5.15.2

Let U be a normed space, $\boldsymbol{u}_0$ be a fixed vector in U and f be a continuous, linear functional on U. Define the operator $T : U \to U$ by

$$T\boldsymbol{u} = f(\boldsymbol{u})\boldsymbol{u}_0$$

Obviously, T is continuous and its range is of dimension 1. Consequently, T is compact. $\square$

There are many continuous operators that are not compact. For instance, the identity operator on infinite-dimensional Banach spaces may not be compact since it maps the unit ball into itself, and while the unit ball is bounded, it need not be totally bounded, and hence is not compact.

The following proposition explains why the linear and compact operators are called completely continuous.

PROPOSITION 5.15.1

A linear and continuous operator T from a reflexive Banach space U to a Banach space V is compact (completely continuous) iff it maps weakly convergent sequences in U into strongly convergent sequences in V, i.e.,

$$u_n \rightharpoonup u \text{ in } U \Longrightarrow Tu_n \longrightarrow Tu \text{ in } V$$

PROOF

Necessity. Let $\boldsymbol{u}_n$ converge weakly to $\boldsymbol{u}$. Suppose that $\boldsymbol{v}_n = T\boldsymbol{u}_n$ does *not* converge strongly to $\boldsymbol{v} = T\boldsymbol{u}$. This implies that there exists an $\varepsilon > 0$ and a subsequence $\boldsymbol{v}_{n_k}$ such that $\|\boldsymbol{v}_{n_k} - \boldsymbol{v}\| \geq \varepsilon$. Now, according to Proposition 5.14.2, $\boldsymbol{u}_{n_k}$ is bounded and therefore enclosed in a sufficiently large ball $B = B(\mathbf{0}, r)$. T being compact implies that $\overline{T(B)}$ is compact and therefore sequentially compact, which means that there must exist a strongly convergent subsequence of $\boldsymbol{v}_{n_k}$, denoted by the same symbol, convergent to an element $\boldsymbol{v}_0$.

However, every linear and continuous operator is also weakly continuous (see Exercise 5.14.2) and therefore $\boldsymbol{v}_{n_k} \rightharpoonup \boldsymbol{v}$. Since, at the same time, $\boldsymbol{v}_{n_k} \rightharpoonup \boldsymbol{v}_0$ (strong convergence implies weak convergence), by the uniqueness of the limit (weak topologies are Hausdorff), $\boldsymbol{v} = \boldsymbol{v}_0$, a contradiction.

Sufficiency. Let A be a bounded set in U. It is sufficient to show that $\overline{T(A)}$ is sequentially compact. Let $\boldsymbol{v}_n = T\boldsymbol{u}_n$ be a sequence from $T(A)$, $\boldsymbol{u}_n \in A$. Since A is bounded and U reflexive, there must exist a weakly convergent subsequence $\boldsymbol{u}_{n_k} \rightharpoonup \boldsymbol{u}$. Consequently, $\boldsymbol{v}_{n_k} = T\boldsymbol{u}_{n_k} \to \boldsymbol{v} = T\boldsymbol{u}$, which finishes the proof. ∎

REMARK 5.15.1. Notice that reflexivity of U was used only in the "sufficiency" part of the proof. ∎

In retrospect, one might inquire as to whether the range of a compact operator must be finite dimensional. This is not true in general; however, compact operators come close to having a finite-dimensional range. Indeed, it can be shown that, for any given $\epsilon > 0$, there exists a finite-dimensional subspace $\mathcal{N}$ of $\mathcal{R}(T)$, T being compact, such that $T(U)$ can be made arbitrarily close to $\mathcal{N}$.

PROPOSITION 5.15.2

Let $T : U \to V$ be a compact operator from a Banach space U into another Banach space V. Then, given $\epsilon > 0$, there exists a finite-dimensional subspace M of $\mathcal{R}(T)$ such that

$$\inf\{\|Tu - v\|_V \ : v \in M\} \leq \epsilon \|u\|_U$$

PROOF Let $\epsilon > 0$ be given, and D be the closed unit ball in U. Since T is compact, $T(D)$ is contained in a compact set, and hence there is an

ϵ-net in $\mathcal{R}(T) \cap T(D)$. Let M be the linear subspace of V generated by this ϵ-net. It follows that M is finite dimensional, and dist $(Tu, M) \leq \epsilon$ for all $u \in D$. Then, if u is any point in U, $u/\|u\|_U \in D$, and

$$\inf \left\{ \left\| T\left(\frac{u}{\|u\|_U}\right) - v' \right\|_V : v' \in M \right\} \leq \epsilon$$

or

$$\inf \{ \|Tu - v\|_V : v = v' \|u\|_U \in M \} \leq \epsilon \|u\|_U$$

This completes the proof. ∎

We conclude this section with a number of simple properties of linear and compact operators.

PROPOSITION 5.15.3

Let U, V, W be Banach spaces and A, B denote linear operators. The following properties hold:

(i) A linear combination of compact operators is compact

$$A, B : U \to V \text{ compact } \Longrightarrow \alpha A + \beta B : U \to V \text{ compact}$$

(ii) Compositions of continuous and compact operators are compact

$$A : U \to V \text{ compact, } B \in \mathcal{L}(V, W) \Longrightarrow B \circ A : U \to W \text{ compact}$$

$$A \in \mathcal{L}(U, V), B : V \to W \text{ compact} \Longrightarrow B \circ A : U \to W \text{ compact}$$

(iii) A limit of a sequence of compact operators is compact

$$A_n : U \to V \text{ compact, } \|A_n - A\|_{\mathcal{L}(U,V)} \stackrel{n\to\infty}{\longrightarrow} 0$$
$$\Longrightarrow A : U \to V \text{ compact}$$

In other words, compact operators form a closed subspace in $\mathcal{L}(U, V)$.

PROOF (i) follows immediately from the definition. (ii) follows from Proposition 5.15.1. To prove (iii), assume that D is a bounded set in U. It is sufficient to prove that $A(D)$ is totally bounded in V. Let D be enclosed in a ball $B(\mathbf{0}, r)$. Pick an $\varepsilon > 0$ and select n such that $\|A_n - A\| \leq \delta = \frac{\varepsilon}{2r}$. Let $V_\varepsilon = \{\boldsymbol{v}_1, \ldots, \boldsymbol{v}_m\}$ be the $\frac{\varepsilon}{2}$-net for $A_n(D)$. Then

$$\begin{aligned}
\inf_i \|A\boldsymbol{u} - \boldsymbol{v}_i\|_V &\leq \inf_i \{\|A\boldsymbol{u} - A_n\boldsymbol{u}\|_V + \|A_n\boldsymbol{u} - \boldsymbol{v}_i\|_V\} \\
&\leq \|A - A_n\| \ \|\boldsymbol{u}\|_U + \inf_i \|A_n\boldsymbol{u} - \boldsymbol{v}_i\|_V \\
&\leq \frac{\varepsilon}{2r} r + \frac{\varepsilon}{2} = \varepsilon
\end{aligned}$$

for every $\boldsymbol{u} \in D$, which proves that V_ε is an ε-net for $A(D)$. ∎

Exercises

5.15.1. Let $T : U \to V$ be a linear continuous operator from a normed space U into a *reflexive* Banach space V. Show that T is *weakly sequentially compact*, i.e., it maps bounded sets in U into sets whose closures are weakly sequentially compact in V.
A is bounded in $U \Longrightarrow \overline{T(A)}$ is weakly sequentially compact in V.

5.15.2. Let U and V be normed spaces. Prove that a linear operator $T : U \to V$ is compact iff $T(B)$ is precompact in V for B—the unit ball in U.

5.15.3. Use the Frechet-Kolmogorov theorem (Theorem 4.9.4) to prove that operator T from Example 5.15.1 with an appropriate condition on kernel $K(x, \xi)$ is a compact operator from $L^p(\mathbb{R})$ into $L^r(\mathbb{R})$, $1 \leq p, r < \infty$.

Closed Range Theorem. Solvability of Linear Equations

5.16 Topological Transpose Operators. Orthogonal Complements

In Chapter 2 we introduced the idea of the transpose of a linear transformation $A : X \to Y$ from a vector space X to a vector space Y. The (algebraic) transpose was defined as an operator $A^T : Y^* \to X^*$ from the algebraic dual Y^* to the algebraic dual X^* by the formula

$$A^T : Y^* \to X^* \ , \ A^T\boldsymbol{y}^* = \boldsymbol{x}^* \text{ where } \boldsymbol{x}^* = \boldsymbol{y}^* \circ A$$

or, in other words,

$$\langle \boldsymbol{y}^*, A\boldsymbol{x} \rangle = \langle A^T\boldsymbol{y}^*, \boldsymbol{x} \rangle \quad \forall\, \boldsymbol{x} \in X, \boldsymbol{y}^* \in Y^*$$

where $\langle \cdot, \cdot \rangle$ stands for the duality pairings.

Topological Transpose. The same concept may be developed for continuous linear operators defined on normed spaces. Let X, Y be two normed spaces and $A \in \mathcal{L}(X, Y)$. The *topological transpose* of A, or, briefly, the *transpose of* A, denoted A', is defined as the restriction of the algebraic transpose A^T to the topological dual Y':

$$A' : Y' \to X' : A' = A^T/_{Y'} \ , \ A'\boldsymbol{y}' = \boldsymbol{x}', \text{ where } \boldsymbol{x}' = \boldsymbol{y}' \circ A$$

or, equivalently,

$$\langle \boldsymbol{y}', A\boldsymbol{x} \rangle = \langle A'\boldsymbol{y}', \boldsymbol{x} \rangle \quad \forall\, \boldsymbol{x} \in X, \boldsymbol{y}' \in Y'$$

Note that A' is well-defined (takes on values in X') since the composition $\boldsymbol{y}' \circ A$ is continuous.

Notice that in the above definition, the duality pairings are defined on different spaces, i.e., more appropriately, we could write

$$\langle \boldsymbol{y}', A\boldsymbol{x} \rangle_{Y' \times Y} = \langle A'\boldsymbol{y}', \boldsymbol{x} \rangle_{X' \times X}$$

The concept of the topological transpose is illustrated in Fig. 5.4.

The following proposition summarizes a number of properties of the transpose (see Proposition 2.11.1).

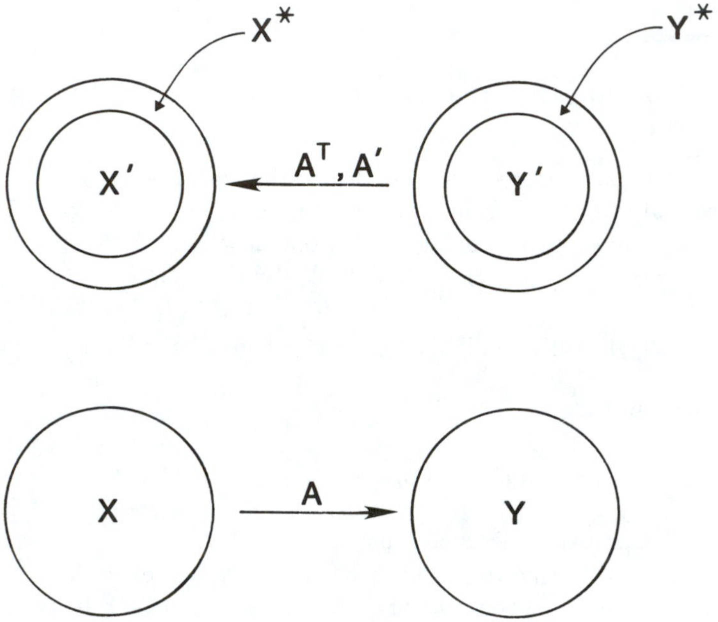

Figure 5.4
Illustration of the algebraic and topological transpose operators.

PROPOSITION 5.16.1

Let X, Y, Z be normed spaces with their topological duals X', Y', Z'. Let $A, A_i \in \mathcal{L}(X, Y)$, $i = 1, 2$, and $B \in \mathcal{L}(Y, Z)$. The following properties hold:

(i) *Transpose of a linear combination of operators is equal to the linear combination of their transpose operators:*

$$(\alpha_1 A_1 + \alpha_2 A_2)' = \alpha_1 A_1' + \alpha_2 A_2'$$

(ii) *Transpose of a composition is equal to the composition of the transpose operators (with inverted order)*

$$(B \circ A)' = A' \circ B'$$

(iii) *Transpose of the identity operator equals the identity operator on the dual space*

$$(id_X)' = id_{X'}$$

(iv) *If the inverse* A^{-1} *exists and is continuous, then* A' *has a continuous inverse, too, and*

$$(A')^{-1} = (A^{-1})'$$

(v) $\|A\|_{\mathcal{L}(X,Y)} = \|A'\|_{\mathcal{L}(Y',X')}$.

PROOF Proof of the first four properties follow precisely lines of the proof of Proposition 2.11.1 and is left as a straightforward exercise. To prove (v) see that

$$|A'\boldsymbol{y}'(\boldsymbol{x})| = |\langle A'\boldsymbol{y}', \boldsymbol{x}\rangle| = |\langle \boldsymbol{y}', A\boldsymbol{x}\rangle|$$

$$\leq \|\boldsymbol{y}'\|_{Y'} \|A\boldsymbol{x}\|_Y$$

$$\leq \|A\| \ \|\boldsymbol{y}'\|_{Y'} \|\boldsymbol{x}\|_X$$

which implies that

$$\|A'\boldsymbol{y}'\|_{X'} \leq \|A\| \ \|\boldsymbol{y}'\|_{Y'}$$

and, consequently, $\|A'\| \leq \|A\|$.

Conversely, by Corollary 5.5.2 to the Hahn-Banach theorem, for every $\boldsymbol{x} \neq \boldsymbol{0}$, there exists a functional $\boldsymbol{y}' \in Y'$ such that $\|\boldsymbol{y}'\|_{Y'} = 1$ and $\langle \boldsymbol{y}', A\boldsymbol{x}\rangle = \|A\boldsymbol{x}\|_Y$. Consequently,

$$\|A\boldsymbol{x}\|_Y = \langle \boldsymbol{y}', A\boldsymbol{x}\rangle = \langle A'\boldsymbol{y}', \boldsymbol{x}\rangle \leq \|A'\| \ \|\boldsymbol{x}\|$$

which proves that $\|A\| \leq \|A'\|$. ∎

Example 5.16.1

Let $A : X \to Y$ be a linear transformation from a finite-dimensional space X into a finite dimensional space Y and $\boldsymbol{e}_i$, $\boldsymbol{f}_j$ be bases for X and Y, respectively. In Chapter 2 we proved that if $\boldsymbol{A} = \{A_{ij}\}$ is the matrix representation for transformation A with respect to these bases, then the transpose matrix $\boldsymbol{A}^T = \{A_{ij}\}$ is the matrix representation for the transpose operator A^T with respect to the dual bases $\boldsymbol{f}_j^*$ and $\boldsymbol{e}_i^*$.

A similar property holds in the case of the integral operator $A: L^2(0,1) \to L^2(0,1)$ of the form

$$Au = v \; , \; v(x) = \int_0^1 K(x,\xi)u(\xi)d\xi$$

where the kernel $K(x,\xi)$ is assumed to be an L^2 function on $(0,1)\times(0,1)$.

Let (compare representation theorem for the duals of L^p spaces)

$$f : v \to \int_0^1 wvdx$$

be an arbitrary linear and continuous functional on $L^2(0,1)$ represented by function $w \in L^2(0,1)$. We have

$$\begin{aligned}
\langle f, Au\rangle &= \int_0^1 wAudx \\
&= \int_0^1 w(x)\left(\int_0^1 K(x,\xi)u(\xi)d\xi\right)dx \\
&= \int_0^1 \left(\int_0^1 K(x,\xi)w(x)dx\right)u(\xi)d\xi \\
&= \langle A^T f, u\rangle
\end{aligned}$$

where $A^T f$ is represented by the L^2 function

$$y(\xi) = \int_0^1 K(x,\xi)w(x)dx = \int_0^1 K^T(\xi,x)w(x)dx$$

where

$$K^T(\xi,x) = K(x,\xi)$$

Identifying this L^2 space with its dual, we see that the transpose of the integral operator A is obtained by interchanging arguments in the kernel $K(x,\xi)$ in much the same way the transpose matrix represents transpose of a linear operator on finite-dimensional spaces. $\square$

Topological Orthogonal Complements. Let X be a vector space and X^* its algebraic dual. In Chapter 2 we defined for a subspace $Z \subset X$ its (algebraic) orthogonal complement as

$$Z^{\perp} = \{x^* \in X^* : \langle x^*, z\rangle = 0 \quad \forall\, z \in Z\}$$

The same concept can now be generalized to a normed space X. By the *topological orthogonal complement* (or simply the *orthogonal complement*) of a subspace $Z \subset X$, denoted $Z^{\perp}$, we mean

$$Z^{\perp} = \{x' \in X' : \langle \boldsymbol{x}', \boldsymbol{z}\rangle = 0 \qquad \forall\, \boldsymbol{z} \in Z\}$$

It is easy to check that $Z^{\perp}$ is a *closed* subspace of X'.

In the same way, we define the orthogonal complement for a subspace $M \subset X'$:

$$M^{\perp} = \{\boldsymbol{z} \in X : \langle \boldsymbol{x}', \boldsymbol{z}\rangle = 0 \quad \forall\, \boldsymbol{x}' \in M\}$$

Again, $M^{\perp}$ is a closed subspace of X. Note that in defining the orthogonal complement of $M \subset X'$, we refer back to the original space X and *not* to the bidual X''.

Let $Z \subset X$ be a linear subspace of a normed space X. For every $\boldsymbol{x}' \in Z^{\perp}$, by definition of $Z^{\perp}$, $\langle \boldsymbol{x}', \boldsymbol{z}\rangle = 0$ and therefore, by definition of $M^{\perp}$, for $M = Z^{\perp}$,

$$Z \subset \left(Z^{\perp}\right)^{\perp}$$

The following proposition formulates a sufficient and necessary condition for the two sets to be equal to each other.

PROPOSITION 5.16.2

Let $Z \subset X$ be a subspace of a normed space X. The following conditions are equivalent to each other:

(i) Z *is closed.*

(ii) $(Z^{\perp})^{\perp} = Z$.

PROOF

(ii) $\Longrightarrow$ (i) follows from the fact that orthogonal complements are closed.

(i) $\Longrightarrow$ (ii) It remains to prove that $(Z^\perp)^\perp \subset Z$. Suppose, contrary to the assertion, that there exists $\boldsymbol{z} \in (Z^\perp)^\perp$ such that $\boldsymbol{z} \notin Z$. By the Mazur separation theorem (Lemma 5.13.1) there exists a linear continuous functional f such that

$$f|_Z = 0 \quad \text{and} \quad f(\boldsymbol{z}) \neq 0$$

which means that $f \in Z^\perp$ and $\langle f, \boldsymbol{z} \rangle \neq 0$ and therefore $\boldsymbol{z} \notin \left(Z^\perp\right)^\perp$, a contradiction. ∎

Exercises

5.16.1. Prove Proposition 5.16.1 (i)–(iv).

5.16.2. Let $A \in \mathcal{L}(U, V)$ be compact. Show that A' is also compact.

5.17 Solvability of Linear Equations in Banach Spaces. The Closed Range Theorem

In this section we shall examine a collection of ideas that are very important in the abstract theory of linear operator equations on Banach spaces. They concern the solvability of equations of the form

$$A\boldsymbol{u} = \boldsymbol{f} \ , \ A : U \longrightarrow V$$

where A is a linear and continuous operator from a normed space into a normed space V. Obviously, this equation can represent systems of linear algebraic equations, partial differential equations, integral equations, etc., so that general theorems concerned with its solvability are very important.

The question about the existence of solutions $\boldsymbol{u}$ to the equation above, for a given $\boldsymbol{f}$, can obviously be rephrased as

$$\text{when does } \boldsymbol{f} \in \mathcal{R}(A)$$

where $\mathcal{R}(A)$ denotes the range of A. The characterization of the range $\mathcal{R}(A)$ is therefore crucial to our problem.

From the definition of the transpose

$$\langle \boldsymbol{v}', A\boldsymbol{u} \rangle = \langle A'\boldsymbol{v}', \boldsymbol{u} \rangle \quad \forall\, \boldsymbol{u} \in U, \boldsymbol{v}' \in V'$$

we have

$$\boldsymbol{v}' \in \mathcal{N}(A') \Longleftrightarrow \boldsymbol{v}' \in \mathcal{R}(A)^{\perp}$$

which implies that

$$\mathcal{R}(A)^{\perp} = \mathcal{N}(A')$$

Combining this observation with Proposition 5.16.2, we arrive at the following theorem:

THEOREM 5.17.1
(The Closed Range Theorem for Continuous Operators)

Let U and V be normed spaces and $A \in \mathcal{L}(U, V)$ a linear and continuous operator from U to V. The following conditions are equivalent to each other:

(i) $\mathcal{R}(A)$ *is closed.*

(ii) $\mathcal{R}(A) = \mathcal{N}(A')^{\perp}$.

PROOF The proof follows immediately from Proposition 5.16.2. ∎

COROLLARY 5.17.1
(Solvability of Linear Equations)

Assume $A \in \mathcal{L}(U, V)$ and that $\mathcal{R}(A)$ is closed in V. Then the linear problem $A\boldsymbol{u} = \boldsymbol{f}$ possesses a solution if and only if

$$\boldsymbol{f} \in \mathcal{N}(A')^{\perp}$$

As we can see, it is essential to determine sufficient (and possibly necessary) conditions for the range $\mathcal{R}(A)$ to be closed.

Bounded Below Operators. A linear operator $A : U \to V$ from a normed space to a normed space V is said to be bounded below iff there exists a constant $c > 0$ such that

$$\|A\boldsymbol{u}\|_V \geq c\,\|\boldsymbol{u}\|_U \qquad \forall\, \boldsymbol{u} \in D(A)$$

This immediately implies that a bounded below operator possesses a continuous inverse on its range $\mathcal{R}(A)$. Indeed, $A\boldsymbol{u} = \boldsymbol{0}$ implies $\boldsymbol{u} = \boldsymbol{0}$ and therefore A is injective, and for $\boldsymbol{u} = A^{-1}\boldsymbol{v}$ we get

$$\|A^{-1}\boldsymbol{v}\|_U \leq \frac{1}{c}\|\boldsymbol{v}\|_V$$

The following theorem establishes the fundamental result showing equivalence of the closed range with the boundedness below for injective operators on Banach spaces.

THEOREM 5.17.2

Let U and V be Banach spaces and let $A \in \mathcal{L}(U, V)$ be injective. Then the range $\mathcal{R}(A)$ of A is closed if and only if A is bounded below.

PROOF

Sufficiency. Suppose $\boldsymbol{v}_n \in \mathcal{R}(A), \boldsymbol{v}_n \to \boldsymbol{v}$. Does $\boldsymbol{v} \in \mathcal{R}(A)$? Let $\boldsymbol{u}_n \in U$ be such that $A\boldsymbol{u}_n = \boldsymbol{v}_n$. But $\|\boldsymbol{v}_n - \boldsymbol{v}_m\|_V = \|A(\boldsymbol{u}_n - \boldsymbol{u}_m)\|_V \geq c\|\boldsymbol{u}_n - \boldsymbol{u}_m\|_U \to 0$ as $m, n \to \infty$. Thus, $\boldsymbol{u}_n$ is Cauchy. But, since U is complete, $\exists \boldsymbol{u}$ such that $\boldsymbol{u}_n \to \boldsymbol{u}$ in U. But A is continuous. Thus, $A\boldsymbol{u}_n \to A\boldsymbol{u} = \boldsymbol{v} \in \mathcal{R}(A)$; i.e., $\mathcal{R}(A)$ is closed.

Necessity. As a closed subspace of Banach space V, the range $\mathcal{R}(A)$ is a Banach space, too. Thus, A is a continuous, injective operator from U onto $\mathcal{R}(A)$ and, by the Banach theorem (Corollary 5.9.2 to the open mapping theorem), A has a continuous inverse A^{-1}, i.e.,

$$\|A^{-1}\boldsymbol{v}\|_U \leq \|A^{-1}\| \; \|\boldsymbol{v}\|_V \quad \forall\, \boldsymbol{v} \in \mathcal{R}(A)$$

But this is equivalent to A being bounded below. ∎

Thus, for injective operators A, the boundedness below is equivalent to the closedness of range $\mathcal{R}(A)$, which in turn is equivalent to the criterion for the existence expressed in terms of the transpose of operator A.

We proceed now with a discussion for noninjective operators A.

Quotient Normed Spaces. Let U be a vector space and $M \subset U$ a subspace of U. In Chapter 2 we defined the quotient space U/M consisting of equivalence classes of $\boldsymbol{u} \in U$ identified as affine subspaces of U of the form

$$[\boldsymbol{u}] = \boldsymbol{u} + M = \{\boldsymbol{u} + \boldsymbol{v} : \boldsymbol{v} \in M\}$$

If, in addition, U is a normed space and M is *closed*, the quotient space U/M can be equipped with the norm

$$\|[\boldsymbol{u}]\|_{U/M} \stackrel{\text{def}}{=} \inf_{\boldsymbol{v}\in[\boldsymbol{u}]} \|\boldsymbol{v}\|_U$$

Indeed, all properties of norms are satisfied:

(i) $\|[\boldsymbol{u}]\| = 0$ implies that there exists a sequence $\boldsymbol{v}_n \in [\boldsymbol{u}]$ such that $\boldsymbol{v}_n \to \boldsymbol{0}$. By closedness of M and, therefore, of every equivalence class $[\boldsymbol{u}]$ (why?), $\boldsymbol{0} \in [\boldsymbol{u}]$, which means that $[\boldsymbol{u}] = [\boldsymbol{0}] = M$ is the zero vector in the quotient space U/M.

(ii)

$$\|\lambda[\boldsymbol{u}]\| = \|[\lambda\boldsymbol{u}]\|$$

$$= \inf_{\lambda\boldsymbol{v}\in[\lambda\boldsymbol{u}]} \|\lambda\boldsymbol{v}\|$$

$$= |\lambda| \inf_{\boldsymbol{v}\in[\boldsymbol{u}]} \|\boldsymbol{v}\| = |\lambda|\ \|[\boldsymbol{u}]\|$$

(iii) Let $[\boldsymbol{u}], [\boldsymbol{v}] \in U/M$. Pick an arbitrary $\varepsilon > 0$. There exist then $\boldsymbol{u}_\varepsilon \in [\boldsymbol{u}]$ and $\boldsymbol{v}_\varepsilon \in [\boldsymbol{v}]$ such that

$$\|\boldsymbol{u}_\varepsilon\| \le \|[\boldsymbol{u}]\|_{U/M} + \frac{\varepsilon}{2} \text{ and } \|\boldsymbol{v}_\varepsilon\| \le \|[\boldsymbol{v}]\|_{U/M} + \frac{\varepsilon}{2}$$

Consequently,

$$\|\boldsymbol{u}_\varepsilon\| \le \|[\boldsymbol{u}]\|_{U/M} + \|[\boldsymbol{v}]\|_{U/M} + \varepsilon$$

But $\boldsymbol{u}_\varepsilon + \boldsymbol{v}_\varepsilon \in [\boldsymbol{u} + \boldsymbol{v}]$ and therefore, taking the infimum on the left-hand side and passing to the limit with $\varepsilon \to 0$, we get the triangle inequality for the norm in U/M.

It also turns out that for a Banach space U, the quotient space U/M is also Banach.

LEMMA 5.17.1

Let M be a closed subspace of a Banach space U. Then U/M is Banach.

PROOF Let $[\boldsymbol{u}_n]$ be a Cauchy sequence in U/M. One can extract a subsequence $\boldsymbol{u}_{n_k}$ such that

$$\|[\boldsymbol{u}_{n_{k+1}}] - [\boldsymbol{u}_{n_k}]\| \leq \frac{1}{2^{k+2}}$$

Select next, for every k, an element $\boldsymbol{v}_k$ such that

$$\boldsymbol{v}_k \in [\boldsymbol{u}_{n_{k+1}}] - [\boldsymbol{u}_{n_k}] = [\boldsymbol{u}_{n_{k+1}} - \boldsymbol{u}_{n_k}]$$

and

$$\|\boldsymbol{v}_k\|_U \leq \|[\boldsymbol{u}_{n_{k+1}}] - [\boldsymbol{u}_{n_k}]\|_{U/M} + \frac{1}{2^{k+2}} \leq \frac{1}{2^{k+1}}$$

and consider the sequence

$$\boldsymbol{v}_0 = \boldsymbol{u}_{n_1}, \ \boldsymbol{v}_1, \boldsymbol{v}_2, \cdots$$

The sequence of partial sums $\boldsymbol{S}_k = \sum_{i=0}^{k} \boldsymbol{v}_i$ is Cauchy and therefore converges to an element $\boldsymbol{v}$ in U.

At the same time,

$$\boldsymbol{S}_k = \boldsymbol{v}_0 + \boldsymbol{v}_1 + \boldsymbol{v}_2 + \cdots \boldsymbol{v}_k \in [\boldsymbol{u}_{n_1}] + [\boldsymbol{u}_{k_2} - \boldsymbol{u}_{n_1}]$$

$$+ \cdots [\boldsymbol{u}_{n_{k+1}} - \boldsymbol{u}_{n_k}] = [\boldsymbol{u}_{n_{k+1}}]$$

which implies

$$\|[\boldsymbol{u}_{n_{K+1}}] - [\boldsymbol{v}]\|_{U/M} \leq \|\boldsymbol{S}_K - \boldsymbol{v}\|_U \to 0$$

and, finally, by the triangle inequality,

$$\|[\boldsymbol{u}_n] - [\boldsymbol{v}]\| \leq \|[\boldsymbol{u}_n] - [\boldsymbol{u}_{n_{k+1}}]\| + \|[\boldsymbol{u}_{n_{k+1}}] - [\boldsymbol{v}]\|$$

which proves that the entire sequence converges to $[\boldsymbol{v}]$. ∎

We continue now with the discussion of sufficient and necessary conditions for the range $\mathcal{R}(A)$ of an operator A, to be closed.

THEOREM 5.17.3

Let U and V be Banach spaces and let $A \in \mathcal{L}(U,V)$ be a linear and continuous operator on U. Then the range $\mathcal{R}(A)$ of A is closed if and only if there exists a constant $c > 0$ such that

$$\|A\boldsymbol{u}\|_V \geq c \inf_{\boldsymbol{w}\in\mathcal{N}(A)} \|\boldsymbol{u}+\boldsymbol{w}\|_U$$

PROOF Let $M = \mathcal{N}(A)$. By continuity of A, M is closed. Consider next the quotient operator

$$\widetilde{A} : U/M \ni [\boldsymbol{u}] \to \widetilde{A}[\boldsymbol{u}] = A\boldsymbol{u} \in V$$

$\widetilde{A}$ is obviously a well-defined injective operator on Banach space. Taking the infimum with respect to $\boldsymbol{w}$ in the inequality:

$$\|A\boldsymbol{u}\|_V = \|A\boldsymbol{w}\|_V \leq \|A\| \ \|\boldsymbol{w}\|_U \quad \forall \boldsymbol{w} \in [\boldsymbol{u}]$$

we see that $\widetilde{A}$ is also continuous.

The inequality in the theorem can now be reinterpreted as boundedness below of operator $\widetilde{A}$:

$$\left\|\widetilde{A}[\boldsymbol{u}]\right\|_V \geq c \,\|[\boldsymbol{u}]\|_{U/M}$$

which reduces the whole case to the previous theorem for injective operators. ∎

COROLLARY 5.17.2
(Solvability of Linear Equations)

Let U and V be Banach spaces and let $A \in \mathcal{L}(U,V)$ be a linear and continuous operator such that

$$\|A\boldsymbol{u}\|_V \geq c \inf_{\boldsymbol{w}\in\mathcal{N}(A)} \|\boldsymbol{u}+\boldsymbol{w}\|_U \ , \ c > 0$$

Then the linear problem $A\boldsymbol{u} = \boldsymbol{f}$, for some $\boldsymbol{f} \in V$, has a solution $\boldsymbol{u}$ if and only if

$$\boldsymbol{f} \in \mathcal{N}(A')^{\perp}$$

The solution $\boldsymbol{u}$ is determined uniquely up to elements from the null space of A, i.e., $\boldsymbol{u}+\boldsymbol{w}$ is also a solution for every $\boldsymbol{w} \in \mathcal{N}(A)$.

We emphasize that the boundedness below of the quotient operator $\widetilde{A}$ provides not only a sufficient condition for the *solvability criterion* above ($\boldsymbol{f} \in \mathcal{N}(A')^{\perp}$), but it is *equivalent* to it, as follows from the presented theorems.

Notice also that the boundedness below is *equivalent* to the continuity of the inverse operator $\widetilde{A}^{-1}$:

$$\widetilde{A}^{-1} : V \ni \boldsymbol{v} \longrightarrow [\boldsymbol{u}] \in U/M$$

which is just another way of saying that the solutions $\boldsymbol{u}$ of $A\boldsymbol{u} = \boldsymbol{f}$ should depend continuously on the data, i.e., the right-hand side, $\boldsymbol{f}$.

5.18 Generalization for Closed Operators

Surprising as it looks, most of the results from the preceding two sections can be generalized to the case of closed operators.

Topological Transpose. Let X and Y be two normed spaces and let $A : X \supset D(A) \to Y$ be a linear operator, not necessarily continuous. Consider all points $(\boldsymbol{y}', \boldsymbol{x}')$ from the product space $Y' \times X'$ such that

$$\langle \boldsymbol{y}', A\boldsymbol{x} \rangle = \langle \boldsymbol{x}', \boldsymbol{x} \rangle \quad \forall \boldsymbol{x} \in D(A)$$

We claim that $\boldsymbol{y}'$ uniquely defines $\boldsymbol{x}'$ iff the domain $D(A)$ of operator A is dense in X. Indeed, assume that $\overline{D(A)} = X$. By linearity of both sides with respect to the first argument it is sufficient to prove that

$$\langle \boldsymbol{x}', \boldsymbol{x} \rangle = 0 \quad \forall\, \boldsymbol{x} \in D(A) \quad \text{implies} \quad \boldsymbol{x}' = \boldsymbol{0}$$

But this follows easily from the density of $D(A)$ in X and continuity of $\boldsymbol{x}'$.

Conversely, assume that $\overline{D(A)} \neq X$. Let $\boldsymbol{x} \in X - \overline{D(A)}$. By the Mazur separation theorem (Lemma 5.13.1) there exists a continuous and linear functional $\boldsymbol{x}'_0$, vanishing on $\overline{D(A)}$, but different from zero at $\boldsymbol{x}$. Consequently, the zero functional $\boldsymbol{y}' = \boldsymbol{0}$ has two corresponding elements, $\boldsymbol{x}' = \boldsymbol{0}$, and $\boldsymbol{x}' = \boldsymbol{x}'_0$, a contradiction.

Restricting thus ourselves to the case of operators A with domains $D(A)$ dense in X, we can identify the collection of $(\boldsymbol{y}', \boldsymbol{x}')$ discussed above (see Proposition 5.10.1) as the graph of a linear operator from Y' to X', denoted A', and called the *transpose* (or *dual*) of operator A. Due to our construction, this definition generalizes the definition of the transpose for $A \in \mathcal{L}(X, Y)$.

The next observation we will make is that the transpose operator A', if it exists, is always *closed*. Indeed, consider a sequence $\boldsymbol{y}'_n \in D(A')$ such that $\boldsymbol{y}'_n \to \boldsymbol{y}'$ and $\boldsymbol{A}'\boldsymbol{y}'_n \to \boldsymbol{x}'$. Passing to the limit in the equality

$$\langle \boldsymbol{y}'_n, A\boldsymbol{x}\rangle = \langle \boldsymbol{A}'\boldsymbol{y}'_n, \boldsymbol{x}\rangle \quad \boldsymbol{x} \in D(A)$$

we conclude immediately that $\boldsymbol{y}' \in D(A')$ and $A'\boldsymbol{y}' = \boldsymbol{x}'$. Consequently, by Proposition 5.10.2, A' must be closed.

We summarize a number of properties for this generalized operator in the following proposition.

PROPOSITION 5.18.1

Let X, Y, Z be normed spaces with their topological duals X', Y', Z'.

(i) *Let $A_i : X \supset D \to Y$, $i = 1, 2$ be two linear operators defined on the same domain D, dense in X. Then*

$$(\alpha_1 A_1 + \alpha_2 A_2)' = \alpha_1 A_1' + \alpha_2 A_2'$$

(ii) *Let $A : X \supset D(A) \to Y$, $B : Y \supset D(B) \to Z$ be linear operators with domains dense in X and Y, respectively, and let $\mathcal{R}(A) \subset D(B)$ (to make sense for the composition $B \circ A$). Then*

$$(B \circ A)' \supset A' \circ B'$$

i.e., the transpose $(B \circ A)'$ exists and is an extension of the composition $A' \circ B'$.

(iii) *If $A : X \supset D(A) \to Y$ is a linear injective operator with domain $D(A)$ dense in X and range $\mathcal{R}(A)$ dense in Y then the transpose operator A' has an inverse and*

$$(A')^{-1} = (A^{-1})'$$

PROOF The proof follows directly from the definitions and is left as a straightforward exercise. ∎

Consider now again the abstract linear equation of the form

$$A\boldsymbol{u} = \boldsymbol{f}\ ,\ A : U \supset D(A) \to V,\ \overline{D(A)} = U$$

where A is a *closed* operator from the dense domain $D(A)$ in a normed space U into another normed space V. We have the following fundamental result due to S. Banach.

THEOREM 5.18.1
(The Closed Range Theorem for Closed Operators)

Let U and V be normed spaces and $A : U \supset D(A) \rightarrow V$, $\overline{D(A)} = U$, be linear and closed. The following conditions are equivalent to each other:

(i) $\mathcal{R}(A)$ *is closed in* V.

(ii) $\mathcal{R}(A) = \mathcal{N}(A')^{\perp}$.

PROOF

(ii) $\Longrightarrow$ (i) follows from the fact that orthogonal complements are always closed.

(i) $\Longrightarrow$ (ii) From the definition of the transpose operator A'

$$\langle \boldsymbol{v}', A\boldsymbol{u} \rangle = \langle A'\boldsymbol{v}', \boldsymbol{u} \rangle \quad \forall\, \boldsymbol{u} \in D(A),\ \boldsymbol{v}' \in D(A')$$

we have

$$\boldsymbol{v}' \in \mathcal{N}(A') \Longleftrightarrow \boldsymbol{v}' \in D(A') \text{ and } A'\boldsymbol{v}' = \boldsymbol{0}$$

$$\Longleftrightarrow \langle \boldsymbol{v}', A\boldsymbol{u} \rangle = 0 \quad \forall\, \boldsymbol{u} \in D(A)$$

$$\Longleftrightarrow \boldsymbol{v}' \in \mathcal{R}(A)^{\perp}$$

Thus, as in the case of continuous operators,

$$\mathcal{R}(A)^{\perp} = \mathcal{N}(A')$$

Applying Proposition 5.16.2, we finish the proof. ∎

As before, we have immediately the following corollary.

COROLLARY 5.18.1
(Solvability of Linear Equations)

Let A be a closed operator discussed above, and let the range $\mathcal{R}(A)$ of A be *closed* in V. Then the linear problem $A\boldsymbol{u} = \boldsymbol{f}$ possesses a solution if and only if

$$\boldsymbol{f} \in \mathcal{N}(A')^{\perp}$$

As in the case of continuous operators, the closedness of the range $\mathcal{R}(\mathcal{A})$ turns out to be equivalent to the boundedness below.

THEOREM 5.18.2

Let U and V be Banach spaces and

$$A : U \supset D(A) \to V$$

denote a closed, linear operator. The following conditions are equivalent to each other:

(i) $\mathcal{R}(A)$ *is closed in* V.

(ii) There exists a positive constant $c > 0$ *such that*

$$\|A\boldsymbol{u}\|_V \geq c \inf_{\boldsymbol{w} \in \mathcal{N}(A)} \|\boldsymbol{u} + \boldsymbol{w}\|_U$$

PROOF

Case 1. A injective.

(ii) $\Longrightarrow$ (i) The inequality implies that A is bounded below and therefore its inverse A^{-1} is continuous. A being closed and bounded below implies that its domain $D(A)$ is closed and therefore its range $\mathcal{R}(A)$ coincides with the inverse image of $D(A)$ through the continuous inverse A^{-1} and therefore must be closed.

(i) $\Longrightarrow$ (ii) If A is closed then A^{-1} is closed as well and is defined on the closed range $\mathcal{R}(A)$ in V, which can be identified as a Banach space itself. By the closed graph theorem, A^{-1} must be continuous which is equivalent to the boundedness below of A.

Case 2. A arbitrary.

As in the proof of Theorem 5.17.3, consider the quotient map

$$\widetilde{A} : U/M \supset D(\widetilde{A}) \ni [\boldsymbol{u}] \to \widetilde{A}[\boldsymbol{u}] = A\boldsymbol{u} \in V$$

where $M = \mathcal{N}(A)$.

A few comments are necessary:

1. Null space of a closed operator is closed. Indeed if

$$D(A) \supset \mathcal{N}(A) \ni \boldsymbol{u}_n \to \boldsymbol{u}$$

then $A\boldsymbol{u}_n = \boldsymbol{0}$ is constant and therefore converges trivially to $\boldsymbol{0}$ which, by Proposition 5.10.2, implies that $\boldsymbol{u} \in D(A)$ and $A\boldsymbol{u} = \boldsymbol{0}$. Consequently, $\boldsymbol{u} \in \mathcal{N}(A)$, which proves that $\mathcal{N}(A)$ is closed.

2. By Lemma 5.17.1, the space U/M is Banach.
3. The domain $D(\widetilde{A})$ of $\widetilde{A}$ is equal to $D(A)/M$.
4. $\widetilde{A}$ is closed. Indeed, let

$$D(\widetilde{A}) \ni [\boldsymbol{u}_n] \rightarrow [\boldsymbol{u}] \; , \; \widetilde{A}[\boldsymbol{u}_n] \rightarrow \boldsymbol{v}$$

By definition of the norm in U/M one can find a sequence $\boldsymbol{w}_n \in [\boldsymbol{u}_n]$ (see Lemma 5.17.1) such that

$$\boldsymbol{w}_n \longrightarrow \boldsymbol{w} \in [\boldsymbol{u}]$$

At the same time, $\widetilde{A}[\boldsymbol{u}_n] = A\boldsymbol{w}_n \rightarrow \boldsymbol{v}$ and therefore, by closedness of A,

$$\boldsymbol{w} \in D(A) \text{ and } A\boldsymbol{w} = \boldsymbol{v}$$

Consequently,

$$[\boldsymbol{u}] = [\boldsymbol{w}] \in D(\widetilde{A}) \text{ and } \widetilde{A}[\boldsymbol{u}] = \boldsymbol{v}$$

which proves that $\widetilde{A}$ is closed.

Finally, it is sufficient to apply the first case result to $\widetilde{A}$. ∎

We conclude this section with the generalization of Corollary 5.17.2.

COROLLARY 5.18.2
(Solvability of Linear Equations)

Let U and V be Banach spaces and let

$$A : \; U \supset D(A) \longrightarrow V, \quad \overline{D(A)} = U$$

be a linear, *closed* operator with the domain $D(A)$ *dense* in U such that

$$\exists c > 0 \; : \quad \|A\boldsymbol{u}\|_V \geq \inf_{\boldsymbol{w} \in \mathcal{N}(A)} \|\boldsymbol{u} + \boldsymbol{w}\|_U \quad \forall \boldsymbol{u} \in D(A)$$

Then the linear problem

$$A\boldsymbol{u} = \boldsymbol{f}, \qquad \boldsymbol{f} \in V$$

has a solution $\boldsymbol{u}$ if and only if

$$\boldsymbol{f} \in \mathcal{N}(A')^{\perp}$$

where A' is the transpose of A

$$A' : V' \supset D(A') \longrightarrow U'$$

The solution $\boldsymbol{u}$ is determined uniquely up to elements from the null space of A.

Note that all the comments concluding the preceding section remain valid.

Exercises

5.18.1. Prove Proposition 5.18.1.

5.19 Examples

In this section, we give two simple examples from mechanics dealing with the solution of a linear problem $A\boldsymbol{u} = \boldsymbol{f}$ and showing the interpretation of the solvability condition $\boldsymbol{f} \in \mathcal{N}(A')^{\perp}$.

Example 5.19.1

Consider the beam equation

$$(EIw'')'' = q \qquad 0 < x < \ell$$

where EI is the stiffness of the beam (product of Young modulus E and cross-sectional moment of inertia I) and $q = q(x)$ the intensity of the load applied to beam (see Fig. 5.5).

The beam is not supported, and both ends are subjected to neither concentrated forces nor concentrated moments which, in view of the formulas for the bending moment M and shear force V:

$$M = -EIw'' \ , \ V = -(EIw'')'$$

translates into boundary conditions

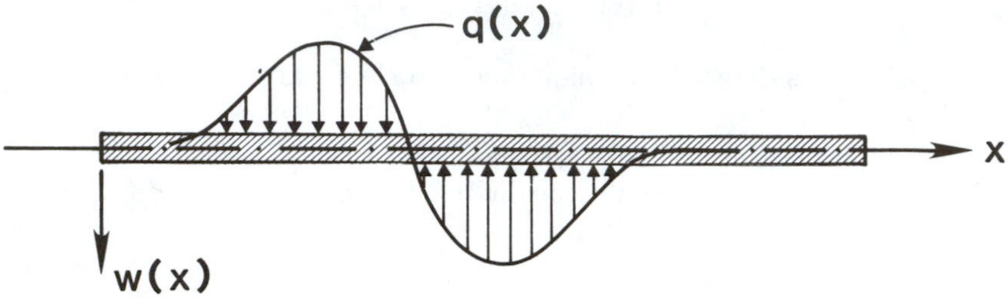

Figure 5.5
A "free" beam loaded with the distributed loading with intensity $q(x)$.

$$w''(0) = w'''(0) = 0 \text{ and } w''(\ell) = w'''(\ell) = 0$$

provided we assume for simplicity that $EI = \text{const.}$

We will formulate now the problem in the operator form. Toward this goal we introduce the space

$$W = \left\{ w \in H^4(0,\ell) : w''(0) = w'''(0) = w''(\ell) = w'''(\ell) = 0 \right\}$$

consisting of all functions w from the Sobolev space of fourth order H^4 satisfying the boundary conditions. As a closed subspace of $H^4(0,\ell)$ (see Exercise 5.19.1), W is itself a Banach (in fact, Hilbert) space. Next, we consider the operator $A : W \to V = L^2(0,1)$ defined as

$$Aw = (EIw'')'' = EIw''''$$

Obviously, A is both linear and continuous and the whole boundary value problem reduces to the operator equation

$$Aw = q$$

provided we assume that the load q is square integrable.

We continue now by determining the transpose of A. First of all, according to the representation theorem for L^p spaces, the dual to $L^2(0,\ell)$ can be identified with itself. Consequently, the duality pairing is replaced with the L^2 product on L^2 (see Chapter 2) and the definition of the conjugate operator A' reads as

$$\int_0^\ell v(EIw'''')dx = (v, Aw) = \langle A'v, w\rangle$$

$$\forall\, w \in W,\ v \in V = L^2(0,1)$$

where $\langle\cdot,\cdot\rangle$ stands for the duality pairing between W and its dual and $(\cdot,\cdot)$ is the L^2 product.

Recall that, for a continuous operator $A : W \to V$ defined on the whole space W, the topological transpose A' is defined on the whole dual space V'. Its value at a particular v is given precisely by the left-hand side of the formula above.

Next, we determine the kernel (null space) of A'. Restricting ourselves first to $w \in C_0^\infty(0,\ell) \subset W$, we get

$$\int_0^\ell v(EIw'''')dx = 0 \qquad \forall\, w \in C_0^\infty(0,\ell)$$

which, by the definition of the distributional derivatives, means that v has a distributional derivative of fourth order, v'''' and that

$$v'''' = 0$$

Integration by parts yields now (see Exercise 5.19.2)

$$\int_0^\ell v(EIw'''')dx = \int_0^\ell v''''EIw\,dx + (v''w' - v'''w)|_0^\ell$$

$$= (v''w' - v'''w)|_0^\ell \quad \forall\, w \in W$$

As there are no boundary conditions on w and w' in the definition of W, both w and w' may take arbitrary values at 0 and ℓ, which implies that

$$v''(0) = v''(\ell) = v'''(0) = v'''(\ell) = 0$$

Consequently,

$$\mathcal{N}(A') = \{v :\ v(x) = \alpha x + \beta\ ,\ \alpha,\beta \in \mathbb{R}\}$$

Notice that the null space $\mathcal{N}(A')$ of the transpose operator coincides with the null space $\mathcal{N}(A)$ of the operator itself, interpreted as the *space of infinitesimal rigid body motions.* Consequently, the necessary and sufficient condition for the existence of a solution $w \in W$

$$q \in \mathcal{N}(A')^{\perp}$$

reduces to

$$\int_0^\ell q(x)dx = 0 \text{ and } \int_0^\ell q(x)x dx = 0$$

The two conditions above are easily recognized as the *global equilibrium* equations for the load q (resultant force and moment must vanish).

Note that the solution u is determined only up to the rigid body motions.
□

REMARK 5.19.1. It may be a little confusing, but it is very illustrative to see how the same example is formulated using the formalism of closed operators. Introducing only one space $V = L^2(0,\ell)$, identified with its dual, we define operator $A : V \to V$ as follows:

$$D(A) = \{u \in L^2(0,\ell) : u'''' \in L^2(0,\ell) \text{ and}$$
$$u''(0) = u''(\ell) = u'''(\ell) = 0\}$$
$$(= W \text{ from the example})$$

$$A = EIu''''$$

It is an easy exercise to prove that A is well defined and *closed.* By the same calculations as before we find out that

$$\int_0^\ell EIv''''udx + (v''u' - v'''u)|_0^\ell = (A'v, u)$$

$$\forall\, u \in D(A),\ v \in D(A')$$

This leads to the transpose (adjoint) operator in the form

$$D(A') = \{v \in L^2(0,\ell) : v'''' \in L^2(0,\ell) \text{ and}$$

$$v''(0) = v''(\ell) = v'''(0) = v'''(\ell) = 0\}$$

$$A'v = EIv''''$$

Thus the transpose operator A' coincides with A itself. Note the difference between the domains of this and previous operators.

The rest of the conclusions are the same. ∎

Example 5.19.2

The solvability condition $\boldsymbol{f} \in \mathcal{N}(A')^\perp$ admits to simple physical interpretation, like in the previous example, in most applications in mechanics. We shall briefly describe now an application to a class of boundary-value problems in linear elasticity.

A two-dimensional version of the situation is illustrated in Fig. 5.6. An elastic body, occupying a domain Ω, is subjected to body forces of density $\boldsymbol{f}$ per unit volume and surface tractions $\boldsymbol{g}$ on a portion Γ_t of the boundary $\Gamma = \partial\Omega$ of Ω. On the remaining portion of the boundary, Γ_u, the displacement vector $\boldsymbol{u}$ is prescribed as zero, $\boldsymbol{u}|_{\Gamma_u} = \boldsymbol{0}$.

We wish to find the displacement vector field $\boldsymbol{u} = \boldsymbol{u}(\boldsymbol{x})$ for which the body will be at rest (in equilibrium) under the action of forces $\boldsymbol{f}$ and $\boldsymbol{g}$. We obtain the familiar boundary-value problem

Find the displacement $\boldsymbol{u}$ such that

$$-(E_{ijk\ell}u_{k,\ell})_{,j} = f_i \qquad \text{in } \Omega$$

subjected to the boundary conditions

$$\begin{aligned} E_{ijk\ell}u_{k,\ell}n_j &= g_i \quad \text{on } \Gamma_t \\ u_i &= 0 \quad \text{on } \Gamma_u \end{aligned}$$

where $E_{ijk\ell}$ is the tensor of elasticities satisfying the customary assumptions, $\boldsymbol{n} = (n_j)$ is the outward normal unit to boundary Γ, and commas denote the partial differentiation.

Our interest here is to interpret the compatibility conditions on the data, and for this purpose we consider a restricted problem for which $\Gamma_u = \emptyset, \Gamma_t = \partial\Omega$, i.e., tractions are prescribed on all of $\partial\Omega$. The operator A is identified as a composite operator (see Section 1.11) prescribing for each displacement field $\boldsymbol{u}$ the corresponding body force in Ω and traction $\boldsymbol{t}$ on the boundary

$$A\boldsymbol{u} = (-(E_{ijk\ell}u_{k,\ell})_{,j} \ ; \ E_{ijk\ell}u_{k,\ell}n_j)$$

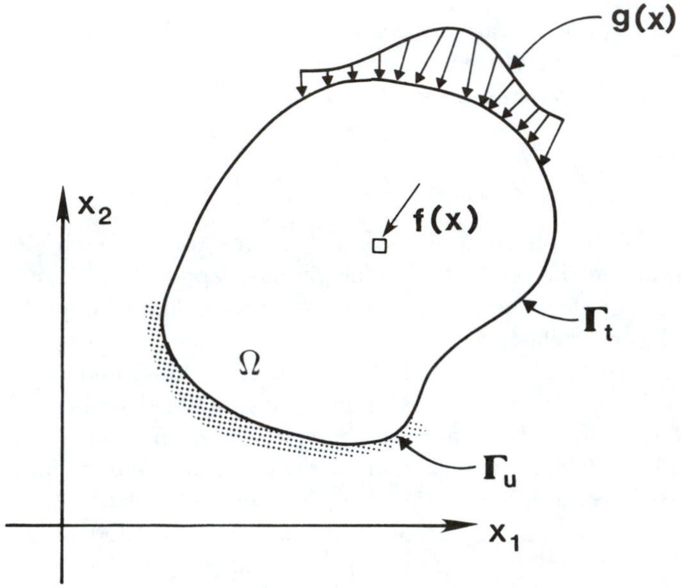

Figure 5.6
An elastic body in equilibrium under the action of external forces.

With an appropriate setting of function spaces it can be proved that the kernel of the transpose operator coincides with that of operator A itself and consists of vector fields $\boldsymbol{v}$ of the form

$$\boldsymbol{v}(\boldsymbol{r}) = \boldsymbol{c} + \boldsymbol{\theta} \times \boldsymbol{r}$$

where $\boldsymbol{c}$ and $\boldsymbol{\theta}$ are constants and $\boldsymbol{r} = (x_i)$ is the position vector with respect to the origin of the system of coordinates. Physically, $\boldsymbol{c}$ is a rigid translation and $\boldsymbol{\theta}$ is an infinitesimal rigid rotation. The data (*load*) is compatible with A iff

$$\int_\Omega \boldsymbol{f}\boldsymbol{v}dx + \int_{\partial\Omega} \boldsymbol{g}\boldsymbol{v}ds = 0 \quad \forall\, \boldsymbol{v} \in \mathcal{N}(A')$$

Setting $\boldsymbol{\theta} = \mathbf{0}$ and arguing that $\boldsymbol{c}$ is arbitrary reveals that

$$\int_\Omega \boldsymbol{f} dx + \int_{\partial\Omega} \boldsymbol{g} ds = \mathbf{0} \tag{a}$$

whereas setting $\boldsymbol{c} = \mathbf{0}$ and using arbitrary $\boldsymbol{\theta}$ gives

$$\int_\Omega \boldsymbol{r} \times \boldsymbol{x} dx + \int_{\partial\Omega} \boldsymbol{r} \times \boldsymbol{g} ds = \mathbf{0} \tag{b}$$

We recognize these compatibility conditions as the *global equations of equilibrium*: (a) is the requirement that the vector sum of external forces vanish and (b) is the requirement that the moment of all external forces about the origin vanish.

We have thus transformed the *local* equilibrium equations into global requirements on the data $\boldsymbol{f}$ and $\boldsymbol{g}$. $\square$

Exercises

5.19.1. Prove that the linear mapping (functional)

$$H'(0,1) \ni w \longrightarrow w(x_0), \quad \text{where} \quad x_0 \in [0,1]$$

is continuous. Use the result to prove that space W in Example 5.19.1 is closed.
Hint: Show first that

$$w(b) - w(a) = \int_a^b w'(x) dx$$

for every $w \in H'(0,1)$.

5.19.2. Let $u, v \in H^1(0,1)$. Prove the integration by parts formula

$$\int_0^1 uv' dx = -\int_0^1 u'v dx + uv|_0^1$$

Hint: Make use of the density of $C^\infty[0,1]$ in $H^1(0,1)$.

5.19.3. Work out all the details of Example 5.19.1 once again, with different boundary conditions:

$$w(0) = w''(0) = 0 \quad \text{and} \quad w''(\ell) = w'''(\ell) = 0$$

(left end of the beam is supported by a pin support).

5.19.4. Prove that operator A from Remark 5.19.1 is closed.

5.19.5. (a) Determine conditions on the data $\boldsymbol{f}$ necessary for solutions to exist to the linear systems of equations that follows.
(b) Determine if the solutions are unique, and if not, describe the null space of the associated operator.

$$\boldsymbol{A}\boldsymbol{u} = \boldsymbol{f}$$

$$\boldsymbol{A} = \begin{bmatrix} 1 & 2 & -1 \\ 4 & 0 & 2 \\ 3 & -2 & -3 \end{bmatrix} \qquad \boldsymbol{A} = \begin{bmatrix} 3 & 1 & -1 & 2 \\ 6 & 2 & -2 & 4 \\ 9 & 3 & -3 & 6 \end{bmatrix}$$

$$\boldsymbol{A} = \begin{bmatrix} 1 & -1 & 0 \\ -1 & 0 & 1 \end{bmatrix}$$

5.20 Equations with Completely Continuous Kernels. Fredholm Alternative

In this last section we would like to study a special class of abstract equations of the form

$$\boldsymbol{x} - T\boldsymbol{x} = \boldsymbol{y}$$

where T is a *completely* continuous operator from a Banach space X into itself, $\boldsymbol{y}$ is a given vector from X, and $\boldsymbol{x}$ is the unknown. Such equations are frequently called *equations of the second type* and they are typical in the theory of integral equations. Results presented in this section are essentially due to Fredholm, a Swedish mathematician who devoted much of his efforts to the study of these equations.

Introducing the identity operator $I : X \to X$ and the corresponding operator $A = I - T$ one could, of course, rewrite the equation in the usual form

$$A\boldsymbol{x} = \boldsymbol{y}$$

However, the point is that then the special properties of operator T are lost and the general theory for linear equations presented in the previous sections cannot deliver such strong results as a direct study of the equation of the second type.

We begin with an essential observation concerning operator $A = I - T$.

LEMMA 5.20.1

The range $\mathcal{R}(A)$ of operator A is closed.

PROOF Let $M = \mathcal{N}(A)$ denote the null space of operator A. According to Theorem 5.17.3, closedness of $\mathcal{R}(A)$ is equivalent to the boundedness below of the quotient operator

$$\widetilde{A}: X/M \longrightarrow X \ , \ \widetilde{A}[\boldsymbol{x}] = A\boldsymbol{x}$$

where $[\boldsymbol{x}] = \boldsymbol{x} + M$ denotes the equivalence class of $\boldsymbol{x}$.

In contrast, assume that $\widetilde{A}$ is *not* bounded below. There exists, then, a sequence of equivalence classes $\widetilde{\boldsymbol{x}}_n \in X/M$ such that

$$\|\widetilde{\boldsymbol{x}}_n\|_{X/M} = 1 \quad \text{and} \quad \left\|\widetilde{A}\widetilde{\boldsymbol{x}}_n\right\|_X \longrightarrow 0$$

It follows from the definition of norm in the quotient space that there exists a corresponding sequence of vectors $\boldsymbol{x}_n \in \widetilde{\boldsymbol{x}}_n$ such that

$$\|\boldsymbol{x}_n\|_X \le \|\widetilde{\boldsymbol{x}}_n\|_{X/M} + \varepsilon \le 1 + \varepsilon,$$

for some $\varepsilon > 0$. Consequently, $\boldsymbol{x}_n$ is bounded, and by the compactness of T, we can extract a subsequence $\boldsymbol{x}_{n_k}$ such that $T\boldsymbol{x}_{n_k} \to \boldsymbol{x}_0$ strongly for some $\boldsymbol{x}_0 \in X$. Consequently,

$$\boldsymbol{x}_{n_k} = (T + A)\boldsymbol{x}_{n_k} \longrightarrow \boldsymbol{x}_0$$

By the continuity of T, $T(\boldsymbol{x}_{n_k}) \to T(\boldsymbol{x}_0)$, which proves that

$$\boldsymbol{x}_0 = T\boldsymbol{x}_0 \Longrightarrow \boldsymbol{x}_0 \in M$$

From the continuity of the map

$$X \ni \boldsymbol{x} \longrightarrow [\boldsymbol{x}] \in X/M$$

it follows that

$$[\boldsymbol{x}_{n_k}] \longrightarrow [\boldsymbol{x}_0] \ \text{in} \ X/M$$

a contradiction since $\|[\boldsymbol{x}_{n_k}]\|_{X/M} = 1$ and $\|[\boldsymbol{x}_0]\|_{X/M} = 0$. ∎

Before we proceed with the study of further properties of operator A, we stop to prove a simple, but fundamental, result which holds in any normed vector space.

LEMMA 5.20.2
(Lemma on Almost Perpendicularity)

Let X be a normed space and X_0 a *closed* subspace of X, different from X. Then, for an arbitrary small $\varepsilon > 0$, there exists a corresponding unit vector $\boldsymbol{x}_\varepsilon$, $\|\boldsymbol{x}_\varepsilon\| = 1$ such that

$$\rho\,(\boldsymbol{x}_\varepsilon, X_0) > 1 - \varepsilon$$

PROOF Recall the definition of the distance between a vector $\boldsymbol{x}$ and set (space) X_0

$$\rho(\boldsymbol{x}, X_0) = \inf_{\boldsymbol{y} \in X_0} \|\boldsymbol{x} - \boldsymbol{y}\|$$

As X_0 is closed and different from X, there must be a vector $\overline{\boldsymbol{x}} \in X$ separated from X_0 by a positive distance d:

$$\rho\,(\overline{\boldsymbol{x}}, X_0) = d > 0$$

(Otherwise X_0 would be dense in X and by closedness would have to coincide with the whole X.) By definition of the distance $\rho(\overline{\boldsymbol{x}}, X_0)$, for every $1 > \varepsilon > 0$ there exists a vector $\boldsymbol{x}' \in X_0$ such that

$$\|\overline{\boldsymbol{x}} - \boldsymbol{x}'\| \leq \frac{d}{1-\varepsilon} \qquad (> d)$$

Define

$$\boldsymbol{x}_\varepsilon = \frac{\overline{\boldsymbol{x}} - \boldsymbol{x}'}{\|\overline{\boldsymbol{x}} - \boldsymbol{x}'\|} = a\,(\overline{\boldsymbol{x}} - \boldsymbol{x}') \ , \ a = \|\overline{\boldsymbol{x}} - \boldsymbol{x}'\|^{-1}$$

Then, for every $\boldsymbol{x} \in X_0$, we have

$$\|\boldsymbol{x}_\varepsilon - \boldsymbol{x}\| = \|a\overline{\boldsymbol{x}} - a\boldsymbol{x}' - \boldsymbol{x}\| = a\left\|\overline{\boldsymbol{x}} - \left(\boldsymbol{x}' + \frac{\boldsymbol{x}}{a}\right)\right\|$$

$$\geq ad > \frac{1-\varepsilon}{d}d = 1 - \varepsilon$$

since $\left(\boldsymbol{x}' + \dfrac{\boldsymbol{x}}{a}\right) \in X_0$. ∎

REMARK 5.20.1. If X were a Hilbert space then, taking any unit vector $\boldsymbol{x}_0$ from the *orthogonal complement* of X_0 (cf. Theorem 6.2.1), we would have

$$\begin{aligned}\rho(\boldsymbol{x}_0, X_0)^2 &= \inf_{\boldsymbol{y}\in X_0} \|\boldsymbol{x}_0 - \boldsymbol{y}\|^2 \\ &= \inf_{\boldsymbol{y}\in X_0} (\boldsymbol{x}_0 - \boldsymbol{y}, \boldsymbol{x}_0 - \boldsymbol{y}) \\ &= \inf_{\boldsymbol{y}\in X_0} \left\{ (\boldsymbol{x}_0, \boldsymbol{x}_0) + \{\boldsymbol{y}, \boldsymbol{y}\} \right\} = 1\end{aligned}$$

where $(\cdot, \cdot)$ is the inner product in X. This explains the name of the lemma. ∎

COROLLARY 5.20.1

Let X be a normed space. The following conditions are equivalent to each other:

(i) X is finite dimensional, $\dim X < \infty$.

(ii) A set $E \subset X$ is compact iff E is closed and bounded.

PROOF Implication (i) $\Rightarrow$ (ii) follows from the famous Heine-Borel theorem (Theorem 4.3.1). To prove (ii) $\Rightarrow$ (i), assume instead that $\dim X = \infty$. Next, take an arbitrary unit vector $\boldsymbol{x}_1$ and consider subspace $X_1 = \mathbb{R}\boldsymbol{x}_1(\mathbb{C}\boldsymbol{x}_1)$. By the lemma on almost perpendicularity there exists a unit vector $\boldsymbol{x}_2$ such that

$$\rho(\boldsymbol{x}_2, X_1) > \frac{1}{2}$$

and by induction we have a sequence of unit vectors $\boldsymbol{x}_n$ such that

$$\rho(\boldsymbol{x}_n, X_{n-1}) > \frac{1}{2}$$

where $X_n = \mathbb{R}\boldsymbol{x}_1 \oplus \ldots \oplus \mathbb{R}\boldsymbol{x}_n$ $(\mathbb{C}\boldsymbol{x}_1 \oplus \ldots \oplus \mathbb{C}\boldsymbol{x}_n)$. As the unit ball is closed and bounded, according to (ii) it must be compact and therefore sequentially compact as well. Consequently, we can extract a converging subsequence $\boldsymbol{x}_{n_k}$ which, in particular, must satisfy the Cauchy condition, i.e.,

$$\lim_{k,l\to\infty} \|\boldsymbol{x}_{n_k} - \boldsymbol{x}_{n_l}\| = 0,$$

a contradiction, since by construction of $\boldsymbol{x}_n$

$$\|\boldsymbol{x}_{n_k} - \boldsymbol{x}_{n_l}\| > \frac{1}{2}$$

∎

We return now to the study of equations of the second type. As a direct consequence of the lemma on almost perpendicularity we get the following further characterization of operator $A = I - T$.

LEMMA 5.20.3

Let $A^n = A \circ \ldots \circ A$ (n times). Then the sequence of null spaces $\mathcal{N}(A^n)$ is increasing:

$$\mathcal{N}(A) \subset \mathcal{N}\left(A^2\right) \subset \ldots \subset \mathcal{N}\left(A^n\right) \subset \mathcal{N}\left(A^{n+1}\right) \subset \ldots$$

and contains only a *finite* number of different sets.

PROOF We have

$$\boldsymbol{x} \in \mathcal{N}\left(A^n\right) \Longleftrightarrow A^n\boldsymbol{x} = \boldsymbol{0} \Longrightarrow A\left(A^n\boldsymbol{x}\right) = \boldsymbol{0} \Longrightarrow \boldsymbol{x} \in \mathcal{N}\left(A^{n+1}\right)$$

which proves the monotonicity.

Denote $X_n = \mathcal{N}(A^n)$. If $X_n = X_{n+1}$ for some n, then $X_{n+1} = X_{n+2}$ and, consequently, $X_m = X_{m+1}$, for any $m \geq n$. Indeed,

$$\begin{aligned}
\boldsymbol{x} \in X_{n+2} \Longrightarrow A^{n+2}\boldsymbol{x} = \boldsymbol{0} \quad &\Longrightarrow A^{n+1}(A\boldsymbol{x}) = \boldsymbol{0} \\
\Longrightarrow A\boldsymbol{x} \in X_{n+1} \quad &\Longrightarrow A\boldsymbol{x} \in X_n \\
\Longrightarrow A^n(A\boldsymbol{x}) = \boldsymbol{0} &\Longrightarrow \boldsymbol{x} \in X_{n+1}
\end{aligned}$$

Finally, assume to the contrary that $X_n \neq X_{n+1}, \quad \forall\ n$. By the lemma on almost perpendicularity, there exists a sequence of unit vectors $\boldsymbol{x}_n$ such that

$$x_{n+1} \in X_{n+1}, \ \|x_{n+1}\| = 1, \ \rho(x_{n+1}, X_n) > \frac{1}{2}$$

Let $m > n$. Then

$$Tx_m - Tx_n = x_m - Ax_m - (x_n - Ax_n) = x_m - \overline{x}$$

where we have denoted

$$\overline{x} = Ax_m + x_n - Ax_n$$

Moreover,

$$A^{m-1}\overline{x} = A^m x_m + A^{m-1}x_n - A^m x_n = \mathbf{0}$$

and therefore $\overline{x} \in X_{m-1}$, which implies that

$$\|Tx_m - Tx_n\| = \|x_m - \overline{x}\| > \frac{1}{2}$$

This leads to a contradiction: we can extract a subsequence x_{n_k} such that Tx_{n_k} converges strongly to some element in X and, in particular, it satisfies the Cauchy condition

$$\lim_{k,l\to\infty} \|Tx_{n_k} - Tx_{n_l}\| = 0$$

a contradiction. ∎

LEMMA 5.20.4

A sequence of range spaces $A^n(X)$ is decreasing

$$A(X) \supset \ldots \supset A^n(X) \supset A^{n+1}(X) \supset \ldots$$

and contains only a finite number of different sets.

PROOF We have

$$x \in A^n(X) \Longrightarrow \exists\ y \in X \quad x = A^n y \Longrightarrow x = A^{n-1}(Ay)$$

$$\Longrightarrow x \in A^{n-1}(X)$$

which proves the monotonicity.

Next, $A^n(X) = A^{n+1}(X)$ implies trivially $A^{n+1}(X) = A^{n+2}(X)$. Finally, assuming to the contrary that $A^n(X) \neq A^{n+1}(X)$, the lemma on almost perpendicularity implies again that

$$\exists\, \boldsymbol{x}_n \in T^n(X),\ \|\boldsymbol{x}_n\| = 1,\ \rho\left(\boldsymbol{x}_n, T^{n+1}(X)\right) > \frac{1}{2}$$

We get for $m > n$

$$T\boldsymbol{x}_n - T\boldsymbol{x}_m = x_n - (A\boldsymbol{x}_n + \boldsymbol{x}_m - A\boldsymbol{x}_m) = \boldsymbol{x}_n - \overline{\boldsymbol{x}}$$

where

$$\overline{x} = A\boldsymbol{x}_n + x_m - A\boldsymbol{x}_m \in A^{n+1}(X)$$

and therefore

$$\|T\boldsymbol{x}_n - T\boldsymbol{x}_m\| = \|\boldsymbol{x}_n - \overline{\boldsymbol{x}}\| > \frac{1}{2}$$

which leads to the same contradiction as in the proof of the previous lemma. ∎

Let m now denote the minimum index n such that $A^n(X) = A^{n+1}(X)$. (If $X = A(X)$, i.e., A is surjective, then $m = 0$.) Denote

$$Y \stackrel{\text{def}}{=} A^m(X) = \mathcal{R}\left(A^m\right),\ Z \stackrel{\text{def}}{=} \mathcal{N}\left(A^m\right)$$

We will continue now with a detailed discussion of the restrictions of operator A to spaces Y and Z.

Step 1.

It follows from the definition of Y that

$$A(Y) = A\left(A^m(X)\right) = A^{m+1}(X) = A^m(X) = Y$$

which proves that operator A takes Y onto Y. It follows also that the restriction of A to Y is one-to-one. Indeed, assume that $A\boldsymbol{y} = \boldsymbol{0}$ for some $\boldsymbol{y} \in Y$. As $\boldsymbol{y} \in A^m(X) = A^n(X), n \geq m$; for every $n \geq m$ there exists a corresponding $\boldsymbol{x}$ such that $\boldsymbol{y} = A^n\boldsymbol{x}$. Consequently, $\boldsymbol{0} = A\boldsymbol{y} = A^{n+1}\boldsymbol{x}$ implies $\boldsymbol{x} \in \mathcal{N}(A^{n+1})$, and, for sufficiently large n such that $\mathcal{N}(A^{n+1}) = \mathcal{N}(A^n)$, $\boldsymbol{x} \in \mathcal{N}(A^n)$, and therefore $A^n\boldsymbol{x} = \boldsymbol{y} = \boldsymbol{0}$, which proves that $A\big|_Y$ is injective.

Operator $A^m = (I - T)^m$ can be represented in the form

$$A^m = (I - T)^m = I - T_1$$

where T_1 is a sum of compositions of T and as such it is completely continuous. Since the restriction of A^m to Z is zero, we have

$$T_1 \boldsymbol{z} = \boldsymbol{z} \qquad \text{for} \qquad \boldsymbol{z} \in Z$$

This implies that any bounded and closed set in Z must be compact. Indeed, if $\boldsymbol{z}_n$ is a bounded sequence in Z, then by the compactness of T_1 we can extract a subsequence $\boldsymbol{z}_{n_k}$ such that $T\boldsymbol{z}_{n_k} \to \boldsymbol{z}_0$ strongly for some $\boldsymbol{z}_0 \in Z$, which implies that $\boldsymbol{z}_{n_k}$ itself converges *strongly* to $\boldsymbol{z}$. Thus, by Corollary 5.20.1, Z must be finite dimensional. When restricted to Z, operator A maps Z into itself. Indeed, if $m = 0$, then A is injective and the assertion is trivial ($Z = \{\mathbf{0}\}$). For $m \neq 0$ and $\boldsymbol{z} \in Z$ ($A^m \boldsymbol{z} = \mathbf{0}$), we have

$$A^m(A\boldsymbol{z}) = A^{m+1}\boldsymbol{z} = A\left(A^m \boldsymbol{z}\right) = \mathbf{0}$$

Step 2.

By Lemma 5.20.1 applied to operator $A^m = (I - T)^n = I - T_1$, space Y must be closed and therefore is a Banach space. By the open mapping theorem then, restriction $A_0 = A|_Y$ has a continuous inverse $A_0^{-1} : Y \to Y$. For any $\boldsymbol{x} \in X$ we define

$$\boldsymbol{y} = A_0^{-m} A^m \boldsymbol{x} \qquad \boldsymbol{z} = \boldsymbol{x} - \boldsymbol{y}$$

By definition, $\boldsymbol{x} = \boldsymbol{y} + \boldsymbol{z}$ and $\boldsymbol{y} \in Y$. Also,

$$A^m \boldsymbol{z} = A^m \boldsymbol{x} - A^m \boldsymbol{y} = \mathbf{0}$$

which proves that $\boldsymbol{z} \in Z$. Due to bijectivity of $A_0 = A|_Y$, the decomposition is unique. Indeed, if there were

$$\boldsymbol{x} = \boldsymbol{y} + \boldsymbol{z} = \boldsymbol{y}_1 + \boldsymbol{z}_1$$

for some other $\boldsymbol{y}_1 \in Y$, $\boldsymbol{z}_1 \in Z$, then it would be

$$\mathbf{0} = (\boldsymbol{y} - \boldsymbol{y}_1) + (\boldsymbol{z} - \boldsymbol{z}_1)$$

and, consequently,

$$A^m \Big((\boldsymbol{y} - \boldsymbol{y}_1) + (\boldsymbol{z} - \boldsymbol{z}_1) \Big) = A^m \left(\boldsymbol{y} - \boldsymbol{y}_1\right) = \mathbf{0}$$

which, by bijectivity of A^m restricted to Y, implies $\boldsymbol{y} = \boldsymbol{y}_1$. Thus space X can be represented as the direct sum of Y and Z as

$$X = Y \oplus Z$$

Step 3.

Let n denote now the smallest integer k such that $\mathcal{N}\left(A^k\right) = \mathcal{N}\left(A^{k+1}\right)$. It turns out that $n = m$.

We first prove that $n \le m$. It is sufficient to show that $\mathcal{N}\left(A^{m+1}\right) \subset \mathcal{N}\left(A^m\right)$. Let $\boldsymbol{x} \in \mathcal{N}\left(A^{m+1}\right)$. Using the just-proved decomposition $\boldsymbol{x} = \boldsymbol{y} + \boldsymbol{z}$, we have

$$\mathbf{0} = A^{m+1}\boldsymbol{x} = A^{m+1}\boldsymbol{y} + A^{m+1}\boldsymbol{z} = A^{m+1}\boldsymbol{y} + A\left(A^m\boldsymbol{z}\right) = A^{m+1}\boldsymbol{y}$$

which implies that $\boldsymbol{y} = \mathbf{0}$ and, consequently, $\boldsymbol{x} = \boldsymbol{z} \in \mathcal{N}\left(A^m\right)$.

To prove that $m \le n$, consider $A^n\boldsymbol{x}$. We have

$$A^n\boldsymbol{x} = A^n(\boldsymbol{y} + \boldsymbol{z}) = A^n\boldsymbol{y} + A^n\boldsymbol{z}$$

$$= A^n\boldsymbol{y} = A^n A A^{-1}\boldsymbol{y} = A^{n+1}\boldsymbol{y}_1$$

because $\mathcal{N}(A^n) = \mathcal{N}(A^m)$ $(m \ge n)$ and where $\boldsymbol{y}_1 = A^{-1}\boldsymbol{y}$. Thus $A^n(X) \subset A^{n+1}(X)$, which proves that $m \le n$ and, consequently, $m = n$.

Step 4.

Let Π_Y and Π_Z denote the (continuous) projections corresponding to the decomposition $X = Y \oplus Z$ (cf. Step 2):

$$\Pi_Y = A_0^{-m}A^m, \qquad \Pi_Z = I - \Pi_Y$$

Defining

$$T_Y \stackrel{\text{def}}{=} T \circ \Pi_Y, \qquad T_Z \stackrel{\text{def}}{=} T \circ \Pi_Z$$

we can decompose T into the sum of completely continuous operators T_Y, T_Z:

$$T = T_Y + T_Z$$

where, according to the Step 1 results, T_Y maps X into Y and T_Z maps X into Z. In particular, both compositions $T_Y T_Z$ and $T_Z T_Y$ are zero,

$$T_Y T_Z = T_Z T_Y = 0$$

Finally, the decomposition of T implies the corresponding decomposition of $A = I - T$

$$A = (I - T_Y) - T_Z$$

The first map, $W \stackrel{\text{def}}{=} I - T_Y = I - T \circ \Pi_Y$, turns out to be an isomorphism of Banach spaces. According to the open mapping theorem, it is sufficient to prove that W is bijective.

Let $W\boldsymbol{x} = \mathbf{0}$. Using decomposition $\boldsymbol{x} = \boldsymbol{y} + \boldsymbol{z}$, $\boldsymbol{y} \in Y$, $\boldsymbol{z} \in Z$, we have

$$\begin{aligned} \mathbf{0} = W\boldsymbol{x} &= \boldsymbol{x} - T_Y\boldsymbol{y} - T_Y\boldsymbol{z} = \boldsymbol{y} - T_Y\boldsymbol{y} + \boldsymbol{z} \\ &= A\boldsymbol{y} + \boldsymbol{z} \end{aligned}$$

which, due to the fact that $A\boldsymbol{y} \in Y$, implies that $A\boldsymbol{y} = \boldsymbol{z} = \mathbf{0}$ and, consequently, $\boldsymbol{y} = \mathbf{0}$ as well. Thus W is injective.

To prove surjectivity, pick $\boldsymbol{x} \in X$ and consider the correponding decomposition

$$\boldsymbol{x} = \boldsymbol{y} + \boldsymbol{z} \ , \ \boldsymbol{y} \in Y, \ \boldsymbol{z} \in Z$$

Next, define

$$\boldsymbol{w} = A_0^{-1}\boldsymbol{y} + \boldsymbol{z}$$

We have

$$\begin{aligned} W\boldsymbol{w} &= \boldsymbol{w} - T\Pi_Y\left(A_0^{-1}\boldsymbol{y} + \boldsymbol{z}\right) = A_0^{-1}\boldsymbol{y} + \boldsymbol{z} - TA_0^{-1}\boldsymbol{y} \\ &= (I - T)A_0^{-1}\boldsymbol{y} + \boldsymbol{z} = AA_0^{-1}\boldsymbol{y} + \boldsymbol{z} = \boldsymbol{y} + \boldsymbol{z} = \boldsymbol{x} \end{aligned}$$

which proves that W is surjective.

We summarize the results in the following theorem.

THEOREM 5.20.1

Let X be a Banach space and $T: X \to X$ a completely continuous operator taking X into itself. Define $A = I - T$, where I is the identity operator on X. Then the following properties hold:

(i) There exists an index $m \geq 0$ such that

$$\mathcal{N}(A) \not\supseteq \ldots \not\supseteq \mathcal{N}\left(A^m\right) = \mathcal{N}\left(A^{m+1}\right) = \ldots$$

$$A(X) \supsetneq \ldots \supsetneq A^m(X) = A^{m+1}(X) = \ldots$$

(ii) *Space X can be represented as a direct sum*

$$X = Y \oplus Z, \quad Y \stackrel{\text{def}}{=} A^m(X), \quad Z \stackrel{\text{def}}{=} \mathcal{N}(A^m)$$

where Z is finite dimensional and the corresponding projections Π_Y and Π_Z are continuous.

(iii) *Operator T admits a decomposition*

$$T = T_Y + T_Z$$

where $T_Y \in \mathcal{L}(X, Y)$, $T_Z \in \mathcal{L}(X, Z)$ are completely continuous and $I - T_Y$ is an isomorphism of Banach spaces.

COROLLARY 5.20.2

The equation $A\boldsymbol{x} = \boldsymbol{x} - T\boldsymbol{x} = \boldsymbol{y}$ is solvable *for every* $\boldsymbol{y} \in Y$, i.e., operator A is surjective if and only if A is injective (compare the case of a finite-dimensional space X).

PROOF Consider the case $m = 0$ in Theorem 5.20.1. ∎

Together with the original equation we can consider the corresponding equation in the dual space X'.

$$\boldsymbol{f} - T'\boldsymbol{f} = \boldsymbol{g} \qquad \boldsymbol{f}, \boldsymbol{g} \in X'$$

We first show that the structure of the transpose operator is the same as the original one.

LEMMA 5.20.5

Let T be a completely continuous operator from a Banach space X into a Banach space Y. Then the transpose operator $T': Y' \to X'$ is completely continuous as well.

PROOF Let $f_n \rightharpoonup f$ be a sequence from Y' converging weakly to an element $f \in Y'$. It is sufficient to show that

$$f_n \circ T \longrightarrow f \circ T \quad \text{strongly in } X'$$

which is equivalent to

$$\sup_{\|\boldsymbol{x}\|_X \le 1} |f_n(T\boldsymbol{x}) - f(T\boldsymbol{x})| \to 0$$

or

$$\sup_{\boldsymbol{y} \in K} |f_n(\boldsymbol{y}) - f(\boldsymbol{y})| \to 0$$

where K is the image of unit ball $\overline{B}(\mathbf{0}, 1)$ in X under transformation T

$$K = T\Big(\overline{B}(\mathbf{0}, 1)\Big)$$

Due to the complete continuity (compactness) of T, $\overline{K}$ is compact in Y.

By the uniform boundedness theorem, f_n are uniformly bounded on a unit ball in Y, which implies that f_n are *uniformly continuous* and due to the boundedness of K, uniformly bounded on K.

Application of the Ascoli-Arzela theorem finishes the proof. ∎

Thus all conclusions for operator T hold for the transpose operator T' as well. We also have

LEMMA 5.20.6

Kernels of operator T and its conjugate T' have the same dimension.

$$\dim \mathcal{N}(T) = \dim \mathcal{N}(T')$$

PROOF Let $A = I - T$ and let $m \ge 0$ be the smallest integer such that $\mathcal{N}(A^m) = \mathcal{N}(A^{m+1})$. Since $\mathcal{N}(A) \subset \mathcal{N}(A^m)$ and, according to Theorem 5.20.1, $\mathcal{N}(A^m)$ is finite dimensional, the kernel of A, $\mathcal{N}(A)$ must be finite dimensional as well. The same applies to the kernel of the transpose operator:

$$(id_X - A)' = id_{X'} - A' = I - A'$$

where we have used the same symbol I to denote the identity operator in X'.

Assume now that $n = \dim \mathcal{N}(A) \le m = \dim \mathcal{N}(A')$, and let $\boldsymbol{x}_1, \ldots, \boldsymbol{x}_n$ be a basis for $\mathcal{N}(A)$ and $\boldsymbol{g}_1, \ldots, \boldsymbol{g}_m$ a basis for $\mathcal{N}(A')$. Let next

$$\boldsymbol{f}_1, \ldots, \boldsymbol{f}_n \in (\mathcal{N}(A))^* = (\mathcal{N}(A))'$$

denote the dual basis to $\boldsymbol{x}_1, \dots, \boldsymbol{x}_n$ in the (finite-dimensional) dual to kernel $\mathcal{N}(A)$. By the Hahn-Banach theorem, functionals $\boldsymbol{f}_1, \dots, \boldsymbol{f}_n$ can be extended to linear and continuous functionals defined on the whole space X. Thus

$$\langle \boldsymbol{f}_i, \boldsymbol{x}_j \rangle = \delta_{ij} \qquad i,j = 1,\dots,n$$

Similarly, let $\boldsymbol{y}_1, \dots, \boldsymbol{y}_m$ be a set of linearly independent vectors in X such that

$$\langle \boldsymbol{g}_i, \boldsymbol{y}_j \rangle = \delta_{ij} \qquad i,j = 1,\dots,m$$

Define now a new operator R as

$$R = T + S \quad \text{where} \quad S\boldsymbol{x} \stackrel{\text{def}}{=} \sum_{k=1}^{n} f_k(\boldsymbol{x})\boldsymbol{y}_k$$

As transformation S is also completely continuous (why ?), R is completely continuous, too.

We now claim that operator $I - R$ is injective. Indeed,

$$\boldsymbol{x} - R\boldsymbol{x} = \boldsymbol{x} - T\boldsymbol{x} - S\boldsymbol{x} = A\boldsymbol{x} - S\boldsymbol{x} = \boldsymbol{0}$$

implies that

$$A\boldsymbol{x} - \sum_{k=1}^{n} f_k(\boldsymbol{x})\boldsymbol{y}_k = \boldsymbol{0}$$

and, consequently,

$$\langle \boldsymbol{g}_i, A\boldsymbol{x} \rangle - \sum_{k=1}^{n} f_k(\boldsymbol{x}) \langle \boldsymbol{g}_i, \boldsymbol{y}_k \rangle = \boldsymbol{0} \quad i = 1,\dots,n$$

or

$$\langle A'\boldsymbol{g}_i, \boldsymbol{x} \rangle - f_i(\boldsymbol{x}) = \boldsymbol{0} \qquad i = 1,\dots,n$$

As $A'\boldsymbol{g}_i = \boldsymbol{0}$, this implies that

$$f_i(\boldsymbol{x}) = 0 \qquad i = 1,\dots,n$$

and, consequently, $A\boldsymbol{x} = \boldsymbol{0}$, i.e., $\boldsymbol{x} \in \mathcal{N}(A)$. But this implies that $\boldsymbol{x}$ can be represented in the form

$$\boldsymbol{x} = \sum_{i=1}^{n} a_i \boldsymbol{x}_i$$

and, since $f_j(\boldsymbol{x}) = a_j = 0$, it follows that $\boldsymbol{x} = \mathbf{0}$. Thus $I - R$ is injective and, by Corollary 5.20.2, surjective as well. In particular, there exists a solution, say $\overline{\boldsymbol{x}}$, to the equation

$$A\overline{\boldsymbol{x}} - \sum_{k=1}^{n} f_k\left(\overline{\boldsymbol{x}}\right) \boldsymbol{y}_k = \boldsymbol{y}_{n+1}$$

Applying $\boldsymbol{g}_{n+1}$ to the left-hand side, we get

$$\langle \boldsymbol{g}_{n+1}, A\overline{\boldsymbol{x}} \rangle - \sum_{k=1}^{n} f_k\left(\overline{\boldsymbol{x}}\right) \langle \boldsymbol{g}_{n+1}, \boldsymbol{y}_k \rangle$$

$$= \langle A' \boldsymbol{g}_{n+1}, \overline{\boldsymbol{x}} \rangle = 0$$

whereas, when applied to the right-hand side, it yields

$$\langle \boldsymbol{g}_{n+1}, \boldsymbol{y}_{n+1} \rangle = 1$$

a contradiction. Thus it must be that $m \leq n$.

Using the same arguments for the adjoint equation, we show that $m \geq n$. ∎

We conclude our study with the general result concerning equations of the second type with completely continuous operators, known as the *Fredholm alternative.*

THEOREM 5.20.2 (Fredholm Alternative)

Let X be a Banach space and $T: X \rightarrow X$ a completely continuous operator from X into itself. Then, either the equations

$$\boldsymbol{x} - T\boldsymbol{x} = \boldsymbol{y} \quad \textit{and} \quad \boldsymbol{g} - T'\boldsymbol{g} = \boldsymbol{f}$$

are solvable for every $\boldsymbol{y}$ and $\boldsymbol{f}$ and, in such a case solutions $\boldsymbol{x}$ and $\boldsymbol{g}$ are unique, or else the homogeneous equations

$$\boldsymbol{x} - T\boldsymbol{x} = \mathbf{0} \quad \textit{and} \quad \boldsymbol{g} - T'\boldsymbol{g} = \mathbf{0}$$

have the same finite number of linearly independent solutions

$$\boldsymbol{x}_1, \ldots \boldsymbol{x}_n \quad \textit{and} \quad \boldsymbol{g}_1, \ldots, \boldsymbol{g}_n$$

In such a case, the necessary and sufficient condition for the solutions to exist is

$$\langle \boldsymbol{g}_i, \boldsymbol{y} \rangle = 0 \qquad i = 1, \ldots, n$$
$$\langle \boldsymbol{f}, \boldsymbol{x}_i \rangle = 0 \qquad i = 1, \ldots, n$$

and, if satisfied, the solutions are determined up to the vectors $\boldsymbol{x}_1, \ldots, \boldsymbol{x}_n$ *and* $\boldsymbol{g}_1, \ldots, \boldsymbol{g}_n$, *i.e., they are in the form*

$$\boldsymbol{x} + \sum_{i=1}^{n} a_i \boldsymbol{x}_i, \ \boldsymbol{g} + \sum_{i=1}^{n} b_i \boldsymbol{g}_i, \ a_i, b_i \in \mathbb{R}(\mathbb{C})$$

where $\boldsymbol{x}$ *and* $\boldsymbol{g}$ *are arbitrary solutions of the original equations.*

PROOF The proof follows immediately from Lemma 5.20.6 and Corollary 5.18.2. $\blacksquare$

Example 5.20.1
(Integral Equations of the Second Type)

Consider the integral equation

$$u(x) - \lambda \int_0^1 K(x, \xi) u(\xi) d\xi = v(x), \ x \in [0, 1]$$

where kernel $K(x, \xi)$ is a real- or complex-valued, continuous function on the (closed) square domain $[0, 1] \times [0, 1]$ and $\lambda \in \mathbb{R}(\mathbb{C})$. Introducing the Banach space $C([0, 1])$ with the Chebyshev metric, we can rewrite the equation in the operator form as

$$u - \lambda T u = v$$

where the corresponding integral operator T, considered in Example 5.15.1, was proved to be completely continuous. $\square$

The same problem can be formulated using space $X = L^2(0, 1)$. In such a case, the assumption on the kernel $K(x, \xi)$ can be weakened to the condition that K is an L^2 function. It can be proved again (see Exercise 5.15.3) that operator T is completely continuous. Moreover, as the dual of space $L^2(0, 1)$ can be identified with the space itself, the transposed problem

$$g - \lambda T' g = f$$

is equivalent to the equation (cf. Example 5.16.1)

$$g(\xi) - \lambda \int_0^1 \overline{K(x,\xi)}\, g(x)dx = f(\xi)$$

According to the Fredholm alternative, either both equations admit unique solutions for every $v, f \in L^2(0,1)$, or the corresponding homogeneous equations have the same number of n linearly independent solutions

$$u_1, \ldots, u_n \;,\; g_1, \ldots, g_n$$

In such a case, a necessary and sufficient condition for the solutions u and g to exist is

$$\int_0^1 v g_i dx = 0 \qquad i- = 1, \ldots, n$$

for the original problem, and

$$\int_0^1 f u_i dx = 0 \qquad i = 1, \ldots, n$$

The same conclusions hold for the case of continuous functional $v(x), f(x)$ and kernel $K(x,\xi)$, except that the second integral equation *cannot* be directly interpreted as the conjugate problem to the original equation, for the dual of $C([0,1])$ does not coincide with the space itself.

Let us finally mention that values $\lambda \in \mathbb{C}$ for which the original and the transpose equations have no unique solutions are called the *characteristic values* of operators T and T'.

Exercises

5.20.1. Complete the proof of Lemma 5.20.6.

6

Hilbert Spaces

Basic Theory

6.1 Inner Product and Hilbert Spaces

Much of functional analysis involves abstracting and making precise ideas that have been developed and used over many decades, even centuries, in physics and classical mathematics. In this regard, functional analysis makes use of a great deal of "mathematical hindsight" in that it seeks to identify the most primitive features of elementary analysis, geometry, calculus, and the theory of equations in order to generalize them, to give them order and structure, and to define their interdependencies. In doing this, however, it simultaneously unifies this entire collection of ideas and extends them to new areas that could never have been completely explored within the framework of classical mathematics or physics.

The final abstraction we investigate in this book is of geometry: We add to the idea of vector spaces enough structure to include abstractions of the geometrical terms *direction, orthogonality, angle between vectors*, and *length of a vector*. Once these ideas are established, we have the framework for not only a geometry of function spaces, but also a theory of linear equations, variational methods, approximation theory, and numerous other areas of mathematics.

We begin by recalling the definition of scalar product (cf. Section 2.13).

Scalar (Inner) Product. Let V be a vector space defined over the complex number field $\mathbb{C}$. A scalar-valued function $p : V \times V \longrightarrow \mathbb{C}$ that associates with each pair $\boldsymbol{u}$, $\boldsymbol{v}$ of vectors in V a scalar, denoted $p(\boldsymbol{u}, \boldsymbol{v}) = (\boldsymbol{u}, \boldsymbol{v})$, is called a *scalar (inner) product* on V iff

(i) $(\boldsymbol{u}, \boldsymbol{v})$ is linear with respect to the first argument
$(\alpha_1\boldsymbol{u}_1 + \alpha_2\boldsymbol{u}_2, \boldsymbol{v}) = \alpha_1(\boldsymbol{u}_1, \boldsymbol{v}) + \alpha_2(\boldsymbol{u}_2, \boldsymbol{v})$

$$\forall\, \alpha_1, \alpha_2 \in \mathbb{C}, \quad \boldsymbol{u}_1, \boldsymbol{u}_2, \boldsymbol{v} \in V$$

(ii) $(\boldsymbol{u}, \boldsymbol{v})$ is symmetric (in the complex sense)

$$(\boldsymbol{u}, \boldsymbol{v}) = \overline{(\boldsymbol{v}, \boldsymbol{u})}, \qquad \forall\, \boldsymbol{u}, \boldsymbol{v} \in V$$

where $\overline{(\boldsymbol{v}, \boldsymbol{u})}$ denotes the complex conjugate of $(\boldsymbol{v}, \boldsymbol{u})$

(iii) $(\boldsymbol{u}, \boldsymbol{v})$ is positive definite, i.e.,

$$(\boldsymbol{u}, \boldsymbol{u}) > 0 \quad \forall\, \boldsymbol{u} \neq \boldsymbol{0}\,, \; \boldsymbol{u} \in V$$

Note that the first two conditions imply that $(\boldsymbol{u}, \boldsymbol{v})$ is *antilinear* with respect to the second argument

$$\begin{aligned}
(\boldsymbol{u}, \beta_1 \boldsymbol{v}_1 + \beta_2 \boldsymbol{v}_2) &= \overline{(\beta_1 \boldsymbol{v}_1 + \beta_2 \boldsymbol{v}_2, \boldsymbol{u})} \\
&= \overline{\beta}_1 \overline{(\boldsymbol{v}_1, \boldsymbol{u})} + \overline{\beta}_2 \overline{(\boldsymbol{v}_2, \boldsymbol{u})} \\
&= \overline{\beta}_1 (\boldsymbol{u}, \boldsymbol{v}_1) + \overline{\beta}_2 (\boldsymbol{u}, \boldsymbol{v}_2)
\end{aligned}$$

for every $\beta_1, \beta_2 \in \mathbb{C}$, $\boldsymbol{v}_1, \boldsymbol{v}_2 \in V$.

In the case of a real vector space V, condition (ii) becomes one of symmetry

(ii) $\qquad (\boldsymbol{u}, \boldsymbol{v}) = (\boldsymbol{v}, \boldsymbol{u}) \qquad \forall\, \boldsymbol{u}, \boldsymbol{v} \in V$

and then $(\boldsymbol{u}, \boldsymbol{v})$ is linear with respect to both arguments $\boldsymbol{u}$ and $\boldsymbol{v}$. Note also that, according to the second condition,

$$(\boldsymbol{u}, \boldsymbol{u}) = \overline{(\boldsymbol{u}, \boldsymbol{u})}$$

is a real number and therefore condition (iii) makes sense.

Inner Product Spaces. A vector space V on which an inner product has been defined is called an *inner product space.* If V is a real vector space, with an inner product, then V is called a real inner product space.

Orthogonal Vectors. Two elements $\boldsymbol{u}$ and $\boldsymbol{v}$ of an inner product space V are said to be *orthogonal* if

$$(\boldsymbol{u}, \boldsymbol{v}) = 0$$

Example 6.1.1

Let $V = \mathbb{C}^n$, the vector space of n-tuples of complex numbers.

$$v \in \mathcal{C}^n \Leftrightarrow v = (v_1, v_2, \ldots, v_n), \quad v_j = \alpha_j + i\beta_j$$

$i = \sqrt{-1}$, $1 \leq j \leq n$. Then the operation $(\cdot,\cdot) : \mathcal{C}^n \times \mathcal{C}^n \to \mathcal{C}$, defined by

$$(\boldsymbol{u}, \boldsymbol{v}) = u_1\bar{v}_1 + u_2\bar{v}_2 + \cdots + u_n\bar{v}_n$$

where $\bar{v}_j = \alpha_j - i\beta_j$ denotes the complex conjugate of v_j, is an inner product on $\mathcal{C}^n$, as is easily verified.

Take $n = 2$, and consider the two vectors

$$\boldsymbol{u} = (1+i, 1+i) \text{ and } \boldsymbol{v} = (-2-2i, 2+2i)$$

These two vectors are orthogonal with respect to the inner product defined previously:

$$(u, v) = (1+i)\,(-2+2i) + (1+i)\,(2-2i) = 0$$

□

Example 6.1.2

Let $V = C(a, b)$ be the vector space of continuous, complex-valued functions defined on an interval (a, b) of the real line. Then

$$(f, g) = \int_a^b f(x)\overline{g(x)}\, dx$$

is an inner product on V, wherein $\overline{g(x)}$ denotes the complex conjugate of $g(x)$.

Let $a = 0$, $b = 1$ and consider the functions

$$f(x) = \sin \pi x + i \sin \pi x, \quad g(x) = -\sin 2\pi x + i \sin 3\pi x$$

These functions are orthogonal; indeed,

$$\begin{aligned}\int_0^1 f(x)\overline{g(x)}dx &= \int_0^1 [-\sin \pi x \sin 2\pi x + \sin \pi x \sin 3\pi x \\ &\quad - i(\sin \pi x \sin 2\pi x + \sin \pi x \sin 3\pi x)]dx \\ &= 0 + i0\end{aligned}$$

□

The essential property of vector spaces with the scalar-product structure is that they form a special subclass of normed spaces as confirmed by the following proposition.

PROPOSITION 6.1.1

Every inner product space V is a normed space. The mapping

$$V \in \boldsymbol{u} \longrightarrow \|\boldsymbol{u}\| \stackrel{\text{def}}{=} (\boldsymbol{u}, \boldsymbol{u})^{\frac{1}{2}}$$

defines a norm on V.

PROOF The first two norm axioms (positive definiteness and homogeneity) are automatically satisfied. The Cauchy-Schwarz inequality (Proposition 2.13.1) can be put to use to verify that $(\boldsymbol{u}, \boldsymbol{u})^{\frac{1}{2}}$ also satisfies the triangle inequality:

$$\begin{aligned}
\|\boldsymbol{u}+\boldsymbol{v}\|^2 &= (\boldsymbol{u}+\boldsymbol{v}, \boldsymbol{u}+\boldsymbol{v}) \\
&= (\boldsymbol{u}, \boldsymbol{u}) + (\boldsymbol{u}, \boldsymbol{v}) + (\boldsymbol{v}, \boldsymbol{u}) + (\boldsymbol{v}, \boldsymbol{v}) \\
&\leq \|\boldsymbol{u}\|^2 + 2\|\boldsymbol{u}\| \; \|\boldsymbol{v}\| + \|\boldsymbol{v}\|^2 \\
&= (\|\boldsymbol{u}\| + \|\boldsymbol{v}\|)^2
\end{aligned}$$

which completes the proof. ∎

It follows that the Cauchy-Schwarz inequality can be rewritten in the form

$$|(\boldsymbol{u}, \boldsymbol{v})| \leq \|\boldsymbol{u}\| \; \|\boldsymbol{v}\|$$

which is reminiscent of the rule for inner products of vectors in the usual Euclidean setting in $\mathbb{R}^3$. In real inner product spaces, this observation prompts us to define the *angle between vectors* by

$$\cos\theta = \frac{(\boldsymbol{u}, \boldsymbol{v})}{\|\boldsymbol{u}\| \; \|\boldsymbol{v}\|}$$

REMARK 6.1.1. It follows immediately from the Cauchy-Schwarz inequality that the inner product is continuous. Indeed, let $\boldsymbol{u}_n \rightarrow \boldsymbol{u}$, $\boldsymbol{v}_n \rightarrow \boldsymbol{v}$. Then

$$|(\boldsymbol{u}, \boldsymbol{v}) - (\boldsymbol{u}_n, \boldsymbol{v}_n)| \leq |(\boldsymbol{u}, \boldsymbol{v}) - (\boldsymbol{u}_n, \boldsymbol{v}) + (\boldsymbol{u}_n, \boldsymbol{v}) - (\boldsymbol{u}_n, \boldsymbol{v}_n)|$$

$$\leq \|\boldsymbol{u} - \boldsymbol{u}_n\| \, \|\boldsymbol{v}\| + \|\boldsymbol{u}_n\| \, \|\boldsymbol{v} - \boldsymbol{v}_n\|$$

and the right-hand side converges to zero. ∎

The existence of the norm also gives meaning to the concept of completeness of inner product spaces.

Hilbert Space. An inner product space V is called a *Hilbert space* if it is complete with respect to the norm induced by the scalar product.

Every finite-dimensional inner product space is a Hilbert space since every finite-dimensional space is complete. Obviously, every Hilbert space is a Banach space. The converse, however, is not true.

Unitary Maps. Equivalence of Hilbert Spaces. Let U and V be two inner product spaces with scalar products $(\cdot, \cdot)_U$ and $(\cdot, \cdot)_V$, respectively. A linear map

$$T : U \longrightarrow V$$

is said to be *unitary* if

$$(T\boldsymbol{u}, T\boldsymbol{v})_V = (\boldsymbol{u}, \boldsymbol{v})_U \qquad \forall\, \boldsymbol{u}, \boldsymbol{v} \in U$$

Note that this implies that T is an isometry

$$\|T\boldsymbol{u}\|_V = \|\boldsymbol{u}\|_U \qquad \forall\, \boldsymbol{u} \in U$$

and therefore, in particular, it must be injective. If, additionally, T is surjective we say that spaces U and V are *unitarily equivalent.* Obviously, both T and T^{-1} are then continuous and $\|T\| = \|T^{-1}\| = 1$. Also, if U and V are unitarily equivalent then U is complete if and only if V is complete.

Example 6.1.3

The space ℓ^2 consisting of square-summable sequences of complex numbers

$$\ell^2 = \left\{ \boldsymbol{x} = \{x_i\}_{i=1}^{\infty} : \sum_{i=1}^{\infty} |x_i|^2 < \infty \right\}$$

is a Hilbert space with the scalar product

$$(\boldsymbol{x}, \boldsymbol{y}) = \sum_{i=1}^{\infty} x_i \bar{y}_i$$

Hölder's inequality with $p = 2$ describes Schwarz's inequality for this space

$$|(\boldsymbol{x}, \boldsymbol{y})| \leq \left(\sum_{i=1}^{\infty} |x_i|^2 \right)^{\frac{1}{2}} \left(\sum_{j=1}^{\infty} |y_i|^2 \right)^{\frac{1}{2}}$$

□

Example 6.1.4

The space $\mathcal{P}^n$ of real polynomials $p = p(x)$ of degree less than or equal to n defined over an interval $a \leq x \leq b$, with the inner product defined as

$$(p, q) = \int_a^b p(x)q(x)dx$$

is an inner product space. Since $\mathcal{P}^n$ is finite dimensional, it is complete. Hence it is a Hilbert space. □

Example 6.1.5

The space $L^2(a, b)$ of equivalence classes of complex-valued functions defined on (a, b) whose squares are Lebesgue integrable is a Hilbert space with inner product

$$(u, v) = \int_a^b u(x)\overline{v(x)}dx$$

The integral form of Hölder's inequality describes the Schwarz inequality for $L^2(a, b)$ if we set $p = 2$. □

Example 6.1.6

A nontrivial example of a unitary map is provided by the Fourier transform in space $L^2(\mathbb{R}^n)$.

We introduce first the *space of rapidly decreasing (at ∞) functions*, denoted $\mathcal{S}(\mathbb{R}^n)$, which contains all $C^\infty(\mathbb{R}^n)$ functions f such that

$$\sup_{\boldsymbol{x}\in\mathbb{R}^n} \left| \boldsymbol{x}^{\boldsymbol{\beta}} D^{\boldsymbol{\alpha}} f(\boldsymbol{x}) \right| < \infty$$

for every pair of multi-indices $\boldsymbol{\alpha}$ and $\boldsymbol{\beta}$. Space $\mathcal{S}(\mathbb{R}^n)$ includes C^∞ functions with compact support, $C_0^\infty(\mathbb{R}^n)$, and such functions as, e.g., $\exp(-|\boldsymbol{x}|^2)$.

Similarly to the space of test functions, $\mathcal{S}(\mathbb{R}^n)$ can be topologized with a locally convex topology. The corresponding dual, denoted $\mathcal{S}'(\mathbb{R}^n)$, is known as *the space of tempered distributions* and can be identified as a subspace of regular distributions.

For a function $f \in \mathcal{S}(\mathbb{R}^n)$, we define the *Fourier transform* $\hat{f}$ as

$$\hat{f}(\boldsymbol{\xi}) = (2\pi)^{-\frac{n}{2}} \int_{\mathbb{R}^n} e^{-i\boldsymbol{\xi}\boldsymbol{x}} f(\boldsymbol{x}) d\boldsymbol{x}$$

where

$$\boldsymbol{\xi}\boldsymbol{x} = \sum_{i=1}^{n} \xi_i x_i$$

The *inverse Fourier transform* $\tilde{g}(x)$ of a function $g \in \mathcal{S}(\mathbb{R}^n)$ is defined as

$$\tilde{g}(x) = (2\pi)^{-\frac{n}{2}} \int_{\mathbb{R}^n} e^{i\boldsymbol{x}\boldsymbol{\xi}} g(\boldsymbol{\xi}) d\boldsymbol{\xi}$$

It can be proved that the Fourier transform defines a linear and continuous map $\mathcal{F}$ from $\mathcal{S}(\mathbb{R}^n)$ into $\mathcal{S}(\mathbb{R}^n)$ with inverse $\mathcal{F}^{-1}$ exactly equal to the inverse Fourier transform, i.e.,

$$\tilde{\hat{f}} = f \text{ and } \hat{\tilde{g}} = g$$

Consequently (substituting $-\boldsymbol{x}$ for $\boldsymbol{x}$ in the inverse transform),

$$\left(\hat{\hat{f}}\right)(\boldsymbol{x}) = f(-\boldsymbol{x}) \text{ and } \left(\tilde{\tilde{g}}\right)(\boldsymbol{\xi}) = g(-\boldsymbol{\xi})$$

Also,

$$\int_{\mathbb{R}^n} f(\boldsymbol{\xi})\hat{g}(\boldsymbol{\xi})\, d\boldsymbol{\xi} = \int_{\mathbb{R}^n} f(\boldsymbol{\xi})(2\pi)^{-\frac{n}{2}} \int_{\mathbb{R}^n} e^{-i\boldsymbol{\xi}\boldsymbol{x}} g(\boldsymbol{x})\, d\boldsymbol{x}\, d\boldsymbol{\xi}$$

$$= \int_{\mathbb{R}^n} (2\pi)^{-\frac{n}{2}} (\int_{\mathbb{R}^n} e^{-i\boldsymbol{\xi}\boldsymbol{x}} f(\boldsymbol{\xi})\, d\boldsymbol{\xi})\, g(\boldsymbol{x})\, d\boldsymbol{x}$$

$$= \int_{\mathbb{R}^n} \hat{f}(\boldsymbol{x}) g(\boldsymbol{x}) \, d\boldsymbol{x}$$

which, upon observing that ($^-$ stands for the complex conjugate)

$$\widehat{(\bar{f})} = \overline{(\hat{f})}$$

leads to the *Parseval relation*

$$\int_{\mathbb{R}^n} f(\boldsymbol{x}) \overline{g(\boldsymbol{x})} \, d\boldsymbol{x} = \int_{\mathbb{R}^n} \hat{f}(\boldsymbol{\xi}) \overline{\hat{g}(\boldsymbol{\xi})} \, d\boldsymbol{\xi}$$

Substituting $g = f$, we get

$$\|f\|_{L^2} = \|\hat{f}\|_{L^2} \text{ for } f \in \mathcal{S}(\mathbb{R}^n)$$

Using the same concept as for the differentiation of distributions, we define next the Fourier transform of a tempered distribution $T \in \mathcal{S}'(\mathbb{R}^n)$ as

$$\langle \hat{T}, \phi \rangle \overset{\text{def}}{=} \langle T, \hat{\phi} \rangle \qquad \forall \phi \in \mathcal{S}(\mathbb{R}^n)$$

and its inverse

$$\langle \tilde{T}, \phi \rangle \overset{\text{def}}{=} \langle T, \tilde{\phi} \rangle \qquad \forall \phi \in \mathcal{S}(\mathbb{R}^n)$$

Again, it can be shown that $\hat{T}$ is an isomorphism between $\mathcal{S}'(\mathbb{R}^n)$ and itself with $\tilde{T}$ being precisely its inverse.

Let f be now an arbitrary L^2 function on $\mathbb{R}^n$ and T_f the corresponding regular distribution, i.e.,

$$\langle T_f, \phi \rangle = \int_{\mathbb{R}^n} f(\boldsymbol{x}) \phi(\boldsymbol{x}) \, d\boldsymbol{x}$$

As space of rapidly decreasing functions $\mathcal{S}(\mathbb{R}^n)$ is continuously imbedded in $L^2(\mathbb{R}^n)$, it follows from the Schwarz inequality that T_f is a tempered distribution as well. Calculating its Fourier transform, we get

$$|\langle \hat{T}_f, \phi \rangle| = |\langle T_f, \hat{\phi} \rangle| = \left| \int_{\mathbb{R}^n} f(\boldsymbol{x}) \hat{\phi}(\boldsymbol{x}) \, d\boldsymbol{x} \right|$$

$$\leq \|f\|_{L^2} \|\hat{\phi}\|_{L^2} = \|f\|_{L^2} \|\phi\|_{L^2}$$

As $\mathcal{S}(\mathbb{R}^n)$ is dense in $L^2(\mathbb{R}^n)$, it follows from the representation theorem for the duals to L^p spaces that there exists a unique function $\hat{f} \in L^2(\mathbb{R}^n)$ such that

$$\langle \hat{T}_f, \phi \rangle = \langle T_{\hat{f}}, \phi \rangle \qquad \forall \phi \in \mathcal{S}(\mathbb{R}^n)$$

and also

$$\|\hat{f}\|_{L^2} \le \|f\|_{L^2}$$

which implies that

$$\|f(\cdot)\|_{L^2} = \|f(-\cdot)\|_{L^2} = \|\hat{\hat{f}}\|_{L^2}$$

$$\le \|\hat{f}\|_{L^2}$$

and therefore, finally,

$$\|\hat{f}\|_{L^2} = \|f\|_{L^2}$$

Function $\hat{f} \in L^2(\mathbb{R}^n)$ is called the *Fourier transform of function* $f \in L^2(\mathbb{R}^n)$ and, consequently, the Fourier transform is identified as the *unitary map* from $L^2(\mathbb{R}^n)$ onto itself.

Note the delicate detail concerning the definition: for $f \in L^1(\mathbb{R}^n)$ the Fourier transform can be defined directly, using the same definition as for the rapidly decreasing functions, but for $f \in L^2(\mathbb{R}^n)$ it *cannot*, because the kernel $e^{-i\boldsymbol{x}\boldsymbol{\xi}}$ is *not* an L^2 function in $\mathbb{R}^n \times \mathbb{R}^n$!

We conclude this example with the fundamental property of the Fourier transform in conjuction with differentiation. We have, by definition

$$\widehat{D^{\boldsymbol{\beta}}\phi}(\boldsymbol{\xi}) = (2\pi)^{-\frac{n}{2}} \int_{\mathbb{R}^n} e^{-i\boldsymbol{\xi}\boldsymbol{x}} D^{\boldsymbol{\beta}}\phi(\boldsymbol{x})\, d\boldsymbol{x}$$

Integrating the right-hand side by parts we arrive at the formula

$$\widehat{D^{\boldsymbol{\beta}}\phi}(\boldsymbol{\xi}) = i^{|\boldsymbol{\beta}|}\boldsymbol{\xi}^{\boldsymbol{\beta}}\hat{\phi}(\boldsymbol{\xi})$$

for $\phi \in \mathcal{S}(\mathbb{R}^n)$ and, consequently, for $T \in \mathcal{S}'(\mathbb{R}^n)$ as well.

In other words, Fourier transform converts derivatives of functions (distributions) into products of transforms and polynomials $\boldsymbol{\xi}^{\boldsymbol{\beta}}$! It is this property which makes the transform a fundamental tool in solving linear differential equations with constant coefficients in the whole $\mathbb{R}^n$. $\square$

Example 6.1.7

A special class of the Sobolev spaces $W^{m,p}(\Omega)$, $m \geq 0$, $1 \leq p \leq \infty$, described in Section 5.11, constitutes one of the most important examples of Hilbert spaces. Let Ω be an open set in $\mathbb{R}^n$. The space

$$H^m(\Omega) \stackrel{\text{def}}{=} W^{m,2}(\Omega) \qquad (p = 2)$$

is a Hilbert space with the scalar product defined as

$$(u, v)_{H^m(\Omega)} = \sum_{|\boldsymbol{\alpha}| \leq m} \left(D^{\boldsymbol{\alpha}} u, D^{\boldsymbol{\alpha}} v\right)_{L^2(\Omega)}$$

$$= \int_\Omega \sum_{|\boldsymbol{\alpha}| \leq m} D^{\boldsymbol{\alpha}} u \cdot \overline{D^{\boldsymbol{\alpha}} v}\, dx$$

with the corresponding norm

$$\|u\|_{H^m(\Omega)} = \left(\int_\Omega \sum_{|\boldsymbol{\alpha}| \leq m} \left|D^{\boldsymbol{\alpha}} u\right|^2 \, dx \right)^{\frac{1}{2}}$$

$\square$

For example, if $\Omega \subset \mathbb{R}^2$,

$$(u, v)_{H^2(\Omega)} = \int_\Omega \left(uv + \frac{\partial u}{\partial x}\frac{\partial v}{\partial x} + \frac{\partial u}{\partial y}\frac{\partial v}{\partial y} + \frac{\partial^2 u}{\partial x^2}\frac{\partial^2 v}{\partial x^2} \right.$$

$$\left. + 2\frac{\partial^2 u}{\partial x \partial y}\frac{\partial^2 v}{\partial x \partial y} + \frac{\partial^2 u}{\partial y^2}\frac{\partial^2 v}{\partial y^2} \right) dx\, dy$$

or if $\Omega = (a, b) \subset \mathbb{R}$,

$$(u, v)_{H^m(a,b)} = \int_b^a \sum_{k=0}^m \frac{d^k u}{dx^k}\frac{d^k v}{dx^k}\, dx$$

Relation between Real and Complex Vector Spaces. For the remainder of this chapter we will select the complex vector spaces as a natural context for developing the concepts of the Hilbert spaces theory. This degree of generality is not only necessary for developing, for instance, the spectral theories, but proves to be absolutely essential in discussing some problems which simply do not admit "real" formulations (e.g., linear acoustics equations; see [9]). Obviously, every complex vector space can be considered as a real space when we restrict ourselves to the real scalars only (compare proofs of the representation theorem in Section 2.11 and the proof of the Hahn-Banach theorem for complex spaces in Section 5.5). Thus, intuitively speaking, whatever we develop and prove for complex spaces should also remain valid for real spaces as a particular case. We devote the rest of this section to a more detailed discussion of this issue.

1. Let us start with an intuitive observation that for most (if not all) of the practical applications we deal with *function spaces*, e.g., $C(\Omega)$, $L^2(\Omega)$, $H^k(\Omega)$, etc. Every space of *real-valued* functions can be immediately generalized to the space of *complex-valued* functions defined on the same domain and possessing the same class of regularity. For instance, a real-valued square-integrable function f defined as an open set Ω

$$f : \Omega \longrightarrow \mathbb{R} \ , \quad \int_\Omega |f(x)|^2 \, dx < \infty$$

can be identified with a real part of a complex-valued L^2 function F

$$F : \Omega \longrightarrow \mathbb{C} \ , \quad F(x) = f(x) + ig(x)$$

$$\int_\Omega |F(x)|^2 \, dx = \int_\Omega \left(f^2(x) + g^2(x)\right) dx < \infty$$

Most of the time extensions like this are done quite naturally by replacing the absolute value of real numbers with modulus of complex ones.

2. Any abstract real vector space X can be extended into a complex space by considering pairs (x, y) of vectors from the real space X. More precisely, we introduce the space

$$Z = X \times X$$

with operations defined as

$$(\boldsymbol{x}_1, \boldsymbol{y}_1) + (\boldsymbol{x}_2, \boldsymbol{y}_2) \stackrel{\text{def}}{=} (\boldsymbol{x}_1 + \boldsymbol{x}_2, \boldsymbol{y}_1 + \boldsymbol{y}_2)$$

$$\lambda(\boldsymbol{x}, \boldsymbol{y}) = (\alpha\boldsymbol{x} - \beta\boldsymbol{y},\ \alpha\boldsymbol{y} + \beta\boldsymbol{x})$$

where $\lambda = \alpha + \beta i$ is an arbitrary complex number. It is easy to check that this abstract extension is linearly isomorphic with the natural extensions of function spaces discussed previously.

3. The complex extension Z of a *normed* real space X may be equipped with a (*not unique*) norm, reducing to the norm on X for real elements of Z. We may set, for instance,

$$\|\boldsymbol{z}\|_Z = \|(\boldsymbol{x}, \boldsymbol{y})\|_Z \stackrel{\text{def}}{=} (\|\boldsymbol{x}\|_X^p + \|\boldsymbol{y}\|_X^p)^{\frac{1}{p}} \qquad 1 \le p < \infty$$

or

$$\|\boldsymbol{z}\|_Z = \|(\boldsymbol{x}, \boldsymbol{y})\|_Z \stackrel{\text{def}}{=} \max_\theta \|\boldsymbol{x}\cos\theta + \boldsymbol{y}\sin\theta\|_X$$

etc. While all these norms are different, they prove to be equivalent and therefore all the corresponding topological properties will be the same. Consequently, any of the presented norms can be used. Again, for function spaces the norms are usually naturally generalized by replacing the absolute value with the modulus.

4. The complex extension Z of a real space X with an inner product $(\cdot,\cdot)_X$ can be equipped with a corresponding product $(\cdot,\cdot)_Z$ reducing to the original one for real elements. More precisely, for $\boldsymbol{z}_1 = (\boldsymbol{x}_1, \boldsymbol{y}_1)$ and $\boldsymbol{z}_2 = (\boldsymbol{x}_2, \boldsymbol{y}_2)$ we define

$$(\boldsymbol{z}_1, \boldsymbol{z}_2)_Z \stackrel{\text{def}}{=} \{(\boldsymbol{x}_1, \boldsymbol{x}_2)_X + (\boldsymbol{y}_1, \boldsymbol{y}_2)_X\} + i\,\{(\boldsymbol{x}_2 \boldsymbol{y}_1)_X - (\boldsymbol{x}_1, \boldsymbol{y}_2)_X\}$$

One can easily check that the above is a well-defined scalar product on complex extension Z. The presented construction is *identical* to the definition of the L^2 scalar product for complex-valued functions f and g

$$(f, g)_{L^2(\Omega)} = \int_\Omega f(x)\overline{g(x)}dx$$

$$= \int_\Omega \{(Re\ f\ Re\ g$$

$$+ Im\ f\ Im\ g) + i(Im\ f\ Re\ g - Re\ f\ Im\ g)\}dx$$

$$= \{(Re\ f, Re\ g)$$

$$+ (Im\ f, Im\ g)\} + i\{(Im\ f, Re\ g) - (Re\ f, Im\ g)\}$$

where $(\cdot,\cdot)$ denotes the L^2 product for real-valued functions.

5. Any linear operator $L : X \to Y$ defined on real spaces X and Y can be naturally extended to their complex extensions by setting

$$\widetilde{L}(x,y) \stackrel{\text{def}}{=} (L(x), L(y))$$

Indeed, $\widetilde{L}$ is trivially additive and is also homogeneous, since

$$\widetilde{L}(\lambda(x,y)) = \widetilde{L}((\alpha x - \beta y, \alpha y + \beta x))$$

$$= (L(\alpha x - \beta y), L(\alpha y + \beta x))$$

$$= (\alpha Lx - \beta Ly, \alpha Ly + \beta Lx)$$

$$= \lambda(Lx, Ly) = \lambda \widetilde{L}(x,y)$$

where $\lambda = \alpha + \beta i$ is a complex number.

Most of the properties of L transfer immediately to its extension $\widetilde{L}$. For instance,

$$
\begin{array}{lll}
L \text{ is continuous} & \Longrightarrow & \widetilde{L} \text{ is continuous,} \\
L \text{ is closed} & \Longrightarrow & \widetilde{L} \text{ is closed,} \\
L \text{ is completely continuous} & \Longrightarrow & \widetilde{L} \text{ is completely continuous,}
\end{array}
$$

etc. For operators L defined on function spaces, the abstract extension $\widetilde{L}$ corresponds to natural extensions of L for complex-valued functions. For example, for a differential operator $L = \frac{d}{dx}$ and $f \in C^2(0,1)$,

$$
\frac{d}{dx}(f) = \frac{d}{dx}(Ref) + i\,\frac{d}{dx}\,(Imf)
$$

or in the case of an integral operator L with *real* kernel $K(x,y)$,

$$
\int_0^1 K(x,y)f(y)dy = \int_0^1 K(x,y)Re\ f(y)dy + i\int_0^1 K(x,y)Im\ f(y)dy
$$

Let us emphasize, however, that *there are* operators which *are not* extensions of real operators, for instance,

$$
L : f \longrightarrow if
$$

We conclude this section with an example emphasizing the importance of complex analysis.

Example 6.1.8

Most of the time when designing a time-marching algorithm for evolution equations we are concerned with the fundamental issue of *linear stability.* As an example, consider a linear convection equation with periodic boundary conditions

$$
\begin{cases}
\text{Find } u(x,t)\ ,\ x \in [0,1],\ t \geq 0 : \\
u_t + cu_x = 0\ ,\ x \in (0,1),\ t > 0 \quad c = \text{const} \\
u(0,t) = u(1,t),\ u_x(0,t) = u_x(1,t),\ t > 0 \\
u(x,0) = u_0(x)
\end{cases}
\tag{6.1}
$$

where u_t and u_x denote the derivatives with respect to time t and spatial coordinate x, respectively. $\square$

As a starting point for discretization in time we assume the following finite difference formula of second order

$$u(t+\Delta t) - \frac{\Delta t^2}{2}\, u_{tt}(t+\Delta t) = u(t) + \Delta t u_t(t) + 0(\Delta t^3) \tag{6.2}$$

where Δt is a time interval and u_{tt} denotes the second order time derivative. Using next the original differential equation, we represent the time derivatives in terms of spatial derivatives

$$\begin{aligned} u_t &= -cu_x \\ u_{tt} &= -(cu_x)_t = -c(u_t)_x = c^2 u_{xx} \end{aligned} \tag{6.3}$$

which leads to a *one-step problem* of the form

$$\begin{cases} u^{n+1} - \dfrac{(c\Delta t)^2}{2}\, u_{xx}^{n+1} = u^n + c\Delta t u_x^n \\ \\ u^{n+1}(0) = u^{n+1}(1)\ ,\ u_x^{n+1}(0) = u_x^{n+1}(1) \end{cases} \tag{6.4}$$

where $u^n = u(n\Delta t, \cdot)$ is an approximate solution at time level $t^n = n\Delta t$ and the initial condition u_0 is used in place of the zeroth iterate u^0. Thus, formally, the time-continuous problem is replaced with a sequence of the equations above solved for iterates u^n, $n = 1, 2, \ldots$.

In order to construct a fully discrete scheme, Equations (6.4) must be next discretized in the space variable x. Probably the simplest approach would be to use a uniformly spaced finite difference grid

$$x_\ell = \ell h,\ \ell = 0, 1, \ldots, N, N+1,\ h = 1/N \tag{6.5}$$

with corresponding discrete solution values u_i (see Fig. 6.1).

Using the finite difference formulas

$$\begin{aligned} u_x(\ell h) &= (u_{\ell+1} - u_{\ell-1})\,/2h + O(h^3) \\ u_{xx}(\ell h) &= (u_{\ell+1} - 2u_\ell + u_{\ell-1})\,/h^2 + O(h^3) \end{aligned} \tag{6.6}$$

we replace the differential equations with their finite difference approximations

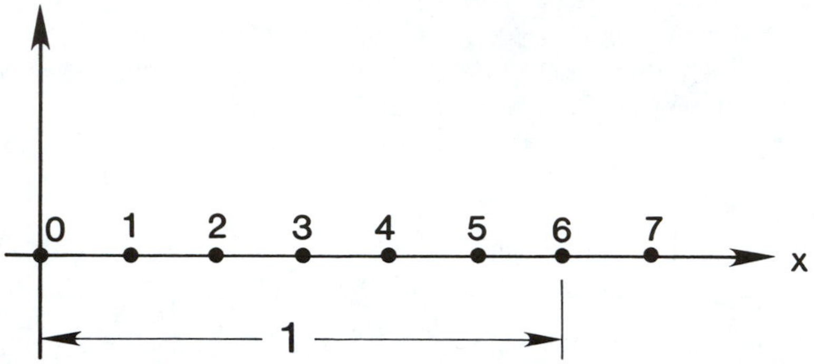

Figure 6.1
Example 6.1.8. A uniform finite difference grid ($N = 6$) on unit interval $(0, 1)$.

$$\begin{aligned} u_\ell^{n+1} - \frac{(c\Delta t)^2}{2}\left(u_{\ell+1}^{n+1} - 2u_\ell^{n+1} + u_{\ell-1}^{n+1}\right)/h^2 \\ = u_\ell^n + (c\Delta t)\left(u_{\ell+1}^n - u_{\ell-1}^n\right)/2h \end{aligned} \tag{6.7}$$

The first boundary condition will translate into

$$u_N = u_0 \tag{6.8}$$

and the second one, after the finite difference approximations

$$u_x(0) = (u_1 - u_0)/h \qquad u_x(1) = (u_{N+1} - u_N)h \tag{6.9}$$

reduces to the condition

$$u_{N+1} = u_1 \tag{6.10}$$

Consequently, solution of one time step reduces to solving the system of N simultaneous linear equations (6.7) for $i = 1, \ldots, N$ with values u_0 and u_{N+1} eliminated by conditions (6.8) and (6.10).

Identifying the finite difference representation

$$u_\ell^n, \; \ell = 1, 2, \ldots, N \tag{6.11}$$

with a vector $\boldsymbol{u}^n \in \mathbb{R}^N$, we introduce a linear operator A prescribing to the solution $\boldsymbol{u}^n$ at time level n, the corresponding solution $\boldsymbol{u}^{n+1}$ at the next time level $n + 1$.

$$A : \mathbb{R}^N \longrightarrow \mathbb{R}^N, \; A\boldsymbol{u}^n = \boldsymbol{u}^{n+1}$$

Obviously, A may be identified with a real $n \times n$ matrix A_{ij}.

We say now that the prescribed method is (*linearly*) *stable* if all eigenvalues of A are bounded in modulus by one, i.e.,

$$|\lambda_j| \leq 1 \quad j = 1, \ldots, N$$

It is at this point where we implicitly replace $\mathbb{R}^N$ with its complex extension $\mathbb{C}^N$ and extend operator A to the complex space $\mathbb{C}^N$

$$\widetilde{A} : \mathbb{C}^N \rightarrow \mathbb{C}^N \; , \; \widetilde{A}\boldsymbol{z}^n = \boldsymbol{z}^{n+1}$$

where

$$\boldsymbol{z}^n = (\boldsymbol{u}^n, \boldsymbol{v}^n) \;\text{ and }\; \widetilde{A}z^n = (A\boldsymbol{u}^n, A\boldsymbol{v}^n)$$

This extension to the complex setting is very essential. It will follow from the general spectral theory prescribed at the end of this chapter that there exists a sequence of unit eigenvectors $\boldsymbol{w}_j \in \mathbb{C}^N$, forming a basis in $\mathbb{C}^N$. Consequently, any vector $\boldsymbol{z} \in \mathbb{C}^N$ can be represented in the form

$$\boldsymbol{z} = \sum_{j=1}^{N} z_j \boldsymbol{w}_j \; , \; z_j \in \mathbb{C} \tag{6.12}$$

Applying operator A to $\boldsymbol{z}$ we get

$$A\boldsymbol{z} = \sum_{j=1}^{N} z_j A\boldsymbol{w}_j = \sum_{j=1}^{N} z_j \lambda_j \boldsymbol{w}_j$$

and after n iterations

$$A^n \boldsymbol{z} = \sum_{j=1}^{N} z_j \lambda_j^n \boldsymbol{w}_j$$

Thus, if any of the eigenvalues λ_j is greater in modulus than one, the corresponding component will grow geometrically to infinity (in modulus) and the solution will "blow up."

It is interesting to see the difference between real and complex eigenvalues. If λ is a real eigenvalue and $\boldsymbol{w}$ denotes the corresponding (complex!) eigenvector then both real and imaginary parts of $\boldsymbol{w}$ (if not zero), $\boldsymbol{u} = \text{Re}\boldsymbol{w}$ and $\boldsymbol{v} = \text{Im}\boldsymbol{w}$ are eigenvectors of operator A in the real sense. Indeed,

$$(A\boldsymbol{u}, A\boldsymbol{v}) = \widetilde{A}(\boldsymbol{u}, \boldsymbol{v}) = \widetilde{A}\boldsymbol{w} = \lambda \boldsymbol{w} = \lambda(\boldsymbol{u}, \boldsymbol{v}) = (\lambda \boldsymbol{u}, \lambda \boldsymbol{v})$$

implies that both $A\boldsymbol{u} = \lambda \boldsymbol{u}$ and $A\boldsymbol{v} = \lambda \boldsymbol{v}$.

If $|\lambda| > 1$, the loss of stability is observed as a rapid (geometrical) growth of an initial value component corresponding to the eigenvector $\boldsymbol{u}$ or $\boldsymbol{v}$. In particular, starting with an initial value $\boldsymbol{u}_0$ equal to the real eigenvector $\boldsymbol{u}$ (or $\boldsymbol{v}$ if both are different from zero and from each other), after n iterations solution $\boldsymbol{u}^n$ takes on the form

$$\boldsymbol{u}^n = A^n \boldsymbol{u} = \lambda^n \boldsymbol{u}$$

The situation is more complicated if λ is complex. Representing λ in the form

$$\lambda = |\lambda| e^{i\theta} = |\lambda|(\cos\theta + i \sin\theta)$$

we get

$$\lambda^n = |\lambda|^n e^{in\theta} = |\lambda|^n (\cos n\theta + i \sin n\theta)$$

and, consequently, if $\boldsymbol{w} = (\boldsymbol{u}, \boldsymbol{v})$ is the corresponding eigenvector,

$$(A^n \boldsymbol{u}, A^n \boldsymbol{v}) = \widetilde{A}^n \boldsymbol{w} = \lambda^n \boldsymbol{w} = (Re\lambda^n \boldsymbol{u} - Im\lambda^n \boldsymbol{v},\ Im\lambda^n \boldsymbol{u} + Re\lambda^n \boldsymbol{v})$$

$$= |\lambda|^n (\cos(n\theta)\boldsymbol{u} - \sin(n\theta)\boldsymbol{v},\ \sin(u\theta)\boldsymbol{u} + \cos(n\theta)\boldsymbol{v})$$

which implies that

$$A^n \boldsymbol{u} = |\lambda|^n (\cos(n\theta)\boldsymbol{u} - \sin(n\theta)\boldsymbol{v})$$

$$A^n \boldsymbol{v} = |\lambda|^n (\sin(n\theta)\boldsymbol{u} + \cos(n\theta)\boldsymbol{v})$$

Starting therefore with the real part $\boldsymbol{u}$ of the eigenvalue $\boldsymbol{w}$, we *do not* observe the simple growth of $\boldsymbol{u}$ as in the case of a real eigenvalue, but a growth coupled with a simultaneous interaction between the real and imaginary parts of $\boldsymbol{w}$. Only for an appropriate phase

$$n\theta \approx k\pi \ , \ k = 1, 2, \ldots ,$$

we have

$$A^n \boldsymbol{u} \approx (-1)^k |\lambda|^n \boldsymbol{u}$$

and a simple amplification of $\boldsymbol{u}$ will be observed.

We conclude this lengthy example intended to show the importance of complex analysis with an evaluation of eigenvalues and eigenvectors for operator $\widetilde{A}$ from our example.

We simply postulate the following form for eigenvectors

$$\boldsymbol{w}_j = \{w_{j,\ell}\}_{\ell=0}^n$$

where

$$\begin{aligned} w_{j,\ell} &= e^{i(j2\pi x_\ell)} = e^{i2\pi j\ell h} = e^{i\beta_j \ell} \\ &= (\cos(\beta_j \ell) \ , \ \sin \beta_j \ell) \end{aligned} \tag{6.13}$$

with

$$\beta_j \overset{\text{def}}{=} 2\pi h j$$

In particular,

$$\begin{aligned} e^{i\beta_j(\ell+1)} + e^{i\beta_j(\ell-1)} &= \left(e^{i\beta_j} + e^{-i\beta_j}\right) e^{i\beta_j \ell} \\ &= 2\cos\beta_j e^{i\beta_j \ell} \end{aligned} \tag{6.14}$$

and

$$\begin{aligned} e^{i\beta_j(\ell+1)} - e^{i\beta_j(\ell-1)} &= \left(e^{i\beta_j} - e^{-i\beta_j}\right) e^{i\beta_j \ell} \\ &= 2i\sin\beta_j e^{i\beta_j \ell} \end{aligned} \tag{6.15}$$

Assuming next

$$\boldsymbol{u}^n = \boldsymbol{w}_j \ , \ \boldsymbol{u}^{n+1} = \widetilde{A}\boldsymbol{w}_j = \lambda_j \boldsymbol{w}_j \tag{6.16}$$

where λ_j is the corresponding eigenvalue, and substituting (6.13)–(6.14) into (6.7) we get

$$\lambda_j \left\{(1+d^2) - d^2 \cos\beta_j\right\} = 1 + di \sin\beta_j$$

where

$$d = \frac{c\Delta t}{h}$$

Thus

$$\lambda_j = \frac{1 + di \sin\beta_j}{1 + d^2(1 - \cos\beta_j)}$$

It is easily checked that the conjugates

$$\bar{\lambda}_j = \frac{1 - di \sin\beta_j}{1 + d^2(1 - \cos\beta_j)}$$

are also eigenvalues with corresponding eigenvectors which are conjugates of vectors $\boldsymbol{w}_j$.

In particular,

$$|\lambda_j|^2 = \lambda_j \bar{\lambda}_j = \frac{1 + d^2 \sin^2\beta_j}{(1 + d^2(1-\cos\beta_j))^2} = \frac{1 + 4d^2 \sin^2 \frac{\beta_j}{2} \cos^2 \frac{\beta_j}{2}}{(1 + 2d^2 \sin^2 \frac{\beta_j}{2})^2}$$

$$= \frac{1 + 4d^2 \sin^2 \frac{\beta_j}{2} - 4d^2 \sin^4 \frac{\beta_j}{2}}{1 + 4d^2 \sin^2 \frac{\beta_j}{2} + 4d^4 \sin^4 \frac{\beta_j}{2}}$$

$$\leq 1$$

which shows that our method is stable for an arbitrary time step Δt. Such schemes are called *unconditionally stable.*

Exercises

6.1.1. Let V be an inner product space. Prove that

$$(\boldsymbol{u}, \boldsymbol{w}) = (\boldsymbol{v}, \boldsymbol{w}) \quad \forall\, \boldsymbol{w} \in V$$

if and only if $\boldsymbol{u} = \boldsymbol{v}$.

6.1.2. (a) Prove the *parallelogram law* for inner product spaces

$$\|\boldsymbol{u}+\boldsymbol{v}\|^2 + \|\boldsymbol{u}-\boldsymbol{v}\|^2 = 2\|\boldsymbol{u}\|^2 + 2\|\boldsymbol{v}\|^2$$

(b) Conversely, let V be a *real* normed space with its norm satisfying the condition above. Proceed with the following steps to prove that

$$(\boldsymbol{u}, \boldsymbol{v}) \stackrel{\text{def}}{=} \frac{1}{4}(\|\boldsymbol{u}+\boldsymbol{v}\|^2 - \|\boldsymbol{u}-\boldsymbol{v}\|^2)$$

is an inner product on V.
Step 1. Continuity

$$\boldsymbol{u}_n \to \boldsymbol{u},\ \boldsymbol{v}_n \to \boldsymbol{v} \Longrightarrow (\boldsymbol{u}_n, \boldsymbol{v}_n) \to (\boldsymbol{u}, \boldsymbol{v})$$

Step 2. Symmetry

$$(\boldsymbol{u}, \boldsymbol{v}) = (\boldsymbol{v}, \boldsymbol{u})$$

Step 3. Positive definiteness

$$(\boldsymbol{u}, \boldsymbol{u}) = 0 \Longrightarrow \boldsymbol{u} = \boldsymbol{0}$$

Step 4. Use the parallelogram law to prove that

$$\|\boldsymbol{u}+\boldsymbol{v}+\boldsymbol{w}\|^2 + \|\boldsymbol{u}\|^2 + \|\boldsymbol{v}\|^2 + \|\boldsymbol{w}\|^2 = \|\boldsymbol{u}+\boldsymbol{v}\|^2 + \|\boldsymbol{v}+\boldsymbol{w}\|^2 + \|\boldsymbol{w}+\boldsymbol{u}\|^2$$

Step 5. Use the Step 4 identity to show additivity

$$(\boldsymbol{u}+\boldsymbol{v}, \boldsymbol{w}) = (\boldsymbol{u}, \boldsymbol{w}) + (\boldsymbol{v}, \boldsymbol{w})$$

Step 6. Homogeneity

$$(\alpha\boldsymbol{u}, \boldsymbol{v}) = \alpha(\boldsymbol{u}, \boldsymbol{v})$$

Hint: Use the Step 5 identity to prove the assertion first for $\alpha = k/m$, where k and m are integers, and use the continuity argument.

(c) Generalize the result to a complex normed space V using the formula (so-called *polarization formula*)

$$(\boldsymbol{u}, \boldsymbol{v}) = \frac{1}{4}(\|\boldsymbol{u}+\boldsymbol{v}\|^2 - \|\boldsymbol{u}-\boldsymbol{v}\|^2 + i\|\boldsymbol{u}+i\boldsymbol{v}\|^2 - i\|\boldsymbol{u}-i\boldsymbol{v}\|^2)$$

Compare the discussion on the extension of a scalar product from a real space to its complex extension.

6.1.3. Use the results of Exercise 6.1.2 to show that the spaces ℓ^p, $p \neq 2$ are *not* inner product spaces.
Hint: Verify the parallelogram law.

6.1.4. Let $\boldsymbol{u}$ and $\boldsymbol{v}$ be nonzero vectors in a real inner product space V. Show that

$$\|\boldsymbol{u}+\boldsymbol{v}\| = \|\boldsymbol{u}\| + \|\boldsymbol{v}\|$$

if and only if $\boldsymbol{v} = \alpha \boldsymbol{u}$ for some real number $\alpha > 0$. Does the result extend to complex vector spaces?

6.1.5. Let $\{\boldsymbol{u}_n\}$ be a sequence of elements in an inner product space V. Prove that if

$$(\boldsymbol{u}_n, \boldsymbol{u}) \longrightarrow (\boldsymbol{u}, \boldsymbol{u}) \text{ and } \|\boldsymbol{u}_n\| \longrightarrow \|\boldsymbol{u}\|$$

then $\boldsymbol{u}_n \longrightarrow \boldsymbol{u}$, i.e., $\|\boldsymbol{u}_n - \boldsymbol{u}\| \longrightarrow 0$.

6.1.6. Show that the sequence of sequences

$$\boldsymbol{u}_1 = (\alpha_1, 0, 0, \ldots)$$

$$\boldsymbol{u}_2 = (0, \alpha_2, 0, \ldots)$$

$$\boldsymbol{u}_3 = (0, 0, \alpha_3, \ldots)$$

etc., where the α_i are scalars, is an *orthogonal sequence* in ℓ^2, i.e., $(\boldsymbol{u}_n, \boldsymbol{u}_m) = 0$ for $m \neq n$.

6.2 Orthogonality and Orthogonal Projections

Orthogonal Complements. Let V be an inner product space and let V' be its topological dual. If M is any subspace of V, recall that (see Section 5.16) we have defined the space

$$M^{\perp} \overset{\text{def}}{=} \{f \in V' : \langle f, \boldsymbol{u} \rangle = 0 \quad \forall\, \boldsymbol{u} \in M\}$$

as the *orthogonal complement of M with respect to the duality pairing* $\langle \cdot, \cdot \rangle$.

Since V is an inner product space, the inner product can be used to construct orthogonal subspaces of V rather than its dual. In fact, we also refer to the space

$$M_V^{\perp} \overset{\text{def}}{=} \{\boldsymbol{v} \in V : (\boldsymbol{u}, \boldsymbol{v}) = 0 \quad \forall\, \boldsymbol{u} \in M\}$$

as the orthogonal complement of M with respect to the inner product $(\cdot, \cdot)$.

The situation is really not as complicated as it may seem, because the two orthogonal complements $M^{\perp}$ and $M_V^{\perp}$ are algebraically and topologically equivalent. We shall take up this equivalence in some detail in the next section. In this section we shall investigate some fundamental properties of orthogonal complements with respect to the inner product $(\cdot, \cdot)$. Taking for a moment the equivalence of two notions for the orthogonal complements for granted, we shall denote the orthogonal complements $M_V^{\perp}$ simply as $M^{\perp}$.

THEOREM 6.2.1
(The Orthogonal Decomposition Theorem)

Let V be a Hilbert space and $M \subset V$ a closed subspace of V. Then:

(i) *$M^{\perp}$ is a closed subspace of V.*

(ii) *V can be represented as the direct sum of M and its orthogonal complement $M^{\perp}$*

$$V = M \oplus M^{\perp}$$

i.e., every vector $\boldsymbol{v} \in V$ can be uniquely decomposed into two orthogonal vectors $\boldsymbol{m}, \boldsymbol{n}$

$$\boldsymbol{v} = \boldsymbol{m} + \boldsymbol{n} \ , \quad \boldsymbol{m} \in M \ , \quad \boldsymbol{n} \in M^{\perp}$$

PROOF

(i) $M^{\perp}$ is trivially closed with respect to vector space operations and therefore is a vector subspace of V. Continuity of scalar product implies also that $M^{\perp}$ is closed. Indeed, let $\boldsymbol{v}_n \in M^{\perp}$ be a sequence converging to a vector $\boldsymbol{v}$. Passing to the limit in

$$(\boldsymbol{u}, \boldsymbol{v}_n) = 0, \quad \boldsymbol{u} \in M$$

we get that $(\boldsymbol{u}, \boldsymbol{v}) = 0$ for every $\boldsymbol{u} \in M$ and therefore $\boldsymbol{v} \in M^\perp$.

(ii) We need to prove that

$$M \cap M^\perp = \{\boldsymbol{0}\}$$

and

$$V = M + M^\perp$$

The first condition is simple. If $\boldsymbol{v} \in M \cap M^\perp$ then $\boldsymbol{v}$ must be orthogonal with itself

$$\|\boldsymbol{v}\|^2 = (\boldsymbol{v}, \boldsymbol{v}) = 0$$

which implies $\boldsymbol{v} = \boldsymbol{0}$.

If $M = V$ then the decomposition is trivial

$$\boldsymbol{v} = \boldsymbol{v} + \boldsymbol{0}$$

and $M^\perp$ reduces to the zero vector $\boldsymbol{0}$.

Let us assume then that M is a proper subspace of V. We will show that there exists an element $\boldsymbol{m} \in M$ realizing the distance between $\boldsymbol{v}$ and the subspace M (see Fig. 6.2), i.e.,

$$\|\boldsymbol{v} - \boldsymbol{m}\| = d = \inf_{\boldsymbol{u} \in M} \|\boldsymbol{v} - \boldsymbol{u}\| \qquad (> 0)$$

To prove it, consider a minimizing sequence $\boldsymbol{u}_n \in M$ such that

$$d = \lim_{n \to \infty} \|\boldsymbol{v} - \boldsymbol{u}_n\|$$

We claim that $\boldsymbol{u}_n$ is Cauchy. Indeed, making use of the *parallelogram law* (Exercise 6.1.2) we have

$$\begin{aligned}
\|\boldsymbol{u}_n - \boldsymbol{u}_m\|^2 &= \|(\boldsymbol{u}_n - \boldsymbol{v}) + (\boldsymbol{v} - \boldsymbol{u}_m)\|^2 \\
&= 2\left(\|\boldsymbol{v} - \boldsymbol{u}_m\|^2 + \|\boldsymbol{v} - \boldsymbol{u}_n\|^2\right) - \|2\boldsymbol{v} - \boldsymbol{u}_n - \boldsymbol{u}_m\|^2 \\
&= 2\left(\|\boldsymbol{v} - \boldsymbol{u}_m\|^2 + \|\boldsymbol{v} - \boldsymbol{u}_n\|^2\right) - 4\left\|\boldsymbol{v} - \frac{\boldsymbol{u}_n + \boldsymbol{u}_m}{2}\right\|^2 \\
&\le 2\left(\|\boldsymbol{v} - \boldsymbol{u}_m\|^2 + \|\boldsymbol{v} - \boldsymbol{u}_n\|^2\right) - 4d^2
\end{aligned}$$

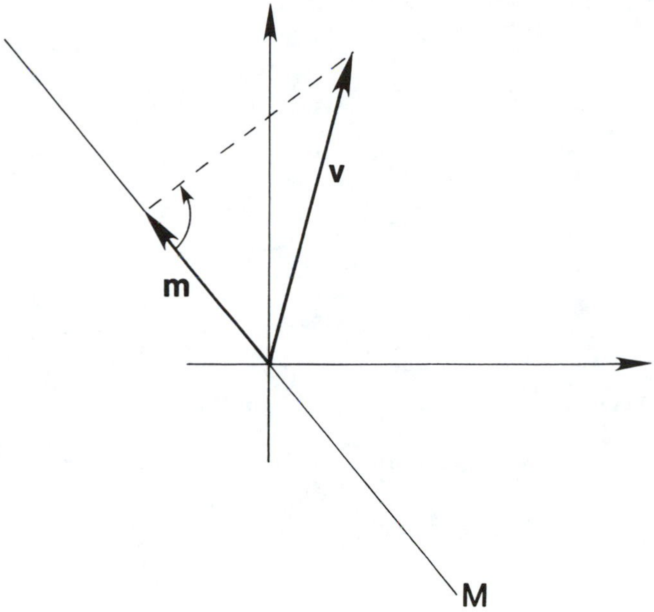

Figure 6.2
Construction of element m in the orthogonal decomposition $v = m + n$.

because $(\boldsymbol{u}_n + \boldsymbol{u}_m)/2$ is an element of M and therefore

$$d \leq \left\| \boldsymbol{v} - \frac{\boldsymbol{u}_n + \boldsymbol{u}_m}{2} \right\|$$

Consequently, if both $n, m \to \infty$ then

$$\|\boldsymbol{u}_n - \boldsymbol{u}_m\|^2 \longrightarrow 2\left(d^2 + d^2\right) - 4d^2 = 0$$

which proves that $\boldsymbol{u}_n$ is Cauchy and therefore converges to a vector $\boldsymbol{m}$. By closedness of M, $\boldsymbol{m} \in M$.

Consider now the decomposition

$$\boldsymbol{v} = \boldsymbol{m} + (\boldsymbol{v} - \boldsymbol{m})$$

It remains to show that $\boldsymbol{n} = \boldsymbol{v} - \boldsymbol{m} \in M^{\perp}$. Let $\boldsymbol{m}'$ be an arbitrary vector of M. For any $\alpha \in I\!\!R$ the linear combination $\boldsymbol{m} + \alpha\boldsymbol{m}'$ belongs to M as well and therefore

$$d^2 \leq \|\boldsymbol{v} - \boldsymbol{m} - \alpha\boldsymbol{m}'\|^2 = \|\boldsymbol{n} - \alpha\boldsymbol{m}'\|^2 = (\boldsymbol{n} - \alpha\boldsymbol{m}', \boldsymbol{n} - \alpha\boldsymbol{m}')$$

$$= \|\boldsymbol{n}\|^2 - \alpha(\boldsymbol{n}, \boldsymbol{m}') - \alpha(\boldsymbol{m}', \boldsymbol{n}) + \alpha^2 \|\boldsymbol{m}'\|^2$$

$$= d^2 - 2\alpha Re(\boldsymbol{n}, \boldsymbol{m}') + \alpha^2 \|\boldsymbol{m}'\|^2$$

Consequently,

$$-2\alpha Re(\boldsymbol{n}, \boldsymbol{m}') + \alpha^2 \|\boldsymbol{m}'\|^2 \geq 0 \quad \forall\, \alpha \in I\!\!R$$

which implies that

$$Re(\boldsymbol{n}, \boldsymbol{m}') = 0$$

At the same time,

$$Im(\boldsymbol{n}, \boldsymbol{m}') = -Re(\boldsymbol{n}, i\boldsymbol{m}') = 0$$

since $i\boldsymbol{m}' \in M$ as well. ∎

COROLLARY 6.2.1

(Recall Proposition 5.16.2.)

Let V be a Hilbert space and M a vector subspace of V. The following conditions are equivalent to each other:

(i) M is closed.

(ii) $(M^{\perp})^{\perp} = M$.

PROOF

(ii) $\Longrightarrow$ (i) follows from the fact that orthogonal complements are closed.

(i) $\Longrightarrow$ (ii) Obviously, $M \subset (M^{\perp})^{\perp}$. To prove the inverse inclusion consider an arbitrary $\boldsymbol{m} \in (M^{\perp})^{\perp}$ and the corresponding unique decomposition

$$\boldsymbol{m} = \boldsymbol{m}_1 + \boldsymbol{n}$$

where $\boldsymbol{m}_1 \in M$ and $\boldsymbol{n} \in M^\perp$. From the orthogonality of $\boldsymbol{m}$ to $M^\perp$ it follows that

$$0 = (\boldsymbol{n}, \boldsymbol{m}) = (\boldsymbol{n}, \boldsymbol{m}_1 + \boldsymbol{n}) = (\boldsymbol{n}, \boldsymbol{n})$$

which implies that $\boldsymbol{n} = \boldsymbol{0}$ and therefore $\boldsymbol{m} = \boldsymbol{m}_1 \in M$. ∎

Orthogonal Projections. Let M be a closed subspace of a Hilbert space V. The linear projection P_M corresponding to the decomposition

$$V = M \oplus M^\perp \ , \ \boldsymbol{v} = \boldsymbol{m} + \boldsymbol{n}$$

and prescribing for any vector $\boldsymbol{v}$ its $\boldsymbol{m}$ component (recall Section 2.7)

$$P_M : V \longrightarrow V \ , \ P_M \boldsymbol{v} \stackrel{\text{def}}{=} \boldsymbol{m}$$

is called the *orthogonal projection onto the subspace* M. Using the nomenclature of Section 2.7, we identify the orthogonal projection on M as the (linear) *projection on* M *in the direction of its orthogonal complement* $M^\perp$.

In general, there are many projections on M corresponding to various (not necessarily orthogonal) decompositions $V = M \oplus N$, but there is only one orthogonal projection on M corresponding to the choice $N = M^\perp$.

From the orthogonality of the decomposition

$$\boldsymbol{v} = \boldsymbol{m} + \boldsymbol{n} \ , \ \boldsymbol{m} \in M \ , \ \boldsymbol{n} \in M^\perp$$

it follows that

$$\|\boldsymbol{v}\|^2 = (\boldsymbol{v}, \boldsymbol{v}) = (\boldsymbol{m} + \boldsymbol{n}, \boldsymbol{m} + \boldsymbol{n}) = (\boldsymbol{m}, \boldsymbol{m}) + (\boldsymbol{n}, \boldsymbol{n})$$

$$= \|\boldsymbol{m}\|^2 + \|\boldsymbol{n}\|^2$$

which implies that

$$\|\boldsymbol{m}\| = \|P_M \boldsymbol{v}\| \le \|\boldsymbol{v}\| \qquad \forall\, \boldsymbol{v} \in V$$

and, consequently, the norm of the orthogonal projection $\|P_M\| \le 1$. But at the same time, if $M \ne \{\boldsymbol{0}\}$, then

$$P_M \boldsymbol{m} = \boldsymbol{m} \qquad \forall\, \boldsymbol{m} \in M$$

and therefore $\|P_M\| = 1$.

We summarize the properties of orthogonal projections in the following proposition.

PROPOSITION 6.2.1

Let M be a closed subspace of a Hilbert space V. There exists a linear, bounded operator P_M with a unit norm, $\|P_M\| = 1$, prescribing for each $\boldsymbol{v} \in V$ a unique element $\boldsymbol{m} \in M$ such that

(i) $\|\boldsymbol{v} - \boldsymbol{m}\| = \inf_{\boldsymbol{m} \in M} \|\boldsymbol{v} - \boldsymbol{u}\|$ *and*

(ii) $\boldsymbol{v} - \boldsymbol{m} \in M^{\perp}$.

Example 6.2.1

Let $V = L^2(-1,1)$ be the space of square-integrable functions on interval $(-1,1)$ and M denote the subspace of even functions on $(-1,1)$

$$u \in M \Longleftrightarrow u(x) = u(-x) \text{ for (almost) all } x \in (-1,1)$$

As the L^2 convergence of a sequence implies the pointwise convergence of a subsequence almost everywhere (Exercises 6.2.5 and 6.2.6), M is closed. From the decomposition

$$u(x) = \frac{u(x) + u(-x)}{2} + \frac{u(x) - u(-x)}{2}$$

it follows that the orthogonal complement $M^{\perp}$ can be identified as the space of odd functions on $(-1,1)$. Indeed, if u is even and v odd, then

$$\begin{aligned}
\int_{-1}^{1} u(x)v(x)dx &= \int_{-1}^{0} u(x)v(x)dx + \int_{0}^{1} u(x)v(x)dx \\
&= \int_{0}^{1} u(-x)v(-x)dx + \int_{0}^{1} u(x)v(x)dx \\
&= 0
\end{aligned}$$

Operators prescribing for any function $u \in L^2(-1,1)$ its even and odd contributions are orthogonal projections. In particular, functions $(u(x) +$

$u(-x))/2$ and $(u(x)-u(-x))/2$ can be interpreted as the *closest* (in the L^2 sense) even and odd functions to function u. $\square$

Exercises

6.2.1. Let V be an inner product space and M, N denote vector subspaces of V. Prove the following algebraic properties of orthogonal complements:

(i) $M \subset N \Longrightarrow N^\perp \subset M^\perp$.

(ii) $M \subset N \Longrightarrow (M^\perp)^\perp \subset (N^\perp)^\perp$.

(iii) $M \cap M^\perp = \{\mathbf{0}\}$.

(iv) If M is dense in V, ($\overline{M} = V$) then $M^\perp = \{\mathbf{0}\}$.

6.2.2. Let M be a subspace of a Hilbert space V. Prove that

$$\overline{M} = (M^\perp)^\perp$$

6.2.3. Two subspaces M and N of an inner product space V are said to be *orthogonal*, denoted $M \perp N$, if

$$(\boldsymbol{m}, \boldsymbol{n}) = 0 \ , \ \forall\, \boldsymbol{m} \in M \ , \ \boldsymbol{n} \in N$$

Let V now be a Hilbert space. Prove or disprove the following:

(i) $M \perp N \Longrightarrow M^\perp \perp N^\perp$.

(ii) $M \perp N \Longrightarrow (M^\perp)^\perp \perp (N^\perp)^\perp$.

6.2.4. Let Ω be an open, bounded set in $\mathbb{R}^n$ and $V = L^2(\Omega)$ denote the space of square integrable functions on Ω. Find the orthogonal complement in V of the space of constant functions

$$M = \left\{ u \in L^2(\Omega) \,:\, u = \text{const a.e. in } \Omega \right\}$$

6.2.5. Let $\Omega \subset \mathbb{R}^N$ be a measurable set and $f_n : \Omega \to \mathbb{R}(\mathbb{C})$ a sequence of measurable functions. We say that sequence f_n *converges in measure* to a measurable function $f : \Omega \to \mathbb{R}(\mathbb{C})$ if, for every $\varepsilon > 0$,

$$\mathrm{m}(\{\boldsymbol{x} \in \Omega \,:\, |f_n(\boldsymbol{x}) - f(\boldsymbol{x})| \geq \varepsilon\}) \to 0 \quad \text{as} \quad n \to 0$$

Let now $\mathrm{m}(\Omega) < \infty$. Prove that $L^p(\Omega)$ convergence, for any $1 \leq p \leq \infty$, implies convergence in measure.

Hint:

$$\mathrm{m}(\{\boldsymbol{x} \in \Omega : |f_n(\boldsymbol{x}) - f(\boldsymbol{x})| \geq \varepsilon\})$$

$$\leq \begin{cases} \frac{1}{\varepsilon}(\int_\Omega |f_n(\boldsymbol{x}) - f(\boldsymbol{x})|^p \, d\boldsymbol{x})^{\frac{1}{p}} & 1 \leq p < \infty \\ \frac{1}{\varepsilon} \operatorname{ess\,sup}_{\boldsymbol{x}\in\Omega} |f_n(\boldsymbol{x}) - f(\boldsymbol{x})| & p = \infty \end{cases}$$

6.2.6. Let $\Omega \subset \mathbb{R}^N$ be a measurable set and $f_n : \Omega \to \mathbb{R}(\mathbb{C})$ a sequence of measurable functions *converging in measure* to a measurable function $f : \Omega \to \mathbb{R}(\mathbb{C})$. Prove that one can extract a subsequence f_{n_k} converging to function f almost everywhere in Ω.

Hint: Follow the steps given below.

Step 1. Show that, given an $\varepsilon > 0$, one can extract a subsequence f_{n_k} such that

$$\mathrm{m}(\{\boldsymbol{x} \in \Omega : |f_{n_k}(\boldsymbol{x}) - f(\boldsymbol{x})| \geq \varepsilon\}) \leq \frac{1}{2^{k+1}} \quad \forall k \geq 1$$

Step 2. Use the diagonal choice method to show that one can extract a subsequence f_{n_k} such that

$$\mathrm{m}(\{\boldsymbol{x} \in \Omega : |f_{n_k}(\boldsymbol{x}) - f(\boldsymbol{x})| \geq \frac{1}{k}\}) \leq \frac{1}{2^{k+1}} \quad \forall k \geq 1$$

Consequently,

$$\mathrm{m}(\{\boldsymbol{x} \in \Omega : |f_{n_k}(\boldsymbol{x}) - f(\boldsymbol{x})| \geq \varepsilon\}) \leq \frac{1}{2^{k+1}}$$

for every $\varepsilon > 0$, and for k large enough.

Step 3. Let $\varphi_k = f_{n_k}$ be the subsequence extracted in Step 2. Use the identities

$$\{\boldsymbol{x} \in \Omega : \inf_{\nu\geq 0} \sup_{n\geq\nu} |\varphi_n(\boldsymbol{x}) - f(\boldsymbol{x})| > 0\}$$

$$= \bigcup_k \{\boldsymbol{x} \in \Omega : \inf_{\nu\geq 0} \sup_{n\geq\nu} |\varphi_n(\boldsymbol{x}) - f(\boldsymbol{x})| \geq \frac{1}{k}\}$$

$$\{\boldsymbol{x} \in \Omega : \inf_{\nu\geq 0} \sup_{n\geq\nu} |\varphi_n(\boldsymbol{x}) - f(\boldsymbol{x})| \geq \varepsilon\}$$

$$= \bigcap_{\nu\geq 0} \{\boldsymbol{x} \in \Omega : \sup_{n\geq\nu} |\varphi_n(\boldsymbol{x}) - f(\boldsymbol{x})| \geq \varepsilon\}$$

to prove that

$$\mathrm{m}(\{\boldsymbol{x} \in \Omega : \limsup_{n\to\infty} |\varphi_n(\boldsymbol{x}) - f(\boldsymbol{x})| > 0\})$$

$$\leq \sum_k \lim_{\nu\to\infty} \mathrm{m}(\{\boldsymbol{x} \in \Omega : \sup_{n\geq\nu} |\varphi_n(\boldsymbol{x}) - f(\boldsymbol{x})| \geq \frac{1}{k}\})$$

Step 4. Use the identity

$$\begin{aligned}\{\boldsymbol{x} \in \Omega \,:\, \sup_{n \geq \nu} |\,\varphi_n(\boldsymbol{x}) - f(\boldsymbol{x})| \geq \frac{1}{k}\} \\ \subset \{\boldsymbol{x} \in \Omega \,:\, \sup_{n \geq \nu} |\varphi_n(\boldsymbol{x}) - f(\boldsymbol{x})| > \frac{1}{k-1}\} \\ \subset \bigcup_{n \geq \nu} \{\boldsymbol{x} \in \Omega \,:\, |\varphi_n(\boldsymbol{x}) - f(\boldsymbol{x})| > \frac{1}{k-1}\}\end{aligned}$$

and the result of Step 2 to show that

$$\mathrm{m}(\{\boldsymbol{x} \in \Omega \,:\, \sup_{n \geq \nu} |\varphi_n(\boldsymbol{x}) - f(\boldsymbol{x})| \geq \varepsilon\}) \leq \frac{1}{2^\nu}$$

for every $\varepsilon > 0$ and ν (ε-dependent !) large enough.
Step 5. Use the results of Step 3 and Step 4 to conclude that

$$\mathrm{m}(\{\boldsymbol{x} \in \Omega \,:\, \lim_{k \to \infty} f_{n_k}(\boldsymbol{x}) \neq f(\boldsymbol{x})\}) = 0$$

REMARK 6.2.1. The Lebesgue dominated convergence theorem establishes conditions under which *pointwise convergence* of a sequence of functions f_n to a limit function f implies the L^p convergence. While the converse, in general, is not true, the results of the last two exercises at least show that the L^p convergence of a sequence f_n implies the pointwise convergence (almost everywhere only, of course) of a subsequence f_{n_k}. ∎

6.3 Orthonormal Bases and Fourier Series

One of the most important features of Hilbert spaces is that they provide a framework for the Fourier representation of functions. We shall now examine this and the related idea of an orthonormal basis in the Hilbert space (recall Example 2.4.9).

Orthogonal and Orthonormal Families of Vectors. A (not necessarily countable) family of vectors $\{\boldsymbol{e}_\iota\}_{\iota \in I}$ is said to be *orthogonal* if

$$(\boldsymbol{e}_\iota, \boldsymbol{e}_\kappa) = 0$$

for every pair of different indices ι, κ. If, additionally, all vectors are unit, i.e., $\|\boldsymbol{e}_\iota\| = 1$, the family is said to be *orthonormal.* As every orthogonal family $\{\boldsymbol{e}_\iota\}_{\iota \in I}$ of nonzero vectors can be turned to an orthonormal one by normalizing the vectors, i.e., replacing $\boldsymbol{e}_\iota$ with $\boldsymbol{e}_\iota/\|\boldsymbol{e}_\iota\|$, there is a limited need for the use of orthogonal families and for most of the time we will talk about the orthonormal ones only.

Every orthonormal family $\{\boldsymbol{e}_\iota\}_{\iota\in I}$ is linearly independent. Indeed, if one of the vectors, say $\boldsymbol{e}_\kappa$, could be represented as a linear combination of a finite subset I_0 of vectors from the family:

$$\boldsymbol{e}_\kappa = \sum_{\iota\in I_0} \alpha_\iota \boldsymbol{e}_\iota \,,\ \#I_0 < \infty \,,\ I_0 \subset I$$

then

$$\|\boldsymbol{e}_\kappa\|^2 = \left(\sum_{\iota\in I_0} \alpha_\iota \boldsymbol{e}_\iota, \boldsymbol{e}_\kappa\right) = \sum_{\iota\in I_0} \alpha_\iota \left(\boldsymbol{e}_\iota, \boldsymbol{e}_\kappa\right) = 0$$

is a contradiction.

Orthonormal Basis. An orthonormal family $\{\boldsymbol{e}_\iota\}_{\iota\in I}$ of vectors in a Hilbert space V is called an *orthonormal basis* of V iff it is maximal, i.e., no extra vector $\boldsymbol{e}_0$ from V can be added such that $\{\boldsymbol{e}_\iota\}_{\iota\in I} \cup \{\boldsymbol{e}_0\}$ will be orthonormal. In other words,

$$(\boldsymbol{e}_\iota, \boldsymbol{v}) = 0 \quad \forall\, \iota \in I \text{ implies } \boldsymbol{v} = \boldsymbol{0}$$

We shall examine now closely the special case when the basis is *countable*, i.e., it can be represented in the sequential form $\boldsymbol{e}_1, \boldsymbol{e}_2, \ldots$, the sequence being finite or infinite.

Let M denote the linear span of vectors $\boldsymbol{e}_1, \boldsymbol{e}_2 \ldots$ forming the basis

$$M = \text{span } \{\boldsymbol{e}_1, \boldsymbol{e}_2, \ldots\}$$

The definition of the orthonormal basis implies that the orthogonal complement of M reduces to the zero vector

$$M^\perp = \{\boldsymbol{0}\}$$

which (recall Exercise 6.2.2) implies that

$$\overline{M} = \left(M^\perp\right)^\perp = \{\boldsymbol{0}\}^\perp = V$$

Thus M is (everywhere) dense in the space V. Consequently, for any vector $\boldsymbol{v} \in V$ there exists a sequence $\boldsymbol{u}_n \in M$ converging to $\boldsymbol{v}$, $\boldsymbol{u}_n \to \boldsymbol{v}$.

In particular, since any finite-dimensional space is automatically closed, we immediately see that the existence of a finite orthonormal basis implies that the space V is finite dimensional. Orthonormal bases then constitute a special subclass of usual (Hamel) bases in a finite-dimensional Hilbert (Euclidean) space.

Let $V_n = \text{span}\,\{\boldsymbol{e}_1, \ldots, \boldsymbol{e}_n\}$ denote now the span of first n vectors from the basis and let P_n be the corresponding orthogonal projection on V_n. We claim that

$$P_n\boldsymbol{v} \longrightarrow \boldsymbol{v}\,, \quad \text{for every } \boldsymbol{v} \in V$$

Indeed, let $\boldsymbol{u}_n \in M$ be a sequence converging to $\boldsymbol{v}$. Pick an arbitrary $\varepsilon > 0$ and select an element $\boldsymbol{u}_k \in M$ such that

$$\|\boldsymbol{u}_k - \boldsymbol{v}\| < \frac{\varepsilon}{2}$$

Let $N = N(k)$ be an index such that $\boldsymbol{u}_k \in V_N$. We have then for every $n \geq N$

$$\begin{aligned}
\|P_n\boldsymbol{v} - \boldsymbol{v}\| &\leq \|P_n\boldsymbol{v} - P_n\boldsymbol{u}_k\| + \|P_n\boldsymbol{u}_k - \boldsymbol{v}\| \\
&\leq \|P_n\|\ \|\boldsymbol{v} - \boldsymbol{u}_k\| + \|\boldsymbol{u}_k - \boldsymbol{v}\| \\
&\leq \|\boldsymbol{v} - \boldsymbol{u}_k\| + \|\boldsymbol{u}_k - \boldsymbol{v}\| \leq 2\frac{\varepsilon}{2} = \varepsilon
\end{aligned}$$

since $\|P_n\| = 1$ and $P_n\boldsymbol{u}_k = \boldsymbol{u}_k$ for $n \geq N$.

Define now

$$\boldsymbol{v}_1 = P_1\boldsymbol{v}\,,\ \boldsymbol{v}_n = P_n\boldsymbol{v} - P_{n-1}\boldsymbol{v}$$

We have

$$\begin{aligned}
\boldsymbol{v} &= \lim_{n\to\infty} P_n\boldsymbol{v} = \lim_{n\to\infty} \{P_1\boldsymbol{v} + (P_2\boldsymbol{v} - P_1\boldsymbol{v}) + \ldots + (P_n\boldsymbol{v} - P_{n-1}\boldsymbol{v})\} \\
&= \lim_{n\to\infty} \sum_{i=1}^{n} \boldsymbol{v}_i = \sum_{i=1}^{\infty} \boldsymbol{v}_n
\end{aligned}$$

Also, representing $P_n\boldsymbol{v} \in V_n$ as a linear combination of vectors $\boldsymbol{e}_1, \ldots, \boldsymbol{e}_n$

$$P_n\boldsymbol{v} = v_n^1\boldsymbol{e}_1 + \ldots + v_n^{n-1}\boldsymbol{e}_{n-1} + v_n^n\boldsymbol{e}_n$$

we see that

$$P_{n-1}\boldsymbol{v} = v_n^1\boldsymbol{e}_1 + \ldots + v_n^{n-1}\boldsymbol{e}_{n-1}$$

and, consequently,

$$\boldsymbol{v}_n = P_n\boldsymbol{v} - P_{n-1}\boldsymbol{v} = v_n^n \boldsymbol{e}_n$$

or, simplifying the notation,

$$\boldsymbol{v}_n = v_n \boldsymbol{e}_n \,, \text{ for some } v_n \in \mathbb{C}(\mathbb{R})$$

Thus, we have found that any vector $\boldsymbol{v} \in V$ can be represented in the form of the series

$$\boldsymbol{v} = \sum_{i=1}^{\infty} v_i \boldsymbol{e}_i$$

The coefficients v_i can be viewed as the *components* of $\boldsymbol{v}$ with respect to the orthonormal basis $\{\boldsymbol{e}_i\}$.

Example 6.3.1

Vectors

$$\boldsymbol{e}_k = \left(0, \ldots, 1_{(k)}, \ldots, 0\right)$$

form a (canonical) orthonormal basis in $\mathbb{C}^n$ with the canonical scalar product. $\square$

Example 6.3.2

(Recall Example 2.4.9.)

Vectors

$$\boldsymbol{e}_k = \left(0, \ldots, 1_{(k)}, \ldots, 0, \ldots\right)$$

form a (canonical) orthonormal basis in ℓ^2.

Indeed, let $\boldsymbol{v} \in \ell^2$, $\boldsymbol{v} = (v_1, v_2, v_3, \ldots)$. Then

$$(\boldsymbol{e}_k, \boldsymbol{v}) = v_k$$

and therefore, trivially, $(\boldsymbol{e}_k, \boldsymbol{v}) = 0$, $k = 1, 2, \ldots$ implies that $\boldsymbol{v} = \boldsymbol{0}$. Also, since

$$\boldsymbol{v} = \sum_{i=1}^{\infty} v_i \boldsymbol{e}_i$$

numbers v_i are interpreted as components of $\boldsymbol{v}$ with respect to the canonical basis. $\square$

Example 6.3.3

We will prove that functions

$$e_k(x) = e^{2\pi i k x},\ k = 0,\ \pm 1,\ \pm 2, \ldots$$

form an orthonormal basis in $L^2(0,1)$. $\square$

Step 1. Orthonormality

$$\begin{aligned}
(\boldsymbol{e}_k, \boldsymbol{e}_\ell) &= \int_0^1 e_k(x)\overline{e}_\ell(x)dx \\
&= \int_0^1 e^{2\pi i k x} e^{-2\pi i \ell x} dx \\
&= \int_0^1 e^{2\pi i (k-\ell) x} dx \\
&= \begin{cases} 0 \text{ if } k \neq \ell \\ 1 \text{ if } k = \ell \end{cases}
\end{aligned}$$

Step 2. Let $f \in L^2(0,1)$ be a real continuous function on $[0,1]$. We claim that

$$\int_0^1 f(x)\overline{e}_k(x)dx = 0\,,\ k = 1, 2, \ldots$$

implies that $f \equiv 0$.

Suppose, to the contrary, that there exists $x_0 \in (0,1)$ such that $f(x_0) \neq 0$. Replacing f with $-f$, if necessary, we can assume that $f(x_0) > 0$. It follows from the continuity of f that there exists $\delta > 0$ such that

$$f(x) \geq \beta > 0 \quad \text{for} \quad |x - x_0| \leq \delta$$

Define now a function

$$\kappa(x) \stackrel{\text{def}}{=} 1 + \cos 2\pi(x - x_0) - \cos 2\pi\delta$$

Obviously, κ is a linear combination (with complex coefficients) of functions $\boldsymbol{e}_\ell$. Due to the properties of the exponential function, the same holds for any power $\kappa^m(x)$ of function κ.

It is easy to check (see the graph of $\kappa(x)$ shown in Fig. 6.3) that

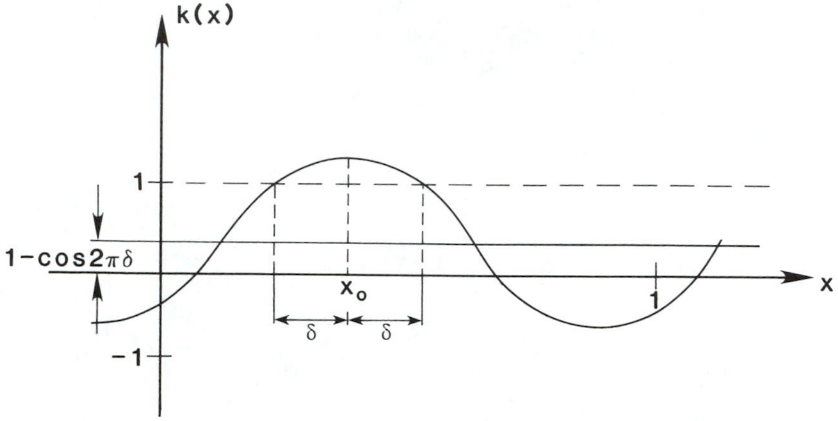

Figure 6.3
Example 6.3.3. Function $\kappa(x)$.

$$\kappa(x) \begin{array}{ll} > 1 & \text{for } |x - x_0| < \delta \\ = 1 & \text{for } |x - x_0| = \delta \\ < 1 & \text{for } |x - x_0| > \delta \end{array}$$

We have therefore

$$(\kappa^m, f)_{L^2} = \int_0^{x_0-\delta} \kappa^m f dx + \int_{x_0-\delta}^{x_0+\delta} \kappa^m f dx + \int_{x_0+\delta}^1 \kappa^m f dx$$

It follows from the Schwarz inequality that

$$\left(\int_0^{x_0-\delta} \kappa^m f dx\right)^2 \le \int_0^{x_0-\delta} \kappa^{2m} dx \int_0^{x_0-\delta} f^2 dx$$

and by the Lebesgue dominated convergence theorem

$$\int_0^{x_0-\delta} \kappa^{2m} dx \longrightarrow 0$$

In the same way, the last integral converges to zero as well, but the middle one

$$\int_{x_0-\delta}^{x_0+\delta} \kappa^m f dx \geq 2\delta\beta > 0$$

At the same time, due to the orthogonality of f with e_ℓ,

$$(\kappa^m, f)_{L^2} = 0$$

is a contradiction.

The assertion of this step is immediately generalized to complex-valued functions f.

Step 3. Let $f \in L^2(0,1)$ be an arbitrary function. By the density of continuous functions in $L^2(0,1)$ there exists a sequence of continuous functions $f_n \in L^2(0,1)$ converging to f. Assume now that

$$(e_k, f) = 0 \qquad k = 0, \pm 1, \pm 2, \ldots$$

In particular, $(k = 0)$ $\int_0^1 f(x)dx = 0$. As

$$\int_0^1 f_n(x)dx \longrightarrow \int_0^1 f(x)dx = 0$$

we can replace the original functions $f_n(x)$ with

$$f_n(x) - \int_0^1 f_n(x)dx$$

and assume that also $\int_0^1 f_n(x)dx = 0$.

Now, for each n define the function

$$g_n(x) = \int_0^x f_n(s)ds$$

From the fundamental theorem of integral calculus it follows that g_n is C^1 and $g_n' = f_n$. Consequently,

$$\int_0^1 g_n(x)v'(x)dx = -\int_0^1 f_n(x)v(x)dx$$

for every C^1 function $v(x)$, since $g_n(0) = g_n(1) = 0$.

Passing to the limit, we get

$$\int_0^1 g(x)v'(x)dx = -\int_0^1 f(x)v(x)dx$$

for the (continuous) function $g(x) = \int_0^x f(s)ds$. Substituting functions $e_k(x)$ for $v(x)$, we draw the conclusion that

$$(e_k, g) = 0 \qquad k = 0, \pm 1, \pm 2, \ldots$$

By the Step 2 result, $g \equiv 0$, which implies that

$$\int_a^b f(x)dx = \left(f, \chi_{[a,b]}\right) = 0 \qquad 0 < a < b < 1$$

for every characteristic function of an interval $(a, b) \subset (0, 1)$. Since the span of such characteristic functions forms a dense subset in $L^2(a, b)$ (see Exercise 4.9.2), f must vanish almost everywhere (Exercise 6.2.1 (iv)).

Example 6.3.4

Functions

$$f_0 \equiv 1$$

$$f_k = \sqrt{2}\cos 2\pi k x \ , \ k = 1, 2, \ldots$$

$$g_k = \sqrt{2}\sin 2\pi k x \ , \ k = 1, 2, \ldots$$

form an orthonormal basis in $L^2(0, 1)$. $\square$

By a straightforward calculation we check that the functions are orthonormal. To prove that they form a maximal set it is enough to see that

$$f_0 = e_0, \ f_k = \frac{e_k + e_{-k}}{\sqrt{2}}, \ g_k = \frac{e_k - e_{-k}}{i\sqrt{2}}, \ k = 1, 2, \ldots$$

where $e_k, k = 1, 2, \ldots$ are the exponential functions from the previous example.

Note that in contrast to the previous example, functions from this example form an orthonormal basis in both *real* and *complex* $L^2(a, b)$ spaces.

Let $\boldsymbol{e}_1, \ldots, \boldsymbol{e}_n$ be a *finite* set of orthonormal vectors in a Hilbert space V and V_n and its linear span. It is easy to derive an explicit formula for the orthogonal projection P_n on the subspace V_n.

Toward this goal, pick an arbitrary vector $\boldsymbol{v} \in V$ and represent $P_n\boldsymbol{v}$ in the form of a linear combination of vectors $\boldsymbol{e}_1, \ldots, \boldsymbol{e}_n$

$$P_n\boldsymbol{v} = v_1\boldsymbol{e}_1 + \ldots + v_n\boldsymbol{e}_n$$

Multiplying both sides by $\boldsymbol{e}_j$ in the sense of the scalar product and using the orthonormality of $\boldsymbol{e}_i$, we get

$$v_j = (P_n \boldsymbol{v}, \boldsymbol{e}_j)$$

According to the orthogonal decomposition theorem, however, $\boldsymbol{v}$ can be represented in the form

$$\boldsymbol{v} = P_n \boldsymbol{v} + (\boldsymbol{v} - P_n \boldsymbol{v})$$

where $\boldsymbol{v} - P_n \boldsymbol{v}$ is orthogonal to V_n and therefore

$$v_j = (\boldsymbol{v}, \boldsymbol{e}_j)$$

Thus we end up with the formula

$$P_n \boldsymbol{v} = \sum_{j=1}^{n} (\boldsymbol{v}, \boldsymbol{e}_j) \boldsymbol{e}_j$$

Gram-Schmidt Orthonormalization. Given an arbitrary sequence of linearly independent vectors $\{\boldsymbol{v}_i\}_{i=1}^{\infty}$ in a Hilbert space V, it is easy to construct a corresponding orthonormal sequence by using the so-called *Gram-Schmidt orthonormalization* procedure.

We begin by normalizing the first vector $\boldsymbol{v}_1$

$$\boldsymbol{e}_1 \stackrel{\text{def}}{=} \frac{\boldsymbol{v}_1}{\|\boldsymbol{v}_1\|}$$

Next, we take the second vector $\boldsymbol{v}_2$ and subtract from it its orthogonal projection $P_1 \boldsymbol{v}_2$ on $V_1 = \text{span}\,\{\boldsymbol{e}_1\}$

$$\widehat{\boldsymbol{e}}_2 \stackrel{\text{def}}{=} \boldsymbol{v}_2 - P_1 \boldsymbol{v}_2 = \boldsymbol{v}_2 - (\boldsymbol{v}_2, \boldsymbol{e}_1)\boldsymbol{e}_1$$

It follows from the linear independence of $\boldsymbol{v}_1$ and $\boldsymbol{v}_2$ that vector $\widehat{\boldsymbol{e}}_2$ is different from zero. We define now $\boldsymbol{e}_2$ by normalizing $\widehat{\boldsymbol{e}}_2$

$$\boldsymbol{e}_2 \stackrel{\text{def}}{=} \frac{\widehat{\boldsymbol{e}}_2}{\|\widehat{\boldsymbol{e}}_2\|}$$

By induction, given $n-1$ vectors $\boldsymbol{e}_1, \ldots, \boldsymbol{e}_{n-1}$, we construct first $\widehat{\boldsymbol{e}}_n$ by subtracting from $\boldsymbol{v}_n$ its orthogonal projection $P_{n-1}\boldsymbol{v}_n$ on $V_{n-1} = \text{span}\,\{\boldsymbol{e}_1, \ldots, \boldsymbol{e}_{n-1}\} = \text{span}\,\{\boldsymbol{v}_1, \ldots, \boldsymbol{v}_{n-1}\}$

$$\widehat{\boldsymbol{e}}_n \stackrel{\text{def}}{=} \boldsymbol{v}_n - P_{n-1}\boldsymbol{v}_n = \boldsymbol{v}_n - \sum_{j=1}^{n-1} (\boldsymbol{v}_n, \boldsymbol{e}_j)\boldsymbol{e}_j$$

and normalize it

$$e_n \stackrel{\text{def}}{=} \frac{\widehat{e}_n}{\|\widehat{e}_n\|}$$

It follows from the construction that vectors $e_i, i = 1, 2, \ldots$ are orthonormal.

Example 6.3.5
(Legendre Polynomials)

By applying the Gram-Schmidt orthonormalization to monomials

$$1,\ x,\ x^2,\ x^3, \ldots$$

in the real $L^2(a, b)$, we obtain the so-called *Legendre polynomials* p_n which can be represented in a concise form as

$$p_n(x) = \frac{1}{\gamma_n} \frac{d^n}{dx^n} \{(x-a)^n (x-b)^n\} \ n = 0, 1, \ldots$$

with constants γ_n chosen to satisfy $\|p_n\| = 1$.

We prove the assertion by induction. For $n = 0$, $p_0 \equiv 1$. Assume now $n > 0$. Obviously, p_n is a polynomial of order n. To prove that it coincides with the function e_n resulting from the Gram-Schmidt orthonormalization, it is sufficient to show that p_n is orthogonal with $p_1, \ldots, p_{n-1}$ or, equivalently, with all monomials x^m of order $m \leq n-1$. We have

$$\int_a^b x^m \frac{d^n}{dx^n} \{(x-a)^n (x-b)^n\} dx$$

$$= - \int_a^b m x^{m-1} \frac{d^{n-1}}{dx^{n-1}} \{(x-a)^n (x-b)^n\} dx$$

$$+ x^m \frac{d^{n-1}}{dx^{n-1}} \{(x-a)^n (x-b)^n\} \Big|_a^b$$

Continuing the integration by parts another $m-1$ times, we conclude that the integral must vanish. $\square$

Let V be an arbitrary Hilbert space. By modifying the proofs of Theorems 2.4.2 and 2.4.3, and using the Kuratowski-Zorn lemma, it can be shown that every Hilbert space V has an orthonormal basis and that every two orthonormal bases in V have precisely the same number of elements (cardinal number). This number is frequently identified as the *dimension of the Hilbert space* V (not the same as the dimension of V treated as a vector space only).

We shall content ourselves here with a much simpler, explicit construction of an orthonormal basis in a *separable* Hilbert space.

THEOREM 6.3.1

Every separable Hilbert space V *has a countable orthonormal basis.*

PROOF Let $\{v_n\}_{n=1}^{\infty}$ be an everywhere dense sequence in V.

Step 1. Select a subsequence of linearly independent vectors

$$\{\boldsymbol{v}_{n_k}\}_{k=1}^{\infty}$$

We proceed by induction. We take first the smallest index n_1 such that $\boldsymbol{v}_{n_1} \neq \mathbf{0}$. Next, assume that k linearly vectors $\boldsymbol{v}_{n_1}, \ldots, \boldsymbol{v}_{n_k}$ have already been selected. Two possible cases may occur:

Case 1. All remaining $\boldsymbol{v}_n$, $n > n_k$ are linearly dependent with the vectors selected so far. In this case the subsequence of linearly independent vectors will be finite.

Case 2. There exists the smallest index $n_{k+1} > n_k$ such that vectors

$$\boldsymbol{v}_{n_1}, \ldots, \boldsymbol{v}_{n_k}, \boldsymbol{v}_{n_{k+1}}$$

are linearly independent.

Step 2. Apply the Gram-Schmidt orthonormalization, yielding vectors $\boldsymbol{e}_1, \boldsymbol{e}_2, \ldots$. Obviously,

$$\text{span}\,(\boldsymbol{e}_1, \boldsymbol{e}_2, \ldots) = \text{span}\,(\boldsymbol{v}_{n_1}, \boldsymbol{v}_{n_2}, \ldots) = \text{span}\,(\boldsymbol{v}_1, \boldsymbol{v}_2, \ldots)$$

and therefore the span of vectors $\boldsymbol{e}_1, \boldsymbol{e}_2, \ldots$ is dense in V, which implies that $\boldsymbol{e}_1, \boldsymbol{e}_2, \ldots$ is an orthonormal basis in V. ∎

COROLLARY 6.3.1

A Hilbert space V is separable iff it posseses a countable basis.

PROOF It remains to show sufficiency. Let $\boldsymbol{e}_1, \boldsymbol{e}_2, \ldots$ be an orthonormal basis in V. Accordingly, any vector $\boldsymbol{v} \in V$ can be represented in the form

$$\boldsymbol{v} = \sum_{i=1}^{\infty} v_i \boldsymbol{e}_i$$

the series being finite if V is finite dimensional. Pick now an arbitrary $\varepsilon > 0$ and select rational numbers $\overline{v}_i$ (complex numbers with rational real and imaginary parts in the complex case) such that

$$|v_i - \overline{v}_i| \leq \frac{\varepsilon}{2^i} \qquad i = 1, 2, \ldots$$

It follows that

$$\left\| \boldsymbol{v} - \sum_{i=1}^{\infty} \overline{v}_i \boldsymbol{e}_i \right\| = \left\| \sum_{i=1}^{\infty} (v_i - \overline{v}_i) \, \boldsymbol{e}_i \right\|$$

$$\leq \sum_{i=1}^{\infty} |v_i - \overline{v}_i| \; \|\boldsymbol{e}_i\| \leq \sum_{i=1}^{\infty} \frac{\varepsilon}{2^i} = \varepsilon$$

which proves that linear combinations of vectors $\boldsymbol{e}_i$, $i = 1, 2, \ldots$ with rational coefficients are dense in V. ∎

Example 6.3.6

We give now without proof three examples of orthonormal bases in different L^2 spaces.

1. The *Legendre polynomials*

$$p_n(x) = \left(\frac{2n+1}{2} \right)^{\frac{1}{2}} \frac{1}{2^n n!} \frac{d^n}{dx^n} \left(x^2 - 1\right)^n$$

form an orthonormal basis in $L^2(-1, 1)$.

2. The *Laguerre functions*

$$\phi_n(x) = \frac{1}{n!} e^{-x/2} L_n(x), \qquad n = 0, 1, \ldots$$

where $L_n(x)$ is the *Laguerre polynomial*

$$L_n(x) = \sum_{i=0}^{n} (-1)^i \binom{n}{i} n(n-1)\dots(i+1)x^i$$

$$= e^x \tfrac{d^n}{dx^n}\left(x^n e^{-x}\right)$$

form an orthonormal basis in $L^2(0,\infty)$.

3. The *Hermite functions*

$$\phi_n(x) = \left[2^n n! \sqrt{\pi}\right]^{-\frac{1}{2}} e^{x^2/2} H_n(x), \qquad n = 0, 1, \dots$$

where $H_n(x)$ is the *Hermite polynomial*

$$H_n(x) = (-1)^n e^{x^2} \frac{d^n}{dx^n}\left(e^{-x^2}\right)$$

form an orthonormal basis for $L^2(-\infty, \infty) = L^2(\mathbb{R})$.

□

Fourier Series. The existence of an orthonormal basis in a separable Hilbert space provides a very useful tool in studying properties of the space because it allows us to represent arbitrary elements of the space as (infinite) linear combinations of the basis functions. Suppose that $\{\boldsymbol{e}_n\}_{n=1}^{\infty}$ is an orthonormal basis for a Hilbert space V. Taking an arbitrary vector $\boldsymbol{v}$ and representing it in the form of the series

$$\boldsymbol{v} = \sum_{n=1}^{\infty} v_i \boldsymbol{e}_i = \lim_{N\to\infty} \sum_{n=1}^{N} v_i \boldsymbol{e}_i$$

we easily find the explicit formula for coefficients v_i. Orthonormality of vectors $\boldsymbol{e}_i$ implies that

$$\left(\sum_{i=1}^{N} v_i \boldsymbol{e}_i \,,\, \boldsymbol{e}_j\right) = v_j \text{ for } N \geq j$$

Passing to the limit with $N \to \infty$, we get

$$v_j = (\boldsymbol{v}, \boldsymbol{e}_j)$$

and, consequently,

$$v = \sum_{i=1}^{\infty} (v, e_i)\, e_i$$

The series is called the (generalized) *Fourier series* representation of $v \in V$, and the scalars $v_n = (v, e_n)$ are called the *Fourier coefficients* of v relative to the basis $\{e_i\}_{i=1}^{\infty}$.

Substituting the Fourier series representation for vectors u and v in the scalar product (u, v), we get immediately

$$\begin{aligned}
(u, v) &= \left(\sum_{i=1}^{\infty} u_i e_i, \sum_{j=1}^{\infty} v_j e_j \right) \\
&= \lim_{N\to\infty} \left(\sum_{i=1}^{N} u_i e_i, \sum_{j=1}^{N} v_j e_j \right) \\
&= \lim_{N\to\infty} \sum_{i=1}^{N} u_i \overline{v}_i \\
&= \sum_{i=1}^{\infty} u_i \overline{v}_i \\
&= \sum_{i=1}^{\infty} (u, e_i)\, \overline{(v, e_i)}
\end{aligned}$$

The formula

$$(u, v) = \sum_{i=1}^{\infty} (u, e_i)\, \overline{(v, e_i)}$$

is known as *Parseval's identity*. Substituting $v = u$ in particular implies that

$$\|u\|^2 = \sum_{i=1}^{\infty} |(u, e_i)|^2$$

Exercises

6.3.1. Prove that every (not necessarily separable) nontrivial Hilbert space possesses an orthonormal basis. *Hint:* Compare the proof of Theorem 2.4.3.

6.3.2. Prove that every separable Hilbert space V is unitary equivalent with the space ℓ^2. *Hint:* Establish a bijective correspondence between the canonical basis in ℓ^2 and an orthonormal basis in V and use it to define a unitary map mapping ℓ^2 onto V.

6.3.3. Let $\{\boldsymbol{e}_n\}_{n=1}^{\infty}$ be an orthonormal family (not necessarily maximal) in a Hilbert space V. Prove *Bessel's inequality*

$$\sum_{i=1}^{\infty} |(\boldsymbol{u}, \boldsymbol{e}_i)|^2 \leq \|\boldsymbol{u}\|^2 \qquad \forall\, \boldsymbol{u} \in V$$

6.3.4. Let $\{\boldsymbol{e}_n\}_{n=1}^{\infty}$ be an orthonormal family in a Hilbert space V. Prove that the following conditions are equivalent to each other.

(i) $\{\boldsymbol{e}_n\}_{n=1}^{\infty}$ is an orthonormal basis, i.e., it is maximal.

(ii) $\boldsymbol{u} = \sum_{n=1}^{\infty} (\boldsymbol{u}, \boldsymbol{e}_n)\, \boldsymbol{e}_n \qquad \forall\, \boldsymbol{u} \in V.$

(iii) $(\boldsymbol{u}, \boldsymbol{v}) = \sum_{n=1}^{\infty} (\boldsymbol{u}, \boldsymbol{e}_n)\, \overline{(\boldsymbol{v}, \boldsymbol{e}_n)}.$

(iv) $\|\boldsymbol{u}\|^2 = \sum_{n=1}^{\infty} |(\boldsymbol{u}, \boldsymbol{e}_n)|^2.$

6.3.5. Prove the *Riesz-Fisher theorem.*
Let V be a separable Hilbert space with an orthonormal basis $\{\boldsymbol{e}_n\}_{n=1}^{\infty}$. Then

$$V = \left\{ \sum_{n=1}^{\infty} v_n \boldsymbol{e}_n : \sum_{n=1}^{\infty} |v_n|^2 < \infty \right\}$$

In other words, elements of $\boldsymbol{v}$ can be characterized as infinite series $\sum_{n=1}^{\infty} v_n \boldsymbol{e}_n$ with ℓ^2-summable coefficients v_n (recall Exercise 6.3.2).

6.3.6. Let $I = (-1, 1)$ and let V be the four-dimensional inner product space spanned by the monomials $\{1, x, x^2, x^3\}$ with

$$(f, g)_V = \int_{-1}^{1} fg\,dx$$

(i) Use the Gram-Schmidt process to construct an orthonormal basis for V.

(ii) Observing that $V \subset L^2(I)$, compute the orthogonal projection Πu of the function $u(x) = x^4$ onto V.

(iii) Show that $(x^4 - \Pi x^4, v)_{L^2(I)} = 0 \quad \forall v \in V$.

(iv) Show that if $p(x)$ is any polynomial of degree ≤ 3, then $\Pi p = p$.

(v) Sketch the function Πx^4 and show graphically how it approximates x^4 in V.

6.3.7. Use the orthonormal basis from Example 6.3.4 to construct the (classical) Fourier series representation of the following functions in $L^2(0, 1)$.

(i) $f(x) = x$

(ii) $f(x) = x + 1$

Duality in Hilbert Spaces

6.4 Riesz Representation Theorem

The properties of the topological dual of a Hilbert space constitute one of the most important collection of ideas in Hilbert space theory and in the study of linear operators. We recall from our study of topological duals of Banach spaces in the previous chapter that the dual of a Hilbert space V is the vector space V' consisting of all continuous linear functionals on V. If f is a member of V' we write, as usual,

$$f(\boldsymbol{v}) = \langle f, \boldsymbol{v} \rangle$$

where $\langle \cdot, \cdot \rangle$ denotes the duality pairing on $V' \times V$. Recall that V' is a normed space equipped with the dual norm

$$\|f\|_{V'} = \sup_{\boldsymbol{v} \neq 0} \frac{\langle f, \boldsymbol{v} \rangle}{\|\boldsymbol{v}\|_V}$$

Now, in the case of Hilbert spaces, we have a ready-made device for constructing linear and continuous functionals on V by means of the scalar product $(\cdot,\cdot)_V$. Indeed, if $\boldsymbol{u}$ is a fixed element of V, we may define a linear functional $f_{\boldsymbol{u}}$ directly by

$$f_{\boldsymbol{u}}(\boldsymbol{v}) \stackrel{\text{def}}{=} (\boldsymbol{v},\boldsymbol{u}) = \overline{(\boldsymbol{u},\boldsymbol{v})} \qquad \forall\, \boldsymbol{v} \in V$$

This particular functional depends on the choice $\boldsymbol{u}$, and this suggests that we describe this correspondence by introducing an operator R from V into V' such that

$$R\boldsymbol{u} = f_{\boldsymbol{u}}$$

We have by the definition

$$\langle R\boldsymbol{u},\boldsymbol{v}\rangle = (\boldsymbol{v},\boldsymbol{u}) = \overline{(\boldsymbol{u},\boldsymbol{v})} \qquad \forall\, \boldsymbol{u},\boldsymbol{v} \in V$$

Now, it is not clear at this point whether or not there might be some functionals in V' that cannot be represented by inner products on V. In fact, all we have shown up to now is that

$$R(V) \subset V'$$

It is a remarkable fact, proven by the Hungarian mathematician Riesz, that *all* functionals in V' can be represented in this way; that is,

$$R(V) = V'$$

The statement of this important assertion is set forth in the following theorem (recall the representation theorems in Section 5.12).

THEOREM 6.4.1
(The Riesz Representation Theorem)

Let V be a Hilbert space and let f be a continuous linear functional on V. Then there exists a unique element $\mathbf{u} \in V$ such that

$$f(\boldsymbol{v}) = (\boldsymbol{v},\boldsymbol{u}) \qquad \forall\, \boldsymbol{v} \in V$$

where $(\cdot,\cdot)$ is the scalar product on V. Moreover,

$$\|f\|_{V'} = \|\boldsymbol{u}\|_V$$

PROOF

Step 1. Uniqueness.

Suppose that

$$f(\boldsymbol{v}) = (\boldsymbol{v}, \boldsymbol{u}_1) = (\boldsymbol{v}, \boldsymbol{u}_2) \qquad \forall\, \boldsymbol{v} \in V$$

and some $\boldsymbol{u}_2$, $\boldsymbol{u}_2 \in V$. Then $(\boldsymbol{v}, \boldsymbol{u}_1 - \boldsymbol{u}_2) = 0$ and upon substituting $\boldsymbol{v} = \boldsymbol{u}_1 - \boldsymbol{u}_2$ we get

$$\|\boldsymbol{u}_1 - \boldsymbol{u}_2\|^2 = 0$$

which implies $\boldsymbol{u}_1 = \boldsymbol{u}_2$.

Step 2. Existence.

The case $f \equiv 0$ is trivial. Assume that $f \not\equiv 0$. Since functional f is continuous, the null space

$$N = f^{-1}\{0\} = \ker f = \{\boldsymbol{u} \in V : f(\boldsymbol{u}) = 0\}$$

is a closed subspace of V and therefore by the orthogonal decomposition theorem, the space V can be represented in the form of the direct, orthogonal sum

$$V = N \oplus N^{\perp}$$

Pick an arbitrary nonzero vector $\boldsymbol{u}_0 \in N^{\perp}$ and define

$$\boldsymbol{u} = \overline{\frac{f(\boldsymbol{u}_0)}{\|\boldsymbol{u}_0\|^2}}\, \boldsymbol{u}_0$$

It follows from the choice of $\boldsymbol{u}$ that both functionals f and $R\boldsymbol{u}$ coincide on $N \oplus \mathbb{C}\boldsymbol{u}_0$. Indeed,

$$f(\boldsymbol{v}) = (\boldsymbol{v}, \boldsymbol{u}) = 0 \ \text{ for } \ \boldsymbol{v} \in N$$

and for $\boldsymbol{v} = \alpha \boldsymbol{u}_0$, $\alpha \in \mathbb{C}$

$$(\boldsymbol{v}, \boldsymbol{u}) = (\alpha \boldsymbol{u}_0, \boldsymbol{u}) = \alpha f(\boldsymbol{u}_0) = f(\alpha \boldsymbol{u}_0)$$

We claim finally that $N^{\perp}$ is one dimensional and it reduces to $\mathbb{C}\boldsymbol{u}_0$. We have for any $\boldsymbol{v} \in V$

$$\boldsymbol{v} = \left(\boldsymbol{v} - \frac{f(\boldsymbol{v})}{f(\boldsymbol{u})}\boldsymbol{u}\right) + \frac{f(\boldsymbol{v})}{f(\boldsymbol{u})}\boldsymbol{u}$$

Now,

$$f\left(\boldsymbol{v} - \frac{f(\boldsymbol{v})}{f(\boldsymbol{u})}\boldsymbol{u}\right) = f(\boldsymbol{v}) - \frac{f(\boldsymbol{v})}{f(\boldsymbol{u})} f(\boldsymbol{u}) = 0$$

and therefore the first vector belongs to N, which proves that

$$V = N \oplus \mathbb{C}\boldsymbol{u} = N \oplus \mathbb{C}\boldsymbol{u}_0$$

the decomposition being orthogonal ($\boldsymbol{u}_0 \in N^{\perp}$). To prove that $\|f\|_{V'} = \|\boldsymbol{u}\|_V$ notice that for $\boldsymbol{v} \neq \boldsymbol{0}$

$$\begin{aligned} |f(\boldsymbol{v})| = |(\boldsymbol{v}, \boldsymbol{u})| &\leq \text{(Schwarz inequality)} \\ &\leq \|\boldsymbol{v}\| \, \|\boldsymbol{u}\| \end{aligned}$$

and therefore $\|f\|_{V'} \leq \|\boldsymbol{u}\|_V$.

But at the same time,

$$\|\boldsymbol{u}\|_V^2 = (\boldsymbol{u}, \boldsymbol{u}) = f(\boldsymbol{u}) \leq \|f\|_{V'} \, \|\boldsymbol{u}\|_V$$

so $\|\boldsymbol{u}\|_V \leq \|f\|_{V'}$. ∎

COROLLARY 6.4.1

Let $R \colon V \to V'$ denote the map from a Hilbert space V onto its dual V' such that

$$\langle R\boldsymbol{u}, \boldsymbol{v} \rangle = (\boldsymbol{v}, \boldsymbol{u}) \qquad \forall\, \boldsymbol{u}, \boldsymbol{v} \in V$$

Then:

(i) R is an *antilinear* map from V onto V'.

(ii) R preserves the norm, i.e.,

$$\|R\boldsymbol{u}\|_{V'} = \|\boldsymbol{u}\|_V$$

In particular, in the case of a *real* Hilbert space V, R is a linear norm-preserving isomorphism (surjective isometry) from V onto V'.

PROOF It remains to show that R is antilinear. But this follows directly from the fact that the scalar product is antilinear with respect to the second variable:

$$(\boldsymbol{v}, \alpha_1 \boldsymbol{u}_2 + \alpha_2 \boldsymbol{u}_2) = \overline{\alpha}_1 \, (\boldsymbol{v}, \boldsymbol{u}_1) + \overline{\alpha}_2 \, (\boldsymbol{v}, \boldsymbol{u}_2)$$

Consequently,

$$R(\alpha_1 \boldsymbol{u}_1 + \alpha_2 \boldsymbol{u}_2) = \overline{\alpha}_1 R(\boldsymbol{u}_1) + \overline{\alpha}_2 R(\boldsymbol{u}_2)$$

∎

The antilinear operator R described above is known as the *Riesz map* corresponding to the Hilbert space V and the scalar product $(\cdot,\cdot)$ and it is frequently used to identify the topological dual of V with itself in much the same way as the representation theorems for the L^p spaces were used to identify their duals with (conjugate) spaces L^q $\left(\frac{1}{p} + \frac{1}{q} = 1\right)$.

The Riesz map can be used to transfer the Hilbert space structure to its dual V' which *a priori* is only a normed (Banach) space. Indeed, by a straightforward verification, we check that

$$(f, g)_{V'} \stackrel{\text{def}}{=} (R^{-1}g, R^{-1}f)_V$$

is a well-defined scalar product on V'. (Note that the inverse of an antilinear map if it exists, is antilinear, too.) Moreover, from the fact that R is norm-preserving it follows that the norm corresponding to the just-introduced scalar product on V coincides with the original (dual) norm on V'.

Applying the Riesz representation theorem to the dual space V', we can introduce the Riesz map for the dual space as

$$R_{V'}\colon V' \ni g \to \{f \to (f, g)_{V'} \in \mathbb{C}\} \in (V')'$$

where $(V')'$ is the bidual of V. Composing $R = R_V$ with the Riesz map for the dual space, we get

$$\begin{aligned}
\Big(R_{V'} R_V(\boldsymbol{u})\Big)(f) &= \Big(f, R_V(\boldsymbol{u})\Big)_{V'} \\
&= \Big(R_V^{-1} R_V(\boldsymbol{u}), R_V^{-1} f\Big)_V \\
&= \Big(\boldsymbol{u}, R_V^{-1} f\Big)_V \\
&= \langle f, \boldsymbol{u} \rangle
\end{aligned}$$

which implies that the evaluation map mapping space V into its bidual $(V')'$ (see Section 5.13) coincides with the composition of the two Riesz maps for V and its dual (notice that the composition of two antilinear maps is linear). Consequently, every Hilbert space is *reflexive*.

We conclude this section with a number of examples illustrating the concept of the Riesz map.

Example 6.4.1

We return one more time to the example of a finite-dimensional inner product space V with a scalar product $(\cdot,\cdot)$ studied previously in Chapter 2.

Let $\boldsymbol{e}_1,\ldots,\boldsymbol{e}_n$ be an arbitrary (not necessarily orthonormal) basis in V. Using Einstein's summation convention, we represent two arbitrary vectors $\boldsymbol{x},\boldsymbol{y}\in V$ in the form

$$\boldsymbol{x} = x^k\boldsymbol{e}_k, \qquad \boldsymbol{y} = y^j\boldsymbol{e}_j$$

Now let $\boldsymbol{f}$ be an arbitrary element of the dual $V^* = V'$. It is natural to represent $\boldsymbol{f}$ in the dual basis $\boldsymbol{e}^{*1},\ldots,\boldsymbol{e}^{*n}$:

$$\boldsymbol{f} = f_i\boldsymbol{e}^{*i}$$

Assume now that $\boldsymbol{f} = R\boldsymbol{x}$ where R is the Riesz map from V into its dual V'. We have

$$\langle \boldsymbol{f},\boldsymbol{y}\rangle = \langle R\boldsymbol{x},\boldsymbol{y}\rangle = (\boldsymbol{y},\boldsymbol{x})$$

$$= \left(y^j\boldsymbol{e}_j, x^k\boldsymbol{e}_k\right) = g_{jk}y^j\overline{x}^k$$

where $g_{jk} \stackrel{\text{def}}{=} (\boldsymbol{e}_j,\boldsymbol{e}_k)$ is the positive definite (so-called *Gram*) matrix corresponding to basis $\boldsymbol{e}_1,\ldots,\boldsymbol{e}_n$. At the same time,

$$\langle \boldsymbol{f},\boldsymbol{y}\rangle = \left\langle f_\ell\boldsymbol{e}^{*\ell}, y^j\boldsymbol{e}_j\right\rangle = f_\ell y^j\left\langle \boldsymbol{e}^{*\ell},\boldsymbol{e}_j\right\rangle$$

$$= f_\ell y^j\delta^\ell_j = f_j y^j$$

Comparing both expressions, we get

$$f_j = g_{jk}\overline{x}^k$$

which is precisely the matrix form representation for the equation

$$\boldsymbol{f} = R\boldsymbol{x}$$

As we can see, the Gram matrix can be interpreted as the matrix representation of the Riesz map with respect to the basis $\boldsymbol{e}_1,\ldots\boldsymbol{e}_n$ and its dual. It can also be explicitly seen that the Riesz map is *antilinear*.

Introducing the inverse Gram matrix g^{kj}, we write

$$g^{kj} f_j = \overline{x}^k$$

or taking into account that $g^{kj} = \overline{g}^{jk}$, we get

$$g^{jk} \overline{f}_j = x^k$$

Consequently, the scalar product in the dual space can be represented in the form

$$\begin{aligned}
(\boldsymbol{f}, \boldsymbol{h})_{V'} &= \left(R^{-1}\boldsymbol{h}, R^{-1}\boldsymbol{f}\right) \\
&= g_{jk} \left(R^{-1}\boldsymbol{h}\right)^j \overline{\left(R^{-1}\boldsymbol{f}\right)}^k \\
&= g_{jk} g^{nj} \overline{h}_n g^{km} f_m \\
&= \delta_j^m g^{nj} \overline{h}_n f_m \\
&= g^{nm} f_m \overline{h}_n
\end{aligned}$$

where $\boldsymbol{h} = h_\ell \boldsymbol{e}^{*\ell}$ and $\boldsymbol{f} = f_\ell \boldsymbol{e}^{*\ell}$. □

Example 6.4.2

Let $V = L^2(\Omega)$ with Ω an open set in $\mathbb{R}^n$. We have

$$\langle Ru, v \rangle = (v, u)_{L^2(\Omega)} = \int_\Omega v\overline{u}\, d\Omega$$

and

$$(Ru, Rv)_{V'} = (v, u)_V = \int_\Omega v\overline{u}\, d\Omega$$

□

Example 6.4.3

Consider the Sobolev space $H^m(\Omega)$ of order m introduced in Example 6.1.5. The closure of functions $C_0^\infty(\Omega)$ (infinitely differentiable functions with compact support contained in Ω) is identified as a subspace of $H^m(\Omega)$

$$H_0^m(\Omega) \stackrel{\text{def}}{=} \overline{C_0^\infty(\Omega)}^{H^m(\Omega)}$$

By definition $H_0^m(\Omega)$ is closed and therefore is a Hilbert space with the scalar product from $H^m(\Omega)$

$$(u,v)_{H^m(\Omega)} = \sum_{|\boldsymbol{\alpha}|\le m} \int_\Omega D^\alpha u \overline{D^\alpha v} d\Omega$$

Intuitively speaking, spaces $H_0^m(\Omega)$ consist of all functions from $H^m(\Omega)$ which vanish on the boundary together with all derivatives of order up to (inclusively) $m-1$. A precise interpretation of this fact is based on *Lions' trace theorem*; see [8,12]. $\square$

Topological duals of spaces $H_0^m(\Omega)$ are identified as Sobolev spaces of negative order

$$H^{-m}(\Omega) \stackrel{\text{def}}{=} (H_0^m(\Omega))'$$

Note that, in general, elements from $H^{-m}(\Omega)$ are only linear and continuous *functionals* defined on $H_0^m(\Omega)$ and cannot be identified with functions.

In this example we would like to take a closer look at a particular case of the real space $H_0^1(I)$, with $I = (-1,1) \subset \mathbb{R}$. According to the *Sobolev imbedding theorem*, see [1], the space $H^1(I)$ can be *imbedded* into the Chebyshev space $C(\overline{I})$ of functions continuous on the closed interval $\overline{I} = [-1,1]$. More precisely, there exists a linear and continuous injection

$$T\colon H^1(I) \longrightarrow C(\overline{I})$$

which for functions continuous on $\overline{I}$ reduces to the identity map. Recall that continuity means that

$$\|Tu\|_{C(\overline{I})} \le C\, \|u\|_{H^1(I)} \qquad \forall\, u \in H^1(I)$$

for some $C > 0$. In particular, this implies that the mapping

$$H^1(I) \ni u \longrightarrow (Tu)(x_0) \qquad x_0 \in \overline{I}$$

is continuous for every point x_0 from $\overline{I}$.

It may be a little confusing, but most of the time we drop the letter T and write that

$$H^1(I) \ni u \longrightarrow u(x_0) \qquad x_0 \in \overline{I}$$

is continuous. This notational simplification is at least partially justified by the fact that T reduces to identity for (equivalence classes of) functions from $C(\overline{I})$.

Equipped with these observations we redefine the space $H_0^1(I)$ as

$$H_0^1(I) \overset{\text{def}}{=} \left\{ u \in H^1(I) \colon u(0) = u(1) = 0 \right\}$$

It is not a trivial exercise, but it can be shown that $H_0^1(I)$ defined above coincides with $H_0^1(\Omega)$ defined before, i.e., the closure of functions $C_0^\infty(I)$ in $H^1(I)$.

Notice that due to the continuity of the imbedding map T, $H_0^1(I)$ is a closed subspace of $H^1(I)$ and therefore it is itself a Hilbert space equipped with the scalar product from $H^1(I)$

$$(u, v) = \int_{-1}^{1} \left(uv + \frac{du}{dx}\frac{dv}{dx} \right) dx$$

Continuity of the imbedding implies also that the *Dirac functional*

$$\delta(v) = v(0)$$

is a continuous and linear map on $H^1(I)$ and therefore on the subspace $H_0^1(I)$ as well and therefore is an element of the dual Sobolev space

$$H^{-1}(I) \overset{\text{def}}{=} \left(H_0^1(I) \right)'$$

The Riesz map R from $H_0^1(I)$ onto $H^{-1}(I)$ is defined by

$$R \colon u \longrightarrow f$$

$$\langle f, v \rangle = (v, u) = \int_{-1}^{1} \left(vu + \frac{dv}{dx}\frac{du}{dx} \right) dx \quad \forall\, v \in H_0^1(I)$$

Restricting ourselves to test functions $v \in C_0^\infty(I)$, we see that u can be interpreted as a solution to the distributional equation

$$-\frac{d^2u}{dx^2} + u = f$$

with boundary conditions $u(-1) = u(1) = 0$ (see Sections 6.6 and 6.7 for a more detailed discussion of related issues).

In particular, for $f = \delta$ the corresponding solution $u = u_\delta$ is given by

$$u_\delta(x) = \begin{cases} \sinh(1+x)/2\cosh 1 & -1 \le x \le 0 \\ \sinh(1-x)/2\cosh 1 & 0 \le x \le 1 \end{cases}$$

It follows that

$$\|\delta\|^2_{H^{-1}(I)} = \|u_\delta\|^2_{H^1(I)} = \int_{-1}^{1} \left[u_\delta^2 + \left(\frac{du_\delta}{dx}\right)^2 \right] dx = \frac{\sinh 2}{4\cosh^2 1}$$

6.5 The Adjoint of a Linear Operator

In Sections 5.16 and 5.18 we examined the properties of the transpose of linear and both continuous and closed operators defined on Banach spaces. In the case of Hilbert spaces those ideas can be further specialized, leading to the idea of (topologically) adjoint operators (recall Section 2.14 for a discussion of the same notion in finite-dimensional spaces).

We set the stage for this discussion by reviewing some notations. Let

U, V be (complex) Hilbert spaces with scalar products $(\cdot,\cdot)_U$ and $(\cdot,\cdot)_V$, respectively,

U', V' denote the topological duals of U and V,

$\langle\cdot,\cdot\rangle_U$ and $\langle\cdot,\cdot\rangle_V$ denote the duality pairings on $U' \times U$ and $V' \times V$, and

$R_U\colon U \to U'$, $R_V\colon V \to V'$ be the Riesz operators for U and V, respectively, i.e.,

$$\langle R_U \boldsymbol{u}, \boldsymbol{w}\rangle = (\boldsymbol{w}, \boldsymbol{u})_U \ \forall\, \boldsymbol{w} \in U \quad \text{and}$$

$$\langle R_V \boldsymbol{v}, \boldsymbol{w}\rangle = (\boldsymbol{w}, \boldsymbol{v})_V \ \forall\, \boldsymbol{w} \in V$$

(Topological) Adjoint of a Continuous Operator. Let $A \in \mathcal{L}(U,V)$, i.e., let A be a linear and continuous operator from U into V. Recall that the topological transpose operator $A' \in \mathcal{L}(V',U')$ was defined as

$$A'v' = v' \circ A \qquad \text{for} \qquad v' \in V'$$

or, equivalently,

$$\langle A'v', \boldsymbol{u}\rangle = \langle v', A\boldsymbol{u}\rangle \qquad \forall\, \boldsymbol{u} \in U \;\; v' \in V'$$

The transpose A' of operator A operates on the dual V' into the dual U'. Existence of the Riesz operators establishing the correspondence between spaces U, V and their duals U', V' prompts us to introduce the so-called (*topological*) *adjoint operator* A^* operating directly on the space V into U and defined as the composition

$$A^* \stackrel{\text{def}}{=} R_U^{-1} \circ A' \circ R_V$$

The relationship between A, A', A^* and the Riesz maps is depicted symbolically in Fig. 6.4.

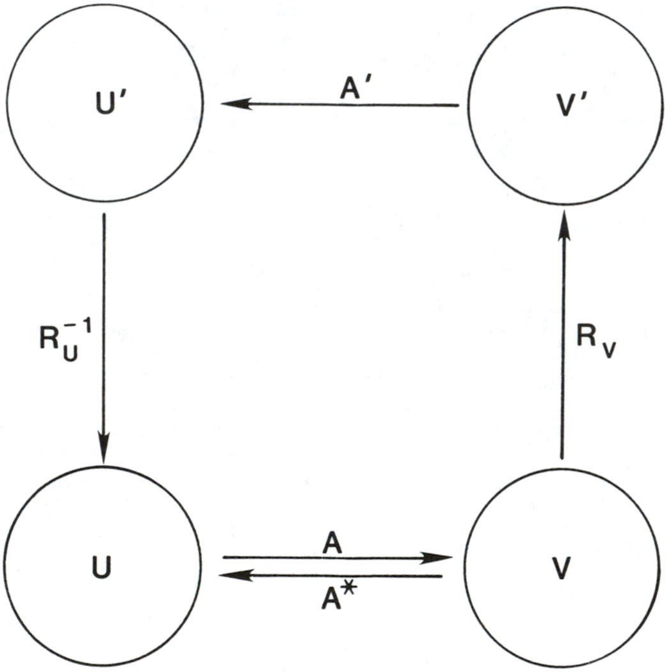

Figure 6.4
Topological adjoint of a continuous operator defined on a Hilbert space.

As in the finite-dimensional case, it follows from the definitions of A', A^* and the Riesz maps that

$$
\begin{aligned}
(\boldsymbol{u}, A^*\boldsymbol{v})_U &= \left(\boldsymbol{u}, R_U^{-1} A' R_V \boldsymbol{v}\right)_U \\
&= \langle A' R_V \boldsymbol{v}, \boldsymbol{u} \rangle_U \\
&= \langle R_V \boldsymbol{v}, A\boldsymbol{u} \rangle_V \\
&= (A\boldsymbol{u}, \boldsymbol{v})_V
\end{aligned}
$$

for every $\boldsymbol{u} \in U$, $\boldsymbol{v} \in V$. Note that even though the Riesz operators are antilinear, the adjoint operator A^* is *linear* as the composition of linear and antilinear maps is antilinear and the composition of two antilinear maps is linear.

Example 6.5.1

The adjoint of the integral operator

$$
Au = v \, , \; v(x) = \int_0^1 K(x,\xi) u(\xi) d\xi
$$

defined on the real space $L^2(0,1)$ with square-integrable kernel $K(x,\xi)$, considered in Example 5.16.1, is equal to the integral operator A^* where

$$
A^* u = v \, , \; v(\xi) = \int_0^1 K(\xi, x) u(x) dx
$$

i.e., the corresponding kernel $K^*(x,\xi) = K(\xi,x)$. Notice that in the complex case a complex conjugate has to be added, i.e.,

$$
K^*(\xi, x) = \overline{K(x,\xi)}
$$

$\square$

Example 6.5.2

Recall that an operator T mapping a Hilbert space U into a Hilbert space V is said to be *unitary* if

$$
(T\boldsymbol{u}, T\boldsymbol{v})_V = (\boldsymbol{u}, \boldsymbol{v})_U \qquad \forall \; \boldsymbol{u} \in U, \; \boldsymbol{v} \in V
$$

Then T is an isometry (in fact, the two conditions are equivalent to each other; cf. Exercise 6.1.2) and in particular injective.

Assume additionally that T is surjective, i.e., $\mathcal{R}(T) = V$. Then the inverse of T coincides with its adjoint. Indeed, substituting $\boldsymbol{w} = T\boldsymbol{v}$ in the above equation we get

$$(T\boldsymbol{u}, \boldsymbol{w})_V = \left(\boldsymbol{u}, T^{-1}\boldsymbol{w}\right)_U \qquad \forall\, \boldsymbol{u} \in U,\ \boldsymbol{w} \in V$$

which implies that $T^{-1} = T^*$.

Conversely, $T^{-1} = T^*$ implies that T is unitary and surjective. $\square$

Reinterpreting properties of the transpose operator (recall Proposition 5.16.1) we get the following result.

PROPOSITION 6.5.1

Let U, V, W be inner product spaces and let A, $A_i \in \mathcal{L}(U, V)$, $i = 1, 2$ and $B \in \mathcal{L}(V, W)$. The following properties hold:

(i) *Adjoint of a linear combination of operators is equal to the linear combination of the corresponding adjoint operators with complex conjugate coefficients*

$$(\alpha_1 A_1 + \alpha_2 A_2)^* = \overline{\alpha}_1 A_1^* + \overline{\alpha}_2 A_2^*$$

(ii) *Adjoint of a composition is equal to the composition of the adjoint operators with inverted order*

$$(B \circ A)^* = A^* \circ B^*$$

(iii) *Adjoint of the identity operator equals the operator itself*

$$(id_U)^* = id_U$$

(iv) *If the inverse A^{-1} exists and is continuous then A^* has a continuous inverse too, and*

$$(A^*)^{-1} = \left(A^{-1}\right)^*$$

(v) *Norm of the adjoint equals norm of the operator*

$$\|A\|_{\mathcal{L}(U,V)} = \|A^*\|_{\mathcal{L}(V,U)}$$

(vi) *Adjoint of the adjoint coincides with the original operator*

$$(A^*)^* = A$$

PROOF All properties follow directly from Proposition 5.16.1 and the definition of the adjoint. Note the difference in the first property. Complex conjugates do not appear in the case of the transpose operators. Indeed,

$$
\begin{aligned}
(\alpha_1 A_1 + \alpha_2 A_2)^* &= R_V^{-1} \circ (\alpha_1 A_1 + \alpha_2 A_2)' \circ R_U \\
&= R_V^{-1} \circ (\alpha_1 A_1' + \alpha_2 A_2') \circ R_U \\
&= R_V^{-1} \circ (\alpha_1 A_1' \circ R_U + \alpha_2 A_2' \circ R_U) \\
&= \overline{\alpha}_1 R_V^{-1} \circ A_1' \circ R_U + \overline{\alpha}_2 R_V^{-1} \circ A_2' \circ R_U \\
&= \overline{\alpha}_1 A_1^* + \overline{\alpha}_2 A_2^*
\end{aligned}
$$

∎

Adjoint of an Operator Defined on a Proper Subspace. As in the case of Banach spaces, the definition of the adjoint operator A^* is more delicate when operator A is defined only on a subspace $D(A)$ of a Hilbert space U. Assuming additionally that $D(A)$ is *dense* in U, we define the adjoint operator A^* again as the composition:

$$A^* = R_U^{-1} \circ A' \circ R_V$$

or, equivalently,

$$(A\boldsymbol{u}, \boldsymbol{v})_V = (\boldsymbol{u}, A^*\boldsymbol{v})_U \qquad \forall\, \boldsymbol{u} \in D(A),\ \boldsymbol{v} \in D(A^*)$$

where

$$D(A^*) = R_U^{-1}\Big(D(A')\Big)$$

can be equivalently characterized as the collection of *all* vectors $\boldsymbol{v}$ for which the above equality holds. It is important to remember the two conditions present in this definition:

1. Domain $D(A)$ must be dense in U (its choice up to a certain extent is up to us when defining the operator).
2. Domain $D(A^*)$ is precisely specified by the definition. This in particular implies that calculating the adjoint operator involves a precise determination of its domain.

Notice finally that the adjoint operators as the compositions of continuous Riesz maps and *closed* transpose operator (recall Section 5.18) are always closed.

Obviously, for continuous operators defined on the whole space U, both notions of the adjoint operator are the same.

Reinterpreting again Proposition 5.18.1, we get

PROPOSITION 6.5.2

Let U, V and W be inner product spaces.

(i) *Let $A_i : U \supset D \to V$, $i = 1, 2$ be two linear operators defined on the same domain D, dense in U. Then*

$$(\alpha_1 A_1 + \alpha_2 A_2)^* = \overline{\alpha}_1 A_1^* + \overline{\alpha}_2 A_2^*$$

(ii) *Let $A: U \supset D(A) \to V$, $B: V \supset D(B) \to W$ be two linear operators with domains dense in U and V, respectively, and let $\mathcal{R}(A) \subset D(B)$. Then*

$$(B \circ A)^* \supset A^* \circ B^*$$

i.e., the adjoint $(B \circ A)^*$ exists and it is an extension of the composition $A^* \circ B^*$.

(iii) *If $A: U \supset D(A) \to V$ is a linear, injective operator with domains $D(A)$ and range $\mathcal{R}(A)$ dense in U and V, respectively, then the adjoint operator A^* has an inverse and*

$$(A^*)^{-1} = \left(A^{-1}\right)^*$$

All theorems involving the transpose operators on Banach spaces, proven in Sections 5.17 and 5.18, may be directly reinterpreted in the context of Hilbert spaces with scalar products replacing duality pairings and the adjoint operators replacing the transpose operators. Reinterpreting, for instance, Corollary 5.18.2, we get

THEOREM 6.5.1
(Solvability of Linear Equations on Hilbert Spaces)

Let U and V be Hilbert spaces and let

$$A: U \supset D(A) \longrightarrow V, \ \overline{D(A)} = U, \ \overline{\mathcal{R}(A)} = \mathcal{R}(A)$$

be a linear, closed operator with the domain $D(A)$ dense in U and range $\mathcal{R}(A)$ closed in V. Then the linear problem

$$Au = f \qquad f \in V$$

has a solution u if and only if

$$f \in \mathcal{N}(A^*)^\perp$$

where A^* is the adjoint of A. The solution u is determined uniquely up to elements from the null space $\mathcal{N}(A)$.

REMARK 6.5.1. Recall (Theorem 5.18.2) that closedness of the range $\mathcal{R}(A)$ in V is *equivalent* to the condition

$$\|Au\|_V \geq c \inf_{w \in \mathcal{N}(A)} \|u + w\|_U \quad \forall\, u \in D(A),\ c > 0$$

This in particular implies that if u is a solution corresponding to f, then

$$\inf_{w \in \mathcal{N}(A)} \|u + w\|_U \leq \frac{1}{c} \|f\|_V$$

∎

Symmetric and Self-Adjoint Operators. An operator A defined on a dense subspace $D(A)$ of a Hilbert space U into itself is said to be *symmetric*, if $A \subset A^*$, i.e.,

$$D(A) \subset D(A^*) \qquad \text{and} \qquad A^*|_{D(A)} = A$$

If, additionally, the domains of both operators are the same, i.e., $A = A^*$, then we say that operator A is *self-adjoint.*

Obviously, every self-adjoint operator is symmetric, but not conversely. There are numerous examples of symmetric operators which are *not* self-adjoint. In the case of a continuous and symmetric operator A defined on the whole space U, however, the adjoint A^* is defined on the whole U as well, and therefore, A is automatically self-adjoint.

Note finally that, since adjoint operators are closed, every self-adjoint operator A is necessarily closed.

Example 6.5.3

Integral operator A discussed in Example 6.5.1 is self-adjoint iff

$$K(x,\xi) = \overline{K(\xi,x)}$$

□

Example 6.5.4

Orthogonal projections in a Hilbert space are self-adjoint. Indeed, let M be a closed subspace of a Hilbert space V and P the corresponding orthogonal projection on M, i.e., if

$$\boldsymbol{u} = \boldsymbol{u}_1 + \boldsymbol{u}_2 \qquad \text{where} \qquad \boldsymbol{u}_1 \in M, \boldsymbol{u}_2 \in M^\perp$$

then $P\boldsymbol{u} = \boldsymbol{u}_1$. Similarly, $P\boldsymbol{v} = \boldsymbol{v}_1$, for $\boldsymbol{v} = \boldsymbol{v}_1 + \boldsymbol{v}_2$, $\boldsymbol{v}_1 \in M$, $\boldsymbol{v}_2 \in M^\perp$. We have

$$\begin{aligned}(P\boldsymbol{u},\boldsymbol{v}) &= (\boldsymbol{u}_1,\boldsymbol{v}) = (\boldsymbol{u}_1,\boldsymbol{v}_1,+\boldsymbol{v}_2)\\ &= (\boldsymbol{u}_1,\boldsymbol{v}_1) = (\boldsymbol{u}_1+\boldsymbol{u}_2,\boldsymbol{v}_1)\\ &= (\boldsymbol{u}_,\boldsymbol{v}_1) = (\boldsymbol{u},P\boldsymbol{v})\end{aligned}$$

for every $\boldsymbol{u},\boldsymbol{v} \in V$.

□

Example 6.5.5

Let $V = L^2(0,1)$ and consider the operator A defined as

$$D(A) = C_0^\infty(0,1), \qquad Au = -\frac{d^2u}{dx^2} = -u''$$

It can be proved that the space of test functions $C_0^\infty(0,1)$ is dense in $L^2(0,1)$ and therefore it makes sense to speak about the adjoint of A. It follows from the definition of the adjoint operator that

$$\int_0^1 (-u'')v dx = \int_0^1 uA^*v dx \qquad \forall u \in C_0^\infty(0,1)$$

which implies that $A^*v = -v''$ in the distributional sense. As the value of A^*v must be in L^2, this implies that for $v \in D(A^*)$, $v'' \in L^2(0,1)$.

It can be proved next that v is a C^1 function. In particular, we can integrate the first integral by parts, which yields

$$\int_0^1 -u''v\,dx = \int_0^1 u'v'dx - u'v|_0^1$$

$$= -\int_0^1 uv''dx - u'v|_0^1$$

Consequently, the domain of operator A^* is identified as

$$D(A^*) = \{v \in L^2(0,1) : v'' \in L^2(0,1), v(0) = v(1) = 0\}$$

As we can see, $D(A) \subset D(A^*)$, but at the same time, $D(A) \neq D(A^*)$ and, therefore, the operator A is symmetric, but *not* self-adjoint. □

For other examples of self-adjoint operators, we refer the reader to the next chapter.

Normal Operators. A continuous linear operator A, defined on a Hilbert space V into itself, is said to be *normal* if it commutes wih its adjoint, i.e.,

$$AA^* = A^*A$$

It follows from the definition that all self-adjoint operators are normal. Also, every unitary and surjective operator is normal as well (see Example 6.5.2).

Example 6.5.6

Let M be a closed vector subspace of a Hilbert space V, and let P be the corresponding orthogonal projection. Operator $A = \lambda P$, where λ is a complex number, is self-adjoint iff λ is real, because

$$A^* = (\lambda P)^* = \overline{\lambda} P^* = \overline{\lambda} P$$

But

$$A^*A = AA^* = \lambda\overline{\lambda} P = |\lambda|^2 P$$

and therefore A is normal for all complex λ. □

The following proposition provides an important characterization of normal operators.

PROPOSITION 6.5.3

Let A be a bounded, linear operator on a Hilbert space V. Then A is normal if and only if

$$\|A\boldsymbol{u}\| = \|A^*\boldsymbol{u}\| \qquad \forall\, \boldsymbol{u} \in V$$

PROOF Assume that A is normal. Then

$$||A\boldsymbol{u}||^2 = (A\boldsymbol{u}, A\boldsymbol{u}) = (\boldsymbol{u}, A^*A\boldsymbol{u}) = (\boldsymbol{u}, AA^*\boldsymbol{u})$$

$$= (A^*\boldsymbol{u}, A^*\boldsymbol{u}) = ||A^*\boldsymbol{u}||^2$$

Conversely, assume that $||A\boldsymbol{u}|| = ||A^*\boldsymbol{u}||$, for every $\boldsymbol{u} \in V$. By a direct calculation, we easily prove that (recall Exercise 6.1.2)

$$(A\boldsymbol{u}, \boldsymbol{v}) = \ \tfrac{1}{4}\,[(A(\boldsymbol{u}+\boldsymbol{v}), \boldsymbol{u}+\boldsymbol{v}) - (A(\boldsymbol{u}-\boldsymbol{v}), \boldsymbol{u}-\boldsymbol{v})$$

$$+i(A(\boldsymbol{u}+i\boldsymbol{v}), u+i\boldsymbol{v}) - i(A(\boldsymbol{u}-i\boldsymbol{v}), \boldsymbol{u}-i\boldsymbol{v})]$$

and, consequently,

$$(A\boldsymbol{u}, \boldsymbol{u}) = 0 \qquad \forall\, \boldsymbol{u} \in V \qquad \text{implies} \qquad (A\boldsymbol{u}, \boldsymbol{v}) = 0 \qquad \forall\, \boldsymbol{u}, \boldsymbol{v} \in V$$

which in turn implies that $A \equiv 0$. But

$$\|A\boldsymbol{u}\|^2 = (A\boldsymbol{u}, A\boldsymbol{u}) = (A^*A\boldsymbol{u}, \boldsymbol{u})$$

and

$$\|A^*\boldsymbol{u}\|^2 = (A^*\boldsymbol{u}, A^*\boldsymbol{u}) = (AA^*\boldsymbol{u}, \boldsymbol{u})$$

imply that

$$((A^*A - AA^*)\boldsymbol{u}, \boldsymbol{u}) = 0 \qquad \forall\, \boldsymbol{u} \in V$$

and by the corollary above, $A^*A - AA^* = 0$. ∎

COROLLARY 6.5.1

Let A be a normal operator on a Hilbert space V. Then

$$\|A^n\| = \|A\|^n$$

PROOF First of all, notice that

$$\|A^n\boldsymbol{u}\| = \left\|AA^{n-1}\boldsymbol{u}\right\| \leq \|A\| \ \left\|A^{n-1}\boldsymbol{u}\right\|$$

$$\leq \ldots \leq \|A\|^n \ \|\boldsymbol{u}\|$$

and therefore, always, $\|A^n\| \leq \|A\|^n$.
We show now that for normal operators the inverse inequality holds as well. We start with the observation that, always,

$$\|A^*A\| = \|A\|^2$$

Indeed,

$$\|A\|^2 = \sup_{\|\boldsymbol{u}\|\leq 1} (A\boldsymbol{u}, A\boldsymbol{u}) = \sup_{\|\boldsymbol{u}\|\leq 1} (A^*A\boldsymbol{u}, \boldsymbol{u}) \leq \sup_{\|\boldsymbol{u}\|\leq 1} (\|A^*A\boldsymbol{u}\| \ \|\boldsymbol{u}\|) \leq \|A^*A\|$$

The inverse inequality follows now from the fact that $\|A^*\| = \|A\|$.

Step 1. A symmetric (self-adjoint), $n = 2^k$. By the result above, we have $\|A^2\| = \|A\|^2$. By induction in k, $\|A^n\| = \|A\|^n$ for $n = 2^k$.

Step 2. A normal, $n = 2^k$. Substituting A^k for A, we have

$$\|(A^k)^*A^k\| = \|A^k\|^2$$

From the commutativity of A and A^*, it follows that arbitrary powers of A and A^* commute, and, therefore,

$$(A^*)^kA^k = (A^*A)^k$$

But A^*A is symmetric, and, by the Step 1 result,

$$\|(A^*A)^k\| = \|A^*A\|^k = \|A\|^{2k}$$

from which the assertion follows.

Step 3. A normal, n arbitrary. One can always find m such that $2^m \leq n < 2^{m+1}$. Denoting $r = 2^{m+1} - n$, we have

$$\|A\|^{n+r} = \|A\|^{2^{m+1}} = \|A^{2^{m+1}}\| = \|A^{n+r}\| \le \|A^n\| \, \|A^r\|$$

and, consequently,

$$\|A\|^n \le \|A^n\|$$

∎

Exercises

6.5.1. Let A be an operator defined on a dense subspace $D(A)$ of a Hilbert space U into a Hilbert space V. Prove that the adjoint operator A^* is closed.

6.5.2. Prove that the composition BA of two self-adjoint continuous operators A and B is self-adjoint iff A and B commute, i.e., $AB = BA$.

6.5.3. Prove that for a self-adjoint operator A, $(A\boldsymbol{u}, \boldsymbol{u})$ is always a real number.

6.5.4. Prove that for a self-adjoint, continuous operator A

$$\|A\| = \sup_{\|\boldsymbol{u}\| \le 1} |(A\boldsymbol{u}, \boldsymbol{u})| = \sup_{\|\boldsymbol{u}\| \le 1} |(\boldsymbol{u}, A\boldsymbol{u})|$$

Hint: Make use of the formula for $(A\boldsymbol{u}, \boldsymbol{v})$ used in the proof of Proposition 6.5.3 and Exercise 6.1.2.

6.5.5. Assuming that for a continuous, self-adjoint operator A, the inverse $(A + iI)^{-1}$ exists and is continuous (see section 6.9), prove that operator

$$Q = (A - iI)(A + iI)^{-1}$$

is unitary.

6.5.6. Let $A : \mathbb{C}^2 \longrightarrow \mathbb{C}^2$ be given by

$$\begin{pmatrix} y_1 \\ y_2 \end{pmatrix} = \begin{pmatrix} a & b \\ c & d \end{pmatrix} \begin{pmatrix} x_1 \\ x_2 \end{pmatrix}$$

What conditions must the complex numbers a, b, c, and d satisfy in order that A be (a) self-adjoint, (b) normal, and (c) unitary.

6.5.7. Prove the Cartesian decomposition theorem: Every linear and continuous operator A on a Hilbert space V can be represented in the form

$$A = B + iC$$

where B and C are self-adjoint. *Hint:* Define

$$B = \frac{1}{2}(A + A^*) \text{ and } C = \frac{1}{2i}(A - A^*)$$

6.5.8. Prove that if A is bijective and normal, then so is A^{-1}.

6.5.9. Determine the adjoints of the following operators in $L^2(I)$, where $I = (0,1)$.

(a) $Au = \frac{du}{dx}$, $D(A) = \{u \in L^2(I) \cap C^1(\overline{I}) : u(0) = 0\}$

(b) $Au = \frac{d^2u}{dx^2} - u$, $D(A) = \{u \in L^2(I) \cap C^2(\overline{I}) : u(0) = u(1) = 0\}$

(c) $Au = -\frac{d}{dx}(x^2\frac{du}{dx}) + x\frac{du}{dx}$
$D(A) = \{u \in L^2(I) \cap C^2(\overline{I}) : u(0) = u(1) = 0\}$

6.6 Variational Boundary-Value Problems

We begin with a simple example of the classical formulation of a boundary problem for the Laplace operator (Example 2.2.3). Given an open set $\Omega \subset \mathbb{R}^2$, we look for a function $u(\boldsymbol{x})$ such that

$$\begin{aligned} -\Delta u = f & \quad \text{in } \Omega \\ u = u_0 & \quad \text{on } \Gamma_u \\ \frac{\partial u}{\partial n} = g & \quad \text{on } \Gamma_t \end{aligned}$$

where Γ_u and Γ_t are two disjoint parts of the boundary Γ. A mathematical formulation of the problem must include a precise specification of the regularity of the solution. Usually, minimum regularity assumptions are desired, admitting the largest possible class of solutions accommodating thus for possible irregular data to the problem, in our case: the domain Ω, boundary $\Gamma = \Gamma_u \cup \Gamma_t$, functions f, u_0, and g specified in Ω and on the two parts of the boundary, respectively. Classical regularity assumptions for the problem above would consist in looking for a solution u in a subspace of $C^2(\Omega)$ consisting of those functions for which the boundary conditions make sense, e.g., the space

$$C^2(\Omega) \cap C^1(\overline{\Omega})$$

It is therefore anticipated that the solution will have second-order derivatives (in the classical sense) continuous in Ω and first-order derivatives and function values continuous on the whole $\overline{\Omega}$, including the boundary.

Classical paradoxes with less regular data to the problem (concentrated on impulse forces in mechanics, resulting in the nonexistence of classical solutions) several decades ago led to the notion of weak or variational solutions.

Variational Formulation. Multiplying the differential equation by a sufficiently regular function v and integrating over the domain Ω, we get

$$-\int_\Omega \Delta u v \, dx = \int_\Omega f v dx$$

Integrating the first integral by parts, we get

$$\int_\Omega \boldsymbol{\nabla} u \boldsymbol{\nabla} v dx - \int_\Gamma \frac{\partial u}{\partial n} v \, dx = \int_\Omega f v \, dx$$

where $\frac{\partial u}{\partial n}$ is the *normal derivative of* u

$$\frac{\partial u}{\partial n} = \sum_{i=1}^{n} \frac{\partial u}{\partial x_i} n_i$$

with n_i the components of the outward normal unit vector $\boldsymbol{n}$.

Substituting g for $\frac{\partial u}{\partial n}$ on Γ_t-boundary and eliminating the unknown normal derivative of u on Γ_u-boundary by restricting ourselves to functions v vanishing on Γ_u, we arrive at the formulation

$$\begin{cases} \text{Find } u(\boldsymbol{x}) \text{ such that} \\ u = u_0 \text{ on } \Gamma_u \text{ and} \\ \displaystyle\int_\Omega \boldsymbol{\nabla} u \boldsymbol{\nabla} v \, dx = \int_\Omega f v \, dx + \int_{\Gamma_t} g v \, ds \\ \text{for every } v = v(\boldsymbol{x}) \text{ such that } v = 0 \text{ on } \Gamma_u \end{cases}$$

The problem above is called the *variational formulation* of the boundary-value problem considered, or *variational boundary-value problem.* Functions $v = v(x)$ are called the *test functions.*

It is easily seen that the regularity assumptions to make sense for the variational formulation are much less demanding than in the classical case. Second-order derivatives of the solution need not exist, and the first-order derivatives can be understood in the distributional sense.

The two formulations are equivalent in the sense that they yield the same solution u in the case when u is sufficiently regular. We have shown so far that every classical solution is a variational solution. It remains to examine when the converse holds, i.e., the variational solution turns out to be the classical one as well.

Toward this goal, we integrate the integral on the left-hand side by parts (it is at this point that we use the assumption that $u \in C^2(\Omega) \times C^1(\overline{\Omega})$), arriving at the identity

$$-\int_\Omega \Delta uv \, dx + \int_{\Gamma_t} \frac{\partial u}{\partial n} v \, ds = \int_\Omega fv \, dx + \int_{\Gamma_t} gv \, ds$$

for every test function vanishing on Γ_u, or, equivalently,

$$\int_\Omega (-\Delta u - f) v dx + \int_{\Gamma_t} (\frac{\partial u}{\partial u} - g) v ds = 0 \qquad \forall\, v, v = 0 \text{ on } \Gamma_u$$

This, in particular, implies that

$$\int_\Omega (-\Delta u - f) v dx = 0$$

for every test function vanishing on the *entire boundary*. Now, if the set (space) of such functions is *dense* in $L^2(\Omega)$, then

$$-\Delta u - f = 0$$

Consequently, the first integral vanishes for *any* test function, this time not necessarily vanishing on the whole boundary, and we arrive at the condition

$$\int_{\Gamma_t} (\frac{\partial u}{\partial n} - g) v \, ds \qquad \forall\, v$$

If, again, the test functions are dense, this time in the space $L^2(\Gamma_t)$, then we conclude that

$$\frac{\partial u}{\partial u} - g = 0$$

We say sometimes that we have recovered both the differential equation and the second (Neumann) boundary condition. Thus, for regular solutions u, the two formulations are equivalent to each other.

Abstract Variational Boundary-Value Problems. A precise formulation of the variational boundary-value problem involves a careful specification of regularity assumptions for both the solution u and the test function v. This usually leads to the selection of function spaces X and Y with a Banach or Hilbert space structure containing the solution u and test functions v. The essential (Dirichlet) boundary conditions on u and v lead next to the introduction of

the set of (kinematically) admissible solutions

$$K = \{u \in X : u = u_0 \text{ on } \Gamma_u\}$$

and

the space of (kinematically) admissible test functions

$$V = \{v \in Y : v = 0 \text{ on } \Gamma_u\}$$

Notice that V is a vector *subspace* of Y, while K is only a *subset* of K, unless $u_0 \equiv 0$ (Example 2.2.3). More precisely, if an element $\hat{u}_0$ from X exists which reduces to u_0 on the boundary Γ_u, then K can be identified as an *affine subspace* or *linear manifold* of X, i.e.,

$$K = \hat{u}_0 + U, \text{ where } U = \{u \in X : u = 0 \text{ on } \Gamma_u\}$$

Finally, the left-hand side of the variational equation in our example is easily identified as a *bilinear form* of solution u and test function v and the right-hand side as a linear form (functional) of test function V.

Introducing symbols

$$B(u, v) = \int_\Omega \nabla u \nabla v \, dx$$

$$L(v) = \int_\Omega f v \, dx + \int_{\Gamma_t} g v \, ds$$

we are prompt to consider an *abstract variational boundary-value problem* in the form

$$\begin{cases} \text{Find } \boldsymbol{u} \in K = \hat{\boldsymbol{u}}_0 + U \text{ such that} \\ B(\boldsymbol{u}, \boldsymbol{v}) = L(\boldsymbol{v}) \qquad \forall\, \boldsymbol{v} \in V \end{cases}$$

The essential question here is what conditions can be imposed so that we are guaranteed that a unique solution exists that depends continuously on the linear functional L. This question was originally resolved for the case $U = V$ (the same regularity assumptions for both solution u and test function v) by Lax and Milgram. We shall prove a more general form of their classic theorem.

THEOREM 6.6.1
(The Generalized Lax-Milgram Theorem)

Let X and Y be two Banach spaces with corresponding closed vector subspaces U and V and let $B : X \times Y \longrightarrow \mathbb{R}$ (or $\mathbb{C}$) be a bilinear functional which satisfies the following three properties:

(i) B *is continuous, i.e., there exists a constant* $M > 0$ *such that*

$$|B(\boldsymbol{u}, \boldsymbol{v})| \leq M \|\boldsymbol{u}\|_X \|\boldsymbol{v}\|_Y \qquad \forall\, \boldsymbol{u} \in X, \boldsymbol{v} \in Y$$

(ii) There exists a constant $\gamma > 0$ *such that*

$$\inf_{\substack{\boldsymbol{u} \in U \\ \|\boldsymbol{u}\|=1}} \sup_{\substack{\boldsymbol{v} \in V \\ \|\boldsymbol{v}\| \leq 1}} |B(\boldsymbol{u}, \boldsymbol{v})| \geq \gamma > 0$$

(iii) $\sup_{\boldsymbol{u} \in U} |B(\boldsymbol{u}, \boldsymbol{v})| > 0 \qquad \forall\, \boldsymbol{v} \neq 0, \boldsymbol{v} \in V.$

Then, for every linear and continuous functional L on V, $L \in V'$, and element $\boldsymbol{u}_0 \in X$ there exists a unique solution to the abstract variational problem:

$$\begin{cases} \textit{Find } \boldsymbol{u} \in \boldsymbol{u}_0 + U \textit{ such that} \\ B(\boldsymbol{u}, \boldsymbol{v}) = L(\boldsymbol{v}) \qquad \boldsymbol{v} \in V \end{cases}$$

Moreover, the solution $\boldsymbol{u}$ depends continuously on the data: functional L and element $\boldsymbol{u}_0$; in fact,

$$\|\boldsymbol{u}\|_X \leq \frac{1}{\gamma} \|L\|_{V'} + \left(\frac{M}{\gamma} + 1\right) \|\boldsymbol{u}_0\|_X$$

PROOF

Step 1. $\boldsymbol{u}_0 = \mathbf{0}$.

For each fixed $\boldsymbol{u} \in U$, $B(\boldsymbol{u}, \cdot)$ defines a linear functional $A\boldsymbol{u}$ on V.

$$\langle A\boldsymbol{u}, \boldsymbol{v} \rangle \stackrel{\text{def}}{=} B(\boldsymbol{u}, \boldsymbol{v}) \qquad \forall\, \boldsymbol{v} \in \boldsymbol{V}$$

and this functional is continuous by virtue of property (i)

$$|\langle A\boldsymbol{u}, \boldsymbol{v}\rangle| \leq M\|\boldsymbol{u}\|_U \ \|\boldsymbol{v}\|_V \ = C\|\boldsymbol{v}\|_V \text{ where } C = M\|\boldsymbol{u}\|$$

where the norms on U and V are those from X and Y, respectively.

Linearity of B with respect to the *first* variable implies also that operator $A : U \longrightarrow V'$ prescribing for each $\boldsymbol{u}$ the corresponding linear and continuous functional $A\boldsymbol{u}$ on V is linear and, by property (i) again, is continuous, i.e.,

$$A \in \mathcal{L}(U, V')$$

Consequently, the variational problem can be rewritten in the operator form as

$$\begin{cases} \text{Find } \boldsymbol{u} \in U \text{ such that} \\ A\boldsymbol{u} = L, L \in V' \end{cases}$$

A simple reexamination of condition (ii) implies that

$$\inf_{\substack{\boldsymbol{u} \in U \\ \|\boldsymbol{u}\|=1}} \|A\boldsymbol{u}\|_{V'} \geq \gamma > 0$$

or in the equivalent form (explain why),

$$\|A\boldsymbol{u}\|_{V'} \geq \gamma\|\boldsymbol{u}\|_U$$

which proves that A is bounded below. As both U and V as closed subspaces of Banach spaces X and Y are the Banach spaces themselves too, boundedness below of A implies that the range $\mathcal{R}(A)$ is closed in V' (recall Theorem 5.17.2).

Finally, condition (iii) implies that the orthogonal complement of $\mathcal{R}(A)$ (equal to the null space of the conjugate operator $A' : V' \longrightarrow U'$) reduces to the zero vector $\mathbf{0}$. Indeed, assume that for some $\boldsymbol{v} \neq \mathbf{0}$

$$< A\boldsymbol{u}, \boldsymbol{v} >= 0 \qquad \forall\, \boldsymbol{u} \in \boldsymbol{U}$$

Then

$$\sup_{\boldsymbol{u} \in U} |B(\boldsymbol{u}, \boldsymbol{v})| = \sup_{\boldsymbol{u} \in U} < A\boldsymbol{u}, \boldsymbol{v} >= 0$$

a contradiction with (iii).

Consequently, A is surjective, and from the boundedness below of A, we have

$$\|\boldsymbol{u}\|_U \leq \frac{1}{\gamma}\|A\boldsymbol{u}\|_{V'} = \frac{1}{\gamma}\|L\|_{V'}$$

where $A\boldsymbol{u} = L$.

Step 2. $\boldsymbol{u}_0 \neq \boldsymbol{0}$. Substituting $\boldsymbol{u} = \boldsymbol{u}_0 + \boldsymbol{w}$, where $\boldsymbol{w} \in U$, we reformulate the variational problem into the form

$$\begin{cases} \text{Find } \boldsymbol{w} \in U \text{ such that} \\ B(\boldsymbol{u}_0 + \boldsymbol{w}, \boldsymbol{v}) = L(\boldsymbol{v}) \qquad \forall\, \boldsymbol{v} \in V \end{cases}$$

or, equivalently,

$$\begin{cases} \text{Find } \boldsymbol{w} \in U \text{ such that} \\ B(\boldsymbol{w}, \boldsymbol{v}) = L(\boldsymbol{v}) - B(\boldsymbol{u}_0, \boldsymbol{v}) \qquad \forall\, \boldsymbol{v} \in V \end{cases}$$

Now, the continuity of B (condition (i)) implies that the right-hand side can be identified as a new, linear and continuous functional on V

$$L_1(\boldsymbol{v}) \stackrel{\text{def}}{=} L(\boldsymbol{v}) - B(\boldsymbol{u}_0, \boldsymbol{v})$$

and

$$\|L_1\|_{V'} \leq \|L\|_{V'} + M\|\boldsymbol{u}_0\|_X$$

Applying the results of Step 1, we prove that there exists a unique solution $\boldsymbol{w}$ and

$$\|\boldsymbol{w}\|_X = \|\boldsymbol{w}\|_U \leq \frac{1}{\gamma}\left(\|L\|_{V'} + M\|\boldsymbol{u}_0\|_X\right)$$

Consequently,

$$\|\boldsymbol{u}\|_X = \|\boldsymbol{u}_0 + \boldsymbol{w}\|_X \leq \frac{1}{\gamma}\|L\|_{V'} + \left(\frac{M}{\gamma} + 1\right)\|\boldsymbol{u}_0\|_X$$

∎

COROLLARY 6.6.1

Let X be a Hilbert space with a closed subspace V and let $B : X \times X \longrightarrow \mathbb{R}$ (or $\mathbb{C}$) be a bilinear functional which satisfies the following two properties:

(i) B is continuous, i.e., there exists a constant $M > 0$ such that

$$|B(\boldsymbol{u}, \boldsymbol{v})| \leq M\|\boldsymbol{u}\| \, \|\boldsymbol{v}\|$$

(ii) B is V-coercive (some authors say V-elliptic), i.e., a constant $\alpha > 0$ exists such that

$$|B(\boldsymbol{u}, \boldsymbol{u})| \geq \alpha\|\boldsymbol{u}\|^2 \qquad \forall\, \boldsymbol{u} \in V$$

Then, for every linear and continuous functional L on V, $L \in V'$, and element $\boldsymbol{u}_0 \in X$, there exists a unique solution to the abstract variational problem

$$\begin{cases} \text{Find } \boldsymbol{u} \in \boldsymbol{u}_0 + V \text{ such that} \\ B(\boldsymbol{u}, \boldsymbol{v}) = L(\boldsymbol{v}) \qquad \forall \boldsymbol{v} \in V \end{cases}$$

and the solution $\boldsymbol{u}$ depends continuously on the data; in fact,

$$\|\boldsymbol{u}\| \leq \frac{1}{\alpha}\|L\|_{V'} + \left(\frac{M}{\gamma} + 1\right)\|\boldsymbol{u}_0\|_X$$

PROOF We show that V-coercivity implies both conditions (ii) and (iii) from Theorem 6.6.1. Indeed,

$$\inf_{\|\boldsymbol{u}\|=1} \sup_{\|\boldsymbol{v}\|\leq 1} |B(\boldsymbol{u}, \boldsymbol{v})| \geq \inf_{\|\boldsymbol{u}\|=1} |B(\boldsymbol{u}, \boldsymbol{u})| \geq \alpha$$

so $\gamma = \alpha$ and

$$\sup_{\boldsymbol{u}\in V} |B(\boldsymbol{u}, \boldsymbol{v})| \ \geq B(\boldsymbol{v}, \boldsymbol{v}) \geq \alpha\|\boldsymbol{v}\|^2 > 0$$

Before we can proceed with examples, we need to prove the classical result known as the *(first) Poincaré inequality.* ∎

PROPOSITION 6.6.1

Let Ω be a bounded, open set in $\mathbb{R}^n$. There exists a positive constant $c > 0$ such that

$$\int_\Omega u^2 dx \leq c \int_\Omega (\boldsymbol{\nabla} u)^2 dx \qquad \forall\, u \in H_0^1(\Omega)$$

PROOF

Step 1. Assume that Ω is a cube in $\mathbb{R}^n$, $\Omega = (-a,a)^n$ and that $u \in C_0^\infty(\Omega)$. Since u vanishes on the boundary of Ω, we have

$$u(x_1, \ldots, x_n) = \int_{-a}^{x_n} \frac{\partial u}{\partial x_n}(x_1, ..., t)dt$$

and by the Schwarz inequality,

$$\begin{aligned} u^2(x_1, \ldots, x_n) &\le \int_{-a}^{x_n} \left(\frac{\partial u}{\partial x_n}(x_1, \ldots, t) \right)^2 dt \ (x_n + a) \\ &\le \int_{-a}^{a} \left(\frac{\partial u}{\partial x_n}(x_1, \ldots, x_n) \right)^2 dx_n \ (x_n + a) \end{aligned}$$

Integrating over Ω on both sides, we get

$$\int_\Omega u^2 dx \le \int_\Omega \left(\frac{\partial u}{\partial x_n} \right)^2 dx \cdot 2a^2$$

Step 2. Ω bounded. $\boldsymbol{u} \in C_0^\infty(\Omega)$. Enclosing Ω in a sufficiently large cube $\Omega_1 = (-a,a)^n$ and extending the function $\boldsymbol{u} \in C_0^\infty(\Omega)$ by zero to the whole Ω_1, we apply the Step 1 results, getting

$$\begin{aligned} \int_\Omega u^2 dx = \int_{\Omega_1} u^2 dx &\le 2a^2 \int_{\Omega_1} \left(\frac{\partial u}{\partial x_n} \right)^2 dx \\ &= 2a^2 \int_\Omega \left(\frac{\partial u}{\partial x_n} \right)^2 dx \end{aligned}$$

Step 3. We use the density argument. Let $u \in H_0^1(\Omega)$ and $u_m \in C_0^\infty(\Omega)$ be a sequence converging to u in $H^1(\Omega)$. Then

$$\int_\Omega u_m^2 dx \le 2a^2 \int_\Omega \left(\frac{\partial u_m}{\partial x_n} \right)^2 dx$$

Passing to the limit, we get

$$\int_\Omega u^2 dx \le 2a^2 \int_\Omega \left(\frac{\partial u}{\partial x_n} \right)^2 dx \le 2a^2 \int_\Omega (\boldsymbol{\nabla} u)^2 dx$$

∎

Example 6.6.1

We now apply Theorem 6.6.1 to establish the uniqueness and continuous dependence upon data results of the variational boundary-value problem for the Laplace operator discussed in the beginning of this section. We select for the space X the (real) Sobolev space $H^1(\Omega)$ and proceed with the verification of the assumptions of the Lax-Milgram theorem.

Step 1. Continuity of the bilinear form follows easily from the Schwarz inequality

$$\begin{aligned} |B(u,v)| &= \left|\int_\Omega \nabla u \nabla v \, dx\right| \\ &\leq \left(\int_\Omega (\nabla u)^2 \, dx\right)^{\frac{1}{2}} \left(\int_\Omega (\nabla v)^2 \, dx\right)^{\frac{1}{2}} \\ &= |u|_1 |v|_1 \leq \|u\|_1 \|v\|_1 \end{aligned}$$

where $|u|_1$ and $\|u\|_1$ denote the first order Sobolev seminorm and norm respectively, i.e.,

$$\begin{aligned} |u|_1^2 &= \int_\Omega (\nabla u)^2 \, dx = \int_\Omega \sum_{i=1}^2 \left(\frac{\partial u}{\partial x_i}\right)^2 \, dx \\ \|u\|_1^2 &= \int_\Omega \left(u^2 + (\nabla u)^2\right) \, dx = \|u\|_0^2 + |u|_1^2 \end{aligned}$$

with $\|u\|_0$ denoting the L^2 -norm.

Step 2. Continuity of the linear functional L follows from *Lions' trace theorem*; see [8,12]. It can be proved that there exists a linear and continuous operator γ, called the *trace operator*, from the space $H^1(\Omega)$ *onto a boundary fractional Sobolev space* $H^{\frac{1}{2}}(\partial\Omega)$ *continuously embedded and dense in* $L^2(\partial\Omega)$ such that for regular functions u, values of γu coincide with the restriction of u to the boundary Γ, i.e.,

$$\gamma : H^1(\Omega) \longrightarrow H^{\frac{1}{2}}(\Gamma), \qquad u \longrightarrow \gamma u$$

$$\|\gamma u\|_{H^{\frac{1}{2}}(\Gamma)} \leq c \|u\|_{H^1(\Omega)} \qquad \forall\, u \in H^1(\Omega), c > 0$$

$$\gamma u = u|_{\partial\Omega} \qquad \forall\, u \in C(\overline{\Omega}) \cap H^1(\Omega)$$

At the same time,

$$\|u\|_{L^2(\Gamma)} \leq C\|u\|_{H^{\frac{1}{2}}(\Gamma)} \qquad \forall\, u \in H^{\frac{1}{2}}(\Gamma), C > 0$$

so, in particular,

$$\|\gamma u\|_{L^2(\Gamma)} \leq C\|u\|_{H^1(\Omega)} \qquad \forall\, u \in H^1(\Omega), C > 0$$

As we can see, a "simple" verification of the assumptions of the Lax-Milgram theorem can get fairly technical. Assuming now regularity assumptions for data f and g as

$$f \in L^2(\Omega), g \in L^2(\Gamma_t)$$

we interpret the linear functional L precisely as follows:

$$Lv = \int_\Omega fv\, dx + \int_\Gamma g\gamma v\, ds$$

where function g has been extended by zero to the whole boundary Γ. It follows now from the Schwarz inequality and the trace theorem that L is continuous.

$$\begin{aligned}|Lv| &\leq |\int_\Omega fvdx| + |\int_\Gamma g\gamma v\, ds| \\ &\leq \|f\|_0\|v\|_0 + C\|g\|_{L^2(\Gamma_t)}\|v\|_1 \\ &\leq \left(\|f\|_0 + C\|g\|_{L^2(\Gamma_t)}\right)\|v\|_1\end{aligned}$$

Step 3. We identify V as the (sub)space of all kinematically admissible functions satisfying the homogeneous kinematic (Dirichlet) boundary conditons on Γ_u.

$$V = \left\{v \in H^1(\Omega) \;:\; \gamma v = 0 \text{ on } \Gamma_u\right\}$$

Assuming that meas $(\Gamma_u) > 0$, it follows now from the continuity of the trace operator γ and the restriction operator

$$L^2(\Gamma) \ni u \longrightarrow u|_{\Gamma_u} \in L^2(\Gamma_u)$$

that V is a *closed* subspace of $H^1(\Omega)$.

Step 4. V-coercivity of the bilinear functional B follows from another very nontrivial result for the Sobolev spaces, the (Rellich) compact imbedding theorem, see [1], which holds under the assumption that meas$(\Gamma_u) > 0$.

In the case when Γ_u coincides with the whole boundary Γ, a simpler argument, based on the Poincaré inequality, can be used. Indeed, by Proposition 6.6.1, we have

$$|u|_1^2 \geq \varepsilon \|u\|_0^2 \qquad \forall\, u \in H_0^1(\Gamma)$$

and, consequently,

$$\begin{aligned}\int_\Omega (\nabla u)^2 dx &= \tfrac{1}{2}|u|_1^2 + \tfrac{1}{2}|u|_1^2 \\ &\geq \tfrac{\varepsilon}{2}\|u\|_0^2 + \tfrac{1}{2}|u|_1^2 \\ &\geq \min\left(\tfrac{\varepsilon}{2}, \tfrac{1}{2}\right) \|u\|_1^2\end{aligned}$$

for every $u \in V = H_0^1(\Omega)$.

Step 5. Postulating finally that function u_0 can be extended to a function $\hat{u}_0$ defined on the whole Ω such that $\hat{u}_0 \in H^1(\Omega)$, we conclude, by the Lax-Milgram theorem, that there exists a unique solution u to the problem

$$\begin{cases}\text{Find } u \in \hat{u}_0 + V \text{ such that} \\ \displaystyle\int_\Omega \nabla u \nabla v dx = \int_\Omega f v dx + \int_{\Gamma_t} g v ds \qquad \forall\, v \in V\end{cases}$$

where, for simplicity, the symbol of trace operator has been omitted.

Solution u depends continuously on the data. There exists positive constants C_1, C_2, C_3, such that

$$\|u\|_{H^1(\Omega)} \leq C_1\|f\|_{L^2(\Omega)} + C_2\|g\|_{L^2(\Gamma_t)} + C_3\|\widehat{u_0}\|_{H^1(\Omega)}$$

□

REMARK 6.6.1.

1. Regularity assumptions on functions f and g are by no means unique! The only condition is that whatever we assume of f and g, it must imply that the corresponding functional L is continuous. In the case of $\Omega \subset \mathbb{R}^2$, it follows, for instance, from the Sobolev imbedding theorems (see [1]) that one can assume that $f \in L^p(\Omega)$ with any $p > 1$.

2. Existence and continuity of the trace operator γ from $H^1(\Omega)$ *into* $L^2(\Gamma)$ (then, it is *not* surjective) can be proved directly, skipping the technical considerations of fractional Sobolev spaces on the boundary Γ (see [12]).

∎

Example 6.6.2

In the case of $\Gamma_u = \emptyset$, i.e., the pure Neumann problem, a solution cannot be unique, as adding an arbitrary constant c to any solution u produces another solution as well. Application of the Lax-Milgram theorem, which implies uniqueness of the solution, requires more caution, and relies on the concept of quotient spaces.

Step 1. We identify the space $V = X$ as the quotient space $H^1(\Omega)/V_0$, where V_0 is the subspace consisting of all constant modes (infinitesimal rigid body motions for the membrane problem). As V_0 is isomorphic with $\mathbb{R}$, we frequently write $H^1(\Omega)/\mathbb{R}$.

As a finite-dimensional subspace, V_0 is closed and therefore, by the results in Sections 5.17 and 5.18, the quotient space $H^1(\Omega)/V_0$ is a Banach space with the norm

$$\|[u]\|_{H^1(\Omega)/V_0} \stackrel{\text{def}}{=} \inf_{c\in\mathbb{R}} \|u + c\|_{H^1(\Omega)}$$

The infimum on the right-hand side is, in fact, attained and by a direct differentiation with respect to c of the function $\|u + c\|^2$ (of one variable c) we find out that

$$c = -\frac{\int_\Omega u\, dx}{\int_\Omega dx}$$

Thus, the norm of the equivalence class of u coincides with the H^1 norm of the representant with a zero mean value.

$$\|[u]\|_{H^1(\Omega)/V_0} = \|u\|_H^1 \text{ where } u \in [u], \int_\Omega u\, dx = 0$$

By the direct verification of the *parallelogram law* or *polarization formula* (cf. Exercises 6.1.2 and 6.6.1) we may check that every quotient space which has been obtained from a *Hilbert* space, is in fact a Hilbert space itself.

Step 2. We define the bilinear form on the quotient space as

$$B([u],[v]) = \int_\Omega \nabla u \nabla v dx, u \in [u], v \in [v]$$

As the right-hand side is independent of the representants u, v, the bilinear form is well-defined. It is also continuous, as follows from taking the infimum with respect to u and v on the right-hand side of the inequality

$$\begin{aligned} \int_\Omega \nabla u \nabla v dx &\leq \|u\|_1 \|v\|_1 \quad \text{implies} \\ |B([u],[v])| &\leq \inf_{u\in[u]} \|u\|_1 \inf_{v\in[v]} \|v\|_1 \\ &= \|u\|_V \|v\|_V \end{aligned}$$

where $V = H^1(\Omega)/V_0$.

Step 3. It follows from the Schwarz inequality that

$$|\int_\Omega u\ dx| \leq \left(\int_\Omega u^2\ dx\right)^{\frac{1}{2}} \left(\int_\Omega dx\right)^{\frac{1}{2}}$$

and, consequently,

$$\begin{aligned} \int_\Omega (\nabla u)^2 dx + (\int_\Omega u\ dx)^2 &\leq \int_\Omega (\nabla u)^2 dx + meas(\Omega) \int_\Omega u^2\ dx \\ &\leq \max(1, meas(\Omega))\ \|u\|^2_{H^1(\Omega)} \end{aligned}$$

for every $u \in H^1(\Omega)$.

For a class of domains Ω (satisfying the so-called segment property) it follows from the Sobolev imbedding theorems that the inverse inequality (sometimes called the *second Poincaré inequality*) holds, i.e., there is a positive number $C > 0$ such that

$$\|u\|^2_{H^1(\Omega)} \leq C \left(\int_\Omega (\nabla u)^2 dx + (\int_\Omega u\ dx)^2 \right)$$

for every $u \in H^1(\Omega)$.

This inequality implies immediately that the bilinear form B is coercive on the quotient space. Indeed, we have

$$\begin{aligned} B([u],[u]) = \int_\Omega (\nabla u)^2 dx &\geq \frac{1}{c} \|u\|^2_{H^1(\Omega)} \\ &= \tfrac{1}{c} \|[u]\|^2_{H^1(\Omega)/V} \end{aligned}$$

provided $u \in [u]$, $\int_\Omega u\ dx = 0$.

The transformation T mapping the closed subspace of functions from $H^1(\Omega)$ with zero average onto the quotient space V is identified as an isomorphism of Banach spaces, and can be used to introduce a scalar product in the quotient space

$$([u],[v])_V \stackrel{\text{def}}{=} (u,v)_{H^1(\Omega)}$$

where $u \in [u], v \in [v]$ and $\int_\Omega u\, dx = \int_\Omega v\, dx = 0$.

Step 4. Continuity of linear functional L. Introducing the linear functional L on the quotient space V as

$$L([v]) = \int_\Omega fv\, dx + \int_\Gamma g\gamma v ds$$

where $v \in [v]$ and γ is the trace operator, we first of all see that, to be well defined, i.e., independent of a particular representant $v \in [v]$, the right-hand side must vanish for $v =$ const. This is equivalent to

$$\int_\Omega f\, dx + \int_\Gamma g\, ds = 0$$

(recall the examples in Section 5.19). With this condition satisfied, functional L is well defined. As in the previous example, we have

$$|L([v])| \le \left(\|f\|_{L^2(\Omega)} + c\|g\|_{L^2(\Gamma)}\right) \|v\|_{H^1(\Omega)}$$

which, upon taking the infimum with respect to $v \in [v]$ on the right-hand side, implies that

$$|L([v])| \le \left(\|f\|_{L^2(\Gamma)}\right) \|[v]\|_V$$

Step 5. Concluding, all assumptions of the Lax-Milgram theorem are satisfied, and therefore there exists a unique solution in the quotient space V to the problem

$$\begin{cases} \text{Find } [u] \in H^1(\Omega)/V_0 \text{ such that} \\ \int_\Omega \boldsymbol{\nabla} u \boldsymbol{\nabla} v\, dx = \int_\Omega fv\, dx + \int_\Gamma gv\, ds \qquad \forall\, v \in H^1(\Omega) \end{cases}$$

where $u \in [u]$.

The continuous dependence upon data (functional L, $\boldsymbol{u}_0 = \boldsymbol{0}$) is interpreted as

$$\|u\|_{H^1(\Omega)} \le C_1\|f\|_{L^2(\Omega)} + C_2\|g\|_{L^2(\Gamma)}$$

where $u \in [u]$ has a zero average:

$$\int_{\Omega} u \, dx = 0$$

□

Example 6.6.3
(The Principle of Virtual Work in Linear Elasticity)

Recall the formulation of the classical boundary-value problem in linear elasticity, considered in Example 5.19. Given a domain $\Omega \subset \mathbb{R}^n (n = 2, 3)$, with boundary Γ consisting of two disjoint parts Γ_u and Γ_t, we are looking for a displacement field $\boldsymbol{u} = \boldsymbol{u}(\boldsymbol{x}), \boldsymbol{x} \in \Omega$, satisfying

- *equilibrium equations:*

  $$-\sigma_{ij,j} = f_i \text{ in } \Omega$$

  where the stress tensor satisfies the constitutive equations

  $$\sigma_{ij} = E_{ijkl} \epsilon_{kl}(\boldsymbol{u})$$

  with elasticities E_{ijkl} satisfying the customary symmetry assumptions and the strain tensor $\epsilon_{kl}(\boldsymbol{u})$ defined as the symmetric part of derivatives $u_{k,l}$

  $$\epsilon_{kl}(\boldsymbol{u}) = \frac{1}{2}(u_{k,l} + u_{l,k})$$

- *kinematic boundary conditions:*

  $$u_i = \hat{u}_i \text{ on } \Gamma_u$$

- *traction boundary conditions:*

  $$t_i = q_i \text{ on } \Gamma_t$$

  where $\boldsymbol{t} = (t_i)$ is the stress vector defined as

  $$t_i = \sigma_{ij} n_j$$

  with $\boldsymbol{n} = (n_j)$ the outward normal unit to boundary Γ.

In order to derive a variational formulation for the problem, we pick an arbitrary test function $\boldsymbol{v} = (v_i)$, multiply both sides of the equilibrium equations by v_i and integrate over Ω, getting

$$-\int_\Omega \sigma_{ij,j} v_i dx = \int_\Omega f_i v_i dx$$

Integrating next first integral by parts, we have

$$-\int_\Omega \sigma_{ij,j} v_i \, dx = \int_\Omega \sigma_{ij} v_{i,j} \, dx - \int_\Gamma \sigma_{ij} n_j v_i \, dS$$

But, due to the symmetry of the stress tensor,

$$\sigma_{ij}(\boldsymbol{u}) v_{i,j} = \sigma_{ij}(\boldsymbol{u}) \epsilon_{ij}(\boldsymbol{v})$$

so

$$\int_\Omega \sigma_{ij}(\boldsymbol{u}) \epsilon_{ij}(\boldsymbol{v}) \, dx = \int_\Omega f_i v_i \, dx + \int_\Gamma t_i(\boldsymbol{u}) v_i dS$$

Finally, restricting ourselves only to test functions vanishing on Γ_u, and using the traction boundary conditions, we arrive at the variational formulation in the form

$$\begin{cases} \text{Find } \boldsymbol{u} = \boldsymbol{u}(\boldsymbol{x}), \boldsymbol{u} = \hat{\boldsymbol{u}} \text{ on } \Gamma_u \text{ such that} \\ \displaystyle\int_\Omega \sigma_{ij}(\boldsymbol{u}) \epsilon_{ij}(\boldsymbol{v}) \, dx = \int_\Omega f_i v_i \, dx + \int_\Gamma q_i v_i \, dS \\ \text{for every } \boldsymbol{v}, \boldsymbol{v} = \boldsymbol{0} \text{ on } \Gamma_u \end{cases}$$

The formulation above is recognized as the classical *principle of virtual work* in mechanics. Test function $\boldsymbol{v} = \boldsymbol{v}(\boldsymbol{x})$ is interpreted as the *virtual displacement*, the integral on the left-hand side as the *virtual work* done by stresses $\sigma_{ij}(\boldsymbol{u})$ on strains $\epsilon_{ij}(\boldsymbol{v})$ corresponding to the virtual displacement, and the integral on the right-hand side as the *work of exterior forces.*

Thus every solution $\boldsymbol{u}$ to the (classical) boundary-value problem is also a solution to the variational formulation, i.e., it satisfies the principle of virtual work.

Conversely, by reversing the entire procedure, we can show that any regular enough solution of the variational formulation is a solution in the classical sense as well.

For a precise analysis of the elasticity problem by means of the Lax-Milgram theorem, we refer the reader to [4].

□

Sesquilinear Forms. In the case of complex-valued functions, when deriving the variational formulation, we frequently prefer to multiply the original equation not by a test funtion $v(\boldsymbol{x})$, but rather its complex conjugate $\bar{v}(\boldsymbol{x})$. In the case of the boundary-value problem for the Laplace operator we get

$$\begin{cases} \text{Find } u(\boldsymbol{x}) \text{such that} \\ u = u_0 \text{ on } \Gamma_u \text{ and} \\ \displaystyle\int_\Omega \nabla u \nabla \bar{v} dx = \int_\Omega f \bar{v} dx + \int_\Gamma g \bar{v} dS \\ \text{for every } v = v(\boldsymbol{x}) \text{ vanishing on } \Gamma_u \end{cases}$$

Consequently, the functional on the left-hand side is not linear, but *antilinear* with respect to the second variable, as in the definition of a scalar product in a complex Hilbert space. Also, the right-hand side is identified as an antilinear functional of v. The particular advantage of such an approach is that, for symmetric sesquilinear forms, i.e.,

$$B(\boldsymbol{u}, \boldsymbol{v}) = \overline{B(\boldsymbol{v}, \boldsymbol{u})}$$

value $B(\boldsymbol{u}, \boldsymbol{u})$, interpreted most of the time in physical applications as an energy, is real.

It is easily verified that antilinear and continuous functionals share all the properties of linear and continuous functionals. In particular, they form a normed vector space with the norm defined as for the linear functionals, i.e.,

$$\| f \| = \sup_{\|\boldsymbol{u}\| \leq 1} | f(\boldsymbol{u}) | \tag{6.17}$$

In fact, many authors prefer to define the algebraic and topological duals of a complex vector space as the space of *antilinear* rather than linear functionals. As a result of such a definition, a few little algebraic changes follow. For instance, the Riesz map is always linear (not antilinear like in our version) and the map prescribing for a linear operator A its transpose A' is antilinear in the same way as it is for adjoint operators on Hilbert spaces.

All these modifications are very cosmetic in nature as there exists a one-to-one correspondence between all linear and antilinear functionals defined through the complex conjugate operation. To see it, define the map J prescribing for each linear functional $f : V \to \mathbb{C}$ on a complex vector space V its complex conjugate $J(f) = \bar{f}$ defined by

$$J(f)(u) = \bar{f}(u) \stackrel{\text{def}}{=} \overline{f(u)} \tag{6.18}$$

It is a straightforward exercise to check that J is *antilinear, bijective* and *norm preserving.*

Using J we can easily reinterpret all results concerning linear functionals in terms of antilinear ones and vice versa. In particular, we have the following reinterpretation of the generalized Lax-Milgram theorem for sesquilinear forms.

COROLLARY 6.6.2
(The Generalized Lax-Milgram Theorem for Sesquilinear Forms)

Let all assumptions of Theorem 6.6.1 or Corollary 6.6.1 hold, except that B is sesquilinear rather than bilinear and L is an antilinear, continuous functional on V. Then, all the conclusions hold as well.

PROOF The proof follows exactly the lines of proof of Theorem 6.6.1. In the first step, we assume $\boldsymbol{u}_0 = \boldsymbol{0}$ and reinterpret the variational formulation in the operator form as

$$A\boldsymbol{u} = L \tag{6.19}$$

where A is the operator corresponding to the sesquilinear form

$$\langle A\boldsymbol{u}, \boldsymbol{v}\rangle \stackrel{\text{def}}{=} B(\boldsymbol{u}, \boldsymbol{v}) \tag{6.20}$$

The *only* difference now is that A takes vectors from U into *antilinear* and continuous functionals on U. Due to linearity of B in $\boldsymbol{u}$, A is still a linear operator and, consequently, the rest of the proof holds without any change. ∎

Exercises

6.6.1. Let X be a Hilbert space and V a closed subspace. Prove that the quotient space X/V, which *a priori* is only a Banach space, is in fact a Hilbert space. *Hint:* Compare Exercise 6.1.2.

6.7 Generalized Green's Formulae for Operators on Hilbert Spaces

As we have seen in the previous chapter, variational boundary-value problems can be treated as generalizations of classical formulations. If a solution to such a variational problem is *additionally* sufficiently regular then it is also a solution to the classical formulation. The question now is whether we can interpret all variational solutions (including those "less" regular as well) as the solutions to the original boundary-value problems, and if the answer is yes, then in what sense.

Trace Property. An abstraction of the idea of boundary values of functions from a Hilbert space, exemplified in the trace theorem for Sobolev spaces, is embodied in the concept of spaces with a trace property. A Hilbert space V is said to have the *trace property* if the following conditions hold:

1. V is continuously imbedded in a larger Hilbert space H

   $$V \hookrightarrow H$$

   Note that this in particular implies that the topology of H, when restricted to V is *weaker* than the original topology of V.

2. There exists a linear and continuous (trace) operator γ that maps V *onto* another (boundary) Hilbert space ∂V such that the kernel of γ, denoted V_0, is everywhere dense in H

   $$\overline{V_0} = H \ , \ V_0 = \ker\gamma = \mathcal{N}(\gamma)$$

   It follows that the original space V is dense in H as well.

Example 6.7.1

Let Ω be a smooth open set in $\mathbb{R}^n$ with boundary Γ and let $V = H^1(\Omega)$ be the first-order Sobolev space. Then V satisfies the trace property where $H = L^2(\Omega), \partial V = H^{\frac{1}{2}}(\Gamma)$ and $\gamma : H^1(\Omega) \rightarrow H^{\frac{1}{2}}(\Gamma)$ is the actual trace operator. $\square$

Let ι denote now the continuous inclusion operator from a Hilbert space V imbedded in a Hilbert space H and let us assume that V is dense in $H, \overline{V} = H$. Its transpose ι^T maps dual H' into dual V' and it may be used to identify H' as a subspace of V', provided it is injective. Since

$$\iota^T(f) = f \circ \iota = f \mid_V$$

this means that, from the fact that two linear functionals continuous in H coincide on subspace V, it should follow that they are equal to each other on the entire H

$$f\,|_V = g\,|_V \quad \Rightarrow \quad f = g \quad f, g \in H'$$

But this follows immediately from continuity of f and g and density of V in H. Indeed, let $\boldsymbol{x} \in H$ and $\boldsymbol{x}_n$ be a sequence in V converging to $\boldsymbol{x}$. Then

$$f(\boldsymbol{x}_n) = g(\boldsymbol{x}_n)$$

and in the limit $f(\boldsymbol{x}) = g(\boldsymbol{x})$, as required.

Frequently, space H is identified with its dual H' using the Riesz map and, in such a case, called the *pivot space*. We shall write shortly then

$$V \hookrightarrow H \sim H' \hookrightarrow V'$$

both imbeddings being continuous.

Example 6.7.2

Let Ω be an open set in $\mathbb{R}^n$. Then

$$H_0^1(\Omega) \hookrightarrow L^2(\Omega) \sim (L^2(\Omega))' \hookrightarrow H^{-1}(\Omega)$$

and the $L^2(\Omega)$ space is a pivot space. $\square$

Formal Operators and Formal Adjoints. Let U and V be now two Hilbert spaces satisfying the trace property, with the corresponding pivot spaces G and H and boundary spaces ∂U and ∂V, i.e.,

$$U \hookrightarrow G \sim G' \hookrightarrow U' \quad V \hookrightarrow H \sim H' \hookrightarrow V'$$

$$\beta : U \twoheadrightarrow \partial U \qquad \gamma : V \twoheadrightarrow \partial V$$

$$U_0 \stackrel{\text{def}}{=} \ker\beta \qquad V_0 \stackrel{\text{def}}{=} \ker\gamma$$

$$U_0 \hookrightarrow G \sim G' \hookrightarrow U_0' \quad V_0 \hookrightarrow H \sim H' \hookrightarrow V_0'$$

All mappings are continuous and symbols $\hookrightarrow$ and $\twoheadrightarrow$ are used to indicate injective and surjective operations, respectively.

Consider now a bilinear and continuous functional B defined on $U \times V$. Restricting first functional B to $U \times V_0$ only (think: the test functions v vanishing on the boundary), we consider the corresponding linear and continuous operator $A : U \to V_0'$ defined as

$$\langle A\boldsymbol{u}, \boldsymbol{v}\rangle_{V_0} = B(\boldsymbol{u}, \boldsymbol{v}) \qquad \forall \boldsymbol{u} \in U, \boldsymbol{v} \in V_0$$

The operator A is called the *formal operator associated with the bilinear form.* Note the difference between the formal operator and operator A corresponding to B and considered in the proof of the Lax-Milgram theorem.

In a similar manner, by inverting the order of arguments, we can consider a corresponding bilinear form B^* on $V \times U$

$$B^*(\boldsymbol{v}, \boldsymbol{u}) \overset{\text{def}}{=} B(\boldsymbol{u}, \boldsymbol{v})$$

with the corresponding formal operator $A^* : V \to U_0'$

$$\langle A^*\boldsymbol{v}, \boldsymbol{u}\rangle_{U_0} \overset{\text{def}}{=} B^*(\boldsymbol{v}, \boldsymbol{u}) = B(\boldsymbol{u}, \boldsymbol{v}) \qquad \forall \boldsymbol{u} \in U_0, \boldsymbol{v} \in V$$

Operator A^* is known as the *formal adjoint* of A.

Green's Formulae. As H' is only a *proper* subspace of V_0', a value of formal operator $A\boldsymbol{u}$ *cannot be identified* in general with a linear and continuous functional on H.

For *some* elements $\boldsymbol{u}$, however, namely for $\boldsymbol{u} \in A^{-1}(H)$, the value of formal operator $A\boldsymbol{u}$ belongs to H' and therefore by the Riesz representation theorem

$$B(\boldsymbol{u}, \boldsymbol{v}) = \langle A\boldsymbol{u}, \boldsymbol{v}\rangle_{V_0} = (\boldsymbol{v}, R_H^{-1}A\boldsymbol{u})_H \qquad \forall \boldsymbol{v} \in V_0$$

As H is identified with its dual, we customary drop the symbol for the Riesz map R_H^{-1} and replace the composition $R_H^{-1}A$ with A itself writing

$$B(\boldsymbol{u}, \boldsymbol{v}) = (\boldsymbol{v}, A\boldsymbol{u})_H \qquad \forall \boldsymbol{v} \in V_0$$

where A is understood now as the operator from a subspace of U into H.

But both $B(\boldsymbol{u}, \cdot)$ and $(\cdot, A\boldsymbol{u})_H$ now are linear and continuous functionals on the whole V and therefore their difference

$$B(\boldsymbol{u}, \cdot) - (\cdot, A\boldsymbol{u})_H$$

can be identified as an element from V', vanishing on V_0.

Consider now the boundary space ∂V and trace operator $\gamma : V \to \partial V$. From the surjectivity of γ follows that the corresponding operator $\tilde{\gamma}$ defined on the quotient (Banach!) space V/V_0 into ∂V

$$\tilde{\gamma} : V/V_0 \rightarrow \partial V$$

is injective and continuous. Consequently, by the corollary to the open mapping theorem, the inverse $\tilde{\gamma}^{-1}$ is continuous as well.

This in particular implies that the boundary space ∂V can be equipped with an equivalent norm of the form

$$|||\boldsymbol{w}|||_{\partial V} \stackrel{\text{def}}{=} \inf\{\|\boldsymbol{v}\|_V : \gamma \boldsymbol{v} = \boldsymbol{w}\}$$

Indeed, taking the infimum in $\boldsymbol{v}$ on the right-hand side of

$$\|\gamma \boldsymbol{v}\|_{\partial V} \leq \|\gamma\| \|\boldsymbol{v}\|_V$$

we get

$$\|\gamma \boldsymbol{v}\|_{\partial V} \leq \|\gamma\| |||\gamma \boldsymbol{v}|||_{\partial V}$$

At the same time, reinterpreting continuity of $\tilde{\gamma}^{-1}$, we get

$$\|\tilde{\gamma}^{-1}(\boldsymbol{w})\|_{V/V_0} = \inf_{\gamma \boldsymbol{v} = \boldsymbol{w}} \|\boldsymbol{v}\|_V = |||\boldsymbol{w}|||_{\partial K}$$

$$\leq \|\tilde{\gamma}^{-1}\| \|\boldsymbol{w}\|_{\partial K}$$

Consider now an arbitrary element $\boldsymbol{w} \in \partial V$ and define a linear functional $\delta \boldsymbol{u}$ on ∂V by

$$\langle \delta \boldsymbol{u}, \boldsymbol{w} \rangle = B(\boldsymbol{u}, \boldsymbol{v}) - (\boldsymbol{v}, A\boldsymbol{u})_H$$

where $\boldsymbol{v} \in V$ is an arbitrary element from V such that $\gamma \boldsymbol{v} = \boldsymbol{w}$. As the right-hand side vanishes for $\boldsymbol{v} \in V_0$, the functional $\delta \boldsymbol{u}$ is well defined, i.e., its value is independent of the choice of $\boldsymbol{v}$.

Introducing now a space

$$U_A = \{\boldsymbol{u} \in U : A\boldsymbol{u} \in H\}$$

with the (so called *operator*) norm

$$\|\boldsymbol{u}\|_{U_A} \stackrel{\text{def}}{=} (\|\boldsymbol{u}\|_U^2 + \|A\boldsymbol{u}\|_H^2)^{\frac{1}{2}}$$

we have immediately

$$|\langle \delta \boldsymbol{u}, \boldsymbol{w} \rangle| \leq M \|\boldsymbol{u}\|_U \|\boldsymbol{v}\|_V + \|A\boldsymbol{u}\|_H \|\boldsymbol{v}\|_H$$

$$\leq (M\|\boldsymbol{u}\|_U + C\|A\boldsymbol{u}\|_H)\|\boldsymbol{v}\|_H$$

Taking infimum in $\boldsymbol{v}$ on the right-hand side we get

$$|\langle \delta \boldsymbol{u}, \boldsymbol{w} \rangle\| \leq (M\|\boldsymbol{u}\|_U + C\|A\boldsymbol{u}\|_H)|||\boldsymbol{w}|||_{\partial V}$$

which proves that $\delta \boldsymbol{u}$ is a continuous functional on ∂V and, at the same time, the operator $\delta : \boldsymbol{u} \to \partial \boldsymbol{u}$ is a linear and continuous operator from U_A into the dual $\partial V'$.

Operator δ is called the *generalized Neumann operator* corresponding to the *trace (Dirichlet) operator* γ. The formula defining $\delta \boldsymbol{u}$ is known as the *generalized* or *abstract Green's formula (of the first type)* for operator A.

Exactly the same results can be obtained for the bilinear form $B^*(\boldsymbol{v}, \boldsymbol{u}) = B(\boldsymbol{u}, \boldsymbol{v})$ and the corresponding formal adjoint operator A^*.

We summarize the results in the following theorem.

THEOREM 6.7.1

Let U and V denote real Hilbert spaces with the trace properties previously described, and let B denote a continuous, bilinear form from $U \times V$ into $\mathbb{R}$ with associated formal operators $A \in \mathcal{L}(U, V_0')$ and $A^* \in \mathcal{L}(V, U_0')$. Moreover, let U_A and V_{A^*} denote the spaces

$$U_A \stackrel{\text{def}}{=} \{\boldsymbol{u} \in U : A\boldsymbol{u} \in H\} \; (H \sim H' \subset V_0')$$

$$V_{A^*} \stackrel{\text{def}}{=} \{\boldsymbol{v} \in V : A^*\boldsymbol{v} \in G\} \; (G \sim G' \subset U_0')$$

with the operator norms

$$\|\boldsymbol{u}\|_{U_A}^2 = \|\boldsymbol{u}\|_U^2 + \|A\boldsymbol{u}\|_H^2$$

$$\|\boldsymbol{v}\|_{V_{A^*}}^2 = \|\boldsymbol{v}\|_V^2 + \|A^*\boldsymbol{v}\|_G^2$$

Then there exist uniquely defined operators

$$\delta \in \mathcal{L}(U_A, \partial V') \ , \ \delta^* \in \mathcal{L}(V_{A^*}, \partial U')$$

such that the following formulas hold:

$$B(\boldsymbol{u}, \boldsymbol{v}) = (\boldsymbol{v}, A\boldsymbol{u})_H + \langle \delta \boldsymbol{u}, \gamma \boldsymbol{v} \rangle_{\partial V} \qquad \boldsymbol{u} \in U_A, \boldsymbol{v} \in V$$

$$B(\boldsymbol{u}, \boldsymbol{v}) = (\boldsymbol{u}, A^* \boldsymbol{v})_G + \langle \delta^* \boldsymbol{v}, \beta \boldsymbol{u} \rangle_{\partial U} \qquad \boldsymbol{u} \in U, \boldsymbol{v} \in V_{A^*}$$

A schematic diagram illustrating the various spaces and operators is given in Fig. 6.5

Green's Formula of the Second Type. As an immediate corollary of Theorem 6.7.1 we get *Green's formula of the second type*

$$(\boldsymbol{u}, A^* \boldsymbol{v})_G = (A\boldsymbol{u}, \boldsymbol{v})_H + \langle \delta \boldsymbol{u}, \gamma \boldsymbol{v} \rangle_{\partial V} - \langle \delta^* \boldsymbol{v}, \beta \boldsymbol{u} \rangle_{\partial U}$$

for every $\boldsymbol{u} \in U_A, \boldsymbol{v} \in V^*_{A^*}$.

The collection of the boundary terms

$$\Gamma(\boldsymbol{u}, \boldsymbol{v}) = \langle \delta \boldsymbol{u}, \gamma \boldsymbol{v} \rangle_{\partial V} - \langle \delta^* \boldsymbol{v}, \beta \boldsymbol{u} \rangle_{\partial U}$$

is called the *bilinear concomitant* of operator $A; \Gamma : U_A \times V_{A^*} \to \mathbb{R}$.

Example 6.7.3

Consider the case in which Ω is a smooth, open, bounded subset of $\mathbb{R}^n$ with a smooth boundary Γ and

$$U = V = H^1(\Omega)$$

$$G = H = L^2(\Omega)$$

$$\partial U = \partial V = H^{\frac{1}{2}}(\Gamma)$$

Let $a_{ij} = a_{ij}(\boldsymbol{x}), b_i = b_i(\boldsymbol{x}), ij = 1, \ldots, n, c = c(\boldsymbol{x})$ be sufficiently regular functions of $\boldsymbol{x}$ (e.g., $a_{ij}, b_i \in C^1(\bar{\Omega})$), and define the bilinear form B : $H^1(\Omega) \times H^1(\Omega) \to \mathbb{R}$ by

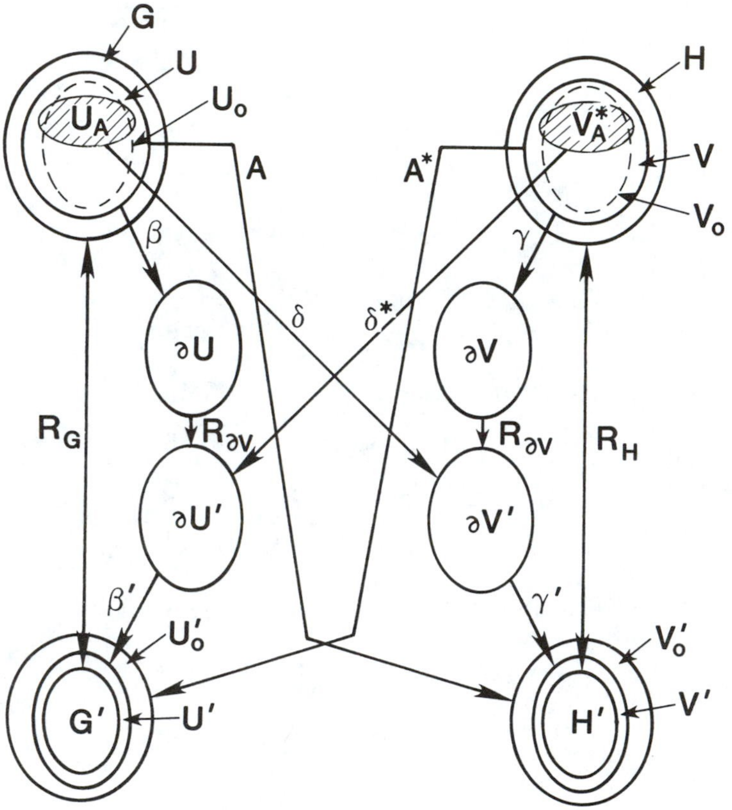

Figure 6.5
Diagram of spaces and operators in the generalized Green's formulae.

$$B(u,v) = \int_\Omega (\sum_{i,j=1}^n a_{ij} \frac{\partial u}{\partial x_j} \frac{\partial v}{\partial x_i} + \sum_{j=1}^n b_j \frac{\partial u}{\partial x_j} v + cuv)\, dx$$

Obviously, if u is sufficiently smooth (e.g., $u \in C^2(\bar{\Omega})$) then

$$B(u,v) = \int_\Omega (-\sum_{i,j=1}^n \frac{\partial}{\partial x_i}(a_{ij} \frac{\partial u}{\partial x_j}) + \sum_{j=1}^n b_j \frac{\partial u}{\partial x_j} + cu) v dx$$

$$+ \int_\Gamma (\sum_{i,j=1}^n a_{ij} \frac{\partial u}{\partial x_j} n_i) v\, dS$$

where $\boldsymbol{n} = (n_i)$ is the outward normal unit to boundary Γ.

This prompts us to consider the formal operator

$$A : H^1(\Omega) \rightarrow (H_0^1(\Omega))' = H^{-1}(\Omega)$$

$$\langle Au, v \rangle = B(u,v) \qquad u \in H^1(\Omega), v \in H_0^1(\Omega)$$

to be a generalization of the classical operator

$$-\sum_{i,j=1}^n \frac{\partial}{\partial x_i}(a_{ij} \frac{\partial}{\partial x_j}) + \sum_{j=1}^n b_j \frac{\partial}{\partial x_j} + c$$

Function u belongs to

$$U_A = \{u \in H^1(\Omega) : Au \in L^2(\Omega)\}$$

if and only if a function $f \in L^2$ exists such that

$$B(u,v) = \int_\Omega fv\, dx \qquad \forall v \in H_0^1(\Omega)$$

or, equivalently,

$$-\sum_{i,j=1}^n \frac{\partial}{\partial x_i}(a_{ij} \frac{\partial u}{\partial x_j}) + \sum_{j=1}^n b_j \frac{\partial u}{\partial x_j} + cu = f$$

Note that from this it *does not* follow that $u \in H^2(\Omega)$!

Finally, the generalized Neumann operator is interpreted as a generalization of the classical operator

$$\delta u = \sum_{i,j=1}^{n} a_{ij} \frac{\partial u}{\partial x_j} n_i \qquad \delta : U_A \to H^{-\frac{1}{2}}(\Gamma)$$

Similarly, for sufficiently smooth v,

$$B(u,v) = \int_\Omega u\Big(-\sum_{i,j=1}^{n} \frac{\partial}{\partial x_j}\Big(a_{ij}\frac{\partial v}{\partial x_i}\Big) - \sum_{j=1}^{n} \frac{\partial}{\partial x_j}(b_j v) + cv\Big)dx$$

$$+ \int_\Gamma u \sum_{j=1}^{n}\Big(\sum_{i=1}^{n} a_{ij}\frac{\partial v}{\partial x_i} + b_j v\Big) n_j \, dS$$

which prompts for the interpretations

$$A^* : H^1(\Omega) \to H^{-1}(\Omega)$$

$$A^* v = -\sum_{i,j=1}^{n} \frac{\partial}{\partial x_j}\Big(a_{ij}\frac{\partial v}{\partial x_i}\Big) - \sum_{j=1}^{n} \frac{\partial}{\partial x_j}(b_j v) + cv$$

$$V_{A^*} = \{v \in H^1(\Omega) : A^* v \in L^2(\Omega)\}$$

$$\delta^* v = \sum_{j=1}^{n}\Big(\sum_{i=1}^{n} a_{ij}\frac{\partial v}{\partial x_i} + b_j v\Big) n_j \ , \ \delta^* : V_{A^*} \to H^{-\frac{1}{2}}(\Gamma)$$

The bilinear concomitant is a generalization of

$$\Gamma(u,v) = \int_\Gamma \Big(\sum_{i,j=1}^{n} a_{ij}\frac{\partial u}{\partial x_j} n_i v - u \sum_{j=1}^{n}\Big(\sum_{i=1}^{n} a_{ij}\frac{\partial v}{\partial xi} + b_j v\Big) n_j\Big)\, dS$$

□

Example 6.7.4
(Interpretation of Solutions to Variational Problems)

Let u be a solution to the variational problem considered in Example 6.6.1

$$\begin{cases} \text{Find } u \in u_0 + V \text{ such that} \\ \int_\Omega \nabla u \nabla v \, dx = \int_\Omega fv \, dx + \int_{\Gamma_t} g\gamma v \, dS \qquad \forall v \in V \end{cases}$$

where

$$V = \{u \in H^1(\Omega) : \gamma u = 0 \;\text{ on }\; \Gamma_u\}$$

and $\gamma : H^1(\Omega) \to H^{\frac{1}{2}}(\Gamma)$ is the trace operator.

The space of traces ∂V, identified as the image of operator γ *does not* coincide with $H^{\frac{1}{2}}(\Gamma_t)$ (unless $\Gamma_u = \emptyset$!) and is frequently denoted as the space $H_{00}(\Gamma_t)$ (functions from $H_{00}(\Gamma_t)$ must decay at an appropriate rate when approaching boundary of Γ_t); see [8].

Taking $v \in H_0^1(\Omega) = \ker\gamma$ in the variational formulation, we get

$$\int_\Omega \nabla u \nabla v \, dx = \int_\Omega fv \, dx \qquad \forall v \in H_0^1(\Omega)$$

Consequently, $-\Delta u = f \in L^2(\Omega)$, which implies that u is in the domain of the generalized Neumann operator

$$\frac{\partial}{\partial n} : H^1(\Delta) \to (H_{00}(\Gamma_t))'$$

where

$$H^1(\Delta) = \{u \in H^1(\Omega) : \Delta u \in L^2(\Omega)\}$$

From the generalized Green's formula, it follows finally that

$$\langle \frac{\partial u}{\partial n}, v \rangle = \int_{\Gamma_t} gv \, dS \qquad \forall v \in H_{00}(\Gamma_t)$$

Summing that up, u being a variational solution implies that

1. u is a solution to the differential equation in the distributional sense,
2. u satisfies the Dirichlet boundary condition in the sense of the trace operator

$$\gamma u = \gamma u_0 \qquad \text{on } \Gamma_u$$

3. u satisfies the Neumann boundary condition in the sense of the generalized Neumann operator

$$\frac{\partial u}{\partial n} = g \qquad \text{on } \Gamma_t$$

Conversely, by reversing the entire procedure, it can be immediately shown that any u satisfying the conditions above is a solution to the variational problem as well. $\square$

All presented results can be immediately generalized to the case of complex Hilbert spaces and sesquilinear forms with the dual spaces redefined as the spaces of *antilinear* functionals.

THEOREM 6.7.2

Let U and V denote two complex Hilbert spaces satisfying the trace property with corresponding pivot spaces G and H, boundary spaces ∂U and ∂V and trace operators $\beta : U \to \partial U$ and $\gamma : V \to \partial V$, respectively. All dual spaces are defined as spaces of antilinear functionals.

Let $B : U \times V \to \mathbb{C}$ be a continuous, sesquilinear form with associated formal operators $A \in \mathcal{L}(U, V_0')$ and $A^* \in \mathcal{L}(V, U_0')$ defined as

$$\langle A\boldsymbol{u}, \boldsymbol{v}\rangle \overset{\text{def}}{=} B(\boldsymbol{u}, \boldsymbol{v}) \qquad \boldsymbol{u} \in U, \boldsymbol{v} \in V_0$$

$$\langle A^*\boldsymbol{v}, \boldsymbol{u}\rangle \overset{\text{def}}{=} B^*(\boldsymbol{v}, \boldsymbol{u}) \overset{\text{def}}{=} \overline{B(\boldsymbol{u}, \boldsymbol{v})} \qquad \boldsymbol{v} \in V, \boldsymbol{u} \in U_0$$

Moreover, let U_A and V_{A^*} denote the spaces

$$U_A \overset{\text{def}}{=} \{\boldsymbol{u} \in U : A\boldsymbol{u} \in H\} \; (H \sim H' \subset V_0')$$

$$V_{A^*} \overset{\text{def}}{=} \{\boldsymbol{v} \in V : A^*\boldsymbol{v} \in G\} \; (G \sim G' \subset U_0')$$

with the operator norms

$$\|\boldsymbol{u}\|_{U_A}^2 = \|\boldsymbol{u}\|_U^2 + \|A\boldsymbol{u}\|_H^2$$

$$\|\boldsymbol{v}\|_{V_{A^*}}^2 = \|\boldsymbol{v}\|_V^2 + \|A^*\boldsymbol{v}\|_G^2$$

Then there exist unique operators

$$\delta \in \mathcal{L}(U_A, \partial V') \ , \ \delta^* \in \mathcal{L}(V_{A^*}, \partial U')$$

such that the following formulas hold:

$$B(\boldsymbol{u}, \boldsymbol{v}) = (A\boldsymbol{u}, \boldsymbol{v})_H + \langle \delta \boldsymbol{u}, \gamma \boldsymbol{v} \rangle_{\partial V} \qquad \boldsymbol{u} \in U_A, \boldsymbol{v} \in V$$

$$\overline{B(\boldsymbol{u}, \boldsymbol{v})} = B^*(\boldsymbol{v}, \boldsymbol{u})$$

$$= (A^*\boldsymbol{v}, \boldsymbol{u})_G + \langle \delta^* \boldsymbol{v}, \beta \boldsymbol{u} \rangle_{\partial U} \qquad \boldsymbol{u} \in U, \boldsymbol{v} \in V_{A^*}$$

or, equivalently,

$$B(\boldsymbol{u}, \boldsymbol{v}) = (\boldsymbol{u}, A^*\boldsymbol{v})_G + \overline{\langle \delta^* \boldsymbol{v}, \beta \boldsymbol{u} \rangle}_{\partial U} \qquad \boldsymbol{u} \in U, \boldsymbol{v} \in V_{A^*}$$

As for the real spaces, Green's formula of the second type follows

$$(\boldsymbol{u}, A^*\boldsymbol{v})_G = (A\boldsymbol{u}, \boldsymbol{v})_H + \langle \delta \boldsymbol{u}, \gamma \boldsymbol{v} \rangle_{\partial V} - \overline{\langle \delta^* \boldsymbol{v}, \beta \boldsymbol{u} \rangle}_{\partial U}$$

for $\boldsymbol{u} \in U_A, \boldsymbol{v} \in V_{A^*}$.

Exercises

6.7.1. Consider the elastic beam equation

$$(EIw'')'' = q \qquad 0 < x < l$$

with the boundary conditions

$$w(0) = w'(0) = 0 \text{ and } w(l) = EIw''(l) = 0$$

(a) Construct an equivalent variational formulation, identifying appropiate spaces.
(b) Use the Lax-Milgram theorem to show that there exists a unique solution to this problem.

6.7.2. Consider again the elastic beam equation

$$(EIw'')'' = q \qquad 0 < x < l$$

with different boundary conditions

$$w(0) = EIw''(0) = 0$$

$$EIw''(l) = -M_l \ , \ (EIw'')'(l) = P_l$$

(a) Construct an equivalent variational formulation, identifying appropriate spaces.
(b) Use the Lax-Milgram theorem to establish existence and uniqueness result in an appropriate quotient space. Derive and interpret the necessary and sufficient conditions for the distributed load $q(x)$, moment M_l and force P_l to yield the existence result.

6.7.3. Let $u, v \in H^2(0, l)$ and $B(\cdot, \cdot)$ denote the bilinear form

$$B(u, v) = \int_0^l (EIu''v'' + Pu'v + kuv)\, dx$$

where EI, P, and k are positive constants. The quadratic functional $B(u, u)$ corresponds to twice the strain energy in an elastic beam of length l with flexural rigidity EI, on elastic foundation with stiffness k and subjected to an axial load P.
(a) Determine the formal operator A associated with $B(\cdot, \cdot)$ and its formal adjoint A^*.
(b) Describe the spaces $U_A, V_{A^*}, G, H, \partial U, \partial V$ for this problem. Identify the trace operators.
(c) Describe the Dirichlet and Neumann problems corresponding to operators A and A^*.
(d) Consider an example of a mixed boundary-value problem for operator A, construct the corresponding variational formulation and discuss conditions under which this problem has a unique solution.

6.7.4. Consider the fourth-order boundary-value problem in two dimensions:

$$\nabla^2\nabla^2 u + u \stackrel{\text{def}}{=} \frac{\partial^4 u}{\partial x^4} + 2\frac{\partial^4 u}{\partial x^2 \partial y^2} + \frac{\partial^4 u}{\partial y^4} + u = f \text{ in } \Omega$$

$$u = 0 \ , \ \frac{\partial u}{\partial n} = 0 \text{ on } \partial\Omega$$

with $f \in L^2(\Omega)$. Construct a variational formulation of this problem, identifying the appropiate spaces, and show that it has a unique solution.
What could be a physical interpretation of the problem?

Elements of Spectral Theory

6.8 Resolvent Set and Spectrum

The spectral analysis of linear operators is basically a geometric study of the behavior of linear operators with special regard to the existence of certain inverses. In particular, if $A \in \mathcal{L}(U, U)$, where U is a Hilbert space, and if λ is a scalar, we are concerned with the existence of the inverse of the operator $(\lambda I - A)$. In finite dimensional spaces, the situation is clear: either $(\lambda I - A)^{-1}$ exists or it does not. If, for a given scalar λ, it does not exist, then λ is called an *eigenvalue* of A, and if $\dim U = n$, there are at most n (distinct) eigenvalues.

However, when U is infinite dimensional, there may be infinitely many, indeed, a continuum of scalars λ such that $(\lambda I - A)^{-1}$ does not exist. If $(\lambda I - A)^{-1}$ exists, the question arises as to whether it is a bounded operator or, moreover, whether its domain, equal to the range $\mathcal{R}(\lambda I - A)$, is dense in U. None of these questions arises in the finite-dimensional case. These questions are in the province of the so-called *spectral theory* of linear operators. The last part of this chapter presents an introductory account of this theory.

Eigenvalues and Characteristic Values of an Operator. Let U be a normed vector space over the complex number field $\mathbb{C}$ and let A be a linear operator from a subspace $D = D(A) \subset U$ into itself. The problem of finding scalars $\lambda \in \mathbb{C}$ such that there exists $\boldsymbol{u} \in D, \boldsymbol{u} \neq \boldsymbol{0}$ satisfying the equation

$$(\lambda I - A)\boldsymbol{u} = \boldsymbol{0}$$

is called an *eigenvalue problem* associated with operator A. Any complex scalar λ such that the equality holds for some nonzero vector $\boldsymbol{u} \in D$ is called an *eigenvalue* of A, and the corresponding nonzero vector $\boldsymbol{u}$ is called an *eigenvector* of A corresponding to λ. The null space of the transformation $\mathcal{N}(\lambda I - A)$ is called the *eigenmanifold* (or *eigenspace*) corresponding to the eigenvalue λ, and the dimension of the eigenspace is called the *multiplicity* of the eigenvalue λ. Note that a scalar λ is an eigenvalue of A if and only if the linear transformation $(\lambda I - A)$ is singular; in other words, if the null space $\mathcal{N}(\lambda I - A)$ is nontrivial.

For *nonzero eigenvalues* λ, we can rewrite the equation in the form

$$(I - \lambda^{-1} A)\boldsymbol{u} = \boldsymbol{0}$$

The inverse λ^{-1} is frequently called the *characteristic value* of operator A.

Resolvent Set. If operator $\lambda I - A$ has a *continuous (bounded)* inverse defined on a dense subset of U, i.e., if $\lambda I - A$ has a range dense in U, operator

$$R_\lambda = (\lambda I - A)^{-1}$$

is called the resolvent of A and λ is said to belong to the *resolvent set* $r(A)$ of operator A.

Note that if A is closed, it follows that $\lambda I - A$ is closed as well, and boundedness of R_λ implies that $\lambda I - A$ has a closed range in U, and therefore the resolvent R_λ is defined on the *whole* space U.

Spectrum. The set of all complex numbers that are not in the resolvent set is called the *spectrum* of the operator A and is denoted by $\sigma(A)$. There are a number of situations in which the operator $\lambda I - A$ has no continuous inverse defined on a dense subset of U. The transformation may not be injective when λ is an eigenvalue of A. Another possibility is that the inverse may not be defined on a dense subset of U or it may not be bounded. It is customary to divide the spectrum $\sigma(A)$ into various categories, depending on which of these circumstances a given scalar λ fails to be in the resolvent set $r(A)$.

Point (or Discrete) Spectrum. The *point spectrum* of A is the subset of all λ's for which $(\lambda I - A)$ is *not* one-to-one. That is, the point spectrum, denoted $\sigma_P(A)$, is exactly the set of all eigenvalues.

Residual Spectrum. This the subset of all λ's for which $(\lambda I - A)$ has no range dense in U. The residual spectrum is denoted by $\sigma_R(A)$.

Continuous Spectrum. The *continuous spectrum* is the subset of all λ's for which $(\lambda I - A)$ is one-to-one and has range dense in U, but for which the inverse defined on its range is not continuous. The continuous spectrum is denoted by $\sigma_c(A)$.

From the definitions, it follows that $\sigma_P(A), \sigma_R(A)$, and $\sigma_c(A)$ are pairwise disjoint sets and that

$$\sigma(A) = \sigma_P(A) \cup \sigma_R(A) \cup \sigma_c(A)$$

Example 6.8.1

Consider the case in which $U = L^2(\mathbb{R})$ and A is the differential operator

$$Au = \frac{du}{dx}$$

with its domain $D(A)$ defined as

$$D(A) = H^1(\mathbb{R})$$

The eigenvalue problem associated with A

$$\lambda u - \frac{du}{dx} = 0$$

has no nonzero solution as the general solution of the differential equation is

$$u(x) = Ce^{\lambda x} \ , C \in \mathbb{C}$$

and $u \in L^2(\mathbb{R})$ only if $C = 0$. Thus the discrete spectrum of A is empty.

To determine the resolvent set of A, assume $\lambda = a + bi$ and consider the equation

$$\lambda u - \frac{du}{dx} = v$$

for $v \in L^2(\mathbb{R})$. Assume the equation above has a solution $u \in H^1(\mathbb{R})$. Applying Fourier transforms to both sides of the equation (cf. Example 6.1.6) yields

$$(\lambda - i\xi)\hat{u}(\xi) = \hat{v}(\xi)$$

or

$$\hat{u}(\xi) = \frac{1}{a - i(\xi - b)}\,\hat{v}(\xi) = \frac{a + i(\xi - b)}{a^2 + (\xi - b)^2}\,\hat{v}(\xi)$$

Consequently,

$$|\hat{u}(\xi)|^2 = \frac{1}{a^2 + (\xi - b)^2}\,|\hat{v}(\xi)|^2$$

and

$$|\widehat{\frac{du}{dx}}(\xi)|^2 = \frac{\xi^2}{a^2 + (\xi - b)^2}\,|\hat{v}(\xi)|^2$$

which allows one to draw the following conclusions.

1. If $a = \mathrm{Re}\lambda \neq 0$, then factors

$$\frac{1}{a^2 + (\xi - b)^2}\ ,\ \frac{\xi^2}{a^2 + (\xi - b)^2}$$

are bounded and therefore $\hat{v} \in L^2(\mathbb{R})$ implies that both $\hat{u}$ and $\widehat{\frac{du}{dx}}$ are L^2 functions, which in turn implies that $u \in H^1(\mathbb{R})$. Consequently, the range of $\lambda I - A$ is equal to the whole $L^2(\mathbb{R})$. It follows from the formula for $\hat{u}(\xi)$ that resolvent R_λ is continuous and therefore the resolvent set $r(A)$ contains all complex numbers λ with nonzero real part.

2. If $a = 0$ then factor

$$\frac{1}{(\xi - b)^2}$$

is *not* bounded and therefore the range of $\lambda I - A$ does *not* coincide with the whole $L^2(\mathbb{R})$. It is, however, *dense* in $L^2(\mathbb{R})$. To see it, pick an L^2 function $\hat{v} \in L^2(\mathbb{R})$ and consider a sequence $\hat{v}_n$

$$\hat{v}_n(\xi) = \begin{cases} 0 & \text{if } |\xi - b| < \frac{1}{n} \\ v(\xi) & \text{otherwise} \end{cases}$$

Obviously, $\hat{v}_n \to \hat{v}$ in $L^2(\mathbb{R})$ and the corresponding inverse transform v_n converges to v. Thus functions of this type form a dense subset of $L^2(\mathbb{R})$. From the formula for $\hat{u}$ and $\widehat{\frac{du}{dx}}$ it follows immediately that v_n is in the range of $\lambda I - A$.

The resolvent R_λ is not, however, continuous. To see it, it is sufficient to consider a sequence of functions $v_n \in L^2(\mathbb{R})$ such that

$$\hat{v}_n(\xi) = \begin{cases} \frac{1}{\sqrt{n}} & \text{for } b - \frac{1}{2n} < \xi < b - \frac{1}{n} \\ 0 & \text{otherwise} \end{cases}$$

Obviously, $\|v_n\|_{L^2} = \|\hat{v}_n\|_{L^2} = 1$ and by inspecting the formula for $\hat{u}$ we see that for the corresponding sequence of functions $\hat{u}_n$

$$\|\hat{u}_n\|_{L^2} \to \infty$$

Summing up, the spectrum of operator A consists only of its continuous part, coinciding with the imaginary axis in the complex plane λ. $\square$

Asymptotic Eigenvalues. Let U be a normed space and let $A : U \supset D(A) \to U$ be a linear operator. A complex number λ is called an *asymptotic eigenvalue* if there exists a sequence of unit vectors $\boldsymbol{x}_n$, $\|\boldsymbol{x}_n\| = 1$, such that

$$(\lambda I - A)\boldsymbol{x}_n \to 0 \text{ for } n \to \infty$$

Obviously, every eigenvalue λ is asymptotic, as one can select $\boldsymbol{x}_n = \boldsymbol{x}$, where $\boldsymbol{x}$ is a unit eigenvector corresponding to λ. The following proposition gives a simple characterization of *essentially* asymptotic eigenvalues.

PROPOSITION 6.8.1

Let U be a normed space and let $A : U \supset D(A) \to U$ be a linear operator. Let λ be a complex number such that $\lambda \notin \sigma_P(A)$, i.e., $\lambda I - A$ is injective. Then the following conditions are equivalent to each other:

(i) λ is an asymptotic eigenvalue of A.

(ii) Resolvent $R_\lambda = (\lambda I - A)^{-1}$ is unbounded.

PROOF

(i) $\Rightarrow$ (ii). Let $\boldsymbol{x}_n$ be a sequence of unit vectors, $\|\boldsymbol{x}_n\| = 1$ such that

$$(\lambda I - A)\boldsymbol{x}_n \to 0$$

Put

$$\boldsymbol{y}_n = \frac{(\lambda I - A)\boldsymbol{x}_n}{\|(\lambda I - A)\boldsymbol{x}_n\|}$$

then $\|\boldsymbol{y}_n\| = 1$ and

$$\|R_\lambda \boldsymbol{y}_n\| = \frac{\|\boldsymbol{x}_n\|}{\|(\lambda I - A)\boldsymbol{x}_n\|} \to \infty$$

which proves that R_λ is unbounded.

(ii) $\Rightarrow$ (i). Unboundedness of R_λ implies that there exists a sequence of unit vectors $\boldsymbol{y}_n$, $\|\boldsymbol{y}_n\| = 1$, such that

$$\|R_\lambda \boldsymbol{y}_n\| \to \infty$$

Put

$$\boldsymbol{x}_n = \frac{R_\lambda \boldsymbol{y}_n}{\|R_\lambda \boldsymbol{y}_n\|}$$

Vectors $\boldsymbol{x}_n$ are unit and

$$(\lambda I - A)\boldsymbol{x}_n = \frac{\boldsymbol{y}_n}{\|R_\lambda \boldsymbol{y}_n\|} \to 0$$

which proves that λ is an asymptotic eigenvalue. ∎

Exercises

6.8.1. Determine spectrum of operator $A : U \supset D(A) \to U$ where

$$U = L^2(\mathbb{R})$$

$$D(A) = H^1(\mathbb{R})$$

$$Au = i\frac{du}{dx}$$

Hint: Use Fourier transforms (cf. Example 6.8.1).

6.9 Spectra of Continuous Operators. Fundamental Properties

In this section we examine some basic properties of operators A taking a (whole) Banach space U into itself. If A were defined only on a subspace $D(A)$ of U, then, by continuity, A could be automatically extended to closure $\overline{D(A)}$ and then, say by zero, to the entire space U. The extension would also have been continuous and the norm of A, $\|A\|$, would have not changed. Thus, whatever we can prove for A defined on the *whole* U could be next reinterpreted for the restriction of A to the original $D(A)$ and therefore it makes a very little sense to "play" with continuous operators A which are defined on proper subspaces only.

We return now to Example 5.7.1 on Neumann series and consider the following sequence of partial sums

$$S_N = I + A + \ldots + A^N$$

The following proposition establishes a simple generalization of the Cauchy criterion of convergence for infinite series of numbers.

PROPOSITION 6.9.1

Let U be a Banach space and $A \in \mathcal{L}(U, U)$ and let S_N be the corresponding sequence of partial sums defined above. The following properties hold:

(i) There exists a limit

$$c = \lim_{n\to\infty} \|A^n\|^{\frac{1}{n}} = \inf_n \|A^n\|^{\frac{1}{n}}$$

(ii) If $c < 1$ then sequence S_N is convergent.
(iii) If $c > 1$ then sequence S_N diverges.

PROOF
(i) Define

$$a = \inf_n \|A^n\|^{\frac{1}{n}}$$

It must be $a \le \|A\|$ since

$$\|A^n\| = \|A \circ \ldots \circ A\| \le \|A\|^n$$

Let $\epsilon > 0$ be now an arbitrary small number. By definition of a, there must be an index m such that

$$\|A^m\|^{\frac{1}{m}} \le a + \epsilon$$

Set

$$M \stackrel{\text{def}}{=} \max\{1, \|A\|, \ldots, \|A^{m-1}\|\}$$

As every integer n can be represented in the form

$$n = k_n m + l_n \ , \ k_n \in \mathbb{Z} \ , 0 \le l_n \le m - 1$$

we obtain

$$a \le \ (\|A^n\|)^{\frac{1}{n}} \le (\|A^{l_n}\| \|A^m\|^{k_n})^{\frac{1}{n}}$$

$$\le M^{\frac{1}{n}} \|A^m\|^{\frac{k_n}{n}} \le M^{\frac{1}{n}} (a + \epsilon)^{\frac{n - l_n}{n}}$$

which, upon passing with $n \to \infty$, proves that

$$a \le \liminf \|A^n\|^{\frac{1}{n}} \le \limsup \|A^n\|^{\frac{1}{n}} \le a + \epsilon$$

from which (i) follows.
(ii) Denote $a_n = \|A^n\|$ and recall the Cauchy convergence test for series of real numbers

$$c = \lim_{n\to\infty} (a_n)^{\frac{1}{n}} < 1 \qquad \Rightarrow \qquad \sum_{n=1}^{\infty} a_n \text{ convergent}$$

Assuming that $c < 1$, we have for $N < M$

$$\begin{aligned} \|S_M - S_N\| &= \|A^{N+1} + \ldots + A^M\| \\ &\leq \|A^N\|(\|A\| + \ldots + \|A^{M-N}\|) \\ &\leq \|A^N\| \textstyle\sum_1^\infty a_n \to 0 \qquad \text{for } N \to \infty \end{aligned}$$

Consequently, S_N is a Cauchy sequence and therefore is convergent.

(iii) Let $c > 1$ and assume to the contrary that S_N is convergent. From $c > 1$ it follows that

$$\exists N \qquad \forall n \geq N \qquad \|A^n\|^{\frac{1}{n}} \geq 1 + \epsilon \;, \; \epsilon > 0$$

Consequently,

$$\|A^n\| \geq (1+\epsilon)^n \geq 1 \text{ for } n \geq N$$

But, at the same time convergence of S_N implies that $\|A^N\| \to 0$, a contradiction. ∎

Spectral Radius of a Continuous Operator. The number

$$\operatorname{spr}(A) = \lim_{n\to\infty} \|A^n\|^{\frac{1}{n}}$$

is called the *spectral radius of operator* A. Obviously,

$$\operatorname{spr}(A) \leq \|A\|$$

Let $\lambda \in \mathbb{C}$ be now an arbitrary complex number. Applying the Cauchy convergence criterion to the series

$$\frac{1}{\lambda}\sum_{k=0}^{\infty} \frac{1}{\lambda^k} A^k \;, \; S_N = \frac{1}{\lambda}\sum_{k=0}^{N} \frac{1}{\lambda^k} A^k$$

we see that S_N converges if $|\lambda| > \operatorname{spr}(A)$ and diverges if $|\lambda| < \operatorname{spr}(A)$. Moreover, passing to the limit in

$$\|(\lambda I - A)S_N - I\| = \|(\frac{A}{\lambda})^{N+1}\|$$

(cf. Example 5.7.1) we prove that for $\lambda < \operatorname{spr}(A)$, S_N converges to a right inverse of $\lambda I - A$.

Similarly, we prove that S_N converges to a left inverse and, consequently, the resolvent

$$R_\lambda = (\lambda I - A)^{-1} = \frac{1}{\lambda}\sum_{k=0}^{\infty}\lambda^{-k}A^k$$

exists and is continuous for $|\lambda| > \text{spr}(A)$. The whole spectrum of A therefore is contained in the closed ball centered at origin with radius equal to the spectral radius.

Later on in this section, we outline a much stronger result showing that the spectral radius $\text{spr}(A)$ is *equal* to the radius of the *smallest* closed ball containing spectrum $\sigma(A)$.

Consider now an arbitrary number λ_0 from the resolvent set of A, $\lambda_0 \in r(A)$ and let λ denote some other complex number. We have

$$\begin{aligned}\lambda I - A &= (\lambda_0 I - A) + (\lambda - \lambda_0)I \\ &= (\lambda_0 I - A)(I - (\lambda_0 - \lambda)R_{\lambda_0}) \\ &= (\lambda_0 - \lambda)(\lambda_0 I - A)((\lambda_0 - \lambda)^{-1}I - R_{\lambda_0})\end{aligned}$$

and, formally,

$$(\lambda I - A)^{-1} = (\lambda_0 - \lambda)^{-1}((\lambda_0 - \lambda)^{-1}I - R_{\lambda_0})^{-1}R_{\lambda_0}$$

Applying the Cauchy convergence criterion to

$$((\lambda_0 - \lambda)^{-1}I - R_{\lambda_0})^{-1}$$

we immediately learn that, if $|\lambda_0 - \lambda|^{-1} > \|R_{\lambda_0}\|$ or, equivalently, $|\lambda - \lambda_0| < \|R_{\lambda_0}\|^{-1}$, then the inverse above exists and is continuous. Consequently, resolvent R_λ exists and is continuous as well. Moreover, the following formula holds

$$R_\lambda = (\lambda_0 - \lambda)^{-1}((\lambda_0 - \lambda)^{-1}I - R_{\lambda_0})^{-1}R_{\lambda_0}$$

It follows that the resolvent set $r(A)$ is open and therefore the spectrum $\sigma(A)$ must be closed. Since it is simultaneously bounded, it must be compact.

We summarize our observations in the following proposition.

PROPOSITION 6.9.2

Let A be a bounded, linear operator from a Banach space U into itself. The following properties hold:

(i) Spectrum of A, $\sigma(A)$, is compact.
(ii) $\sigma(A) \subset \bar{B}(0, \text{spr}(A))$.
(iii) For every $|\lambda| > \text{spr}(A)$ *the corresponding resolvent is a sum of the convergent Neumann series*

$$R_\lambda = \frac{1}{\lambda} \sum_{k=0}^{\infty} \lambda^{-k} A^k$$

(iv) For $\lambda, \mu \in r(A)$

$$R_\lambda - R_\mu = (\mu - \lambda) R_\mu R_\lambda$$

In particular, resolvents are permutable.
(v) Resolvent set of A is contained in the resolvent set of the transpose operator A'

$$r(A) \subset r(A')$$

(vi) In the case of a Hilbert space U, resolvent of the adjoint operator A^*, $r(A^*)$, is equal to the image of $r(A)$ under the complex conjugate operation, i.e.,

$$\lambda \in r(A) \Leftrightarrow \bar{\lambda} \in r(A^*)$$

PROOF It remains to prove (iv), (v) and (vi).
(iv) We have

$$R_\lambda - R_\mu = (\lambda I - A)^{-1} - (\mu I - A)^{-1}$$

Multiplying by $(\lambda I - A)$ from the right-hand side and by $(\mu I - A)$ from the left-hand side, we get

$$(\mu I - A)(R_\lambda - R_\mu)(\lambda I - A) = (\mu I - A) - (\lambda I - A) = (\mu - \lambda) I$$

which proves the assertion.
(v) We have

$$((\lambda I - A)^{-1})' = ((\lambda I - A)')^{-1} = (\lambda I - A')^{-1}$$

Thus, if $(\lambda I - A)^{-1}$ exists and is continuous, then $(\lambda I - A')$ exists and is continuous as well.

Note that for reflexive spaces $r(A') \subset r(A'') = r(A)$ and therefore $r(A) = r(A')$.

(vi) follows from the identity

$$((\lambda I - A)^{-1})^* = (\bar{\lambda} I - A^*)^{-1}$$

∎

We conclude this section with an important geometrical characterization of spectral radius. Only an outline of the proof is provided as the proof uses essentially means of complex analysis exceeding the scope of this book.

PROPOSITION 6.9.3

Let A be a bounded, linear operator from a Banach space U into itself. The following characterization of the spectral radius holds

$$\text{spr}A = \lim_{n\to\infty} \|A^n\|^{\frac{1}{n}} = \max_{\lambda\in\sigma(A)} |\lambda|$$

i.e., spectral radius is equal to the maximum (in modulus) number from the spectrum of A.

PROOF

Step 1. We define the *characteristic set* of A, denoted $\rho(A)$, as the set of characteristic values of A

$$\rho(A) \overset{\text{def}}{=} \{\rho \in \mathbb{C} : \exists \boldsymbol{u} \neq \boldsymbol{0} : (I - \rho A)\boldsymbol{u} = \boldsymbol{0}\}$$

Obviously, for $\lambda \neq 0$

$$\lambda^{-1} \in \rho(A) \qquad \Leftrightarrow \qquad \lambda \in \sigma(A)$$

The characteristic set $\rho(A)$, as an inverse image of spectrum $\sigma(A)$ through the continuous map $\lambda \to \lambda^{-1}$, is closed and obviously does not contain 0.

Step 2. For $\rho^{-1} = \lambda \in r(A)$ we introduce the resolvent (of the second kind) B_ρ, defined as

$$B_\rho = (I - \rho A)^{-1}$$

A direct calculation reveals the relation between the two types of resolvents

$$R_\lambda = \lambda^{-1} I + \lambda^{-2} B_{\lambda^{-1}}$$

Step 3. Property (iv) proved in Proposition 6.9.2 implies that

$$B_\rho - B_\mu = (\rho - \mu) B_\rho B_\mu$$

Step 4. It follows from the Step 3 that resolvent $B_\rho \in \mathcal{L}(U,U)$ is a continuous function of ρ.

Step 5. It follows from the Step 3 and Step 4 results that for any $\boldsymbol{x} \in U$ and $f \in U'$ function

$$\phi(\rho) = \langle f, B_\rho(\boldsymbol{x}) \rangle$$

is holomorphic in ρ (analytic in the complex sense). Indeed, it is sufficient to show that ϕ is differentiable (analyticity in the complex sense is *equivalent* to the differentiability!). But

$$\phi'(\rho) = \lim_{\mu \to \rho} \langle f, \tfrac{B_\mu - B_\rho}{\mu - \rho}(\boldsymbol{x}) \rangle$$

$$= \lim_{\mu \to \rho} \langle f, B_\mu B_\rho(\boldsymbol{x}) \rangle = \langle f, B_\rho^2(\boldsymbol{x}) \rangle$$

Step 6. Consequently, $\phi(\rho)$ can be expanded into its Taylor's series at $\rho = 0$:

$$\phi(\rho) = \sum_{k=0}^{\infty} \frac{\phi^{(k)}(0)}{k!} \rho^k$$

and, at the same time, from the definition of the spectral radius (cf. Proposition 6.9.1) follows that

$$\phi(\rho) = \sum_{k=0}^{\infty} \langle f, A^{k+1} \boldsymbol{x} \rangle \rho^k$$

Both series, as the *same* representations of the *same* function, must converge (uniformly!) in the ball with the *same* radius. The second of the series converges for

$$|\rho| < (\mathrm{spr}(A))^{-1} \qquad (|\lambda| = |\rho^{-1}| > \mathrm{spr}(A))$$

while the first one (a standard result from complex analysis) converges for all ρ from a ball containing no singular points of $\phi(\rho)$, i.e.,

$$|\rho| < (\max_{\lambda \in \sigma(A)} |\lambda|)^{-1}$$

Consequently, it must be

$$\max_{\lambda \in \sigma(A)} = \mathrm{spr}(A)$$

∎

Exercises

6.9.1. Let X be a real normed space and $X \times X$ its complex extension (cf. Section 6.1). Let $A : X \to X$ be a linear operator and let $\tilde{A}$ denote its extension to the complex space defined as

$$\tilde{A}((u, v)) = (Au, Av)$$

Suppose that $\lambda \in \mathbb{C}$ is an eigenvalue of $\tilde{A}$ with a corresponding eigenvector $w = (u, v)$. Show that the complex conjugate $\bar{\lambda}$ is an eigenvalue of $\tilde{A}$ as well with the corresponding eigenvector equal $\bar{w} = (u, -v)$.

6.9.2. Let U be a Banach space and let λ and μ be two different eigenvalues ($\lambda \neq \mu$) of an operator $A \in \mathcal{L}(U, U)$ and its transpose $A' \in \mathcal{L}(U', U')$ with corresponding eigenvectors $\boldsymbol{x} \in U$ and $\boldsymbol{g} \in U'$. Show that

$$\langle \boldsymbol{g}, \boldsymbol{x} \rangle = 0$$

6.10 Spectral Theory for Compact Operators

In this section we focus on the special class of compact (completely continuous) operators on Banach and Hilbert spaces.

Let T be a compact operator from a Banach space X into itself and λ a nonzero complex number. According to the Fredholm alternative (cf. Section 5.20), operator $\lambda I - T$ or, equivalently, $I - \lambda^{-1}T$ has either a continuous inverse (bijectivity and continuity of $A = I - \lambda^{-1}T$ implies continuity of $A^{-1} = R_\lambda$!) or it is not injective and its null space

$$X_\lambda = \mathcal{N}(I - \lambda^{-1}T) = \mathcal{N}(\lambda I - T)$$

has a finite dimension. Consequently, the whole spectrum of T, except for $\lambda = 0$, reduces to the point spectrum $\sigma_P(T)$ consisting of eigenvalues λ with corresponding finite-dimensional eigenspaces X_λ.

The following theorem gives more detailed information on $\sigma_P(T)$.

THEOREM 6.10.1

Let T be a compact operator from a Banach space X into itself. Then $\sigma(T) - \{0\}$ consists of, at most, a countable set of eigenvalues λ_n. If the *set is infinite then* $\lambda_n \to 0$ *as* $n \to \infty$.

PROOF It is sufficient to prove that for every $r > 0$ there exists at most a *finite* number of eigenvalues λ_n such that $|\lambda_n| > r$. Assume, to the contrary, that there exists an infinite sequence of distinct eigenvalues $\lambda_n, |\lambda_n| > r$ with a corresponding sequence of unit eigenvectors $\boldsymbol{x}_n$:

$$T\boldsymbol{x}_n = \lambda_n \boldsymbol{x}_n, \qquad \|\boldsymbol{x}_n\| = 1$$

We claim that $\boldsymbol{x}_n$ are linearly independent. Indeed, from the equality

$$\boldsymbol{x}_{n+1} = \sum_{k=1}^{n} \alpha_k \boldsymbol{x}_k$$

it follows that

$$\lambda_{n+1}\boldsymbol{x}_{n+1} = T\boldsymbol{x}_{n+1} = \sum_{k=1}^{n} \alpha_k T\boldsymbol{x}_k = \sum_{k=1}^{n} \alpha_k \lambda_k \boldsymbol{x}_k$$

and, consequently,

$$\boldsymbol{x}_{n+1} = \sum_{k=1}^{n} \alpha_k \frac{\lambda_k}{\lambda_{n+1}} \boldsymbol{x}_k$$

As the coefficients α_k are unique, there must be $\frac{\lambda_k}{\lambda_{n+1}} = 1$ for some k, a contradiction.

Let X_n denote now the span of the first n eigenvectors $\boldsymbol{x}_k$

$$X_n = \operatorname{span}\{\boldsymbol{x}_1, \dots, \boldsymbol{x}_n\}$$

By the lemma on almost perpendicularity, there exists a sequence of unit vectors $\boldsymbol{y}_n \in X_n$ such that

$$\rho(\boldsymbol{y}_{n+1}, X_n) > \frac{1}{2} \qquad , \qquad \|\boldsymbol{y}_{n+1}\| = 1$$

Let now $\boldsymbol{x} \in X_n$, i.e., $\boldsymbol{x} = \sum_{k=1}^{n} \alpha_k \boldsymbol{x}_k$. Then

$$T\boldsymbol{x} = \sum_{k=1}^{n} \alpha_k T\boldsymbol{x}_k = \sum_{k=1}^{n} \alpha_k \lambda_k \boldsymbol{x}_k \in X_n$$

and, at the same time, denoting $B_n = \lambda_n I - T$

$$B_n \boldsymbol{x} = \sum_{k=1}^{n} \alpha_k (\lambda_n I - T) \boldsymbol{x}_k$$

$$= \sum_{k=1}^{n} \alpha_k (\lambda_n - \lambda_k) \boldsymbol{x}_k \in X_{n-1}$$

Thus, for $m > n$

$$\|T(\frac{\boldsymbol{y}_m}{\lambda_m}) - T(\frac{\boldsymbol{y}_n}{\lambda_n})\| = \|\boldsymbol{y}_m - B_m(\frac{\boldsymbol{y}_m}{\lambda_m}) - \boldsymbol{y}_n + B_n(\frac{\boldsymbol{y}_n}{\lambda_n})\| > \frac{1}{2}$$

since

$$-B_m(\frac{\boldsymbol{y}_m}{\lambda_m}) - \boldsymbol{y}_n + B_n(\frac{\boldsymbol{y}_n}{\lambda_n}) \in X_n$$

At the same time, sequence $\frac{\boldsymbol{y}_n}{\lambda_n}$ is bounded ($|\lambda_n| > r$!) and therefore we can extract a strongly convergent subsequence from $T(\lambda_n^{-1} \boldsymbol{y}_n)$, satisfying in particular the Cauchy condition, a contradiction. ∎

For the rest of this section, we shall restrict ourselves to a more specialized class of compact operators—the *normal* and compact operators. We begin by recording some simple observations concerning all normal and continuous operators (not necessarily compact) on a *Hilbert space* U.

PROPOSITION 6.10.1

Let U be a Hilbert space and $A \in \mathcal{L}(U, U)$ be a normal operator, i.e., $AA^* = A^*A$. The following properties hold:

(i) For any eigenvalue λ of A and a corresponding eigenvector $\boldsymbol{u}$, $\bar{\lambda}$ is an eigenvalue of A^* with the same eigenvector $\boldsymbol{u}$.

(ii) For any two distinct eigenvectors $\lambda_1 \neq \lambda_2$ of A, the corresponding eigenvectors $\boldsymbol{u}_1$ and $\boldsymbol{u}_2$ are orthogonal

$$(\boldsymbol{u}_1, \boldsymbol{u}_2) = 0$$

(iii) $\operatorname{spr}(A) = \|A\|$.

PROOF

(i) If A is a normal operator then $A - \lambda I$ is normal as well and Proposition 6.5.3 implies that

$$\|(A - \lambda I)\boldsymbol{u}\| = \|(A^* - \bar{\lambda} I)\boldsymbol{u}\|$$

Consequently,

$$(A - \lambda I)\boldsymbol{u} = \boldsymbol{0} \qquad \Leftrightarrow \qquad (A^* - \bar{\lambda} I)\boldsymbol{u} = \boldsymbol{0}$$

which proves the assertion.

(ii) We first prove that if λ_1 is an eigenvalue of *any* operator $A \in \mathcal{L}(U, U)$ with a corresponding eigenvector $\boldsymbol{u}_1$, and λ_2 is an eigenvalue of adjoint A^* with corresponding eigenvector $\boldsymbol{u}_2$, then (cf. Exercise 6.9.2)

$$\lambda_1 \neq \overline{\lambda}_2 \qquad \text{implies} \qquad (\boldsymbol{u}_1, \boldsymbol{u}_2) = 0$$

Indeed, for $\lambda_1 \neq 0$ we have

$$(\boldsymbol{u}_1, \boldsymbol{u}_2) = (A(\tfrac{\boldsymbol{u}_1}{\lambda_1}), \boldsymbol{u}_2) = (\frac{\boldsymbol{u}_1}{\lambda_1}, A^*\boldsymbol{u}_2)$$

$$= \tfrac{\bar{\lambda}_2}{\lambda_1}(\boldsymbol{u}_1, \boldsymbol{u}_2)$$

which implies that $(\boldsymbol{u}_1, \boldsymbol{u}_2) = 0$.

We proceed similarly for $\lambda_2 \neq 0$. Finally, (ii) follows from the orthogonality result just proved and property (i).

(iii) follows immediately from Corollary 6.5.1 and the definition of spectral radius. ∎

COROLLARY 6.10.1

Let A be a normal, compact operator from a Hilbert space U into itself. Then the norm of A is equal to the maximum (in modulus) eigenvalue of A.

PROOF The proof follows immediately from Proposition 6.9.3, Theorem 6.10.1 and Proposition 6.10.1 (iii). ∎

We are ready now to state our main result for compact and normal operators on Hilbert spaces.

THEOREM 6.10.2
(Spectral Decomposition Theorem for Compact and Normal Operators)

Let U be a Hilbert space and let $T \in \mathcal{L}(U,U)$ be a compact and normal operator. Let

$$|\lambda_1| \geq |\lambda_2| \geq \dots \qquad (\to 0 \textit{ if infinite})$$

denote the finite or infinite sequence of eigenvalues of T and $P_1, P_2, \dots$ the corresponding orthogonal projections on finite-dimensional eigenspaces

$$N_i \stackrel{\text{def}}{=} \mathcal{N}(\lambda_i I - A)$$

Then

$$T = \sum_{i=1}^{\infty} \lambda_i P_i$$

and for the adjoint operator

$$T^* = \sum_{i=1}^{\infty} \bar{\lambda}_i P_i$$

PROOF Define

$$T_1 = T - \lambda_1 P_1$$

The following properties hold:

(i) T_1 is normal, since both T and P_1 are normal (cf. Example 6.5.6) and linear combinations of normal operators are normal.

(ii) T_1 is compact, since both T and P_1 (the eigenspace is finite-dimensional!) are compact, and linear combinations of compact operators are compact.

(iii) Eigenvalues of operator T

$$|\lambda_2| \geq |\lambda_3| \geq \dots$$

are also eigenvalues of T_1. Indeed, due to the orthogonality of eigenvalues (Proposition 6.10.1)

$$(T - \lambda_1 P_1)\boldsymbol{u}_i = T\boldsymbol{u}_i - \lambda_i P_1 \boldsymbol{u}_i = T\boldsymbol{u}_i = \lambda_i \boldsymbol{u}_i$$

for any eigenvector $\boldsymbol{u}_i \in N_i, i = 2, 3, \dots$.

(iv) T_1 vanishes on N_1 (definition of eigenvalue) and takes on values in $N_1^{\perp}$. Indeed, for any $\boldsymbol{u}_1 \in N_1$ and $\boldsymbol{u} \in U$

$$
\begin{aligned}
(T_1\boldsymbol{u}, \boldsymbol{u}_1) = ((T - \lambda_1 P_1)\boldsymbol{u}, \boldsymbol{u}_1) \\
= \quad (\boldsymbol{u}, T^*\boldsymbol{u}_1) - \lambda_1(\boldsymbol{u}, P_1\boldsymbol{u}_1) \\
= \quad (\boldsymbol{u}, T^*\boldsymbol{u}_1 - \bar{\lambda}_1\boldsymbol{u}_1) \quad = 0
\end{aligned}
$$

Assume now that $\lambda \neq 0$ is an eigenvalue of T_1 with a corresponding eigenvector $\boldsymbol{u}$. Making use of the decomposition

$$
\boldsymbol{u} = \boldsymbol{u}_1 + \boldsymbol{u}_2, \text{ where } \quad \boldsymbol{u}_1 \in N_1, \boldsymbol{u}_2 \in N_1^{\perp}
$$

we have

$$
T_1\boldsymbol{u} = \lambda\boldsymbol{u}_1 + \lambda\boldsymbol{u}_2
$$

and therefore $\boldsymbol{u}_1 = \boldsymbol{0}$. Consequently,

$$
T\boldsymbol{u} = (T_1 + \lambda_1 P_1)\boldsymbol{u} = T_1\boldsymbol{u}_2 = \lambda\boldsymbol{u}_2
$$

which means that λ is also an eigenvalue of T. In other words, there are *no* eigenvalues of T_1 other than the original eigenvalues $\lambda_1, \lambda_2, \ldots$ of T.

(v) Properties (i) to (iv) imply that

$$
\|T_1\| = |\lambda_2|
$$

By induction

$$
\|T - \sum_{i=1}^{\infty} \lambda_i P_i\| = |\lambda_{i+1}|
$$

where the whole process stops if the sequence λ_i is finite or $|\lambda_{i+1}| \to \infty$ in the infinite case. ∎

We will need yet the following lemma.

LEMMA 6.10.1

Let P_n be a sequence of mutually orthogonal projections in a Hilbert space U, i.e.,

$$
P_m P_n = \delta_{mn} P_n
$$

Then:

(i) The series $\sum_{n=1}^{\infty} P_n\boldsymbol{u}$ converges for every $\boldsymbol{u} \in U$ and

$$Pu = \sum_{n=1}^{\infty} P_n u$$

is an orthogonal projection on U.

(ii) $\mathcal{R}(P) = \overline{\text{span}(\cup_n \mathcal{R}(P_n))}$.

PROOF For any $u \in U$,

$$\sum_{k=1}^{n} \|P_k u\|^2 = \|\sum_{k=1}^{n} P_k u\|^2 \leq \|u\|^2$$

which proves that $\sum_{k=1}^{\infty} \|P_k u\|^2$ is convergent. This in turn implies that

$$\|\sum_{k=n}^{m} P_k u\|^2 = \sum_{k=n}^{m} \|P_k u\|^2 \to 0$$

as $n, m \to \infty$, which proves that $\sum_{k=1}^{n} P_k u$ is (strongly) convergent to a limit Pu.

Passing to the limit with $n \to \infty$ in

$$P_m \sum_{k=1}^{n} P_k u = P_m u \qquad m \leq n$$

we get

$$P_m P u = P_m u$$

and upon summing up in m

$$PPu = Pu$$

Thus P is a projection.

In the same way, passing with $n \to \infty$ in

$$(\sum_{k=1}^{n} P_k u, u - \sum_{k=1}^{n} P_k u) = 0$$

we prove that P is an orthogonal projection.

Finally, condition (ii) follows from definition of P. ∎

THEOREM 6.10.3

Let U be a Hilbert space and T be a compact and normal operator from U into itself. Let

$$|\lambda_1| \geq |\lambda_2| \geq \ldots \qquad (\to 0 \textit{ if infinite})$$

denote the sequence of its eigenvalues with corresponding eigenspaces N_i and let N denote the null space of operator T (eigenspace of $\lambda = 0$ eigenvalue!). Let $P_i, i = 1, 2, \ldots$ denote the orthogonal projections on N_i and P_0 the orthogonal projection on N.

Then the following holds:

$$\boldsymbol{u} = \sum_{i=0}^{\infty} P_i \boldsymbol{u}$$

PROOF Let $\boldsymbol{u} \in N, \boldsymbol{v} \in U$. Then

$$0 = (T\boldsymbol{u}, \boldsymbol{v}) = (\boldsymbol{u}, T^*\boldsymbol{v})$$

implies that $\mathcal{R}(T^*) \subset N^\perp$ and, consequently,

$$N \subset \mathcal{R}(T^*)^\perp$$

(cf. Exercises 6.2.1 (i) and 6.2.2).

At the same time, for $\boldsymbol{y} \in \mathcal{R}(T^*)^\perp$ we have

$$0 = (\boldsymbol{y}, T^*\boldsymbol{x}) = (T\boldsymbol{y}, \boldsymbol{x}) \qquad \forall \boldsymbol{x} \in U$$

and, consequently, $T\boldsymbol{y} = 0$, i.e., $\boldsymbol{y} \in N$, which all together proves that

$$N = \mathcal{R}(T^*)^\perp$$

As $\bar{\lambda}_i$ are eigenvalues of T^* with the same corresponding eigenspaces N_i and the range of $T^*, \mathcal{R}(T^*)$, is closed, applying Lemma 6.10.1 we have

$$\boldsymbol{u} = \sum_{i=1}^{\infty} P_i \boldsymbol{u} \qquad \text{for } \boldsymbol{u} \in \mathcal{R}(T^*)$$

and, finally,

$$\begin{aligned} \boldsymbol{y} &= P_0\boldsymbol{u} + (\boldsymbol{u} - P_0\boldsymbol{u}) \\ &= P_0\boldsymbol{u} + \textstyle\sum_{i=1}^{\infty} P_i(\boldsymbol{u} - P_0\boldsymbol{u}) \\ &= \textstyle\sum_{i=0}^{\infty} P_i\boldsymbol{u} \end{aligned}$$

∎

REMARK 6.10.1. The decomposition formula for $\boldsymbol{u}$ is frequently rewritten in the operator form as

$$I = \sum_{i=0}^{\infty} P_i$$

and called the *resolution of identity* (see the definition at the end of this section). The essential difference between the resolution of identity and spectral representation for compact and normal operators

$$A = \sum_{i=1}^{\infty} \lambda_i P_i$$

is the underlying kind of convergence. The first formula is understood in the sense of the *strong convergence* of operators, i.e.,

$$\sum_{i=0}^{n} P_i \boldsymbol{u} \to \boldsymbol{u} \qquad \forall \boldsymbol{u} \in U$$

whereas the second one is in the *operator norm*

$$\| \sum_{i=1}^{n} \lambda_i P_i - A \| \to 0$$

and, in particular, implies the uniform convergence of operator values on bounded sets! ∎

COROLLARY 6.10.2

Let U be a Hilbert space and suppose that a bounded, normal and compact operator T from U into itself exists such that the null space of T is finite dimensional. Then U admits an orthonormal basis.

PROOF Let $\phi_1, \ldots, \phi_{n_0}$ be an orthonormal basis for N, $\phi_{n_0+1}, \ldots, \phi_{n_0+n_1}$ an orthonormal basis for N_1, etc. ∎

COROLLARY 6.10.3

Let $\phi_1, \phi_2, \ldots$ be an orthonormal basis selected in the previous corollary. Then

$$T\boldsymbol{u} = \sum_{k=1}^{\infty} \lambda_k (\boldsymbol{u}, \phi_k) \phi_k$$

where λ_k repeat themselves if dim $N_k > 1$.

PROOF The proof follows immediately from the Fourier series representation. ∎

Spectral Representation for Compact Operators. Let U, V be two Hilbert spaces and T be a compact (not necessarily normal!) operator from U into V. As operator T^*T is compact, self-adjoint and semipositive-definite, it admits the representation

$$T^*T\boldsymbol{u} = \sum_{k=1}^{\infty} \alpha_k^2(\boldsymbol{u}, \phi_k)\phi_k$$

where α_k^2 are the positive eigenvalues of T^*T (cf. Exercise 6.10.1) and ϕ_k are the corresponding eigenvectors

$$T^*T\phi_k = \alpha_k^2\phi_k \qquad k = 1, 2, \dots \qquad \alpha_1 \geq \alpha_2 \geq \dots > 0$$

Set

$$\phi_k' = \alpha_k^{-1}T\phi_k$$

Vectors ϕ_k' form an orthonormal family in V, since

$$\begin{aligned}(\alpha_k^{-1}T\phi_k, \alpha_l^{-1}T\phi_l) &= (\alpha_k^{-1}\alpha_l^{-1}T^*T\phi_k, \phi_l) \\ &= (\tfrac{\alpha_k}{\alpha_l}\phi_k, \phi_l) = \delta_{kl}\end{aligned}$$

We claim that

$$T\boldsymbol{u} = \sum_{k=1}^{\infty} \alpha_k(\boldsymbol{u}, \phi_k)\phi_k'$$

Indeed, the series satisfies the Cauchy condition as

$$\begin{aligned}\sum_{k=n}^{m} \|\alpha_k(\boldsymbol{u}, \phi_k)\phi_k'\|^2 &= \textstyle\sum_{k=n}^{m} \alpha_k^2|(\boldsymbol{u}, \phi_k)|^2 \\ &\leq \alpha_n^2 \textstyle\sum_{k=n}^{m} |(\boldsymbol{u}, \phi_k)|^2 \leq \alpha_n\|\boldsymbol{u}\|^2\end{aligned}$$

Moreover, both sides vanish on $\mathcal{N}(T) = \mathcal{N}(T^*T)$ (explain why the two sets are equal to each other) and on eigenvectors ϕ_l,

$$\sum_{k=1}^{\infty} \alpha_k(\phi_l, \phi_k)\phi'_k = \sum_{k=1}^{\infty} \alpha_k \delta_{lk} \phi'_k = \alpha_l \phi'_l = T\phi_l$$

and therefore on the whole space U.

Resolution of Identity. A sequence of orthogonal projections $\{P_n\}$ on a Hilbert space U is said to be a *resolution of identity* if

(i) P_n is orthogonal to $P_m, m \neq n$ ($P_m P_n = 0,$ for all $m \neq n$) and

(ii) $I = \sum_n P_n$ (strong convergence of the series is assumed).

The series may be finite or infinite.

Thus, according to Theorem 6.10.3, every compact and normal operator in a Hilbert space generates a corresponding resolution of identity of orthogonal projections on its eigenspaces N_λ.

Example 6.10.1

Consider the space $U = L^2(0,1)$ and the integral operator A defined as

$$(Au)(x) \stackrel{\text{def}}{=} \int_0^x u(\xi)d\xi$$

Rewriting it in the form

$$(Au)(x) \stackrel{\text{def}}{=} \int_0^1 K(x,\xi)u(\xi)d\xi$$

where

$$K(x,\xi) = \begin{cases} 1 \text{ for } \xi \leq x \\ 0 \text{ for } \xi > x \end{cases}$$

we easily see that A falls into the category of compact operators discussed in Example 5.15.1 and Exercise 5.15.3. As

$$Au = 0 \qquad \text{implies} \qquad \frac{d}{dx}(Au) = u = 0$$

the null space of A and, consequently, A^*A reduces to the zero vector. The adjoint A^* (cf. Example 5.16.1) is given by the formula

$$(A^*u)(x) = \int_x^1 u(\xi)d\xi$$

The eigenvalue problem for A^*A and $\lambda^2 \neq 0$ reduces to solving the equation

$$\lambda^2 u(y) = \int_y^1 \int_0^x u(\xi)\, d\xi\, dx$$

or, equivalently,

$$-\lambda^2 u'' = u$$

with boundary conditions

$$u(1) = 0 \qquad \text{and} \qquad u'(0) = 0$$

This leads to the sequence of eigenvalues

$$\lambda_n^2 = (\frac{\pi}{2} + n\pi)^{-2} \qquad n = 0, 1, 2, \ldots$$

with the corresponding (normalized) eigenvectors

$$u_n = \frac{1}{\sqrt{2}} \cos((\frac{\pi}{2} + n\pi)x) \qquad n = 0, 1, 2, \ldots$$

Consequently, u_n form an orthonormal basis in $L^2(0,1)$ and we have the following representation for the integral operator A

$$(Au)(x) = \int_0^x u(\xi)\, d\xi = \sum_{n=0}^{\infty} a_n \sin(\frac{\pi}{2} + n\pi)x$$

where

$$a_n = \frac{1}{2}(\frac{\pi}{2} + n\pi)^{-1} \int_0^1 u(x) \cos(\frac{\pi}{2} + n\pi)x\, dx \qquad n = 0, 1, \ldots$$

□

Exercises

6.10.1. Let T be a compact operator from a Hilbert space U into a Hilbert space V. Show that:
(i) T^*T is a compact, self-adjoint, positive definite operator from a space U into itself.
(ii) All eigenvalues of a self-adjoint operator on a Hilbert space are real.

Use Theorem 6.10.3 to conclude that all eigenvalues of T^*T are real and positive.

6.11 Spectral Theory for Self-Adjoint Operators

We conclude our presentation of elements of spectral theory with a discussion of the very important case of self-adjoint operators in Hilbert spaces. Most of the presented results exceed considerably the scope of this book and are presented without proofs. For a complete presentation of the theory we refer the reader to [6].

Let U be a Hilbert space. Recall that an operator A defined on a (dense) domain $D(A) \subset U$ into U is called *self-adjoint* iff it coincides with its adjoint operator, i.e., $A = A^*$. For A defined on a proper subspace only, the equality of operators involves the equality of their domains!, i.e., $D(A) = D(A^*)$. As adjoint operators are always closed, every self-adjoint operator is necessarily closed. If domain of A, $D(A)$ equals the whole space U then, by the closed graph theorem, A must be continuous. Thus only two cases are of interest: the case of continuous, i.e., bounded operators defined on the whole space U and the case of closed operators defined on a proper (dense) subspace $D(A)$ of U. We discuss first the bounded operators.

Spectral Theory for Self-Adjoint Bounded Operators

First of all, as the self-adjoint operators fall into the category of normal operators, all the results concerning compact and normal operators, studied in the previous section, remain valid. Additionally, all eigenvalues of A are real. Indeed, if λ is an eigenvalue of A with a corresponding eigenvector $\boldsymbol{u}$ then

$$\lambda\|\boldsymbol{u}\|^2 = \lambda(\boldsymbol{u},\boldsymbol{u}) = (\lambda\boldsymbol{u},\boldsymbol{u}) = (A\boldsymbol{u},\boldsymbol{u})$$

$$= (\boldsymbol{u},A\boldsymbol{u}) = (\boldsymbol{u},\lambda\boldsymbol{u}) = \bar{\lambda}(\boldsymbol{u},\boldsymbol{u}) = \bar{\lambda}\|\boldsymbol{u}\|^2$$

and, consequently, $\lambda = \bar{\lambda}$.

Thus every self-adjoint and compact operator A admits the representation

$$A = \sum_{i=1}^{\infty} \lambda_i P_i$$

where $|\lambda_1| \geq |\lambda_2| \geq \ldots$ is a decreasing (possibly finite) series of real eigenvalues and P_i are the corresponding orthogonal projections on eigenspaces $N_i = \mathcal{N}(\lambda_i I - A)$.

The observation concerning the eigenvalues of self-adjoint operators can be immediately generalized to the case of asymptotic eigenvalues, which must be real as well. To see it, let λ be an asymptotic eigenvalue of a self-adjoint operator A and $\boldsymbol{u}_n$ a corresponding sequence of unit vectors $\boldsymbol{u}_n, \|\boldsymbol{u}_n\| = 1$, such that

$$(\lambda I - A)\boldsymbol{u}_n \to \mathbf{0} \qquad \text{as } n \to \infty$$

We have

$$((\lambda I - A)\boldsymbol{u}_n, \boldsymbol{u}_n) \to 0 \qquad \text{and} \qquad (\boldsymbol{u}_n, (\lambda I - A)\boldsymbol{u}_n) \to 0$$

Consequently,

$$\lambda = \lambda(\boldsymbol{u}_n, \boldsymbol{u}_n) = (\lambda \boldsymbol{u}_n, \boldsymbol{u}_n) = \lim_{n\to\infty} (A\boldsymbol{u}_n, \boldsymbol{u}_n)$$

and

$$\bar{\lambda} = \bar{\lambda}(\boldsymbol{u}_n, \boldsymbol{u}_n) = (\boldsymbol{u}_n, \lambda \boldsymbol{u}_n) = \lim_{n\to\infty} (\boldsymbol{u}_n, A\boldsymbol{u}_n)$$

both limits on the right-hand side being equal, which proves that $\lambda = \bar{\lambda}$.

It can be proved that the asymptotic eigenvalues consitute the whole spectrum of A, i.e., if λ is not an asymptotic eigenvalue of A, then $(\lambda I - A)^{-1}$ is defined on the *whole* U (i.e., $\mathcal{R}(\lambda I - A) = U$) and is bounded (cf. Proposition 6.8.1). This result has two immediate consequences:

- Self-adjoint operators have no residual spectrum, i.e., $\sigma(A)$ may consist of point and continuous spectrum only.
- Spectrum $\sigma(A)$ is real.

Define now

$$m \overset{\text{def}}{=} \inf_{\|\boldsymbol{u}\|=1} \langle A\boldsymbol{u}, \boldsymbol{u} \rangle \qquad M \overset{\text{def}}{=} \sup_{\|\boldsymbol{u}\|=1} \langle A\boldsymbol{u}, \boldsymbol{u} \rangle$$

Both quantities are finite since $\|A\| = \max\{|m|, |M|\}$ (cf. Exercise 6.5.4). It follows immediately from the definition of an asymptotic eigenvalue that

$$\sigma(A) \subset [m, M]$$

The following theorems formulate the main result concerning spectral representation of self-adjoint operators.

THEOREM 6.11.1

Let U be a Hilbert space and A a bounded, self-adjoint operator from U into itself. There exists then a one-parameter family of orthogonal projections $I(\lambda) : U \to U, \lambda \in \mathbb{R}$, which satisfies the following conditions:

(i) $\lambda \le \mu \Rightarrow I(\lambda) \le I(\mu)$ (the family is increasing).

(ii) $I(\lambda) = 0$ for $\lambda < m$ and $I(\lambda) = I$ for $\lambda > M$.

(iii) function $\lambda \to I(\lambda)$ is right-continuous, i.e.,

$$I(\lambda) = \lim_{\mu \to \lambda^+} I(\mu)$$

(iv) $\lambda \in r(A)$ (resolvent set of A) iff λ is a point of constancy of A, i.e., there exists a constant $\delta > 0$ such that $I(\lambda - \delta) = I(\lambda + \delta)$.

(v) $\lambda \in \sigma_P(A)$ (is an eigenvalue of A) iff λ is a discontinuity point of $I(\lambda)$, i.e.,

$$\lim_{\mu \to \lambda^-} I(\mu) \neq I(\lambda)$$

The inequality of projections in condition (i) of the theorem makes sense for any self-adjoint operators A and B and is understood as

$$A \ge B \overset{\text{def}}{=} A - B \ge 0 \text{ (positive definite)}$$

The family of projections $I(\lambda), \lambda \in \mathbb{R}$ is known as the *spectral family of A*.

The following example explains the relation between the spectral family and the resolution of identity defined in the previous section.

Example 6.11.1

In the case of a compact and self-adjoint operator A on a Hilbert space U, the spectral family $I(\lambda)$ can be represented in terms of orthogonal projections $P(\lambda)$ corresponding to eigenvalues λ as

$$I(\lambda) = \sum_{\mu < \lambda} P(\lambda)$$

where the sum on the right-hand side is finite for $\lambda < 0$ and is to be understood in the sense of the strong convergence of operators for $\lambda \geq 0$ if A has infinitely many eigenvalues. $\square$

Given an arbitrary partition $\mathcal{P}_n$

$$\lambda_0 < \lambda_1 < \ldots < \lambda_n < \lambda_{n+1}$$

where $\lambda_0 < m$ and $\lambda_{n+1} > M$, we construct now two approximate Riemann-like sums

$$s_n = \sum_{k=0}^{n-1} \lambda_k [I_{\lambda_{k+1}} - I_{\lambda_k}]$$

and

$$S_n = \sum_{k=0}^{n-1} \lambda_{k+1} [I_{\lambda_{k+1}} - I_{\lambda_k}]$$

THEOREM 6.11.2

Let all the assumptions of Theorem 6.11.1 hold. Then for any sequence of partitions $\mathcal{P}_n$, such that

$$r(\mathcal{P}_n) = \max |\lambda_{i+1} - \lambda_i| \to 0$$

the corresponding lower and upper sums s_n and S_n converge (in the operator norm!) to operator A.

The approximate sums are interpreted as the Riemann-Stieltjes approximation sums and the result is stated symbolically as

$$A = \int_{-\infty}^{\infty} \lambda \, dI_\lambda \qquad \left(= \int_m^M \lambda \, dI_\lambda\right)$$

Spectral Theory for Self-Adjoint Closed Operators

We turn now our attention to unbounded operators. Let us begin with a simple result concerning any linear operator A on a Hilbert space U.

PROPOSITION 6.11.1

Let A be a linear operator on a Hilbert space U. If there is a complex *number* λ_0 *in the resolvent set of* A *for which the resolvent* $(\lambda_0 I - A)^{-1}$ *is compact and normal, then*

(i) spectrum of A *consists of at most countable set of eigenvalues*

$$|\lambda_1| \le |\lambda_2| \le \dots \qquad (\to \infty \text{ if infinite})$$

(ii) A *can be represented in the form*

$$A = \sum_{i=1}^{\infty} \lambda_i P_i$$

where P_i *are the orthogonal projections on eigenspaces* N_i *corresponding to* λ_i *and the convergence of operators is to be understood in the strong sense, i.e.,*

$$A\boldsymbol{u} = \sum_{i=1}^{\infty} \lambda_i P_i \boldsymbol{u} \qquad \forall \boldsymbol{u} \in D(A)$$

PROOF Let

$$|\mu_1| \ge |\mu_2| \ge \dots \qquad (\to 0 \text{ if infinite})$$

be a sequence of eigenvalues of the resolvent $(\lambda_0 I - A)$ and let P_i denote the orthogonal projections on eigenspaces corresponding to μ_i. By the spectral theorem for compact and normal operators

$$(\lambda_0 I - A)^{-1} = \sum_{i=1}^{\infty} \mu_i P_i$$

Let $\boldsymbol{u} \in D(A)$. We claim that

$$(\lambda_0 I - A)\boldsymbol{u} = \sum_{i=1}^{\infty} \frac{1}{\mu_i} P_i \boldsymbol{u}$$

where the series converges in the strong (norm) sense. Indeed, if

$$\boldsymbol{u} \in D(A) = D(\lambda_0 I - A) = \mathcal{R}((\lambda_0 I - A)^{-1})$$

then $\boldsymbol{u}$ can be represented in the form

$$\boldsymbol{u} = (\lambda_0 I - A)^{-1}\boldsymbol{v} = \sum_{i=1}^{\infty} \mu_i P_i \boldsymbol{v}$$

for some $\boldsymbol{v} \in U$.

Consequently, for $m > n$

$$\sum_{i=n}^{m} \frac{1}{\mu_i} P_i \boldsymbol{u} = \sum_{i=n}^{m} \frac{1}{\mu_i} P_i \Big(\sum_{j=1}^{\infty} \mu_j P_j \boldsymbol{v}\Big) = \sum_{i=n}^{m} P_i \boldsymbol{v}$$

which converges to $\mathbf{0}$ as $n, m \to \infty$ since P_i form a resolution of identity (null space of $(\lambda_0 I - A)^{-1}$ reduces to the zero vector!). Thus the sequence converges (only in the strong sense!). Passing to the limit with $n \to \infty$ in

$$\sum_{i=1}^{n} \frac{1}{\mu_i} P_i \Big(\sum_{i=1}^{\infty} \mu_i P_i \boldsymbol{u}\Big) = \sum_{i=1}^{n} P_i \boldsymbol{u}$$

we prove that $\sum_{i=1}^{\infty} \frac{1}{\mu_i} P_i$ is a left inverse of $(\lambda_0 I - A)$ and, in a similar way, that it is a right inverse as well.

Finally, using again the fact that P_i form a resolution of identity we get

$$A\boldsymbol{u} = \sum_{i=1}^{\infty} \Big(\lambda_0 - \frac{1}{\mu_i}\Big) P_i \boldsymbol{u}$$

or denoting $\lambda_i = \lambda_0 - \mu_i^{-1}$,

$$A\boldsymbol{u} = \sum_{i=1}^{\infty} \lambda_i P_i \boldsymbol{u}$$

It is easy to see that λ_i are eigenvalues of A and P_i, the corresponding eigenprojections. Finally, for any $\lambda \neq \lambda_i$,

$$\lambda I - A = \sum_{i=1}^{\infty} (\lambda - \lambda_i) P_i \qquad \text{(strong convergence)}$$

and using the same reasoning as before, we can prove that

$$(\lambda I - A)^{-1} = \sum_{i=1}^{\infty} (\lambda - \lambda_i)^{-1} P_i$$

where the boundedness of $(\lambda - \lambda_i)^{-1}$ implies the boundedness of $(\lambda I - A)^{-1}$. Thus eigenvalues λ_i are *the only* elements from the spectrum of A. ∎

The proposition just proved not only has a practical importance, but also indicates the kind of convergence in the spectral representation we can expect in the general case for unbounded operators.

Example 6.11.2

Consider the space $U = L^2(0,1)$ and the differential operator

$$Au = -\frac{d^2u}{dx^2}$$

defined on the subspace

$$D(A) = \{u \in H^2(0,1) \ : \ u'(0) = 0, u(1) = 0\}$$

(A is actually self-adjoint !)

The inverse of A, equal to the integral operator

$$(A^{-1}u)(y) = \int_y^1 \int_0^x u(\xi)\, d\xi\, dx$$

(cf. Example 6.10.1) was proved to be compact with the corresponding sequence of eigenvalues

$$\mu_n = (\frac{\pi}{2} + n\pi)^{-2} \qquad n = 0, 1, 2, \ldots$$

and eigenvectors

$$u_n = \frac{1}{\sqrt{2}} \cos((\frac{\pi}{2} + n\pi)x) \qquad n = 0, 1, 2, \ldots$$

Consequently,

$$\lambda_n = (\frac{\pi}{2} + n\pi)^2 \qquad n = 0, 1, 2, \ldots$$

are the eigenvalues of A and the spectral decomposition takes form

$$-u''(x) = \sum_{n=0}^{\infty} a_n \cos((\frac{\pi}{2} + n\pi)x)$$

where

$$a_n = \frac{1}{2}(\frac{\pi}{2} + n\pi)^2 \int_0^1 u(x) \cos((\frac{\pi}{2} + n\pi)x)\, dx$$

$\square$

Continuing our discussion on self-adjoint operators, we first notice that, in the initial considerations in this section, concerning asymptotic eigenvalues of self-adjoint operators, we have nowhere used the assumption that the operator was bounded. Moreover, as in the case of bounded operators, one can show that the spectrum of a self-adjoint operator consists of the asymptotic eigenvalues only and therefore the same conclusions hold: the residual spectrum is empty and the whole spectrum is real.

We are now ready to state the final result concerning the self-adjoint, unbounded operators (cf. Theorems 6.11.1 and 6.11.2).

THEOREM 6.11.3
(The Spectral Theorem for Self-Adjoint Unbounded Operators)

Let U be a Hilbert space and A an unbounded self-adjoint operator defined on a subspace $D(A) \subset U$ into U. There exists then a one-parameter family of orthogonal projections $I(\lambda) : U \to U, \lambda \in \mathbb{R}$, satisfying the following conditions:

(i) $\lambda \leq \mu \quad I(\lambda) \leq I(\mu)$ *(monotonicity).*

(ii) $\lim_{\lambda \to -\infty} I(\lambda) = 0$ *and* $\lim_{\lambda \to \infty} I(\lambda) = I$ *(in the strong sense).*

(iii) function $\lambda \to I(\lambda)$ *is right-continuous, i.e.,*

$$I(\lambda) = \lim_{\mu \to \lambda^+} I(\mu)$$

(iv) $\lambda \in r(A)$ *iff* λ *is a point of constancy of* A.

(v) $\lambda \in \sigma_P(A)$ *iff* λ *is a discontinuity point of* I.

Moreover,

$$A\boldsymbol{u} = \sum_{\infty}^{\infty} \lambda \, dI(\lambda)\boldsymbol{u} \stackrel{\text{def}}{=} \lim_{M \to \infty} \int_{-M}^{M} \lambda \, dI(\lambda)\boldsymbol{u} \qquad \textit{for } \boldsymbol{u} \in D(A)$$

where the convergence is understood in the strong sense and the finite integral is understood in the sense of the Riemann-Stieltjes integral discussed in Theorem 6.11.2.

Additionally, we have the following characterization for the domain of operator A*:*

$$D(A) = \{\boldsymbol{u} \in U \, ; \int_{\infty}^{\infty} \lambda^2 \, d\|I(\lambda)\boldsymbol{u}\|^2 < \infty\}$$

As previously, $I(\lambda)$ is called the *spectral family of operator* A.

Functions of Operators. Given the spectral representation of a linear operator A, and a real function $\phi(\lambda)$, we may define *functions of* A as

$$\phi(A) = \int_{-\infty}^{\infty} \phi(\lambda) \, dI(\lambda)$$

Domain of $\phi(A)$ will consist of only those vectors $\boldsymbol{u}$ for which the integral converges (in the appropriate sense). The same observation applies to the compact and normal operators. Note that, due to the properties of the spectral family $I(\lambda)$, operator $\phi(A)$ is insensitive to the behavior of ϕ outside of spectrum $\sigma(A)$. Thus, in this sense, all the information about operator A is stored in its spectrum.

Exercises

6.11.1. Determine the spectral properties of the integral operator

$$(Au)(x) = \int_0^x \int_0^\eta u(\xi)\, d\xi\, d\eta$$

defined on the space $U = L^2(0,1)$.

6.11.2. Determine the spectral properties of the differential operator

$$Au = -u''$$

defined on the subspace $D(A)$ of $L^2(0,1)$

$$D(A) = \{u \in H^2(0,1) \; : \; u(0) = u(1) = 0\}$$

References

[1] R. A. Adams, *Sobolev Spaces*, Academic Press, New York, 1976.

[2] A. Alexiewicz, *Functional Analysis* (in Polish), Państwowe Wydawnictwo Naukowe, Warszawa, 1969.

[3] J. P. Aubin, *Applied Functional Analysis*, John Wiley & Sons, New York, 1979.

[4] G. Duvaut and J. L. Lions, *Les Inequations en Mécanique et en Physique*, Editions Dunod, Paris, 1972.

[5] L. Kantorovitch and G. Akilov, *Functional Analysis in Normed Spaces*, Moscow, 1955.

[6] T. Riesz, *Perturbation Theory for Linear Operators*, Springer-Verlag, New York, 1966.

[7] M. A. Krasnosel'skii, *Topological Methods in the Theory of Nonlinear Integral Equations*, Pergamon Press, New York, 1964.

[8] J. L. Lions and E. Magenes, *Non-homogeneous Boundary Value Problems and Applications*, Vol. 1, Springer-Verlag, New York, 1972.

[9] R. Leis, *Initial Boundary Value Problems in Mathematical Physics*, John Wiley & Sons, New York, 1986.

[10] S. Łojasiewicz, *Introduction to Real Functions* (in Polish), Państwowe Wydawnictwo Naukowe, Warszawa, 1973.

[11] F. Riesz and B. Sz.-Nagy, *Functional Analysis*, Frederick Ungar Publishing Co., New York, 1955.

[12] R. E. Showalter, *Hilbert Space Methods for Partial Differential Equations*, Pitman Publishing Limited, London, 1977.

[13] I. Stakgold, *Green's Functions and Boundary Value problems*, John Wiley & Sons, New York, 1979.

[14] C. Truesdell and W. Noll, *The Non-Linear Field Theories of Mechanics*, in *Handbuch Der Physik*, Band III/3, Springer-Verlag, New York, 1965.

[15] K. Yosida, *Functional Analysis*, 4th ed., Springer-Verlag, New York, 1974.

Index

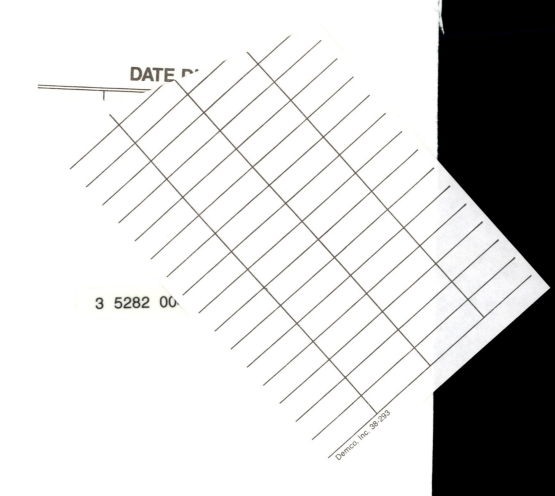